Nanomedicine in
Health and Disease

Nanomedicine in Health and Disease

Editors

Ross J. Hunter
Cardiology Research Fellow
St Bartholomew's Hospital
London
UK

Victor R. Preedy
Professor of Nutritional Biochemistry
School of Biomedical & Health Sciences
King's College London
and
Professor of Clinical Biochemistry
King's College Hospital
UK

CRC Press
Taylor & Francis Group
Boca Raton London New York

CRC Press is an imprint of the
Taylor & Francis Group, an **informa** business
A SCIENCE PUBLISHERS BOOK

CRC Press
Taylor & Francis Group
6000 Broken Sound Parkway NW, Suite 300
Boca Raton, FL 33487-2742

First issued in paperback 2017

© 2011 by Taylor & Francis Group, LLC
CRC Press is an imprint of Taylor & Francis Group, an Informa business

ISBN 13: 978-1-138-11286-5 (pbk)
ISBN 13: 978-1-57808-725-9 (hbk)

Cover Illustrations: Reproduced by kind courtesy of the undermentioned authors:
Figure No. 3 from Chapter 4 by Kerriann R. Greenhalgh
Figure No. 5 from Chapter 9 by Rehab Amin
Figure No. 4 from Chapter 12 by Moritz Beck-Broichsitter et al.

Library of Congress Cataloging-in-Publication Data

Nanomedicine in health and disease / editors, Ross J. Hunter, Victor
R. Preedy.
       p. ; cm. -- (Nanoscience applied to health and medicine)
   Includes bibliographical references and index.
   ISBN 978-1-57808-725-9 (hardcover)
   1. Nanomedicine. 2. Nanotechnology--Health aspects. I. Hunter,
Ross, 1977- II. Preedy, Victor R. III. Series: Nanoscience applied
to
health and medicine.
   [DNLM: 1. Nanomedicine--methods. 2. Nanostructures--therapeutic
use. QT 36.5]
   R857.N34N356 2011
   610.28--dc23
                                                          2011016513

Printed in the United States of America

Preface

The nanosciences are a rapidly expanding field of research with a wide applicability to all areas of health. They encompass a variety of technologies ranging from particles to networks and nanostructures. For example, nanoparticles have been proposed to be suitable carriers of therapeutic agents, while nanostructures provide suitable platforms for sub micro bioengineering. However, understanding the importance of the nanoscience and technology is somewhat problematic as much of the literature on the subject is highly technical, written with little consideration for the novice. In this series *Nanosciences Applied to Health and Biomedical Sciences* we aim to disseminate the information in a readable way by having unique sections for the novice and expert alike. This enables readers to transfer their knowledge base from one discipline to another, or from one academic level to another. Each chapter has an abstract, key facts, applications to other areas of health and disease, and a mini-dictionary of key terms and phrases used within that chapter. Finally, each chapter has a series of summary points. In this book we focus on nanomedicine in health and disease. We cover the regulation of nanomedicine, nanotubes, topical applications of nanoparticles, nanocrystals, antioxidant nanoparticles, lipid nanocapsules, nanotheragnostic colloids, nanotechnology in the control of infectious disease, virus-based nanoparticles and the safety of nanoparticles. We also cover nanomedicine in relation to pulmonary drug delivery, the control of infectious disease, radiation protection, arthritis, cancer nanomedicine, blood diseases, neurodegenerative disorders, and tissue and implant engineering.

Contributors to *Nanomedicine in Health and Disease* are all international or national experts or leading authorities or are carrying out ground-breaking and innovative work on their subject. The book is essential reading for research scientists, medical doctors, health care professionals, pathologists, biologists, biochemists, chemists and physicists, and general practitioners as well as those interested in disease and nano sciences in general.

The Editors

Contents

Section 2: Applications to Health

Section 1: Introduction and Technology

Nanomedicine: An Introduction

Lajos P. Balogh

ABSTRACT

Nanoscience and nanotechnology is closing the gap in our knowledge between molecular and bulk properties. This process has enormous significance in understanding and changing our world. It is expected that this field will expand greatly as advances are being made in using nanoscience and nanotechnology for basic, translational, and clinical medical research. In the next decade, nanomedicine will forever change how we diagnose and treat patients.

The general relationship between science, technology, medicine, business and society is always the same, whether it is "nano" or not "nano". However, the emergence of nanomedicine has a significant practical future in medicine and human health. The market for life sciences and health care products can currently be measured in hundreds of billions of U.S. dollars. A market of many billions more is predicted within ten years, primarily in medical devices, implants, imaging and diagnostics enabled by nanoscience and nanotechnology. As a consequence, the quality of life is expected to improve considerably.

Nanomedicine research is reforming our understanding of biological processes, it will transform clinical medicine, and it will change the way pharmaceutical commercialization and business is done.

364 Ocean Ave, Ocean Shores Tower, Suite 702, Revere, MA 02151; Email: baloghl@prodigy.net; baloghlp@gmail.com

A strategy to accelerate nanoscience and nanoengineering for medicine and health reflects increasing societal need and technology pull, as well as basic science push. Apart from the great promises of the relevant sciences and developing technologies, there are several challenges present for this paradigm-changing field, especially in the areas of communication, commercialization, safety, regulations, standardization, education, and public policies. The success of nanomedicine ultimately will depend on whether or not we can transform and harmonize commercialization processes to allow personalized medicine, whether nanoscience becomes a successful "nanobusiness," and whether these new tools and devices will reach the clinic and improve public health. In reality, at the clinical level, the practitioner does not have to know that it is nanotechnology at work—only that the new procedure or "medicine" works better, and it is more affordable.

INTRODUCTION

A few terms are defined at the outset.

A *nanometer (nm)* is 1×10^{-9} m; 1000 nm equals 1 μm.

Science is the study of the physical world and its manifestations, aimed at discovering ruling principles of the phenomenal world by employing scientific methods, especially systematic observation and experiment. (See also: http://encarta.msn.com/encnet/features/dictionary/dictionaryhome.aspx.)

Engineering is the process of designing and making tools, machines, apparatus, and systems to exploit phenomena for practical human purposes. It translates basic and applied science into technology. Mathematics and predictive models are distinct components of engineering, which is always quantitative.

Technology is the study, development, and application of engineered devices, machines, and techniques for manufacturing and productive processes; a method or methodology that applies technical knowledge or tools; and the sum of a society's or culture's practical knowledge, especially with reference to its material culture. Technology has to satisfy additional requirements such as utility, usability and safety. (See also: http://en.wikipedia.org/wiki/Technology#cite_note-12.)

Medicine is the study and the cycle of diagnosis, treatment, monitoring, prediction, and prevention of diseases. Following up with patients, collecting and evaluating data, and other tasks are also considered part of medicine. The term "medicine" itself could also mean a *substance* that promotes healing. Potential sub-categories of medicine may include diagnostics, therapeutics, regenerative medicine, prosthetics, public health, toxicology, point-of-care monitoring, nutrition, medical devices, prosthetics, biomimetics, and bioinformatics.

Medical research together with monitoring and follow-up of patients improves our health-related knowledge, which is realized in education and practiced through our public health systems. Knowing the cause and mechanism of an illness may lead to its temporary or permanent prevention.

Science, engineering, technology, commercialization, business and society are in a special relationship with each other. Successful new products have to pass the whole sequence from science through business to serve society's needs. These fields are vertically interrelated and embedded in each other (Fig. 1) and are connected by communication and dynamic interactions. Science generates new knowledge, engineering creates new tools, and technology develops processes to manufacture goods. When fundamental business requirements are met, technologies can be commercialized. Standards and appropriate regulations are prerequisites to successful commercialization. Commercialization requires sufficient scientific knowledge (in the form of intellectual property secured by patents) and needs creative engineering, a technology that is competitive on the market; the product must also satisfy society's needs.

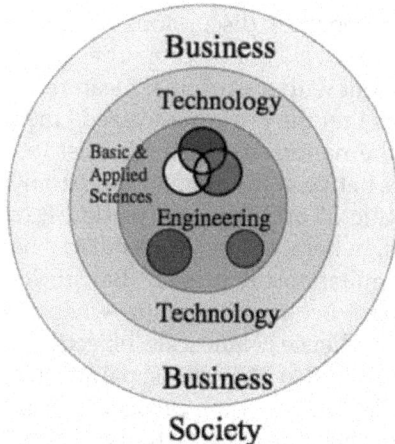

Fig. 1. The {{{{Science}Engineering}Technology}Business} hierarchy. Progress on these fields could be driven by hypothesis, need, and/or opportunity. Efficient communication between these layers is of utmost importance for successful commercialization.

Color image of this figure appears in the color plate section at the end of the book.

Nanoscience is the study of nanoscale substances, objects, structures, and materials applying scientific methods to understand and control matter at the nanoscale (Fig. 2). As such, it is part of all disciplines of natural science. Nanomaterials are not new materials with magical properties; rather, nanoscale materials are made of well-known materials and building blocks that display properties that do not exist outside the nanorange. Measurements performed at that level to understand how things work are also considered part of nanoscience.

Nanotechnology is the "application of scientific knowledge to manipulate and control matter in the nanoscale in order to make use of size- and structure-dependent properties and phenomena, as distinct from those associated with individual atoms or molecules or with bulk materials. Manipulation and control includes material synthesis" (ISO/TS 80004-1:2010, http://db.iso.org).

Nanomedicine exploits and builds upon novel research findings in nanoscience, nanotechnology, biology, and medicine; it unifies the efforts of scientists, engineers, and physicians determined to apply their latest research results to translational and clinical medicine, ultimately developing novel and better approaches and paradigm-changing solutions to health-related issues, ultimately improving the quality of life (Balogh 2009).

Nanoscience, nano-engineering and nanotechnology are the science, engineering, and technology of nanoscale materials and nanosized systems.

Although there are ongoing discussions whether "nanoscale" is 1–100 nm (where most of the novel "nano" properties are usually observed), or 1–1000 nm (in agreement with the original meaning of "nano" (10^{-9}), there is a general consensus that the term "nanosized" means that the structure of a material is on the nanometer scale, at least in one dimension. The nanoscopic range lies between the atomic "angstrom" (10^{-10} m) range and the bulk state (the "micro" and "macro" world). It connects "molecular" and "bulk" regimes, where atomic/molecular, "nano", and materials characteristics are simultaneously present (Balogh 2010a) (Fig. 2).

People interact and behave differently when they are alone, are in groups, or form an audience. Nanoscale objects, atoms and molecules, also behave differently when they are separate from the others, or are held together in small groups, or are present in bulk. In general, "nano" properties are due to the direct interactions of small numbers of atoms and/or molecules keeping them together in a given environment. Emergence of "nano" properties may be due to *quantum effects* between the atoms or molecules (Brus 1984; Alivisatos 1998), the *increased role of surface interactions* (Nel et al. 2006), and the *collective properties* of the particles formed from more than one primary object. Figure 3 depicts the

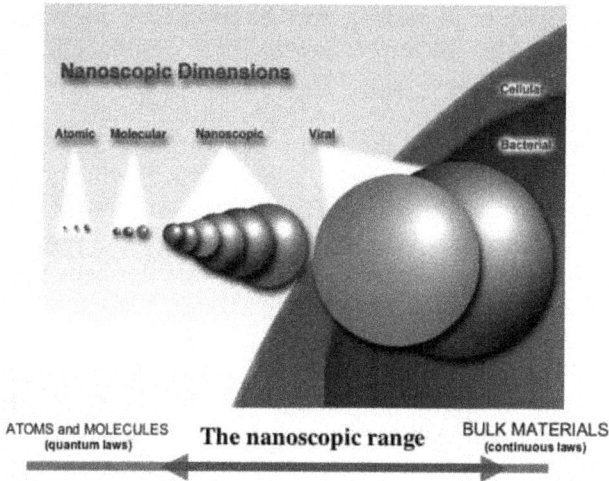

Fig. 2. The nanoscopic range is between "molecular" and "bulk" regimes, where molecular, "nano", and materials characteristics are present simultaneously.

Color image of this figure appears in the color plate section at the end of the book.

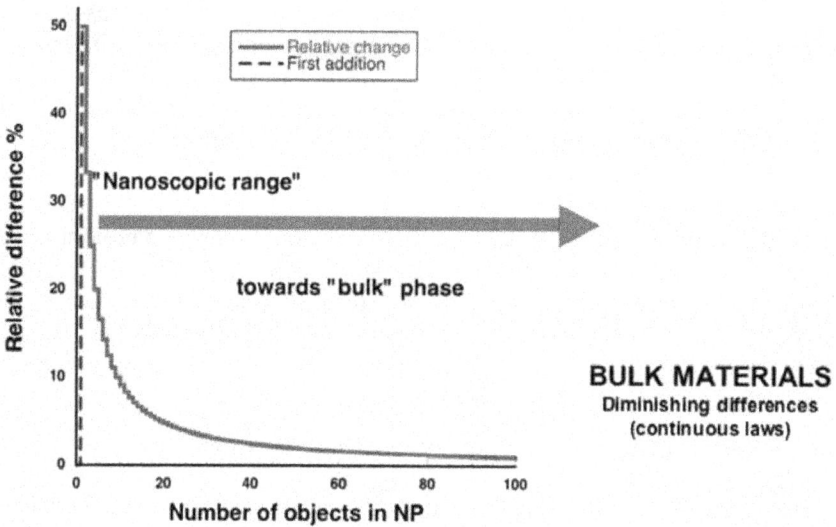

Fig. 3. Change of relative differences in mass as nanoparticles (NP) during transition towards bulk materials. There is no sharp limit towards the bulk phase (Balogh 2010b, Fig. 2., with permission).

relative change in mass of particles formed by adding one more unit to an atom or molecule as a function of the number of original objects. It is clear that as the particle becomes bigger its average diameter increases, and the measurable property differences gradually diminish until the differences become unmeasurable (Balogh 2010b). Scientifically it is impossible to point out the precise length limits of "nanoscale properties", because there is no sharp limit on the "bulk" side. For different substances and materials and even for different (e.g., optical or electromagnetic) properties of the same material, the nanorange (transitional properties between molecular and bulk behavior) may lie somewhere else on the nanometer length-scale.

Size is the most obvious independent variable of nanoparticles that can be measured and nanoscale properties are often attributed to changes in size only. However, size is just one of the factors determining properties, and composition, shape, and architecture also play a definitive role. In principle, "nano" characteristics of materials are determined by the individual properties of the participating constituents (composition) and the means by which they interact with each other and with their immediate environment. Size (i.e., L linear parameter) is directly related to surface (which is proportional to L^2), and increase of the surface area increases everything related to surface interactions (e.g., surface forces, solubility, penetration/scattering of light, biodistribution, degree of opsonization). Similarly, surface/volume ratio ($\sim L^2/\sim L^3$) increases with decreasing size, and more and more atoms and molecules will be on the surface of a particle (Nel 2006).

For medical nanodevices combining inorganic, organic and biological materials the picture is much more complex. Here the primary components comprising a single nanodevice may be objects of different nature (e.g., organic or inorganic molecules, synthetic macromolecules, peptides, proteins, antibodies, nanoparticles). This complexity is unavoidable to successfully address complicated biological processes. We believe that better understanding of nanoparticles, nanoscale materials, and nanoscale laws will lead to a seamless integration of theory and models, which sooner or later will enable the design of better nanodevices and *a priori* prediction of their actions and biological properties.

In summary, "nano" properties are due to the following:

- quantum effects,
- the increased role of surface interactions, and
- collective properties of the objects, which depend on architecture.

"Nano" properties of materials are determined by the individual properties of the constituents and the way primary components interact with each other *and* with their environment. Properties of individual nanoparticles in a nanomaterial may differ slightly; therefore, properties of nanomaterials have a distribution around an average value for every distinct property (the only exception is chemical composition).

Size is just one of the factors determining properties of nanoscale materials. It is impossible to point out the precise length limits of the range of "nanoscale properties" without specifying the material system itself first. Nanoscale properties (transitional between molecular and bulk state) could be different for different properties of the *same* material. Either way, the resulting novel "nano" properties can effectively be used to create innovative tools and design new processes that lead to technologies offering better solutions to medical problems (Balogh 2010a).

THE NEED FOR NEW SOLUTIONS IN HEALTH CARE

The world longs for better health care. In the U.S. the cost of health care is skyrocketing and threatens to bankrupt the health care system. As an example, the direct medical cost in the U.S. was about $90 billion for cancer and about $116 billion for diabetes in 2007. The annual medical care cost for spinal cord injury in the U.S. is about $1.5 billion; the full annual costs are estimated at about $10 billion. In order to remain physically active, approximately 200,000 people receive hip implants and 300,000 people receive knee implants in the U.S. The average lifetime of current orthopedic implants is only 10–15 years, and revision surgeries and their recoveries are not as successful as the first operation. The cost of an implant varies, but is roughly $20,000.

At the same time, in developing countries, especially in Africa, thousands are dying every day because they cannot afford medication. Asian countries are looking at traditional cures as alternatives to western-style pharmaceutics. People are hopeful for change, but they also naturally fear the unknown.

Everybody is looking for new solutions, and most of these new approaches may be offered by nanomedicine.

Nanomedicine connects chemistry, materials science, physics, biology, and engineering to basic and preclinical medical research as well as clinical medicine. While the most developed areas for nanoscale materials are chemistry, materials science, and physics, the last few years have seen unprecedented advances in the field of biology. The decoding of the human genome coupled with gene transfection technologies offer great opportunities for treating illnesses. It has long been recognized by health care practitioners that individual patients respond differently to the same drug, in terms of both efficacy and safety. Drug pharmacological response is underlined by two interrelated dynamic processes: physiologically based pharmacodynamics is associated with what the drug does to the body (i.e., drug-receptor interactions), whereas pharmacokinetics describes what an organism does to the drug (i.e., the processes of drug absorption, distribution, metabolism and elimination) (Viziriakis 2010).

Pharmacogenomic testing and individualized adjustment of drug selection and dosage schemes will also improve treatments and eliminate adverse response effects caused by the differential pharmacological response of different patients to the same drug(s). Understanding the genetic origin and molecular mechanism of diseases will significantly improve the outcome of clinical treatments (Balogh 2009).

There are many reasons to consider nanosized substances for medicine. Decreasing particle size results in increased surface interactions for both nanocrystals and emulsions, which may result in increased solubility for both hydrophilic and hydrophobic drugs. Liposomes, polymers, surface compatibilized nanoparticles, dendrimers and other vesicles are usually able to deliver larger payloads, and chemical modifications of surfaces can accommodate to receptor-specific targeting moieties. Interactions of nanodevices with cells are different from those of small molecules: pinocytosis and receptor-mediated endocytosis are the major uptake mechanisms (Fahmy 2005). Nanoscale devices have different biodistribution, pharmacodynamic and pharmacokinetic behavior compared to traditional drugs (Li 2010) and offer unusual pharmacodynamics and pharmacokinetics that are not accessible for small molecules and traditional substances. Functional and multifunctional nanodevices are assembled from various nanosized components that carry out the various desired functions. For medical nanodevices, there are a few critical basic requirements to meet: they must be stabile in body fluids, lack toxicity, have appropriate biodistribution, and have predictable and ideal pharmacokinetic behavior without significant immunogenicity. In addition to functional interactions such as binding to receptors, blocking or internalization, initiating signaling events and so on, all of these requirements need to be considered in their design. These are typical steps in nanomedical engineering to design, analyze, and manufacture nanodevices.

For example, an ideal anticancer drug must have a low level of systemic toxicity, it should target the appropriate cells at the desired sites of action, its location and effects should be observable by existing imaging techniques, and it should deliver the correct dose of therapeutic agents for the desired length of time. In addition, it may use a non-invasive external trigger to release therapeutic molecules and document response to therapy by identifying residual cells. In this case, *seven* different actions would be needed to accomplish this task. Although modern drugs can very effectively address one or two of these actions, they are simply not complex enough to perform all of the above listed tasks. Once we fine-tune the design and engineering of functional nanodevices, personalized medical devices will become a true possibility that will forever change the face of modern medicine.

Medical devices using nanomaterials and/or nanostructures could be micro- or macro-sized devices, or truly nanosized active systems. Examples of micro- and/or macro-sized devices are diagnostic arrays that use non-bleaching quantum dots as fluorescence markers and microfluidic devices that contain nanoscopic structures in their inner surface to improve detection or catalytic processes. Nanoscience permits the integration of biological and physical systems at the nanoscale allowing the fabrication of real nanoscale devices or *in situ* fabrication of nanostructures.

Nanosized active systems ("nanodevices") don't use electric energy, but rather work on chemical, physical and biological principles. When in action, they deliver drugs, deliver active or passive imaging agents that emit or absorb radiation, and may deliver siRNA, various vectors, or growth factors. These nanodevices have designed and predetermined surface functionality, regulated polarity, narrow size distribution, and specific binding moieties that are targeted to receptors (Bryne 2008).

Safety is one of the primary concerns of both researchers and regulators. Current interests in nanomedicine involve understanding the issues related to toxicity and environmental impact of nanoscale materials. However, taking appropriate and responsible measures cannot precede the formulation of relevant science. For example, it is impossible to judge toxicity of a certain nanomaterial before validated toxicology methods and/or quality control are developed. In addition to toxicity, life-cycle analysis, environmental impact, and effects on human health are all very important issues. That is why the U.S. Environmental Protection Agency initiated the Nanoscale Materials Stewardship Program in 2008, to provide a firmer scientific foundation for regulatory decisions. The program has released its interim report (http://epa.gov/oppt/nano/nmsp-interim-report-final.pdf).

NANOMEDICINE'S PRESENT MAJOR DIRECTIONS

Subfields of medicine that are particularly impacted include drug delivery, novel biomaterials, *in vivo* functional imaging, *in vitro* diagnostics, and active implants. New achievements in these areas will considerably improve our biological and medical understanding.

However, nanomedicine has a very broad potential scope and eventually all aspects of medicine will be involved, as has happened with microtechnology and computers. Computer technology has also found its way into medicine and computers now are being considered a natural part of medical improvements (e.g., in imaging and 3D image reconstructions) without particularly pointing to advances in computer chip and memory technologies as part of the reasons for having better images.

Some outstanding achievements of nanomedicine are analytical applications, where lab-on-a-chip methods have surpassed earlier *ex vivo* and *in vivo* detection methods and provided greatly improved diagnostic accuracy and convenience. Further examples include the scenarios of accelerated hemostasis and wound healing (Greehalgh and Turos 2009; Wong et al. 2006), tissue engineering using autologous donor cells (Guo 2009), cancer cell–specific markers that allow surgeons to visually identify the tissue that needs to be removed during the procedure (Jiang 2004; Nguyen 2010), replacement of surgical brachytherapy procedures by a single injection (Khan 2008), and development of personalized and disease-specific intelligent nanodrugs (Kawasaki and Player 2005; Dutta 2009).

CURRENT CHALLENGES

The most demanding challenges for nanomedicine are not presented by science, engineering, or technology, but rather by social and business aspects: communication, education, standardization, regulatory rules, and the need for new business models.

1. Nanomedicine is developing in a fast-paced and multilingual global theater.

All elements of nanomedicine are developing simultaneously in different spheres and countries of the world. In the beginning, the U.S., Europe, and Japan had the advantage, but in recent years India, China, and other countries of the Pacific are catching up. Smaller countries with more centralized governments have put forward efforts by forming large research centers that are well funded and attract high-level scientists from the West. In general, regulations in developing countries are less stringent, and many translational experiments and clinical trials are outsourced.

A unique problem is the simultaneous development of terminology and nomenclature in different languages, where meaning of words is not necessarily equivalent. Terminology and nomenclature discrepancies are usually unimportant issues for science, because scientists create their own terms that fit their own field. However, it is impossible to patent and secure intellectual property without a globally accepted legal vocabulary. Similarly, insecure or uncertain intellectual property status is not good for business.

2. There are conflicts of interest among participants.

One of the major challenges of every product development is that motives of the participants in the hierarchy are different: scientists want to study, invent and publish in journals with good impact factors to advance their careers; engineers want to develop methods, processes and technology; companies want to own intellectual property and control new markets; businessmen and inventors want to make profit. Thus, development of new drugs, medications and devices based on nanoscience or nanotechnology has to be considered as a system consisting of good science, creative engineering, and effective technology with a competitive edge that provides a solid profit for business.

To be successful, for any large-scale project these interests have to be harmonized, which requires efforts from all stakeholders. Governments and other organizations often directly support basic and developmental nanomedicine to cross the funding gap that exists between research and commercialization (NNI 2010, http://www.nano.gov, the European Technology Platform Nanomedicine (http://www.etp-nanomedicine.eu/public).

3. Communication is fuzzy and terminology and nomenclature are still not generally accepted.

Nanomedicine research has intensified since 2000 (Fig. 4). The increase in "nano" publications is more than exponential. To establish efficient interdisciplinary communication between participants educated in single disciplines may take a long time. Historically, terminology of mathematics,

QuickTime™ and a
decompressor
are needed to see this picture.

Fig. 4. Number of journal publications reporting on nanomedicine research 2000 to present. (Originally presented in the NSF Workshop "Re-Engineering Basic and Clinical Research to Catalyze Translational Nanoscience"), courtesy Lux Research (Murday et al. 2009.)

chemistry, biology and medicine developed in only somewhat overlapping directions. In addition, vocabulary and nomenclature may differ considerably in different disciplines (e.g., "substrate" has a fundamentally different meaning for a materials scientist and for a biologist.) Mutual understanding takes time, even though the participants are eager and motivated to learn as well as teach and coach each other. Genuinely, terms of engineering and technology are closer to business, but it still requires an extra effort from everybody to meaningfully communicate between participating categories and stakeholders.

4. Standards for commercialization (safety, environment, and regulations) will take years to catch up with the research.

Nanoscience, nanotechnology, and nanomedicine are developing at such a pace that patent offices, standardization and regulatory mechanisms lag considerably behind discoveries and technology development. Meaningful information exchange is possible only when relevant nomenclature has been developed and is accepted by the parties.

The International Standard Organization and its national committees and other organizations are feverishly working on the definition and unification of terms and methods. Ongoing projects of the International Standard Organization Technical committee 229–Nanotechnologies (http://www.ansi.org/isotc229tag/) are pursuing development of terminology and nomenclature in topical areas such as basic definitions of nanotechnology, the bio-nano interface and terminology for health care professionals (http://www.nanolawreport.com). However, because of their methodical approach based on consensus of stakeholders, it may take several years before standards catch up with the research.

For regulatory purposes there are several additional complications. One of them is that the classical Composition → Efficacy relationship must be replaced with a (Composition + Structure + Architecture + Controlled function) → Efficacy relationship for nanodevices, which is highly problematic using present regulatory practice (especially since the related scientific understanding is not yet fully available). Clinical trials would require nanomedicines that were manufactured according to good laboratory practices and/or good manufacturing practices, but how is this possible until such procedures and tests are described for specific nanomaterials? Both under-regulation and over-regulation would be a mistake, and agencies are struggling with this issue.

The generally accepted view is that development of nanomedicine will lead to better and more affordable therapeutics. At the same time, the public is worried about the gap between the incredible rate at which new materials are created in this field, and the slow response of agencies overseeing the commercialization process. In 2006, the International Center for Technology Assessment and a coalition of consumer, health, and environmental groups filed a legal petition challenging government agencies for their failure to regulate human health and environmental hazards from nanomaterials in consumer products, specifically calling for the recall of nanosunscreens until reviewed as new drug products rather than as cosmetics (http://www.icta.org/doc/Nano%20FDA%20 petition%20final.pdf).

They requested comprehensive nanomaterial-specific regulations, the development of new paradigms of nano-specific toxicity testing, the classification of nanomaterials as "new" substances, and mandatory nanomaterial product and ingredient labeling, among other things. However, after thorough investigation, the U.S. Food and Drug Administration Nanotechnology Task Force concluded in July 2007 that the Food and Drug Administration need not develop a new regulatory framework or special regulations for nanotech at the current time and that no new labeling was necessary (http://www.fda.gov/ScienceResearch/ SpecialTopics/Nanotechnology/NanotechnologyTaskForceReport2007/ default.htm).

5. It takes time for medical education to catch up with the latest science and technology.

Clinical experience over the last decade with biologically targeted anticancer agents has exposed the increasing inadequacies of many current classifications of malignant disease (Parkinson and Cesano 2009). For example, cancer institutes, clinics, and clinical training are still organized by disease sites and rely much on the knowledge and experience of the individual clinician, who classifies and treats patients according to empirical guidelines. There is a need for greater patient-specific biological characterization of cancer, because conclusions based on statistics may not apply to the individual patient. The recent opportunity of identifying biochemical/genetic fingerprints together with the systematic classification of malignant diseases would create a common language among the participants involved in diagnosis and treatment. Our improving knowledge of molecular mechanisms and understanding of the reasons of biological diversity will give us much better therapeutics than the presently used classification groups based on histology. Of course, this approach will have to be taught at medical universities and for health professionals, too, not to mention the questions regarding the new desired structure of an ideal cancer clinic.

As far as the public is concerned, as the knowledge becomes available we must keep educating people about advantages and risks of nanotechnology. The public is embracing the potential benefits, but worried about the risks, including environmental consequences. These concerns must be addressed at all possible levels.

6. There is a discrepancy between research and traditional drug development.

There is a discrepancy between popular academic research directions and the global problems of pharmaceutical companies. Increasing costs are forcing the companies to merge, cut costs, and outsource research, while medical (and, of course, nanomedical) research is more and more about the details of a specific area. These directions are aimed towards more specific, less general, solutions, i.e., at smaller and smaller markets. Academic ideas such as researching disease mechanisms of rare diseases (which is the declared goal and roadmap of the U.S. National Institutes of Health), targeted anticancer technology, personalized medicine and theranostics (i.e., substances/compounds that allow diagnosis and therapy simultaneously) do not fit well the existing pharmaceutical business models based on identifying effective compounds (e.g., from large libraries), then manufacturing drugs in large volumes and marketing them under similar trademark names (to cut marketing costs). Personalized medicines and theranostics would be impossible under the present regulatory practices, not to mention insurance policies and billing practices. New business and regulatory models are needed to address these new and smaller areas.

Development cost of a new successful product is presently very high, around $2.4 billion per product. Personalized medicine will change the way the pharmaceutical industry works, and instead of thousands of tons of generalized drugs, small amounts will have to be delivered by your nanopharmacist, who uses sophisticated synthetic methods.

7. Nanomedicine must become a successful business.

Recent advances in nanomedicine and nanotechnology and the accumulation of disease- and drug-related genomic data are creating the opportunity for personalized medicine to emerge as the new direction in diagnosis and drug therapy. Personalized medicine may mean two things: one, personalized application of generic drugs and/or their combinations (pharmacotyping). This may involve possible individualized drug selection and dosage adjustment based on assessment of molecular biological, pharmacological, clinical and genomics data coupled with the patient's individual characteristics.

The greatest opportunity for personalized and evidence-based medicine is literally the individual synthesis and application of drugs specific to the particular disease the actual patient has. The rationale behind this argument is that *all treatments are either general or specific*; they cannot be both. A general systemic intervention (e.g., reducing body temperature) requires a general drug (say aspirin), but, in principle, the more specific a treatment, the more efficient it is. Thus, ideally the most effective treatment would be a personalized one.

Unfortunately, a one-person medicine would ultimately reduce the market size to one individual, which is not possible presently. This logic also says that the more specific a drug, the smaller the expected market, which is not a rosy perspective for business. Personalized medicine based on the individual biomedical profile of a particular patient will require manufacturing of small quantities of nanomedicines in a short time, but shrinking market size and constantly rising development costs presently negate this direction. The rightfully existing requirement of statistically validated clinical trials does not work for this kind of individualized and highly effective personal medicines that otherwise technologically are possible. The appropriate business and regulatory mechanisms are presently missing, and the solution is unclear.

SUMMARY

Nanomedicine is not a separate discipline. Rather, nanomedicine exists where nanoscience and nanotechnology overlap and interact with life sciences. Sub-categories of medicine may include diagnostics, therapeutics, regenerative medicine, prosthetics, public health, toxicology, point-of-care monitoring, nutrition, medical devices, prosthetics, biomimetics, and bioinformatics. These areas are either already impacted or will be impacted by the ongoing nano-revolution. Nanomedicine will not replace traditional medicines. It will add new and more efficient tools, substances and procedures to the existing ones. Future success depends on whether or not nanomedicine is helping the public. In reality, at the clinical level, the practitioner does not have to know that it is nanotechnology at work, only that the new procedure works faster, cheaper, and better.

The major challenges are the following:

- Nanomedicine is developing in a fast-paced and multilingual global theater.
- There are conflicts of interests among participants.
- Communication is fuzzy and terminology and nomenclature are still not generally accepted.
- Standards for commercialization will take years to catch up with the research.
- It takes time for medical education to catch up with the latest science and technology.

- There is a discrepancy between research and traditional drug development.
- Nanomedicine *must* become a successful business.

New technologies have always had their enthusiasts and those who have rallied for and against them. However, nay-sayers have never stopped development. Nanotechnology and nanomedicine have the potential to solve many problems, but may cause great harm as well. As with any new science and/or technology, achievements have to be evaluated and responsibly used for the well-being of the global society.

References

Alivisatos, A.P. 1996. Perspectives on the physical chemistry of semiconductor nanocrystals. J. Phys. Chem. 100: 13226–13239.

Balogh, L.P. 2009. The future of nanomedicine and the future of *Nanomedicine: NBM* (Editorial). Nanomedicine: Nanotechnol. Biol. Med. 5: 1.

Balogh, L.P. 2010a. Why do we have so many definitions for nanoscience and nanotechnology? Nanomedicine: Nanotechnol. Biol. Med. 6: 397–398.

Balogh, L.P. 2010b. The nanoscopic range and the effect of architecture on nanoproperties (Editorial). Nanomedicine: Nanotechnol. Biol. Med. 6: 501–503.

Brus, L.E. 1984. Electron-electron and electron-hole interactions in small semiconductor crystallites: the size dependence of the lowest excited electronic state. J. Chem. Phys. 80: 4403–4409.

Byrne, J.D., T. Betancourt, and L. Brannon-Peppas. 2008. Active targeting schemes for nanoparticle systems in cancer therapeutics. Adv. Drug Delivery Rev. 60: 1615–1626.

Dutta, T., M. Burgess, N.A.J. McMillan, and H.S. Parekh. 2009. Dendrosome-based delivery of siRNA against E6 and E7 oncogenes in cervical cancer, Nanomedicine: Nanotechnol. Biol. Med. DOI: 10.1016/j.nano.2009.12.001.

Fahmy, T.M., P.M. Fong, A. Goyal, and W.M. Saltzman. 2005. Targeted for drug delivery. Nano Today, 18 August: 18–26.

Guo, J.S., K.K.G. Leung, H.X. Su, Q. Yuan, L. Wang, T.H. Chu, W.M. Zhang, J.K.S. Pu, G.K.P. Ng, W.M. Wong, X. Dai, and W.L. Wu. 2009. Self-assembling peptide nanofiber scaffold promotes the reconstruction of acutely injured brain. Nanomedicine: Nanotechnol. Biol. Med. 5: 345–351.

Jiang, T., E.S. Olson, Q.T. Nguyen, M. Roy, P.A. Jennings, and R.Y. Tsien. 2004. Tumor imaging by means of proteolytic activation of cell-penetrating peptides. Proc. Natl. Acad. Sci. USA 101(51): 17867–72.

Kawasaki, E.S., A. Player. 2005. Nanotechnology, nanomedicine, and the development of new, effective therapies for cancer, Nanomedicine: Nanotechnol. Biol. Med. 1: 101–109.

Kerriann G., E. Turos. 2009. *In vivo* studies of polyacrylate nanoparticle emulsions for topical and systemic applications, Nanomedicine: Nanotechnol. Biol. Med. 5: 46–54.

Khan, M.K., L.D. Minc, S.S. Nigavekar, M.S.T. Kariapper, B.M. Nair, M. Schipper, A.C. Cook, W.G. Lesniak, and L.P. Balogh 2008. Fabrication of {198Au0} radioactive composite nanodevices by radiation polymerization and their use for nanobrachytherapy. Nanomedicine: Nanotechnol. Biol. Med. 4: 57–69.

Mingguang Li, M.G., T. Khuloud, K.T. Al-Jamal, K. Kostas Kostarelos, and J. Reineke. 2010. Physiologically based pharmacokinetic modeling of nanoparticles, ACS Nano, Article ASAP DOI: 10.1021/nn1018818, publication date (web): October 14, 2010.

Murday, J.S., R.W. Siegel, J. Stein, and J.F. Wright. 2009. Translational nanomedicine: status assessment and opportunities. Nanomedicine: Nanotechnol. Biol. Med. 5: 251–273.

Nel, A., T. Xia, L. Maëdler, and N. Li. 2006. Toxic potential of materials at the nanolevel. Science 311: 622–627.

Nguyen, Q.T., E.S. Olson, T.A. Aguilera, T. Jiang, M. Scadeng, L. Ellies, and R.Y. Tsien. 2010. Surgery with molecular fluorescence imaging using activatable cell penetrating peptides decreases residual cancer and improves survival Proc. Natl. Acad. Sci. USA (accepted for publication).

Parkinson, D.R., and A. Cesano. 2009. Patient-specific classifications of human malignant disease. Curr. Opin. Mol. Ther. 11: 252–259.

Viziriakis, I. 2010. Nanomedicine and personalized medicine: The improvement of drug delivery outcomes by pharmacogenomics leading to pharmacotyping in clinical practice. Nanomedicine workshop, NN10 Symposium, W3-I7.2, Chalkidiki, Greece.

Wong, K.K.Y., J. Tian, C.M. Ho, C.N. Lok, C.M. Che, J.F. Chiu, and P.K.H. Tam. 2006. Topical delivery of silver nanoparticles reduces systemic inflammation of burn and promotes wound healing. Nanomedicine: Nanotechnol. Biol. Med. 2: 306.

2

The Regulation of Nanomedicine

Joel D'Silva[1], and *Diana Bowman[2]*

ABSTRACT

Nanomedicine holds enormous promise for the improved prevention, detection and treatment of a range of diseases and medical conditions. While the potential benefits of nanomedicine appear—at this nascent period of development—to be enormous, uncertainty surrounds the potential risks of many nanomedical products. Concerns about the adequacy of regulatory oversight also threatens to impede the development and commercialization of nanomedicine-based products. It is becoming increasingly clear, therefore, that appropriate and effective regulatory structures will be fundamental to the successful implementation and commercialization of nanomedicine if the technology is to fulfill its promise.

The first section of this chapter explores nanomedicine and regulation and the implications regulation will have in the areas of health and disease. The second and third sections analyse the regulation of nanomedicine in the European Union and the United States respectively. The last section of this chapter addresses the common threads and various issues raised.

[1]Research Fellow, Department of International and European Law, Faculty of Law, K.U. Leuven, Tiensestraat 41, 3000 Leuven, Belgium; Email: joel.dsilva@telenet.be

[2]Senior Research Fellow, Melbourne School of Population Health, The University of Melbourne, Victoria 3010 Australia and Department of International and European Law, Faculty of Law, K.U.Leuven, Tiensestraat 41, 3000 Leuven, Belgium.

*Corresponding author

List of abbreviations after the text.

INTRODUCTION: NANOMEDICINE, REGULATION AND APPLICATIONS TO AREAS OF HEALTH AND DISEASE

Nanomedicine, the application of nanotechnologies to the health care sector, offers numerous improvements to conventional medical diagnosis, drug delivery, therapy and implants. The European Science Foundation has defined nanomedicine as "the science and technology of diagnosing, treating and preventing disease and traumatic injury, of relieving pain, and of preserving and improving human health, using molecular tools and molecular knowledge of the human body" (ESF 2005). Medical nanotechnology involves the engineering, design, fabrication and application of drugs and medical devices that are about 1-100 nm in size or incorporate nanomaterials of this size into their structure. A somewhat speculative projection is that by 2014 the market for pharmaceutical applications of nanotechnology alone will be close to US$18 billion per year. Moreover, commentators have suggested that the demand for nanotechnology-based medical products within the US market alone will increase by 17% a year to US$53 billion in 2011 and US$110 billion in 2016 (Harris and Bawa 2007).

Though it is foreseen that nanomedicine will radically change the health care sector and the way in which patients interact with the medical profession, nanomedicine is also expected to challenge existing perceptions, dynamics and standards relating to ethics, patient and environmental safety and governance (D'Silva and Van Calster 2009; Marchant et al. 2009; Wagner et al. 2006). At this early stage of development and commercialization, there is considerable uncertainty about potential human health and safety risks, especially in relation to the long-term risk of certain "free" nanomaterials and their use in the human body (De Jong and Borm 2008). In addition to the potential risks associated with the nanomedical products themselves, a number of commentators have raised concerns about the potential occupational and environmental risk associated with the manufacture and disposal of nanomedicines, drugs and devices (Linkov et al. 2008; Marchant et al. 2009). Regulators are increasingly being faced by the need to adequately regulate this promising field, and soon, under the current regulatory structures. Knowledge gaps, questions of expertise and definitional issues (Chowdhury 2010: 135) all pose crucial challenges to the effective regulation of nanomedicine.

Despite the perceived benefits of various nanomedical products, consumer reactions and acceptance will be crucial to the commercial "success" of those products. If, for example, one or more of these products fails spectacularly or is associated with widespread health and/or environmental damage, consumer trust and confidence in the technology as a whole may be adversely affected. The widespread reporting of such events, even if limited to one such product, could subsequently undermine further development of these promising products (Throne-Holst and Strandbakken 2009: 394). Hence, it is becoming increasingly clear that appropriate and effective regulatory structures will be fundamental to ensuring that the potential of the technology is fulfilled.

REGULATING NANO-MEDICINE IN THE EUROPEAN UNION

There are no specific nanomedical laws in the European Union (EU), nor indeed in any other jurisdiction, at this time; this does not mean that they are "unregulated" (Van Calster 2006). Rather, such products are regulated in the same way as their non-nanotechnology counterparts, with existing legislation on medicinal products and devices, tissue engineering and other advanced therapies being applicable to nanomedicine-based products and processes.

The Regulation of Medicinal Products

Medicinal products for paediatric use, orphan, herbal medicinal products and advanced therapies are governed by specific rules. The general medical legislation is supported by a series of guidelines.

Table 1. Regulatory bodies and legislation in the European Union relevant to nanomedicine.

Regulatory bodies
Primary Body
European Medicines Agency (EMEA)
Committees and Groups
Committee for Medicinal Products for Human Use (CHMP)
Innovation Task Force (ITF)
New and Emerging Technologies (N&ET) Working group
Legislation
Medical products for Human Use—Directive 2001/83/EC
Authorisation and supervision of medicinal products—Regulation (EC) No 726/2004
Medical Devices Directive—93/42/EEC
Active Implantable Medical Devices Directive—90/385/EEC
In Vitro Diagnostic Medical Devices Directive—98/79/EC
Advanced Therapy Medicinal Products (ATMP)—Regulation 1394/2007/EC
Registration, Evaluation, Authorisation and Restriction of Chemical Substances (REACH)—Regulation (EC) 1907/2006

Regulating Medical Products for Human Use—Directive 2001/83/EC[1]

According to Article 6 of the Directive, no medicinal product can be placed on the market of a Member State unless a marketing authorization has been issued by the competent authorities of that Member.

Authorization and Supervision of Medicinal Products—Regulation (EC) No. 726/2004

Regulation No. 2309/93 set down procedures for the authorization and supervision of medicinal products and established a European Agency for the Evaluation of Medicinal Products.[2] The regulation was updated in 2004 by Regulation 726/2004, at which time the name of the agency was changed to the European Medicines Agency (EMA).[3] The EMA is mandated to perform a number of different functions, the two main functions being:

1. to provide scientific advice to the Community institutions and the Member States in the field of medicinal products in relation to the authorization and supervision of medicinal products; and
2. to co-ordinate the activities of the Member States in the monitoring of adverse reactions to medicinal products (pharmacovigilance).

The EMA comprises five scientific committees: Committee for Medicinal Products for Human Use (CHMP), Committee for Medicinal Products for Veterinary Use (CVMP), Committee for Orphan Medicinal Products (COMP), Committee on Herbal Medicinal Products (HMPC), Paediatric Committee (PDCO) and Committee for Advanced Therapies (CAT).

The Committee for Medicinal Products for Human Use (CHMP)

The CHMP, established under Regulation (EC) No. 726/2004, sits under the umbrella of EMA (Article 5). It has been proactive in the approach to nanotechnologies, having released a reflection paper on the topic in 2006. In the report, the EMA concluded that the evaluation and prevention of potential hazards related to the use of any nanomedicinal products is adequately regulated under the existing, technology-neutral, pharmaceutical legislation. The CHMP, however, accepted that the development and commercialization of increasingly sophisticated nano-products that span regulatory boundaries between medicinal products and medical devices may challenge the current criteria used by EMA

---

[1] OJ [2001] L311/67.
[2] OJ [1993] L 214/1.
[3] OJ [2004] L 136/1.

in relation to classification and evaluation. Should such products be developed, appropriate expertise and guidelines may be needed for their evaluation so as to ensure quality, safety, efficacy and risk-management (CHMP 2006).

The Innovation Task Force (ITF)

The EMA established the Innovation Task Force (ITF) in 2006 in order to provide the EMA with expert regulatory advice on whether new medicinal products for emerging therapies and borderline products are eligible for EMA procedures. The ITF is organized in specialized groups: cell therapy products, gene therapy products, nanomedicines, genomics, and borderline combination products including medical devices and medicinal products. In the absence of specific guidance, potential applicants developing nanomedicinal products are essentially encouraged to interact with the EMA from the early stages of development.

EU Regulation of Medical Devices

Medical devices are regulated by Directive 93/42/EEC (The Medical Devices Directive)[4], Directive 90/385/EEC (The Active Implantable Medical Devices Directive)[5] and Directive 98/79/EC (The In Vitro Diagnostic Medical Devices Directive)[6]. Directive 2007/47/EC[7] amended Council Directive 90/385/EEC but did not make any amendments to the In Vitro Diagnostic Directive. Each of these legislative instruments will play an important role in regulating nanomedicines within the EU as the technology evolves.

The Medical Devices Directive (MDD) 93/42/EEC

All products which fall within the scope of the Directive must meet certain essential safety and administrative requirements and are to be CE marked (Conformité Européenne). Pursuant to the MDD, "medical devices" are classified under four distinct classes:

- Class I: devices with low potential risk,
- Class IIa: with moderate potential risk,
- Class IIb: devices with high potential risk, and
- Class III: devices with critical potential risk.

---

[4] OJ [1993] L169/1.
[5] OJ [1990] L189/17.
[6] OJ [1998] L331/1.
[7] OJ [2007] L247/21.

The categorization of a medical device is important as each class has its own assessment procedures. Manufacturers of medical devices are also under a legal duty to carry out a risk assessment, demonstrate the effectiveness of the device, implement a procedure for post-market surveillance and comply with the essential requirements of Annex 1 of the MDD pertaining to design, use and safety (MDD, Article 3). Compliance with essential requirements of Annex 1 is presumed if the device is in conformity with relevant national standards adopted pursuant to harmonized standards published in the Official Journal of the European Union (MDD, Article 5).

The Active Implantable Medical Devices Directive (AIMDD) 90/385/EEC

A product falls within the scope of the Active Implantable Medical Devices Directive (AIMDD) if it complies with the definition of an AIMD. That means it must be a "medical device" which is, at the same time, both "active" and "implantable". Examples include implantable cardiac pacemakers, implantable nerve stimulators and implantable active drug administration devices. Annex 1 of the Directive sets out the general requirements for compliance as well as certain requirements regarding design and construction. The conformity assessment requirements for devices within the scope of the Directive are essentially the same as for devices classified as class III under the MDD. These include essential safety requirements in terms of function, sterility, material compatibility, marking, "user" instructions, design documentation and CE marking as well as requirements for type approval, production quality management, clinical investigation and manufacturer registration. When an AIMD is intended to administer a substance defined as a medicinal product, that substance shall be subject to the system of marketing authorization provided for a medical product (AIMDD, Article 1(3)). Where an active implantable medical device incorporates, as an integral part, a substance which, if used separately, may be considered to be a medicinal product, that device must be evaluated and authorized in accordance with the provisions of the AIMDD (Article 1(4)).

In Vitro Diagnostic Medical Devices Directive (IVDD) 98/79/EC

The IVDD covers the placement and subsequent service requirements associated with *in vitro* diagnostic medical devices. Devices used *in vitro* for the examination of a specimen derived from the human body, including reagents, instruments and specimen receptacles, fall within the scope of the Directive. All *in vitro* diagnostic devices, including those that incorporate nanomaterials or are manufactured using nanotechnology-based processes, must meet the applicable "essential requirements" on safety, performance and labelling as outlined in Annex I of the IVDD. The fulfilment of the essential requirements has to be demonstrated by the manufacturer for all devices. This includes both new devices and those that have been previously available on the market.

The Use of Harmonized Standards and Risk Management in Nanomedical Regulation

The web of EU Directives outlined above do not prescribe specific features of a medical device to ensure safety, but rather incorporate a series of "essential requirements"[8] that are designed to eliminate or reduce risks as far as possible. Compliance with these requirements is necessary before CE marking may take place. These requirements can be complied with by means of quality systems and by adhering to published harmonized product standards. Products that are subsequently manufactured in conformity with harmonized standards are presumed to conform to essential requirements of the various Directives, and meet the relevant safety and risk assessment requirements: this is often referred to as presumption of compliance or conformity (Delaney and Van de Zande, 2000: 10–11). Another method of compliance is third party testing and product approval. Reliance on this avenue is, however, uncommon because the burden of re-testing after any product change. In the EU, standards are developed by several private bodies including the European Committee for Electrotechnical Standardisation (CENELEC), which deals with all electrical-related standards, and the European Committee for Standardisation (CEN), which deals with standards not covered by CENELEC. While many of these standards are *prima facie* voluntary standards and have little, if any, legal force, they are often seen by manufacturers as being market requirements. Where safety is critical, as in the case of medical devices and products, the means the manufacturer uses to control their production must also be independently assessed. This is generally done by quality management standards established by the ISO9000 series. Standards may only be used to support CE marking on a product if they have been harmonized and their number and title have been published in the Official Journal of the European Communities.

As noted above, these standards and risk management strategies provide guidance for manufacturers on assessing broad risks associated with their products, but are primarily voluntary in nature. Moreover, most of the standards were designed prior to the emergence of nanotechnologies, and are therefore not designed to specifically address any additional risks

---

[8]Generally expressed in Annex I of the directives.

and/or challenges that may be posed by the use of nanomaterials and nanoparticles in such products.

Recognition of this potential shortcoming, in concert with the growing appreciation of the importance and potential market size of nanotechnology-based products, resulted in the establishment by CEN of CEN/TC 352–Nanotechnologies in 2005. The Technical Committee was charged with the task of developing a set of standards to address several aspects of nanotechnologies, including the following:

1. classification, terminology and nomenclature,
2. characterization of physical and chemical properties of nanomaterials,
3. safety and risk assessment, and
4. development of methods, equipment and systems to measure basic characteristics of nano-products.

While the Technical Committee still has much work to do in this area, its work to date has without doubt assisted in the harmonization of nano-standards. Conformity assessments are still in the development stage and their effectiveness and adequacy is yet to be ascertained (CEN 2005).

The New and Emerging Technologies (N&ET) Working Group

New and Emerging Technologies (N&ET), a sub-group of the Medical Devices Expert Group, was invited by the EC in 2007 to assess new and emerging risks posed by nanotechnologies in medical devices. The N&ET Working Group, in a report published in July 2007, considered the adequacy of the existing medical devices regulatory regime in relation to nanotechnologies and recommended the following (N&ET 2007: 6–11):

1. Introduction of a new classification rule in the MDD and AIMDD: "All devices incorporating or consisting of particles, components or devices at the nanoscale are in Class III unless they are encapsulated or bound in such a manner that they cannot be released to the patient's organs, tissues, cells or molecules" (N&ET 2007: 10).
2. In relation to risk assessment and management in the MDD and AIMDD, "special attention should be paid to specific physico-chemical characteristics and toxicological and toxicokinetic properties associated with free nanoparticles" (N&ET 2007: 7).
3. In the case of nano-coated medical devices, risks of nanoparticles coming free are no different from traditional implants (e.g., hip prostheses), which can also generate wear particles, hence current risk assessment is considered to be adequate.
4. A range of measures for ensuring safe commercialization of nanotechnology-based products in the EU market. These include

the introduction of new, nano-specific standards for such products, further development of regulatory guidance on risks, post-marketing guidance, and a voluntary reporting scheme.

The N&ET Working Group did not, however, go as far as suggesting new, nanotechnology-specific legislative instruments. Rather, it concluded that the existing legislations were, in principle, suitable for effectively regulating nanomedical devices as they address any associated risks that may arise as a result of the nano-component in the product. However, it was argued that current procedures may require modification in the short to medium term in order to specifically address issues associated with the use of nanoparticles (N&ET 2007: 5).

Recast of the Medical Devices Directives—Public Consultation 2008

A revision of the legal framework in Europe for medical devices was commenced by the EC in 2008 and was still ongoing in late 2010. Some of the "emerging weaknesses" in the medical devices framework that the Commission has sought to address through the recast of the instrument are the following:

1. gaps and potential loopholes in the current regulatory framework, particularly the scarcity of expertise needed to assess such technologies (MDD Public Document 2008: 2);
2. gaps in regulating "quasi-medical products"—products that are on the market or likely to be marketed that do not fall under the definition of a "medicinal product" or "medical device" nor are covered by "cosmetic product" or similar definitions as they are implanted or injected into the body (MDD 2008: 5);
3. the need for more precise or mandatory standards (MDD 2008: 15); and
4. the challenges posed by "borderline cases", which need expertise and further clarification.

Borderline (Nano)medicines and the Blurring of Boundaries

The boundaries between medicinal products, devices and therapies are likely to become increasingly blurred with advances in medical technology and advanced sciences like nanoscience. The convergence between these traditionally discrete product classes is commonly referred to as "borderline products". To date, the most frequent borderline products have occurred between medical devices and medicinal products. In general, if a product acts pharmacologically, immunologically or metabolically it is unlikely to be considered a medical device. In order to determine whether a product is a device or a medicine, the definitions of both need to be considered, along with the claims for the product, the mode of action on the human body and the intended purpose of the product. Drug delivery systems such as BioSilicon, which involve pharmaceutical substances, are an example of a borderline product, with overlap occurring between drugs and devices. BioSilicon is a nanostructured form of elemental silicon that is engineered to create a "honeycomb" structure of pores. This structure allows silicon to biodegrade while also allowing the retention of various drugs and vaccines within the honeycomb matrix (D'Silva and Van Calster 2009).

Another consideration is the jurisdictional line between the IVDD and the MDD. This determination is of fundamental importance because the two Directives are mutually exclusive (EC IVD Guidance 2004). Article 1.5(a) of the MDD specifically excludes its application to *in vitro* diagnostic medical devices. However, advances in medical technology generally, and nanoscience more specifically, are likely to blur this distinction and make it harder to determine under which regime the product *prima facie* falls. The traditional difference between a medical device and an *in vitro* device is that the former is intended to be used in contact with the patient, while the latter is to be used on specimens taken from the human body. This distinction becomes somewhat blurred in the case of a product that collects specimens for examination and/or has other ancillary functions associated with coming in contact with the patient. To date such products have primarily been treated as *in vitro* devices based on definition and primary application, and regulated as such. The commercialization of more advanced and sophisticated nanodevices that offer a range of different applications, while using various materials and coatings, is likely to blur this distinction and may open up the opportunity for manufacturers to challenge what has been considered to be the generally accepted approach.

Depending on the classification, regulatory requirements for pre-market and/or post-market processes kick in accordingly, with variation in the regulatory pathway existing for medical devices, medical products or therapies. The essential requirements and conformity assessment procedures of the MDD, which were developed and adopted on the premise that the Directive shall not be applicable to *in vitro* diagnostics, is a case in point here. Several harmonized standards have been designed for a certain type of device, as in the case of standards for *in vitro* devices; these would hence not give rise to the legal presumption of conformity if applied to devices that are not clearly *in vitro* devices. It can also be argued that the blurring of these traditional boundaries has the potential to challenge the competencies of notified bodies that are charged with the responsibility to assess and authorize these borderline products.

The issues raised by the increasing number and sophistication of products that sit on and/or straddle these traditional jurisdictional borders are likely to be more prevalent in the nanomedical area than in other sectors such as food or cosmetics (D'Silva and Van Calster 2009: 259). While there are certainly concerns as to whether the current regulatory framework in the EU suffices for the use of nanotechnologies in foods, food contact materials and cosmetics, the majority of these concerns relate primarily to the need for revised risk assessment methodologies and risk management approaches than to the blurring of regulatory boundaries (Bowman and Van Calster 2008; Gergely 2010). That is not to say, however, that such blurring will not occur within these sectors, as already illustrated by the increasing production of, for example, functional foods and nutraceuticals. However, it has been argued that an update of the scientific trimmings of the current regulatory frameworks for these areas is likely to suffice.

By contrast, were the challenges associated with the regulation of borderline products likely to increase through developments in nanomedicine, D'Silva and Van Calster (2009) have argued that a complete overhaul of the underlying regulatory framework may be required. The most recent regulation on advanced therapy medicinal products (ATMP[9]) has been viewed by some commentators as being an attempt to address several issues pertaining to tissue and cell engineering, combination therapies and advanced therapy medical products and could offer regulatory guidance that could be adopted in nanomedicines (Chowdhury 2010). However, much of the ATMP regulation has been "inherited" from existing legislation, standardization, testing and evaluation regimes that do not contain all of the nuances required to effectively tackle the specific concerns relating to borderline nanomedicines and nano-manipulation in general. It also remains to be seen how effective the accountability, transparency and convergence of regulatory models as set out in the ATMP regulation will be (D'Silva and Van Calster 2009: 259).

Regulation of Nanoparticles and Nanomaterials in Medicine

The identification of potential health risks arising from nanosized residues is largely speculative at this point in time because of the embryonic nature of many nanomedicine applications. Studies have reported data on the toxicity of certain nanomaterials under certain laboratory conditions, but to date many of these studies have been inconclusive and contradictory and must therefore be considered within the context of these limitations (Berube 2006). Studies have, for example, examined the toxicological effects of nanoparticles (Oberdorster et al. 2004), quantum dots (Hardman 2006; Monteiro-Riviere et al. 2006), fullerenes (Oberdorster 2004) and carbon nanotubes (Foldvari 2008; Poland et al. 2008; Kane and Hurt 2008). This

---

[9]Regulation 1394/2007/EC - OJ [2007] L 324/ 121.

research has indicated that the small size of engineered nanomaterials and nanoparticles can imbue them with novel properties that are potentially useful across many fields including nanomedicine. However, the comparatively higher reactivity, mobility and/or other properties of some nanomaterials when compared to their larger-scale counterparts may, in some instances, alter their toxicity (Nel et al. 2006; Maynard 2006). Free, insoluble and/or biopersistent nanoparticles are of particular concern because, under certain circumstances, some commentators believe that there is a higher likelihood of such particles entering the body, reacting with cells and giving rise to tissue damage (Holsapple et al. 2005). In the UK, the Royal Society and the Royal Academy of Engineering report *Nanoscience and nanotechnologies: opportunities and uncertainties* concluded that, because of their novel chemical properties, nanoparticles and nanotubes should be treated as new chemicals under UK and European legislation, in order to trigger appropriate safety tests and clear labelling (Royal Society 2004: 79–81).

In March 2006, the Scientific Committee on Emerging and Newly Identified Health Risks (SCENIHR) highlighted three areas of concern and action: appropriateness of existing risk assessment methodologies, the need for new and modified methodologies, and existing information gaps (SCENIHR 2006). The expert body reiterated that, given the uncertainties pertaining to novel properties, hazard evaluation, exposure evaluation and overall risk assessment, current procedures are likely to require modification for nanoparticles (SCENIHR 2006: 60). In June 2007, SCENIHR once again observed that there was a general lack of appropriate standardized protocols and scientific data required for risk and toxicity management (SCENIHR 2007: 9) and recommended that a staged or tiered approach be adopted in order to identify different adverse effects and exposure data. It recommended that evaluation should be carried out on a case-by-case basis with due consideration to emerging knowledge on translocation of nanoparticles, for example within the cardiovascular system or following passage across the blood-brain barrier (SCENIHR 2007: 10).

Registration, Evaluation, Authorisation and Restriction of Chemical Substances Regulation (REACH)[10] and Nanomedicine

European chemicals regulation has been consolidated and integrated with the creation of REACH. Provisions relating to the classification and labelling of substances are dealt with in a separate Regulation on Classification, Labelling and Packaging (CLP)[11] of substances. REACH and CLP are expected to play a critical role in addressing environmental,

---

[10]Regulation (EC) 1907/2006, OJ [2006] L 396/1–849.
[11]Regulation (EC) 1272/2008.

health and safety risks of nanomaterials not least because many such substances enter the market as chemicals for use in a variety of industrial process and products (Breggin et al. 2009: 38). The REACH Regulation contains broad exemptions for pharmaceuticals and medical products (Articles 2(5) and 60(2)) but there remain important regulatory obligations pertaining to disclosure of information and acceptable use. Since the main exemption for pharmaceuticals covers only substances that are explicitly "used in" medicinal products, other substances used in the manufacturing process are likely to be subject to the full scope of REACH. Also, even if the substances are not subject to most of the REACH requirements when intended for medicinal products, pharmaceutical companies are likely to be affected by the restrictions applying to their suppliers (Covington and Burling 2007; Breggin et al. 2009).

There are no provisions in REACH that refer explicitly to nanomaterials (Bowman and van Calster 2007). However, nanomaterials are covered by the "substance" definition in Article 3 of REACH (Breggin et al. 2009: 44–47). In June 2008, the European Commission (EC Communication 2008) concluded that with respect to potential health, safety and environmental risks posed by nanomaterials, the current EU legislative framework was, in principle, adequate to effectively regulate nanoscale substances. In coming to this conclusion, however, the Commission did acknowledge the need for fundamental scientific research to be undertaken in order to determine if the existing standards and risk assessment methodology underpinning the regime, as set out in the Annexes of the REACH Regulation, are appropriate for nanoscale substances. Until the state of the art has evolved adequately so as to be able to answer such questions, testing will continue to be carried out in accordance with the existing guidelines (Breggin et al. 2009: 47).

REGULATING NANOMEDICINE IN THE UNITED STATES

The FDA and Nanomedicine

The Food and Drug Administration (FDA) is the nodal agency responsible for the regulation of food products, human and animal drugs, therapeutic agents, medical devices, radiation-emitting products, cosmetics and animal feed in the US. It consists of several centres, of which the Center for Drug Evaluation and Research (CDER), Center for Devices and Radiological Health (CDRH) and the Center for Biologics Evaluation and Research (CBER) are responsible for the regulation of drugs, devices and biologics respectively. These centres are also primarily responsible for regulating nanomedical products (Miller 2003: 13; Boucher 2008: 103). For purposes of regulation, the FDA classifies medical products as drugs, devices, biologics or combination products (Harris 2009).

The CDER regulates drugs under the Federal Food, Drug, and Cosmetic Act of 1938 and its amendments. A manufacturer of a new drug first files an Investigational New Drug application to get approval for research on human subjects. The purpose of the application is to provide the FDA with information about the drug's active ingredients and structural formula, as well as its chemistry, manufacturing, and controls information (Harris 2003). CDER must then approve and monitor the clinical trials. Upon completion the CDER may then approve a New Drug Application (NDA). The manufacturer must also comply with labelling requirements and a set of manufacturing regulations called the current Good Manufacturing Practices (Miller 2003: 14).

Table 2. Regulatory bodies and legislation in the U.S. relevant to nanomedicine

Regulatory bodies
Primary Body
Food and Drug Administration (FDA)
Centres and Groups
Center for Drug Evaluation and Research (CDER)
Center for Devices and Radiological Health (CDRH)
Center for Biologics Evaluation and Research (CBER)
NanoTechnology Interest Group (NTIG)
Nanotechnology Task Force (NTF)
Legislation
Federal Food, Drug, and Cosmetic Act, 1938
Public Health Service Act, 1944
Toxic Substances Control Act (TSCA), 1976

The CDRH is responsible for regulating medical devices, which are classified as Class I, Class II, or Class III. Class I devices are considered to present the lowest risk and are subject to general controls. Class II devices are subject to special controls, while Class III devices, which are deemed to present the greatest risk to human health, are subject to review for safety and effectiveness. In order to obtain FDA approval for clinical trials, a manufacturer must submit an Investigation Device Exception. In order to market the device, a manufacturer must submit a Premarket Approval Application (PMA), which imposes strict conditions on the manufacturing and labelling of the device. Importantly, a new device that is considered to be "substantially equivalent" to a device already in the market is not subject to review as a Class III device if the manufacturer obtains 510(k) approval.

While drugs and devices are regulated under the Food, Drug, and Cosmetic Act, biologics are regulated by the CBER primarily under the Public Health Service Act. CBER is responsible for regulating a wide variety of "biologics": blood and blood components, devices, allergenic extracts, vaccines, tissues, somatic cell and gene therapies, biotech-derived therapeutics, and xenotransplantation (Miller 2003: 15). Approval must be granted for clinical testing and, in order to obtain a license to market a biologic, the agency must determine that the product is "safe, pure, potent, and manufactured accordingly".

The FDA's position, at least to date, in relation to the application of the current regulatory regime has been to consider particle size as being immaterial and that the safety of the large-particle version (bulk counterparts) of an active ingredient can be used to predict the safety of the nanoscale version of the same ingredient. According to the FDA, if large-particle versions of a product are considered to be safe, then it can be presumed that the nano versions of the product are also safe. Therefore, under the current regime it would appear that nano-scale ingredients are presumed to be bioequivalent to their bulk counterparts. Thus, manufacturers of nanoproducts are neither required to obtain premarket approval from the FDA nor required to list nanoscale ingredients on a product's label.

The FDA has similarly stated that, based on the evidence it has reviewed, existing health and safety tests that it uses to assess the safety of normal-size materials are adequate to assess the health effects of nanoproducts (Bawa 2010). The FDA's position on this is somewhat surprising given that other US regulatory agencies, for example, the Environmental Protection Agency (EPA), have concluded that some nanoparticles may exhibit toxicological properties that differ from their bulk counterparts (Bawa 2010). This can be seen in recent EPA studies conducted on nanoscale silver (2010) and nanoscale titanium dioxide (2009).

In applying its current regulations to nanomedicine products it would appear that the FDA is likely to encounter two other issues that may impact on its ability to effectively regulate such products:

1. The issue of appropriately placing those products into its present classification scheme, and
2. The maintenance of an adequate level of scientific expertise in the nanomedicine field (Miller 2003; Harris 2009).

The FDA uses a product's primary mode of action to determine its classification within the context of the regime and to decide which of its Centres will have primary jurisdiction over the product.

As medical products become smaller, classification will become increasingly difficult for two reasons. First, the ability to operate at the nano-level will increasingly enable manufacturers to combine different types of components in producing a single therapy. Second, in the long run, sophisticated nanomedical products will blur the distinction between "mechanical", "chemical", and "biological" and make it difficult to determine whether a product is a drug, device, biologic, or combination product (Miller 2003: 24, 25).

The second issue posed by nanomedical products is the maintaining of scientific expertise at the FDA (Miller 2003; Harris 2009). Although the agency has taken steps to acquire the technical abilities necessary for effective regulation of other emerging technologies, the FDA, like many other analogous regulatory bodies around the world, is likely to face unique problems in obtaining the aptitude to effectively regulate nanomedical products.

The NTIG and the FDA Task Force

To facilitate the effective regulation of nanoproducts, the FDA formed an internal NanoTechnology Interest Group (NTIG), which is composed of representatives from all of its regulatory Centres. A Nanotechnology Task Force (NTF) was also formed in 2007 and was charged with the responsibility of looking at the Agency's capacity to deal with nanotechnologies. In their publicly available report, the Task Force concluded that existing regulations in the US were sufficient and comprehensive to ensure the safety of nanoproducts since these products would undergo premarket testing and approval either as new drugs under the NDA process, or in the case of medical devices, under the Class III premarket approval process (NTF Report 2007; Bawa 2010). This conclusion is based on the assumption that current regulatory requirements would be sufficient to detect any toxicity or safety issues. Some commentators have criticized this inaccurate extrapolation especially since most nanoproducts approved by the FDA have obtained approval based in whole or in part on studies of non-nano versions (based on their bulk counterparts). In this regard, the approvals were granted on the basis of safety data of equivalent non-nano versions; the nanoproducts did not undergo the full PMA or NDA (Bawa 2010).

The guiding principle behind the FDA's position is that it regulates end products rather than technologies *per se*. As such, the Agency does not regulate nanomaterials or manufacturing processes, but rather the end products that may or may not contain nanomaterials or nanoparticles (Bawa 2010). The NTF Report does allude to the need for regulatory oversight of some nanoproducts but offers no regulatory remedy, stating instead that "the Task Force believes it is important that manufacturers and sponsors be aware of the issues raised by nanoscale materials and the possible change in the regulatory status/pathway when products contain nanoscale materials" (NTF Report 2007: 32).

The Regulation of Nanomaterials and Nanoparticles in Medicine

Pharmaceutical companies will undoubtedly be affected by the restrictions and regulations applied to their suppliers, i.e., chemical companies. In this context in the US, the Toxic Substances Control Act 1976 (TSCA) is the most significant legislative instrument. Until recently, however, only minimal information had been made available to the public about what, if any, regulatory actions the EPA might pursue in relation to nanomaterials. While the EPA has been more forthcoming in recent years, partly through the establishment of a voluntary stewardship program and the publication of two Significant New Use Rules, information is still limited because of claims of confidential business information (Breggin et al. 2009: xiv).

TSCA and REACH are similar as they require companies to determine prior to manufacture whether a chemical is subject to the regulatory requirements set out in the respective instruments. Both US and EU regulators consider their authorities under TSCA and REACH broad enough to cover the current generation of nanomaterials, although both regulatory schemes provide for exemptions and the standards and processes that govern whether a particular nanoscale chemical will be subject to government requirements differ considerably. A principal difference is that REACH eliminates the distinction between new and existing chemicals, in an effort to subject all chemicals to the same regulatory oversight. In contrast, TSCA distinguishes between new and existing chemicals for purposes of the pre-manufacture obligations imposed on manufacturers and the corresponding regulatory tools available to the EPA. "New" chemicals are automatically subject to pre-manufacture notification and review, enabling the EPA to determine whether restrictions should be imposed prior to allowing the chemical to be manufactured. However, the information that companies are required to submit is typically limited in comparison to that required by REACH (Breggin et al. 2009).

The approaches and authorities granted to regulators to require manufacturers to produce information, including environmental, health and safety data, differ significantly under the two systems. Nevertheless, in theory both are science-based approaches that seek to assess the risk of chemicals (Breggin et al. 2009). Hence, US and EU regulators face fundamentally similar challenges in regulating nanomaterials, including limited knowledge of human health and ecotoxicological effects and the need to adjust or develop new test methods.

CONCLUSION: SECURING *NANO RX*

Legislators in both the EU and the US are likely to continue regulating nanomedical products and devices under the existing regulatory structures wherever possible rather than create a *sui generis* regime for nanomedicines. This would arguably seem to be a logical step in the short term as the regimes for medical products and devices regulation are relatively elaborate and involve not only pre-market authorization processes but also post-market surveillance and compliance requirements.

However, in the long term none of the legislations were written with nanomedicinal applications in mind. It therefore seems likely that these legislations will need to be revised in due course in order to take account of specific risks and challenges (EGE 2007; N&ET 2007; SCHENIR 2007; Wagner et al. 2007; Bawa 2010).

The EU has already begun the process of updating regimes in order to incorporate nano-specific provisions as evidenced by the inclusion of such provisions in EU Cosmetics Regulation in 2009; special treatment for nanomaterials and nanoparticles in the Novel Foods Regulation is also envisaged as part of the recast of that instrument (Bowman et al. 2010). The entry into force of the Cosmetic Regulation is the first attempt by a national or supranational body to specifically regulate nanotechnologies. Definitions set out in legislative instruments are fundamental to their operation as well as in establishing the subject matter and scope (Bowman et al. 2010). With a wide array of definitions, finding a precise and feasible definition of nanomedicine and inter-related applications and devices is also likely to be a challenge. For example, the inclusion of a definition of "nanomaterials" in both the Cosmetics Regulation and the proposed Novel Foods Regulation will arguably have a significant impact. It will thus have to be seen whether such nano-specific aspects or definitions will be introduced in the nanomedical sphere in the future.

With the advancement of nanomedicine, the borderline between medicines, devices and therapies is becoming increasingly blurred, which will in turn blur the demarcation between the regulatory systems. Moreover, these products, which have the potential to span the regulatory boundaries between medicinal products, devices and therapies, appear likely to challenge current criteria for classification and evaluation. We would argue that it is therefore essential that steps are taken at this early stage to address these issues as the classification system establishes the regulatory pathway, and subsequent regulatory hurdles, that the product or device is required to clear before it may be placed on the market.

Risk assessment, safety and quality requirements are to be fulfilled by conformity to established quality systems and product standards which—based on the current scientific state of the art—may or may not be suitable to address various concerns relating to nanomedicine. As progress in medicine and nanoscience accelerates in the manufacture and characterization of nanoscale materials and nano-enabled products, it will become increasingly important to researchers, manufacturers, regulators and other stakeholders to have agreed nanotechnology standards. Such standards will have to address the evaluation of the quality, safety, efficacy and risk management of nanomedicinal products. Such standards, if and when introduced, could be introduced initially in guidance documents in order to maximize flexibility and subsequently in legally binding instruments as knowledge gaps are addressed.

Concerns over the current gaps in knowledge, expertise and scientific data concerning nanoparticle characterization, their detection and measurement, the persistence of nanoparticles in humans and in the environment, as well as toxicology related to nanoparticles are not unique to EU or US stakeholders. Yet, despite the many unknowns, products incorporating nanomaterials continue to make their way into the marketplace in many jurisdictions. Given this, and current concerns, it would seem prudent that further toxicological and risk studies be performed as a matter of urgency and that, where appropriate, guidelines be introduced by regulatory agencies for nanomedicinal products.

Though modification of existing regulations and standards to incorporate nanomedicine seems to be one potential recourse, such action will not address the more fundamental issue—that existing risk assessment methodologies may be inadequate primarily because of their reliance on mass metrics. Hence, if existing regulations are modified to make them more nano-conversant, existing risk methodologies will also have to be adapted to introduce agglomeration, particle size, shape and surface reactivity into the assessment criteria (Handy et al. 2008), and steps must be initiated to bolster expertise. Cooperation between, for example, the EU, the US, other OECD members and other key stakeholders on these issues would also help expedite the process. With nanomedicine comes great promise; however, it also introduces a number of regulatory challenges. The direction and application of nanomedicine in the long term will depend on how these challenges are currently addressed.

Key Facts

- Nanomedicine is expected to challenge existing perceptions, dynamics and standards relating to ethics, safety and governance.
- Uncertainty about the risks, as well as concerns about the adequacy of current regulatory frameworks, threatens to impede the development of nanomedicine.
- Regulators are increasingly being faced with the need to regulate this promising field that challenges existing regulatory structures and testing methods.
- Legislators in the EU and the US are likely to continue regulating nanomedical products and devices under the existing regulatory

structures for the time being, though recent developments within the EU in relation to the regulation of cosmetics and foods may lead to a rethink.

Definitions

Medical device: Any instrument, apparatus, appliance, material or other article, whether used alone or in combination, including the software necessary for its proper application intended by the manufacturer to be used for human beings.

Medicinal product: Any substance or combination of substances having properties for treating or preventing disease in human beings. Any substance or combination of substances that may be used in or administered to human beings with a view to restoring, correcting or modifying physiological functions by exerting a pharmacological, immunological or metabolic action, or to making a medical diagnosis.

Nanomedicine: The science and technology of diagnosing, treating and preventing disease and of preserving and improving human health, using molecular tools and molecular knowledge of the human body.

Summary Points

- Though the potential benefits of nanomedicine are enormous, uncertainty about the risks and concerns about the adequacy of regulatory oversight threaten to impede its development.
- It is becoming increasingly clear that appropriate and effective regulatory structures will be critical in the successful implementation of nanomedicine and the fulfillment of its promise.
- Nanomedicine challenges existing regulatory structures and testing methods, while knowledge gaps, questions of expertise and definitional issues all pose crucial challenges.
- There are currently no specific pieces of nanomedical legislation; this does not mean that such products are "unregulated". Existing legislation is and will continue to be applied despite concerns regarding its adequacy.
- The boundaries between medicinal products, devices and therapies are likely to become increasingly blurred, thereby challenging the competencies of regulatory bodies.
- Given the uncertainties, current procedures are likely to require modification, while legislations will need to be revised to take into account specific risks and challenges.

Abbreviations

AIMDD	:	Active Implantable Medical Devices Directive
CBER	:	Center for Biologics Evaluation and Research
CDER	:	Center for Drug Evaluation and Research
CDRH	:	Center for Devices and Radiological Health
CHMP	:	Committee for Medicinal Products for Human Use
EMA	:	European Medicines Agency
EPA	:	Environmental Protection Agency
FDA	:	Food and Drug Administration
ITF	:	Innovation Task Force
IVDD	:	In Vitro Diagnostic Medical Devices Directive
MDD	:	Medical Devices Directive
N&ET	:	New and Emerging Technologies Working Group
NDA	:	New Drug Application
NTF	:	Nanotechnology Task Force
NTIG	:	NanoTechnology Interest Group
PMA	:	Premarket Approval Application
REACH	:	Registration, Evaluation, Authorisation and Restriction of Chemical Substances
SCENIHR	:	Scientific Committee on Emerging and Newly Identified Health Risks
TSCA	:	Toxic Substances Control Act

References

Bawa, R. 2010. Regulating nanomedicine—Can the FDA handle it? Available at:< https://medical.wesrch.com/User_images/Pdf/XXF_1278627200.pdf>

Berube, M.D. 2006. Nano-Hype: The truth behind the nanotechnology buzz, Prometheus Books, NY, 288–300.

Boucher, P. 2008. Nanotechnology—Legal Aspects. CRC Press, Boca Raton, Florida.

Bowman, D.M., and G. van Calster. 2007. Does REACH go too far? Nature Nanotechnol. 1: 525–526.

Bowman, D.M., and G. van Calster. 2008. Flawless or fallible? A review of the applicability of the European Union's Cosmetics Directive in relation to nanocosmetics. Stud. Ethics, Law Technol. 2(3): Article 6.

Bowman, D., J. D'Silva, and G. van Calster. 2010. Defining nanomaterials for the purpose of regulation within the European Union. Eur. J. Risk Reg. 2: 115–122.

Breggin, L. et al. 2009. Securing the Promise of Nanotechnologies—Towards Transatlantic Regulatory Cooperation. Report, Royal Institute of International Affairs.

CHMP. 2006. Reflection paper on Nanotechnology-Based Medicinal Products for Human Use.

Chowdhury, N. 2010. Regulation of nanomedicines in the EU: distilling lessons from the paediatric and the advanced therapy medicinal products approaches. Nanomedicine 5(1): 135–142.

Communication from the Commission to the European Parliament, the Council and the European Economic and Social Committee—Regulatory aspects of nanomaterials, 2008.

Covington and Burling, LLP. 2007. REACH and its impact on pharmaceuticals, available at:< http://www.cov.com/files/Publication/ed03bc97-89de-471d-9163-4f1a26899ba2/ Presentation/PublicationAttachment/e60742d7-e808-45a9-bec6-590676c12a9a/735.pdf>

D'Silva, J., and G. Van Calster. 2009.Taking temperature—A review of European Union regulation in nanomedicine. Eur. J. Health Law 16(3): 249–269.

De Jong, W., and P. Borm. 2008. Drug delivery and nanoparticles: Applications and hazards. Int. J. Nanomedicine. 3(2): 133–149.

Delaney, H., and R. van de Zande. 2000. NIST Special Publication 51—A Guide to EU Standards and Conformity Assessment, 10–11.

European Group on Ethics in Science and New Technologies (EGE)—Opinion 21—On the ethical aspects of nanomedicine, 2007.

European Science Foundation. 2005. ESF Forward Look on Nanomedicine.

Foldvari, M., and M. Bagonluri. 2008. Carbon nanotubes as functional excipients for nanomedicines: Drug delivery and biocompatibility issues, Nanomedicine: Nanotechnol. Biol. Med. 12 June 2008 (10.1016/j.nano.2008.04.003).

Gergely, A., D.M. Bowman, and Q. Chaudhry. 2010. Small ingredients in a big picture: Regulatory perspectives on nanotechnologies in foods and food contact materials. In: Q. Chaudhry, L. Castle and R. Watkins (eds.). Nanotechnologies in Food. The Royal Society of Chemistry, London, pp. 150–181.

Handy, D.R., R. Owen, and E. Jones Valsami. 2008. The ecotoxicology of nanoparticles and nanomaterials: current status, knowledge gaps, challenges and future needs. Ecotoxicology 17: 315–325.

Hardman, R. 2006. A toxicological review of quantum dots: Toxicity depends on physico-chemical and environmental factors. Environ. Health Persp. 114(2): 165–172.

Harris L.D., and R. Bawa. 2007. The carbon nanotube patent landscape in nanomedicine: an expert opinion. Expert Opin. Ther. Patents 17(9): 1–11.

Harris, S. 2009. The regulation of nanomedicine: Will the existing regulatory scheme of the FDA suffice? Richmond J. Law Technol. XVI(4), http://law.richmond.edu/jolt/v16i2/ article4.pdf

Holsapple et al. 2005. Research strategies for safety evaluation of nanomaterials, Part II: Toxicological and safety evaluation of nanomaterials, current challenges and data needs. Toxicol. Sci. 88(1): 12–17.

Kane, A., and R.H. Hurt. 2008. The asbestos analogy revisited, Nature Nanotechnol. 3: 378–379.

Linkov, I., F.K. Satterstrom, and L.M. Corey. 2008. Nanotoxicology and nanomedicine: making hard decisions. Nanomedicine: Nanotechnol. Biol. Med. 4: 167 –171.

Marchant, G.E., D.J. Sylvester, K.W. Abbott, and T.L. Danforth. 2009. International harmonization of regulation of nanomedicine. Stud. Ethics, Law, Technol. 3(3), Article 6.

Maynard, A. 2006. Nanotechnology: The next big thing, or much ado about nothing? Ann. Occupational Hyg. 51(1): 4–7.

Miller, J. 2003. Beyond biotechnology: FDA regulation of nanomedicine. Columbia Sci. Technol. Law Rev. IV: E5.

Monteiro-Riviere, N., et al. 2006. Penetration of intact skin by quantum dots with diverse physicochemical properties. Toxicol. Sci. 91(1): 159–165.

N&ET Working Group. 2007. Report on nanotechnology to the medical devices expert group findings and recommendations.

Nel, A., et al. 2006. Toxic potential of materials at the nanolevel. Science 311: 622–627.

NTF Report. 2007. Nanotechnology: A Report of the US Food and Drug Administration Nanotechnology Task Force.

Oberdörster, E. 2004. Manufactured nanomaterials (Fullerenes, C60) induce oxidative stress in the brain of juvenile largemouth bass. Environ. Health Persp. 112(10): 1058–1062.

Oberdorster, G., Z. Sharp, V. Atudorei, A. Elder, R. Gelein, W. Kreyling et al. 2004. Translocation of inhaled ultrafine particles to the brain. Inhal. Toxicol. 16: 437–445.

Poland, C.A. et al. 2008. Carbon nanotubes introduced into the abdominal cavity of mice show asbestos like pathogenicity in a pilot study. Nature Nanotechnol. 3: 423–428.

Recast of the Medical Devices Directives—Public Consultation document, 2008, available at: <http://ec.europa.eu/enterprise/medical_devices/recast_docs_2008/Public_consultation_en.pdf>

SCENIHR. 2006. Modified Opinion (after public consultation) on the appropriateness of existing methodologies to assess the potential risks associated with engineered and adventitious products of nanotechnologies, adopted by the SCENIHR during the 10th plenary meeting of 10 March 2006.

SCENIHR. 2007. The appropriateness of the risk assessment methodology in accordance with the Technical Guidance Documents for new and existing substances for assessing the risks of nanomaterials.

The Royal Society. 2004. Nanoscience and nanotechnologies: opportunities and uncertainties—Report. Latimer Trend Ltd, UK.

Throne-Holst, H., and P. Strandbakken. 2009. "Nobody told me I was a Nano-consumer": How nanotechnologies might challenge the notion of consumer rights. J. Consumer Pol. 32(4): 393–402.

Van Calster, G. 2006. Regulating nanotechnology in the European Union. Nanotechnol. Law Bus. 3(2): 359–372.

Wagner, V., et al. 2006. The emerging nanomedicine landscape. Nature Biotechnol. 24: 1211–1217.

3

Nanotubes and Their Application to Nanomedicine

Gianni Ciofani

ABSTRACT

Nanomedicine is the application of nanotechnology to medicine and is based on three mutually overlapping and progressively more powerful technologies: (1) nanoscale-structured materials and devices; (2) benefits of molecular medicine via genomics, proteomics, and artificially engineered microorganisms; and (3) molecular machine systems such as nanorobots that will allow instant diagnosis with destruction of cause of pathology. Artificial nanostructures, such as nanoparticles and nanodevices, being of the same size as biological entities, can readily interact with biomolecules on both the cell surface and within the cell. Nanomedical developments range from nanoparticles for molecular diagnostics, imaging and therapy to integrated medical nanosystems, which may perform complex repair actions at the cellular level inside the body. This chapter will focus on applications of nanotubes in nanomedicine, with particular emphasis on drug delivery and cell therapy. Major examples of recent developments of carbon nanotube research will be reviewed. A second section will be dedicated to a highly innovative nanomaterial, boron nitride nanotubes, that have emerged in recent years as promising bio-nanovectors. A brief

Italian Institute of Technology, Smart Materials Lab, Center of MicroBioRobotics c/o Scuola Superiore Sant'Anna, Viale Rinaldo Piaggio, 34 - 56025 Pontedera (Pisa), Italy; Email: g.ciofani@sssup.it; gianni.ciofani@iit.it

List of abbreviations after the text.

overview of other nanostructured tubular materials will conclude the chapter, stressing the necessity to exploit both the chemical and physical properties of these nanomaterials for the development of active and smart nanovectors.

INTRODUCTION

Nanotechnology refers to the research and technology development at atomic, molecular, and macromolecular scales that leads to the controlled manipulation and study of structures and devices with length scales in the range of 1–100 nm (Gao and Xu 2009).

Nanoscale structures and materials have been explored in many biological applications because their novel properties and functions differ drastically from their bulk counterparts. Particularly, their high volume/surface ratio, surface tailorability, improved solubility, and multifunctionality open many new possibilities for biomedicine. Moreover, the intrinsic optical, magnetic, and biological properties of nanomaterials offer remarkable opportunities to study and regulate complex biological processes for biomedical applications in an unprecedented manner.

This chapter is dedicated to the applications in nanomedicine of inorganic nanotubes, with particular attention to carbon and boron nitride nanotubes, and their exploitation as intracellular carriers. At the end, a brief introduction of extremely innovative materials will be given. Carbon nanotubes (CNTs) are molecular-scale tubes of graphitic carbon with outstanding properties. They are among the stiffest and strongest fibres known and have remarkable electronic properties and many other unique characteristics (Sinha and Yeow 2005). For these reasons, they have attracted huge academic and industrial interest, with thousands of papers on nanotubes being published each year. Boron nitride nanotubes (BNNTs) are of significant interest for the scientific community. In fact, like CNTs, they have attracted wide attention because they have potentially unique and important properties for structural and electronic applications (Terrones et al. 2007). A BNNT is a structural analogue of a carbon nanotube: alternating B and N atoms entirely replace C atoms in a graphitic-like sheet with almost no change in atomic spacing; despite this, CNTs and BNNTs have many different properties.

CARBON NANOTUBES

Biocompatibility Issues

An initial review on the safety risk induced by CNTs was presented in 2003 (Gogotsi et al. 2003). In later years, studies on health risks related to nanoparticles found that they penetrate the human body via the lungs and the intestines, depending on their size, surface properties and point of contact. The most extensive analysis of the research devoted to the toxicology of CNTs was collected in a Special Issue of the journal *Carbon* (issue no. 44) in 2006. Data of the latest five years on this topic are often conflicting, pointing out the need for standards for CNTs characterization, dispersion/solubilization, treatment, and biological testing.

Positive results were presented by Muller et al. (2006). Rat peritoneal macrophages were incubated in media containing purified MWCNTs and "ground" MWCNTs at different concentrations (20 to 100 µg/ml) up to 24 h. Ground MWCNTs had a capacity for inducing dose-dependent cytotoxicity and up-regulating pro-inflammatory cytokine tumor necrosis factor-α expression similar to asbestos and carbon black, but lower than purified MWCNTs.

Some studies pointed out that CNT cytotoxicity is related to the degree of dispersion as well as of functionalization. Sayes et al., for example, performed *in vitro* cytotoxicity screens on cultured human dermal fibroblasts with derivatized and surfactant-dispersed SWCNTs: as the degree of sidewall functionalization increases, the SWCNT sample becomes less cytotoxic than surfactant-stabilized SWCNTs (Sayes et al. 2006).

Interestingly, Bardi et al. very recently showed that MWCNTs coated with Pluronic F127 (PF127) surfactant can be injected in the mouse cerebral cortex without causing degeneration of the neurons surrounding the site of injection (Fig. 1). They also showed that, contrary to previous reports on lack of PF127 toxicity on cultured cell lines, concentrations of PF127 as low as 0.01% can induce apoptosis of mouse primary cortical neurons *in vitro* within 24 h. However, the presence of MWCNTs can avoid PF127-induced apoptosis, demonstrating that PF127-coated MWCNTs do not induce apoptosis of cortical neurons. Moreover, the presence of MWCNTs can reduce PF127 toxicity (Bardi et al. 2009).

Functionalized CNTs as Nanocarriers for Drugs, Peptides and Nucleic Acids

The search for new and effective drug delivery systems is a research field in rapid expansion. Many different methodologies have been developed according to the different classes of bioactive molecules to be delivered (e.g., peptides, proteins, nucleic acids and small organic molecules) and the characteristics of the target tissues. Liposomes, emulsions, polymers, micro- and nanoparticles are the most commonly studied vehicles, but new approaches are emerging, mainly due to significant advances in nanotechnology and nanofabrication.

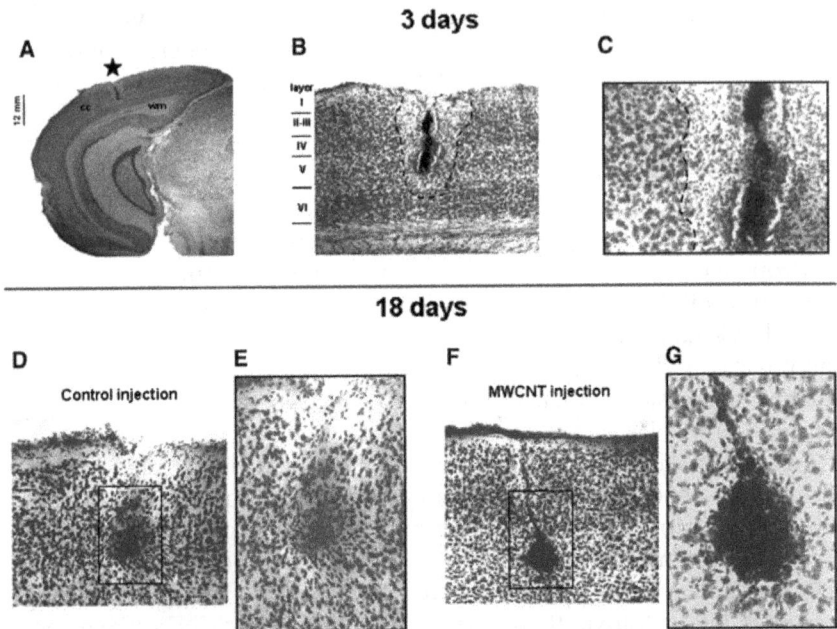

Fig. 1. *In vivo* testing of carbon nanotubes: injection in mouse brain. Sections of mouse brain at the site of CNT injection. (A) The injection site is labeled with a star. *cc*: cerebral cortex; *wm*: white matter. (B) The injection site in mice treated for 3 d at higher magnification. Normal neuronal density and tissue layering is present outside the lesioned site (broken line). (C) Higher magnification of the transition between the lesioned and the unlesioned area (on the right, broken line). Control mice (D, E) and MWCNT-injected mice (F, G) brain cortices present a scar of glial cells surrounding the injection site. Higher magnifications of (E) the control and (G) the MWCNT-injected mice are shown. Reprinted from Bardi et al. (2009) with permission from Elsevier.

Color image of this figure appears in the color plate section at the end of the book.

The ability of f-CNT (functionalized CNTs) to penetrate cells makes them potentially useful as vehicles for the delivery of small drug molecules. The development of delivery systems able to carry one or more therapeutic agents with recognition capacity, optical signals for imaging and/or specific targeting is of fundamental advantage, for example, in the treatment of cancer and of different types of infectious diseases. CNT-based systems can help to solve transport problems for pharmacologically relevant compounds that need to be internalized, and may have several potential therapeutic applications.

The development of functionalized carbon nanotubes to target and to be up-taken by specific cell populations without collateral consequences for healthy tissues would be of paramount importance also in cancer treatment. The molecular targeting of CNT-based delivery systems derivatized with a therapeutic agent is possible if an active recognition moiety is simultaneously present at the surface of the nanocarrier. In addition, attachment of a fluorescent molecule would provide optical signals for imaging and localization of the CNT–drug conjugates. Therefore, multiple functionalization of CNTs is of particular interest for multimodal delivery of anticancer agents. Pastorin et al. (2006) described a straightforward methodology for the introduction of two orthogonally protected amino groups on the sidewalls of CNT, subsequently derivatized with fluoresceinisothiocyanate (FITC) and methotrexate (MTX). MTX is a widely used drug against cancer; however, it suffers from low cellular uptake. Its conjugation to CNT represents a promising approach to overcome its limited cellular uptake by enhancing its internalization via the f-CNT. A similar approach was followed by Wu et al. (2005) that also performed multiple functionalization of CNTs with different types of molecules. A fluorescent probe (fluorescein) for tracking the cellular uptake of the material, and an antibiotic moiety (amphotericin B) as the active molecule were covalently linked to the MWCNTs.

Other researchers have shown alternative mechanisms of interaction between CNTs and peptides. Various proteins, in fact, spontaneously adsorb on the sidewalls of acid-oxidized CNTs. This simple non-specific binding scheme can be used to afford non-covalent protein-nanotube conjugates. The combined treatment of refluxing and sonication in nitric and/or sulfuric acid is known to produce short individual (50–500 nm) or small bundles of CNTs with carboxylic groups (e.g., -COOH) along the sidewalls and ends of the tubes. These functional groups impart hydrophilicity to the nanotubes and make them stable in aqueous solutions. In the work proposed by Shi Kam and Dai (Shi Kam et al. 2005a), it was found that simple mixing of oxidized SWCNTs with protein solutions led to non-specific binding of proteins to the nanotubes, as it can be deduced from AFM data, due to electrostatic forces between functional groups on SWCNTs and to positively charged domains on proteins and/ or to hydrophobic interactions. Bovine serum albumin (BSA), protein A, and human IgG, all of them labeled with a fluorescent probe, and finally cytochrome c (cyt-c, an apoptosis inducer) were tested *in vitro*.

Concerning cancer therapy, it is worthwhile to mention an example of *in vivo* application reported by Liu et al. (2008). They conjugated paclitaxel (PTX), a widely used cancer chemotherapy drug, to branched polyethylene glycol chains on SWCNTs, to obtain a water-soluble SWCNT-PTX conjugate. SWCNT-PTX affords higher efficacy in suppressing tumor growth than clinical Taxol in a murine 4T1 breast cancer model, owing to prolonged blood circulation and 10-fold higher tumor PTX uptake. Drug molecules carried into the reticuloendothelial system are released from SWCNTs and excreted via biliary pathway without causing toxic effects to normal organs.

Ou and colleagues (2009) also proposed a chemical functionalized vector based on CNTs for cancer therapy. In this study, a novel functional SWCNT based on an integrin $\alpha_v\beta_3$ monoclonal antibody was developed and was used for cancer cell targeting *in vitro*. SWCNTs were non-covalently wrapped by phospholipid-bearing polyethylene glycol. *In vitro* study revealed that SWCNT-PEG-mAb presented a high targeting efficiency on integrin $\alpha_v\beta_3$-positive U87MG cells with low cellular toxicity.

Heister *et al.* (2009) presented a method for a triple functionalization of oxidized SWCNTs with the anti-cancer drug doxorubicin, a monoclonal antibody, and a fluorescent marker at non-competing binding sites (Fig. 2). The proposed methodology allowed for the targeted delivery of the anti-cancer drug to cancer cells and the visualization of the cellular uptake of SWCNTs by confocal microscopy. The complex was efficiently taken up by cancer cells with subsequent intracellular release of doxorubicin.

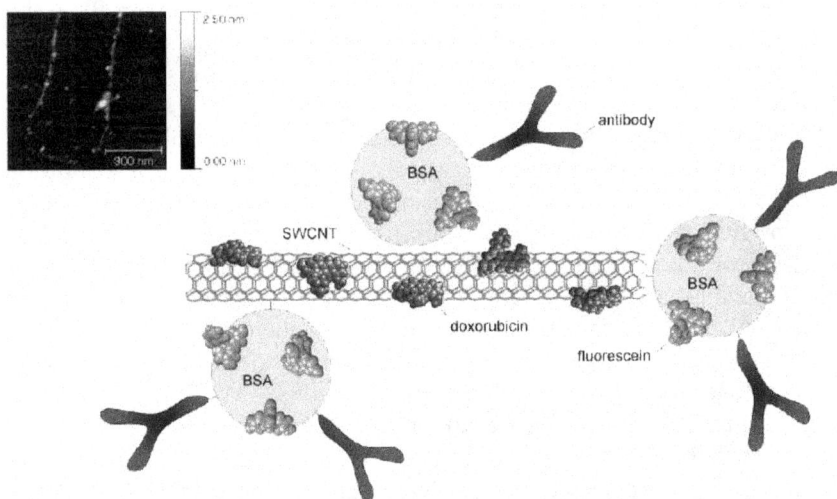

Fig. 2. Example of carbon nanotubes multiple functionalization. Schematic illustration of the doxorubicin–fluorescein–BSA–antibody-SWCNT complexes (red = doxorubicin, green = fluorescein, light blue = BSA, dark blue = antibodies). Inset: AFM image of doxorubicin–fluorescein–BSA–SWCNT complexes (without antibodies). Reprinted from Heister et al. (2009) with permission from Elsevier.

Color image of this figure appears in the color plate section at the end of the book.

The use of CNTs as a new carrier system for nucleic acids is another area of research currently under investigation. The most commonly used DNA carriers are based on viral vectors, liposomes, cationic lipids, polymers and nanoparticles. Although viral vectors are very efficient in gene expression, their use is limited because of the concerns about their safety. Non-viral vectors offer various alternatives regarding the size and type of vectors, they are chemically controllable, and they display a reduced immunogenicity. However, they often present low gene expression efficiency rates because of their poor capability in reaching and crossing the nuclear membrane. In this context, CNTs seem to be very promising because they do not inherently trigger an immune response (Klumpp et al. 2006).

Singh et al. (2004) explored the potential of CNTs as delivery vehicles of nucleic acids, showing as different kinds of functionalization approaches exhibited upregulation of marker gene expression over naked DNA using a human cell line.

Dwyer et al. (2002) proposed an alternative functionalization method for binding CNTs with DNA. Their study represents a step toward the DNA-guided assembly of carbon nanotubes, by demonstrating that the well-known chemical pathway already discovered for attaching amino-terminal compounds to carbon nanotubes is also compatible with DNA functionalization.

Non-covalent binding DNA-CNTs were investigated by Liu et al. (2005) exploiting polyethylenimine (PEI) grafted CNTs. DNA was securely immobilized onto the surface of PEI-g-MWCNTs as demonstrated by the total inhibition of the migration of DNA in gel electrophoresis, and PEI-g-MWCNTs showed transfection efficiency similar to or even several times higher than that one of PEI alone, and several orders of magnitude higher than that one of naked DNA.

A frontier in the field of gene and protein therapy is RNA interference (RNAi) for gene silencing by short interfering RNA (siRNA) delivered to mammalian cells. Efficient intracellular transport and delivery of siRNA are critical to RNAi potency. As an application of cleavable functionalization of SWCNTs, Shi Kam et al. showed transport, release, and delivery of siRNA in mammalian cells by SWCNT carriers, and achieved highly efficient lamin A/C gene silencing compared to existing transfection agents (Shi Kam et al. 2005b).

Exploitation of Physical Properties

We have shown that many CNT functionalizations have been successfully achieved, but no applications for drug delivery have been approved, nor have drugs using this technology entered the market yet. At the moment, traditional nanoplatforms such as liposomes or polymer-based carriers are still preferred because of their assessed biosafety *in vivo* and *in vitro*. Nevertheless, we believe that CNTs should be much more deeply investigated for their potential impact in nanomedicine, in particular combining their impressive physical properties with the chemical approach.

A study by Shi Kam et al., for example, shows that while biological systems are transparent to 700–1,100 nm near-infrared (NIR) light, the strong absorbance of SWCNTs in this window can be used for optical stimulation of nanotubes inside living cells, to afford various useful functions (Shi Kam et al. 2005c). Oligonucleotides transported inside cells by nanotubes, for example, can translocate into cell nucleus upon endosomal rupture triggered by NIR laser pulses. Continuous NIR radiation can cause cell death because of excessive local heating: selective cell destruction can be therefore achieved by functionalization of SWCNTs with a folate moiety, selective internalization of SWCNTs inside cells over-expressing folate receptor, and NIR-triggered cell death, without harming receptor-free normal cells.

A recent study made use of the unique properties of CNTs also for electroporation (Rojas-Chapana et al. 2004). The "lightning rod effect" of CNTs, mixed within cell suspensions, was used to create localized high field regions at the tips, creating pores in the membrane, allowing the uptake of extracellular substances. When placed in an electric field, the CNTs strongly enhance the electric field at their ends by a factor of 10–100, making them ideal for localized electroporation. It has been shown that the transport of gold nanoparticles across the cell wall following microwave irradiation is much more enhanced in the presence of CNTs, without affecting the cell viability.

Also, Raffa et al. (2010) proposed a new modality for cell electro-permeabilization based on the use of carbon nanotubes, but exploiting external static electric fields and the strong anisotropic electrical properties of MWCNTs. The obtained experimental data indicate that this method provides an effective means of lowering the electric field voltage required for reparable cell electro-permeabilization to below 50 V/cm and with an efficiency exceeding 80%. This CNT-enhanced electroporation, in combination with recent advances in CNT functionalization chemistry for transporting and targeting therapeutic molecules, appears to show considerable promise for a new generation of artificial vectors for gene, cancer and vaccine therapy, characterized by high efficiency, cell viability, selectivity and minimal damage to healthy tissues.

Cai et al. have designed an alternative physical method of gene transfer called nanotube "spearing", capable of inducing cell internalization of plasmid DNA (Fig. 3). Nanotubes grown from plasma-enhanced chemical vapor deposition contain nickel particle catalysts trapped in their tips, allowing them to respond to a magnetic field. The tubes were functionalized with a DNA strain containing the sequence coding for the enhanced green fluorescent protein. Dividing and non-dividing cells like Bal17, B-lymphoma, *ex vivo* splenic B cells and primary neurons were incubated with magnetic pDNA/CNTs. A rotating magnetic field first drove the nanotubes to mechanically spear the cells. In a subsequent step, a static magnetic field pulled the tubes into the cells. The cells were efficiently transfected as confirmed by fluorescent microscopy measurements and it was demonstrated that both spearing steps are necessary for an efficient transduction (Cai et al. 2005).

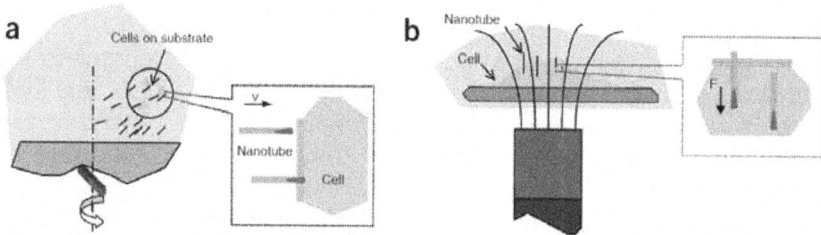

Fig. 3. DNA transfection via carbon nanotubes spearing. A two-step procedure of CNT spearing. (a) In the first step, a rotating magnetic field drives nanotubes (short black lines) to spear the cells (yellow) on a substrate. Inset: Close-up of one cell with the CNT penetrating the membrane. The plasma membrane is illustrated as an assembly of red circles. (b) In the second step, a static field persistently pulls CNT into the cells. *F*: magnetic force; *v*: velocity of CNTs. Reprinted from Cai et al. (2005) with permission from Macmillan Publishers Ltd.

Color image of this figure appears in the color plate section at the end of the book.

BORON NITRIDE NANOTUBES

First Data of Cytocompatibility

As BNNTs are insoluble and hydrophobic nanoparticulates, the first and most essential step to accomplish for an efficient biological implementation is their homogeneous water-dispersion, which necessarily passes through the use of biocompatible detergent-like agents, such as amphyphilic or positively charged biomolecules.

In our laboratories we obtained long-term stable water-dispersions of BNNTs using a non-covalent coating of the nanotube walls with either poly-L-lysine (PLL, Ciofani et al. 2010a) or glycol-chitosan (G-chitosan, Ciofani et al. 2010b).

BNNT effects in biological application have scarcely been explored. The published data are derived from studies of BNNTs incubated with different cell types such as human embryonic kidney cells HEK 293, Chinese hamster ovary cells CHO (Chen et al. 2009), human osteoblasts or mouse macrophages (Lahiri et al. 2010), human neuroblastoma cells SH-SY5Y (Ciofani et al. 2010b), and mouse myoblasts C2C12 (Ciofani et al. 2010a).

Chen and collaborators incubated HEK 293 with BNNTs and MWCNTs using a concentration of 1 µg/ml. They observed that BNNTs did not influence cell growth, which was comparable to that of the control, while MWCNTs inhibited cell proliferation also inducing apoptosis.

A fundamental contribution to cytotoxicity studies of BNNTs was derived from our experiments carried out on C2C12 (Ciofani et al. 2010a). In this study, PLL-coated BNNTs were cultured at different concentrations and for several culture times. Cell metabolic activity was evaluated by MTT test, revealing no statistically significant decrease of C2C12 metabolic activity up to 72 h of incubation with a BNNT concentration of 10 μg/ml. Using higher concentration of BNNTs, a decrease of C2C12 metabolic activity was observed both in PLL-BNNTs treated samples and in cells incubated with PLL alone at the same concentrations, indicating that the surfactant is the main responsible for toxicity. BNNT internalization was confirmed by both transmission electron microscopy and confocal microscopy, the latter demonstrating an energy-dependent internalization mechanism.

In the same study, the effects of PLL-BNNTs on C2C12 differentiation were analyzed, showing that the number of myotubes and their shape were similar in the BNNT-treated samples and in the controls. The production and the cellular distribution of two myogenic differentiation markers such as MyoD and Connexin 43 (Cx43) were also investigated, and the obtained results indicate that both MyoD and Cx43 are present in all the tested samples as both mRNA and expressed protein. All these data demonstrate that PLL-BNNTs do not interfere with myogenic differentiation.

Nevertheless, even though poly-L-lysine allowed interesting preliminary results on cytotoxicity of BNNTs to be obtained, it did not enable the researchers to assess concentration higher than 15–20 μg/ml of dispersed BNNTs in the culture medium, because of its intrinsic toxicity. In a subsequent study (Ciofani et al. 2010b), we showed the excellent results in terms of cytocompatibility and stability of the dispersions obtained using glycol-chitosan as wrapping polymer. WST-1 (2-(4-iodophenyl)-3-(4-nitrophenyl)-5-(2,4-disulfophenyl)-2H-tetrazoilium monosodium salt) and Picogreen tests demonstrated that BNNT-treated human neuroblastoma cells are comparable to controls in terms of viability, metabolic activity, proliferative capability and cell density at BNNT concentrations much higher than 20 μg/ml (Fig. 4). Optimal compatibility was confirmed by several other independent assays, both quantitative and qualitative (viability/toxicity, early apoptosis, and ROS production detection).

Potential Applications in Nanomedicine

In our laboratories we have proposed and investigated, for the first time, the possibility of exploiting BNNTs as boron atom carriers in boron neutron capture therapy (BNCT). Folic acid was used as a targeting ligand to functionalize BNNTs against tumor cells (Ciofani et al. 2009a).

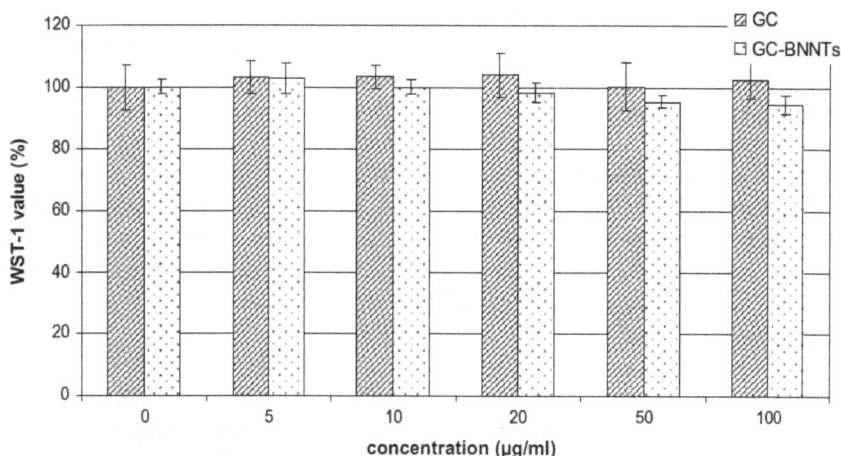

Fig. 4. Cytocompatibility testing of boron nitride nanotubes. WST-1 assay on SH-SY5Y cell cultures incubated for 48 h with different concentrations of glycol-chitosan–coated BNNTs (GC-BNNTs) and, as controls, with the respective concentrations of glycol-chitosan alone (GC). Results are presented as mean value ± standard error; $n = 6$. Reprinted from Ciofani et al. (2010b) with permission from Elsevier.

Folate conjugation was achieved by linking the amino-groups of PLL with the γ carboxylic-groups of the folate molecules, which are not involved in the folate receptor interaction. Human glioblastoma multiforme T98G cells and primary human gingival fibroblasts were used for *in vitro* experiments. The cells were treated with quantum dot labeled PLL-BNNT and folate PLL-BNNT (F-PLL-BNNT) modified culture medium. The images of glioblastoma cells after a 90 min incubation in the presence of F-PLL-BNNTs showed intensive fluorescence, indicating significant cellular uptake of nanotubes. In sharp contrast, fluorescence was considerably weaker when the cells were incubated in the presence of non-folate-functionalized PLL-BNNTs. The same experiments were carried out on healthy human gingival fibroblasts, which acted as controls. In these experiments, similar but much weaker fluorescence intensity was detected in cultures treated with both F-PLL-BNNTs and PLL-BNNTs.

The obtained results indicate that folate-functionalized BNNTs have the potential to act as boron delivery agents to malignant cells, as these preliminary *in vitro* studies have confirmed a strong and selective uptake of these nanotube vectors by glioblastoma multiforme cells but not by normal human fibroblasts. Obviously, extensive *in vivo* and pre-clinical tests are needed to explore the use of functionalized BNNTs as boron target agents for clinical boron neutron capture therapy, but these innovative nanomaterials are really promising for applications in brain diseases.

One issue to be addressed, for example, is the targeting and localization of the nanovectors towards the diseased site. This could be achieved, or at least favored, by the magnetic properties of certain types of boron nitride nanotubes. In a recent study, for example, we reported for the first time on the exploitation of magnetic properties of BNNTs for drug targeting (Ciofani et al. 2009b). Energy dispersive spectroscopy revealed in fact a content of metal catalysts (Fe and Cr) of about 3–5% wt., responsible for the magnetic properties. The magnetic behavior was confirmed by the SQUID analysis, which showed the typical trend of superparamagnetic materials, with almost zero coercivity and remanence and a saturation magnetization of about 8 emu/g for a field of about 15,000 Oe.

In vitro tests performed on human neuroblastoma SH-SY5Y cells demonstrated that cellular uptake of fluorescent labeled BNNTs can be easily modulated with an external magnetic field, thus suggesting that the magnetic properties of BNNTs could be exploited for physical drug targeting, or just to concentrate the nanotubes themselves at a desired target site with the drive of a magnetic field.

Chen et al. also investigated BNNTs as vectors for intracellular delivery of ss-DNA obtained by passive adsorption. FITC-labeled ss-DNA (FITC-DNA) was conjugated to BNNTs in aqueous solution. Fluorescence microscopy revealed that FITC-DNA–loaded BNNTs were internalized by the cells in a carrier-dependent manner.

BNNT functionalization through non-covalent adsorption appears to be a simple and effective method that enables the BNNT surface to display plenty of molecules (e.g., glycodendrimers, G-chitosan, DNA, and RNA) and should facilitate applications of BNNTs in biosensing and bioimaging without cytotoxicity-induced limitations. More generally, the observed properties of BNNTs are highly encouraging for their application in biocompatible materials, also for tissue engineering applications.

We have already mentioned the excellent mechanical properties of BNNTs, which are a potential candidate as a structural reinforcement. Lahiri et al. first reported about a poly-ε-caprolactone (PLC)/BNNT composite for bone tissue engineering applications. Composite films were obtained by casting of acetone dispersion of BNNTs at different concentrations (0, 2 and 5%) and PLC (Lahiri et al. 2010). PLC-BNNT composite showed a highly significant improvement in the elastic modulus (up to 1370%) and tensile strength (up to 109%) with increasing BNNT concentration, maintaining up to 240% elastic elongation.

In the same study, they also confirmed biocompatibility of BNNTs, investigating lactate dehydrogenase release for both osteoblasts and macrophages. Biocompatibility of BNNT-PCL composites was also tested with the Live/Dead assay, which revealed a 30% increase in living cells with respect to PCL alone, demonstrating a positive effect due to BNNTs

(Fig. 5). The genic expression of Runx 2, a key factor in osteoblastogenesis, was also evaluated and, in osteoblasts seeded onto BNNT-PCL film, a four- or seven-fold higher level was observed with respect to cultures seeded onto PCL alone. Such acceleration in terms of cell proliferation and differentiation could be explained with natural affinity between some proteins responsible for osteogenesis, such as Runx2, and BNNTs.

All these results confirm the high potential of BNNTs as nanovectors and nanomaterials for many applications in biomedicine.

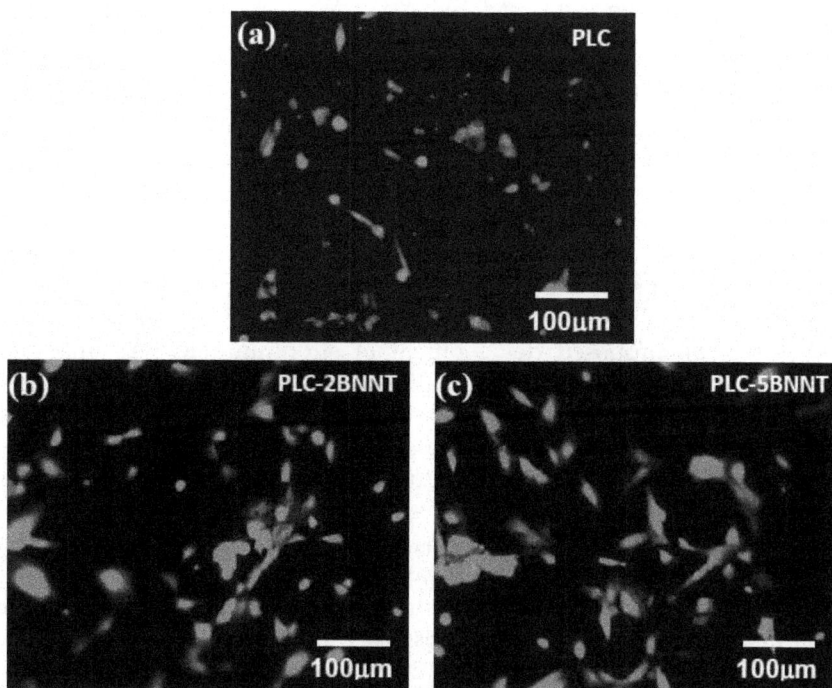

Fig. 5. Testing of boron nitride nanotube doped scaffolds. Fluorescent images of live (green) and dead (red) osteoblast cells obtained through Live/Dead staining after 2.5 d of growth on (a) PLC, (b) PLC-BNNT 2% and (c) PLC-BNNT 5% films. Reprinted from Lahiri et al. (2010) with permission from Elsevier.

Color image of this figure appears in the color plate section at the end of the book.

OTHER TYPES OF NANOTUBES

In this last section, extremely new materials, prepared in the shape of nanotubes/nanowires, are briefly introduced. Applications in nanomedicine are of course still scant, but some of them present extremely interesting and promising properties.

Xiao et al. (2009) prepared TiO_2 nanotube by anodic oxidation of a Ti sheet, and carbon-deposited TiO_2 nanotube arrays by annealing TiO_2 nanotube arrays in carbon atmosphere. The biocompatibility of the as-prepared NT arrays was investigated by observing the growth of osteosarcoma (MG-63) cells on the NT arrays. Both the TiO_2 NTs and the carbon-modified TiO_2 NTs showed good biocompatibility, supporting the normal growth and adhesion of MG-63 cells with no need of extracellular matrix protein coating. The TiO_2 NTs with one end open can easily be filled with drugs, working as an efficient drug delivery vehicle of antineoplastic agents.

In another study, Wang and collaborators (2009) investigated TiO_2 NTs as potential drug vehicles and carried out a preliminary study of the interaction between cells and TiO_2 NTs. FITC-conjugated titania nanotubes (FITC-TiO_2-NTs), internalized in mouse neural stem cells (line C17.2), have been directly imaged by confocal microscopy, showing a strong internalization: after a 24 h incubation, NTs were localized around the cell nucleus without crossing the karyotheca, but nucleus penetration occurred after a further 24 h, opening interesting perspectives in the exploitation of this nanomaterial for gene and DNA-targeting drug delivery.

Xin et al. proposed development of strontium-releasing implants capable of stimulating bone formation and inhibiting bone resorption for curing osteoporosis, based on well-ordered $SrTiO_3$ nanotube arrays (Xin et al. 2009). These devices are capable of Sr release at a slow rate and for a long time, and are successfully fabricated on titanium by simple hydrothermal treatment of anodized titania nanotubes. This surface architecture combines the functions of nanoscaled topography and Sr release to enhance osteointegration, while at the same time leaving space for loading of other functional substances. *In vitro* experiments revealed that the $SrTiO_3$ nanotube arrays possess good biocompatibility and can induce precipitation of hydroxyapatite from simulated body fluids. This Ti-based implant with $SrTiO_3$ nanotube arrays could represent a valid candidate for osteoporotic bone implants, and the method can also be extended to load other biologically useful elements such as Mg and Zn.

Finally, an important study is the first cellular-level study on the biocompatibility and biosafety of ZnO nanowires (NWs), by Li et al. (2008) (Fig. 6). HeLa cell line showed a complete biocompatibility with ZnO nanostructures from low to high NW concentrations beyond a couple of production periods. The L929 cell line showed a good production behavior at lower NW concentration, but when the concentration was close to 100 µg/ml, the viability dropped to about 50% (Li et al. 2008). This study shows the biosafety of ZnO NWs in well-defined concentration ranges, thus demonstrating feasibility of their applications in *in vitro* and *in vivo* biomedical science and engineering.

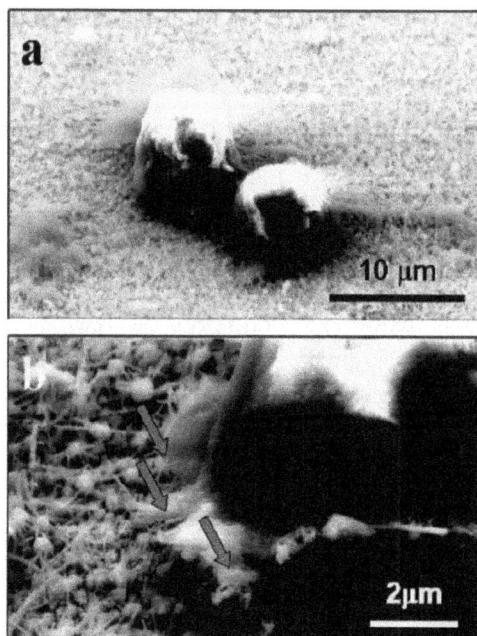

Fig. 6. Study of interaction between HeLa cells and ZnO nanowires. SEM images of HeLa cells on ZnO NWs arrays. (a) Two HeLa cells are growing on the surface of ZnO NWs arrays. (b) Cells are upheld by the NWs. Some ZnO NWs are phagocytosed into the HeLa cell (pointed out by the red arrow). The diameter and the length of the nanowires are about 100 nm and 1.5 μm, respectively. Reprinted from Li et al. (2008) with permission from the American Chemical Society.

CONCLUSION

The introduction and delivery of DNA, proteins, or drug molecules into living cells is an important challenge of biomedical research. Inorganic nanomaterials, including nanocrystals, nanotubes, and nanowires, exhibit advanced physical properties that are promising for various biological applications, including new molecular transporters.

Here we have introduced and discussed nanomedicine applications of the main classes of materials available in "nanotube" form, with particular attention to carbon and boron nitride nanotubes. Others kind of extremely innovative materials have been also introduced.

In conclusion, there is an urgent need to exploit the rapid advances in nanomaterials for biomedical applications. The merging of different disciplines such as bioengineering, materials science, chemistry, physics, and biology, as well as medicine, will be essential for the successful exploration of the applications of nanomaterials inside cells, and for effective and realistic applications in clinical practice.

APPLICATIONS TO OTHER AREAS OF HEALTH AND DISEASE

This chapter has been particularly focused on applications of nanotubes as nanocarriers for drug delivery and cell therapy. Many other nanomedicine fields have seen interesting applications of these nanomaterials. Their use is becoming relevant, for example, in neuroscience research and in tissue engineering (Harrison and Atala 2007).

CNTs are, moreover, promising materials for sensing applications due to several intriguing properties. In particular, their high aspect ratio provides for high surface-to-volume ratio. Moreover, CNTs have an outstanding ability to mediate fast electron-transfer kinetics for a wide range of electroactive species, such as hydrogen peroxide or nicotinamide adenine dinucleotide. In addition, CNT chemical functionalization can be used to attach almost any desired chemical species. CNTs have therefore been used as new platforms to detect antibodies associated with human autoimmune diseases with high specificity (Balasubramanian and Burghard 2006). These findings pave the way for development of CNT-based diagnostic devices for the discrimination and identification of different proteins from serum samples, and in the fabrication of microarray devices for proteomic analyses. In a similar context, CNTs covalently modified at their open ends with DNA have led to innovative systems for hybridization of complementary DNA strands, allowing for ultrasensitive DNA detection (Niu et al. 2009).

Key Facts

- Applications of CNTs in the field of biotechnology have recently started to emerge, raising great hopes.
- CNTs have emerged as a new alternative and efficient tool for transporting and translocating therapeutic molecules: they can be functionalized with bioactive peptides, proteins, nucleic acids and drugs, and used to deliver their cargos to cells, tissues, and organs.
- The discovery of CNTs has the potential to revolutionize biomedical research, as they can show superior performance over other nanoparticles. The advantage lies in a unique, unprecedented combination of electrical, magnetic, optical and chemical properties, which is greatly promising for the development of a new class of CNT-based drugs and therapies.

- A BNNT is a structural analogue of a carbon nanotube: alternating B and N atoms entirely replace C atoms in a graphitic-like sheet with almost no change in atomic spacing.
- While BNNTs possess a very high Young modulus similar to CNTs, they also possess superior chemical, thermal and electrical properties.
- Recently, it has been shown that BNNTs have excellent piezoelectric properties, superior to those of piezoelectric polymers.
- All these properties make BNNTs potentially attractive candidates for a wide range of applications in the nano domain. In the last years, many applications of CNTs in the field of biotechnology have been proposed, but biomedical applications of BNNTs are yet unexplored.

Definitions

Boron nitride nanotubes: A BNNT is a structural analogue of a CNT in nature: alternating B and N atoms entirely replace C atoms in a graphitic-like sheet with almost no change in atomic spacing; despite this, carbon and boron nitride nanotubes present many different chemical and physical properties.

Carbon nanotubes: CNTs are allotropes of carbon with a cylindrical nanostructure. They are among the stiffest and strongest fibres known and have remarkable electronic properties and many other unique characteristics. Nanotubes have been constructed with length-to-diameter ratio of up to 132,000,000:1, which is significantly larger than any other material. These cylindrical carbon molecules have novel properties that make them potentially useful in many applications in nanotechnology, electronics, optics, and other fields of materials science, as well as potential uses in architectural fields.

Cell surgery: Surgery performed via devices of sub-cellular size having the ability to interact with biological matter at the molecular, cellular, and tissue level.

Cell therapy: A technology by which drugs or genes are delivered to human cells, tissues or organs to correct a genetic defect, or to provide new therapeutic functions for the ultimate purpose of preventing or treating diseases.

Drug delivery system: Drug delivery is the method or process of administering a pharmaceutical compound to achieve a therapeutic effect in humans or animals. Nanomedical approaches to drug delivery focus on developing nanoscale particles or molecules to improve drug bioavailability. Bioavailability refers to the presence of drug molecules where they are needed in the body and where they will be more active. Drug delivery focuses on maximizing bioavailability both at specific places in the body and over a period of time. This can potentially be achieved by molecular or physical targeting of nanoengineered devices.

Nanomedicine: Nanomedicine is the medical application of nanotechnology. This relatively new discipline ranges from the medical applications of nanomaterials to nanoelectronic biosensors, and even possible future applications of molecular nanotechnology. It comprises the process of diagnosing, treating, and preventing disease and traumatic injury, of relieving pain, and of preserving and improving human health, using molecular tools and molecular knowledge of the human body. Nanomedicine can actually address many important medical problems by using nanoscale-structured materials and simple nanodevices, including the interaction of nanostructured materials with biological systems.

Nanotoxicology: Nanotoxicology is the study of the toxicity of nanomaterials. With the rapid development of nanotechnology and its applications, a wide variety of nano-structured materials are now used in commodities, pharmaceutics, cosmetics, biomedical products, and industries. While nanoscale materials possess more novel and unique physicochemical properties than bulk materials, they also have an unpredictable impact on human health. Nanotoxicology is a branch of bionanoscience that deals with the study and application of toxicity of nanomaterials.

Summary Points

- Nanoscale structures and materials have been explored in many biological applications because their novel properties and functions differ drastically from their bulk counterparts.
- CNTs are allotropes of carbon with a cylindrical nanostructure; they are among the stiffest and strongest fibres known and have remarkable electronic properties and many other unique characteristics.
- The ability of CNT to penetrate cells offers the potential of using CNTs as vehicles for the delivery of drug molecules and nucleic acids.
- It is of paramount importance to conjugate the concept of CNT-mediated drug delivery to the exploitation of their ability to be manipulated with external magnetic, electric or optical means.
- A BNNT is a structural analogue of a carbon nanotube: alternating B and N atoms entirely replace C atoms in a graphitic-like sheet with almost no change in atomic spacing.
- Preliminary investigations of BNNT biocompatibility revealed excellent results on many different cell lines.
- Biomedical applications of BNNTs range from boron carriers for BNCT to nanovectors for drug delivery up to structural fillers for tissue engineering.

- Extremely innovative nanotube-based materials (TiO$_2$, SrTiO$_3$, ZnO) have started to emerge and are greatly promising for nanomedicine applications.

Abbreviations

AFM : Atomic force microscopy
BNCT : Boron neutron capture therapy
BNNT : Boron nitride nanotube
BSA : Bovine serum albumin
CNT : Carbon nanotube
DNA : Deoxyribonucleic acid
FITC : Fluorescein isothiocyanate
MTT : Dimethyl thiazolyl diphenyl tetrazolium salt
MTX : Methotrexate
MWCNT : Multi-walled carbon nanotube
NT : Nanotube
NW : Nanowire
NIR : Near infrared
PCL : Poly-ε-caprolactone
PEG : Polyethylene glycol
PEI : Polyethylenimine
PLL : Poly-L-lysine
PTX : Placlitaxel
RNA : Ribonucleic acid
SQUID : Superconducting quantum interference device
SWCNT : Single-walled carbon nanotube
WST-1 : 2-(4-iodophenyl)-3-(4-nitophenyl)-5-(2,4-disulfophenyl)-2H-tetrazoilium monosodium salt

References

Balasubramanian, K., and M. Burghard. 2006. Biosensors based on carbon nanotubes. Anal. Bioanal. Chem. 385: 452–468.

Bardi, G., P. Tognini, G. Ciofani, V. Raffa, M. Costa, and T. Pizzorusso. 2009. Pluronic-coated carbon nanotubes do not induce degeneration of cortical neurons *in vivo* and *in vitro*. Nanomed. Nanotechnol. 5: 96–104.

Cai, D., M.J. Mataraza, Z.H. Qin, Z. Huang, J. Huang, T.C. Chiles, D. Carnahan, K. Kempa, and Z. Ren. 2005. Highly efficient molecular delivery into mammalian cells using carbon nanotube spearing. Nat. Met. 2: 449–454.

Chen, X., P. Wu, M. Rousseas, D. Okawa, Z. Gartner, A. Zettl, and C.R. Bertozzi. 2009. Boron nitride nanotubes are noncytotoxic and can be functionalized for interaction with proteins and cells. J. Am. Chem. Soc. 131: 890–891.

Ciofani, G., V. Raffa, A. Menciassi, and A. Cuschieri. 2009a. Folate functionalised boron nitride nanotubes and their selective uptake by glioblastoma multiforme cells: implications for their use as boron carriers in clinical boron neutron capture therapy. Nanoscale Res. Lett. 4: 113–121.

Ciofani, G., V. Raffa, J. Yu, Y. Chen, Y. Obata, S. Takeoka, A. Menciassi, and A. Cuschieri A. 2009b. Boron nitride nanotubes: A novel vector for targeted magnetic drug delivery. Curr. Nanosci. 5: 33–38.

Ciofani, G., L. Ricotti, S. Danti, S. Moscato, C. Nesti, D. D'Alessandro, D. Dinucci, F. Chiellini, A. Pietrabissa, M. Petrini, and A. Menciassi. 2010a. Investigation of interactions between poly-l-lysine-coated boron nitride nanotubes and C2C12 cells: up-take, cytocompatibility, and differentiation. Int. J. Nanomedicine 5: 285–298.

Ciofani, G., S. Danti, D. D'Alessandro, S. Moscato, and A. Menciassi. 2010b. Assessing cytotoxicity of boron nitride nanotubes: Interference with the MTT assay. Biochem. Biophys. Res. Commun. 394: 405–411.

Dwyer, C., M. Guthold, M. Falvo, S. Washburn, R. Superfine, and D. Erie. 2002. DNA functionalized single-walled carbon nanotubes. Nanotechnology 13: 601–604.

Gao, J., and B. Xu. 2009. Applications of nanomaterials inside cells. Nano Today 4: 37–51.

Gogotsi, Y. 2003. How safe are nanotubes and other nanofilaments. Mat. Res. Innovat. 7: 192–1949.

Harrison, B.S., and A. AtalaA. 2007. Carbon nanotube applications for tissue engineering. Biomaterials 28: 344–353.

Heister, E., V. Neves, C. Tilmaciu, K. Lipert, V.S. Beltrán, H.M. Coley, S.R.P. Silva, and J. McFadden. 2009. Triple functionalisation of single-walled carbon nanotubes with doxorubicin, a monoclonal antibody, and a fluorescent marker for targeted cancer therapy. Carbon 47: 2152–2160.

Klumpp, C., K. Kostarelos, M. Prato, and A. Bianco. 2006. Functionalized carbon nanotubes as emerging nanovectors for the delivery of therapeutics. BBA–Biomembranes 1758: 404–412.

Lahiri, D., F. Rouzaud, T. Richard, A.K. Keshri, S.R. Bakshi, L. Kos, and A. Agarwal. 2010. Boron nitride nanotube reinforced polylactide-polycaprolactone copolymer composite: mechanical properties and cytocompatibility with osteoblasts and macrophages *in vitro*. Acta Biomater. 6: 3524–3533.

Li, Z., R. Yang, M. Yu, F. Bai, C. Li, and Z.L. Wang. 2008. Cellular level biocompatibility and biosafety of ZnO nanowires. J. Phys. Chem. C. 112: 20114–20117.

Liu, Y., D.C. Wu, W.D. Zhang, X. Jiang, C.B. He, T.S. Chung, S.H. Goh, and K.W. Leong. 2005. Polyethylenimine-grafted multiwalled carbon nanotubes for secure noncovalent immobilization and efficient delivery of DNA. Angew. Chem. Int. Ed. 44: 4782–4785.

Liu, Z., K. Chen, C. Davis, S. Sherlock, Q. Cao, X. Chen, and H. Dai. 2008. Drug delivery with carbon nanotubes for *in vivo* cancer treatment. Cancer Res. 68: 6652–6660.

Muller, J., F. Huaux, and D. Lison. 2006. Respiratory toxicity of carbon nanotubes: How worried should we be? Carbon 44: 1028–1033.

Niu, S., M. Zhao, R. Ren, and S. Zhang. 2009. Carbon nanotube-enhanced DNA biosensor for DNA hybridization detection using manganese(II)-Schiff base complex as hybridization indicator. J. Inorg. Biochem. 103: 43–49.

Ou, Z., B. Wu, D. Xing, F. Zhou, H. Wang and Y. Tang. 2009. Functional single-walled carbon nanotubes based on an integrin $\alpha_v\beta_3$ monoclonal antibody for highly efficient cancer cell targeting. Nanotechnology 20, art. no. 105102.

Pastorin, G., W. Wu, S. Wieckoski, J.P. Briand, K. Kostarelos, M. Prato, and A. Bianco. 2006. Double functionalisation of carbon nanotubes for multimodal drug delivery. Chem. Commun. 11: 1182–1184.

Raffa, V., G. Ciofani, O. Vittorio, V. Pensabene, and A. Cuschieri. 2010. Carbon nanotube-enhanced cell electropermeabilisation. Bioelectrochemistry 79: 136–141.

Rojas-Chapana, J.A., M.A. Correa-Duarte, Z. Ren, K. Kempa, and M. Giersig. 2004. Enhanced introduction of gold nanoparticles into vital acidothiobacillus ferrooxidans by carbon nanotube-based microwave electroporation. Nano Lett. 4: 985–988.

Sayes, C.M., F. Liang, J.L. Hudson, J. Mendez, W. Guo, J.M. Beach, V.C. Moore, C.D. Doyle, J.L. West, W.E. Billups, K.D. Ausman, and V.L. Colvin. 2006. Functionalization density dependence of single-walled carbon nanotubes cytotoxicity *in vitro*. Toxicol. Lett. 161: 135–142.

Shi Kam, N.W., and H. Dai. 2005a. Carbon nanotubes as intracellular protein transporters: generality and biological functionality. J. Am. Chem. Soc. 127: 6021–6026.

Shi Kam, N.W., Z. Liu, and H. Dai. 2005b. Functionalization of carbon nanotubes *via* cleavable disulfide bonds for efficient intracellular delivery of siRNA and potent gene silencing. J. Am. Chem. Soc. 127: 12492–12493.

Shi Kam, N.W., M. O'Connell, J.A. Wisdom, and H. Dai. 2005c. Carbon nanotubes as multifunctional biological transporters and near-infrared agents for selective cancer cell destruction. Proc. Natl. Acad. Sci. USA 102: 11600–11605.

Singh, R., D. Pantarotto, D. McCarthy, O. Chaloin, J. Hoebeke, C.D. Partidos, J.P. Briand, M. Prato, A. Bianco, and K. Kostarelos. 2004. Binding and condensation of plasmid DNA onto functionalized carbon nanotubes: toward the construction of nanotube-based gene delivery vectors. J. Am. Chem. Soc. 127: 4388–4396.

Sinha, N., and T.W. Yeow. 2005. Carbon nanotubes for biomedical applications. IEEE T. Nanobiosci. 4: 180–195.

Terrones, M., J.M. Romo-Herrera, E. Cruz-Silva, F. López-Urías, E. Muñoz-Sandoval, J.J. Velázquez-Salazar, H. Terrones, Y. Bando, and D. Golberg. 2007. Pure and doped boron nitride nanotubes. Mater. Today 10: 30–38.

Wang, Y., J. Wang, X. Deng, J. Wang, H. Wang, M. Wu, Z. Jiao, and Y. Liu. 2009. Direct imaging of titania nanotubes located in mouse neural stem cell nuclei. Nano Res. 2: 543–552.

Wu, W., S. Wieckowski, G. Pastorin, C. Klumpp, M. Benincasa, J.P. Briand, R. Gennaro, M. Prato, and A. Bianco. 2005. Targeted delivery of amphotericin B to cells using functionalized carbon nanotubes. Angew. Chem. Int. Ed. 44: 6358–6362.

Xiao, X., L. Yang, M. Guo, C. Pan, Q. Cai, and S. Yao. 2009. Biocompatibility and *in vitro* antineoplastic drug-loaded trial of titania nanotubes prepared by anodic oxidation of a pure titanium. Sci. China Ser. B-Chem. 52: 2161–2165.

Xin, Y., J. Jiang, K. Huo, T. Hu, and P.K. Chu. 2009. Bioactive $SrTiO_3$ nanotube arrays: strontium delivery platform on Ti-based osteoporotic bone implants. ACS Nano 3: 3228–3234.

4

Nanoparticles for Topical Application

Kerriann R. Greenhalgh

ABSTRACT

Topical application of nanopharmaceuticals has focused on two main areas in recent years: ocular and transdermal applications. A third and less frequently studied application is in topical delivery to compromised skin, such as chronic ulcers and dermal wound injuries. The majority of these nanopharmaceuticals have been designed with the intention of improving the biocompatibility and bioavailability of pharmaceuticals. Nanopharmaceuticals have had success as topically applied treatments because their small size permits increased transport across the epithelium and intracellular penetration. These characteristics allow drugs with steric and absorptive issues to be delivered to areas previously out of reach. This chapter will focus on the most common forms of nanopharmaceuticals, including metallic and polymeric nanocarriers, and their application in ocular and wound care treatment.

INTRODUCTION

Nanopharmaceuticals have been highly successful in topical application in medicine, especially in the realm of infection control. Escalating drug

---

KeriCure Inc., 26620 Easy St. Wesley Chapel, FL 33544, USA; Email: kgreenha@gmail.com

List of abbreviations after the text.

resistance among pathogenic bacteria has set the stage for new classes of antibacterial agents and alternative therapies, and nanopharmaceuticals have the potential to fulfill this need. Nanotechnology has recently been employed for investigating new drug delivery platforms and new therapeutic agents in the form of nanopharmaceuticals. Studies into the conjugation of both novel and classic antibiotics with various nanoparticle systems have established that these systems present improved performance, sustained or controlled release, improved solubility and stability, lower cytotoxicity, and, in some cases, targeted drug delivery. Since the first application of nanoparticles in therapeutics in 1976 by Kreuter and Speiser, nanotechnology has been applied to many areas in the medical field, including imaging, anti-cancer therapies, and antimicrobial therapy. This chapter focuses on the impact that nanotechnology has had in areas of topical treatments of infection and disease.

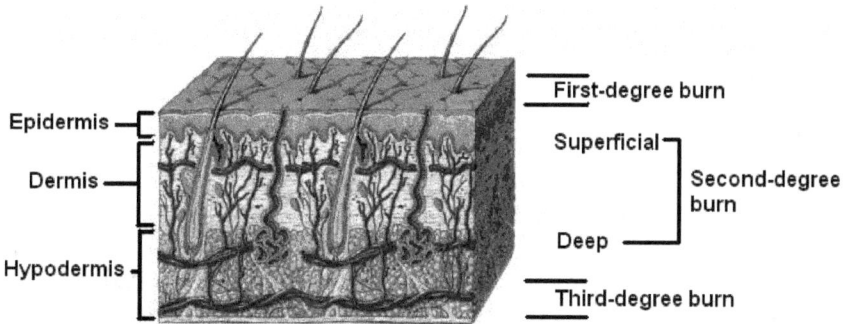

Fig. 1. Anatomy of human skin and subdermal layers. The image shows the three main layers of the human skin. Labels to the right indicate the depth of and area of skin affected by each of the three levels of burn wound severity. Image obtained from Dassinger, M.S. National Library of Medicine (2006) (Steinstraesser et al. 2002).

BENEFITS OF NANOPARTICLES IN TOPICAL APPLICATION

One of the largest advantages of nanoparticles, particularly for topical administration, is the ability to allow hydrophobic or water-insoluble therapeutics to treat areas where they otherwise could not. Many of the commercially available topical creams and ointments use lyposomal, micelle, or colloidal drug conjugates to deliver treatments to skin and soft tissues. All three of these carriers are within the micrometer size range and have proven effective at providing a bulky hydrophobic drug carrier to the otherwise hydrophilic antibiotics, allowing them some degree of permeation (Lamb et al. 1991; Lance et al. 1995; Sorkine et al. 1996; Souza et al. 1993; Washington et al. 1993; Yarkoni and Rapp 1978; Turner and

Wooley 2004). However, these products have many limitations, including a shallow depth of treatment and limited bioavailability of the drug in the tissue. This is especially true when there is a large degree of damage to the skin, as is the case with third degree burns, which penetrate down to the subcutaneous or fatty tissue below the skin. Most topical treatments are not effective here and must be combined with systemic administration of a like drug as a preventive measure. In terms of infection control, this method of treatment can lead to the development of bacterial resistance and is often ineffective at treating the infection.

Table 1. Ideal properties for biological-based temporary wound dressings (Morones et al. 2005).

Biocompatibility
Water permeability and heat retention
Protective barrier against pathogens and trauma
Reduces pain
Adherent, yet flexible and conforming
Applied in one operation
Grows with patient
Low cost
Long shelf life/simple storage

NANOPHARMACEUTICAL RESEARCH

Currently, nanoparticles are among the most heavily researched areas of drug delivery. Many improvements have been made in areas of cancer therapy, targeted delivery of therapeutic compounds, and medical imaging (Turner and Wooley 2004; Couvreur et al. 1986; Kim et al. 2006; Kumar et al. 2003; Duncanson et al. 2007; Pankhurst et al. 2003), yet little effort has been made towards using nanoparticles in topical infection control (Turos et al. 2007a, b; Wang 2006; Tian et al. 2007; Couvreur et al. 1979, 1995; Gu et al. 2003; Maincent et al. 1986; Tom et al. 2004; Morones et al. 2005; Allemann et al. 1993; Cavallaro et al. 1994). Antibiotic-coated gold and silver metallic nanoparticles (Gu et al. 2003; Tom et al. 2004; Morones et al. 2005), antibiotic encapsulated polymeric nanoparticles and liposomes (Couvreur et al. 1995; Allemann et al. 1994), and biodegradable nanospheres (Couvreur et al. 1979; Cavallaro et al. 1994) have all been explored, yet little commercial development has been seen thus far. Couvreur et al. developed a series of poly(alkyl cyanoacrylate) nanoparticles that are bioresorbable and are currently being used as surgical glues (Couvreur et al. 1986) and more recently as carriers of antibiotics (specifically ciprofloxacin) (Couvreur et al. 1979, 1995; Maincent et al. 1986). This drug delivery system allowed the antibiotics to overcome problems associated with oral administration, including stabilizing the drug in the gastrointestinal tract and providing adequate adsorption into the bloodstream. Couvreur et al. has also shown that encapsulation of the ciprofloxacin monomer within the poly(cyanoacrylate) nanoparticle has heightened the *in vitro* activity of this antibiotic against *S. aureus* (Couvreur et al. 1979, 1995; Maincent et al. 1986).

One of the critical aspects in use of nanoparticles for drug delivery is quantifying the percent association of drug to nanoparticle. Multiple studies have investigated improving association efficacies of drugs to nanocarriers. The strength of the drug-nanocarrier association has many ramifications on the overall success of the formulation. Stability of the drug, retention of drug activity, and release profiles *in vivo* are all key factors when designing the nanopharmaceutical. Each factor must be tailored to the specific topical application desired for maximum effectiveness.

NANOPARTICLES IN TOPICAL INFECTION CONTROL

Nanoparticles have played a key role in altering the way doctors approach topical treatments. There are a number of products currently on the market that use nanocrystalline silver or nanosized silver colloids for topical treatment of infection. While these products have seen moderate commercial success, there are some significant drawbacks to these silver-based products, namely toxicity and skin irritation. Some natural polymers, including PEG and PLA, have been investigated as nanocarriers; however, a stand-apart nano-based topical product has not yet seen commercial success as a topical treatment for infection. Research efforts continue to grow in this area, with recent publications on polymer and chitosan-conjugated antibiotics making clear headway, and hopefully will gain traction in the pharmaceutical arena.

Some *in vivo* studies have looked into the effects of various nanoparticles, including chitosan and metal-based nanoparticles, for topical treatment of dermal wounds and have found positive results with increased wound healing and enhanced cosmetic results (Tian et al. 2007; Atiyeh et al. 2007). Advances in nanotechnology have allowed these nanopharmaceuticals to provide topical treatments that penetrate deep within the compromised tissue and extend the activity of the antimicrobials. Nano-based topical infection control also has the potential to decrease the occurrence of nosocomial infections by attacking the infection from the outside in, instead of the typical inside-out approach with systemically administered anti-infectives.

SILVER NANOPARTICLES IN WOUND CARE

Research to improve antimicrobial wound dressings have continued to be explored by a select number of pharmaceutical companies and researchers, mainly smaller startup companies, yet products have not been significantly improved over the contemporary metal-doped bandages. The mainstay of this area of topical wound care has revolved around silver-doped bandages as temporary wound dressings, which has been a major area of wound dressing research for the past 20 years. These bandages contain antibacterial properties through the silver constituent, where silver sulfadiazine is the most commonly used and is considered the benchmark for silver-containing dressings (Atiyeh et al. 2007). Such metal-doped bandages reduce bacterial infection at the site of the wound and also help prevent sepsis (Atiyeh et al. 2007). Since the re-emergence of silver in medical practices, many companies have begun manufacturing wound

Table 2. Microorganisms commonly found in dermal wounds.

Microbe classification	Pathogen
Gram Positive	*S. aureus:* MRSA, MSSA
	S. epidermidis
	Streptococcus spp.
	Pneumococcus spp.
	Haemolytic streptococci
	Bacillus spp.
	Vancomycin-resistant *Enterococcus*
	Enterococcus faecalis
	Corynebacterium spp.
	Neisseria spp.
Gram Negative	*P. aeruginosa*
	Escherichia coli
	Klebsiella spp.
	Enterobacter cloacae
	Serratia marcescens
	Proteus mirabilis
	Acinetobacter calcoaceticus
	Bacteroides spp.
	Citrobacter fruendi
Fungi	*Candida* spp.
	Aspergillus spp.
	Fusarium spp.
	Alternaria spp.
	Rhizopus spp.
Viruses	Herpes simplex virus
	Cytomegalovirus
	Varicella-zoster virus

Compiled from publications over the past 20 years.

dressing products containing various forms of silver salts. Some of these products include silver nylon cloth/activated charcoal (Actisorb®) (Silver et al. 2006), silver sulfadiazine with cerium nitrate (Flammacerium®) (Atiyeh et al. 2007), silver absorbent wound dressing (SilvaSorb®) (Silver et al. 2006), silver-impregnated polyurethane (PolyMem Silver®) (Silver et al. 2006), and silver sulfadiazine-impregnated lipidocolloid wound dressing (Urgotul SSD®) (Atiyeh et al. 2007; Silver et al. 2006).

Two of the best examples of commercial nanopharmaceuticals as topical products are Acticoat™ and Silveron wound dressings. These dressings use nanocrystalline silver for topical application to dermal wounds (Atiyeh et al. 2005). According to the manufacturers' website, the dressings consist of two sheets of high-density polyethylene coated with nanocrystalline silver. The nanocrystalline silver affords moderate penetration of unexcised burn eschar, a highly lipophilic and extremely difficult tissue to penetrate, and controlled release over a prolonged time frame (Atiyeh et al. 2005). Studies have shown that the dressings allow controlled and prolonged release of the silver constituent into the wound area, thereby decreasing the amount of dressing changes required over time (Atiyeh et al. 2007; Rustogi et al. 2005; Holder et al. 2003). Silver substances, especially nanocrystalline silver as used in Acticoat™ and Silverlon®, promote wound healing by decreasing matrix metalloproteinases (MMP) that cause tissue destruction and by enhancing cellular apoptosis (Atiyeh et al. 2007). However, not all silver-based complexes have these beneficial properties. Recent studies have established that the use of Acticoat™ with cultured keratinocytes and other cultured skin substitutes can be cytotoxic within one day and can inhibit re-epithelialization (Atiyeh et al. 2007), while Silvazine® has been shown to have the highest degree of cytotoxicity among the silver-containing wound dressings (Atiyeh et al. 2007).

A second and equally serious issue with these nanopharmaceuticals is that the levels of silver ion delivered to the wound are typically below the bactericidal levels necessary to prevent resistance formation (Atiyeh et al. 2007). The amount of silver available in the wound has an inverse relationship with the concentration of silver required for antibacterial activity in these products. Bacterial resistance to the nanocrystalline silver has been found in clinical studies, despite no observation of resistance *in vitro*, through one of two molecular interactions: bacteria can bind the silver once inside the cell forming an intracellular complex, or the silver can be excreted from the cells prior to invoking its antibacterial activity through the cell's efflux systems (Atiyeh et al. 2007). While there is no question that silver-based topicals are effective antimicrobials, the evolution of nanotechnology appears to be moving towards more robust nanocarriers that are capable of utilizing synthetic and natural products as the drug carrier with stronger activity and biocompatibility profiles.

POLYMERIC NANOPARTICLES

Polymeric drug delivery was established over 30 years ago and is still a heavily researched area (Gombotz and Pettie 1995; Sinha and Khosla 1998; Langer 1998; Polymeric Drug Delivery 2004-2005). However, only recently have polymers gained attention as potential drug-carrying nanoparticles. Couvreur et al. developed a drug delivery system that uses a classic cyanoacrylate polymer prepared through emulsion polymerization for entrapment of antimicrobials, specifically ciprofloxacin. Their work has shown that the poly(cyanoacrylate) nanoparticle system increases the biostability of the antibiotic and has also strengthened the drug's antimicrobial activity *in vitro*.

Most non-biodegradable polymeric nanoparticles use some form of emulsion or microemulsion polymerization for preparing the particles. In some cases, initial micelle formation occurs in an aqueous solution using hydrophobic monomers and surfactants followed by polymerization initiated within the micelle through a radical initiator (Turos et al. 2007a, b; Wang 2006). In other instances, the emulsion is a reverse water-in-oil emulsion, as is the case with Acrymed's hydrogel technology used in their advanced wound care products. Emulsions allow tight control of the resulting particle size (see Fig. 2) and a fairly even particle distribution throughout the prepared solution. Polymeric nanoparticles are typically stable in solution and are capable of purification through dialysis methods (Turos et al. 2007a, b; Wang 2006).

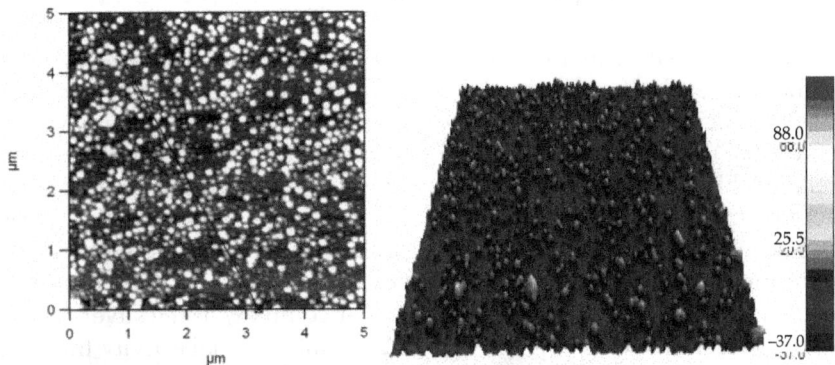

Fig. 2. Atomic force microscopy (AFM) image of polyacrylate-based nanoparticles. Left image is a flat scan of a polyacrylate emulsion. Right image is a 3D depiction of the left image. Particles are 19–23 nm in diameter based on the computer-generated scale to the right. Published with permission from Kerriann Greenhalgh (2008).

Other polymers that have received a great deal of attention as nanocarriers are the biodegradable polymers such as poly(caprolactone), poly(alkyl cyanoacrylate), poly(ethylene glycol) (PEG), poly(glycolic acid) (PGA), poly(lactic acid) (PLA), and various combinations of the polymers. The benefit of biodegradable polymers over the acrylate-based systems is the ability to provide a faster delivery of the drug over a short period of time and then leave the body's system. Depending on the method of delivery, these biodegradable nanopharmaceuticals have shown increase drug concentrations in tissue with sustained or increased drug release (Leroux et al. 1996).In terms of topical applications, this property is most beneficial in areas where a permanent polymer could cause long-term issues and a high concentration of drug is required for treatment, such as in ocular applications. Other applications, such as control of infections in compromised skin, benefit more from a non-degradable nanocarrier where consistent levels of drug over extended time is required. There are a wide range of commercially available biodegradable polymers, allowing multiple configurations of biodegradable nanocarriers that can be tailored to the specific topical application desired. The numerous variables and combinations that biodegradable polymers present have sparked the vast research effort observed for biodegradable nanocarriers, an area that continues to grow today (Greenhalgh and Turos 2006).

The bulk of the research done on polymer-based nanopharmaceuticals, both degradable and non-degradable, has looked at encapsulation of the drug monomers within the core of the nanoparticle. However, Couvreur and Turos have both successfully bound antimicrobials and other bioactive drugs to these polymeric nanoparticles, either ionically or covalently (Turos et al. 2007a, b; Wang 2006; Couvreur et al. 1979, 1995). Studies have shown a drastic increase in overall drug activity and also a rejuvenation of antimicrobial activity against drug-resistant microbes when covalently bound to the nanocarrier (Turos et al. 2007a). Methicillin-resistant *Staphylococcus aureus* (MRSA) is a common drug-resistant pathogen that has plagued many hospitals and patients around the world and is often associated with topical infections in wound beds and in the eye. The work performed by Turos et al. (2007) has shown that by covalently binding penicillin to a polyacrylate nanoparticle system, they are able to regain activity of penicillin against resistant MRSA strains. The ability to revive antibiotics such as penicillin would be a safer and simpler method of infection control. Research efforts continue to grow in this area with the initial success *in vitro* being examined in pre-clinical settings.

CHITOSAN NANOPARTICLES

Chitosan nanoparticles have received much attention since early 2000 as potential nanocarriers. Chitosan, a natural polymer commonly found in the hard skeleton of shellfish, has been used in a number of nanoparticle formulations and has played a critical role in improving delivery of many pharmaceuticals by providing increased bioavailability, drug stability, targeted delivery, and improved cytostatic activity over longer periods of time. Chitosan is a positively charged polysaccharide that is capable of forming nanoparticles through ionic gelation, although other methods of nanoparticle formation have been investigated.

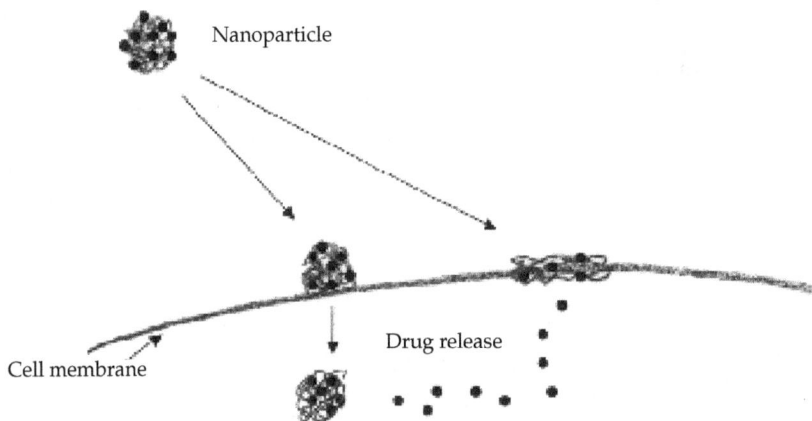

Fig. 3. Schematic image of drug-carrying nanoparticle interactions with cellular membranes. The depiction above is a schematic illustration of how some polymer-based nanoparticles interact with cells. Studies have shown that the nanoparticles may be imported into the cell through phagocytosis or endocytosis, or they may collapse on the surface of the cell, exposing the contained drug to the cell's membrane. Published with permission from Kerriann Greenhalgh (2008).

There are many advantages of chitosan-based nanocarriers in topical applications, particularly in the eye. The biocompatibility and biodegradability of these polymers has been well characterzed since the early 1990s. The polymer itself is well known for enhancing paracellular transport of drugs and thus has been a common choice for investigation of potential nanocarriers. The positively charged chitosan polymer has shown mucoadhesive properties when presented with a negatively charged tissue such as the cornea and conjunctiva (De Campo et al. 2001). Research efforts have used this interaction to increase the bioavailability of topical antimicrobials by coordinating the nanopharmaceutical to the corneal tissue, as well as enhance intraocular penetration of other therapeutics for inner eye treatment (De Campo et al. 2001). Each of these characteristics has enhanced the effectiveness of the therapeutic agent incorporated in the nanopharmaceutical. Success has been demonstrated in topical ocular treatments using drugs such as idebenone, a potent antioxidant that has similar behavior to coenzyme Q10, doxorubicin, an anthracycline drug, and many hydrophobic antimicrobials including cyclosporine A and doxycycline (De Campo et al. 2001).

APPLICATIONS TO AREAS OF HEALTH AND DISEASE

Ophthalmic Application

One of the largest challenges in treatment of extraocular disease is providing long-term treatment without overexposure, which often leads to compromised intraocular structures and potential systemic drug exposure. Drug penetration into the ocular tissue is especially difficult and bioavailability in the tissue is extremely low due to anatomical and physicochemical limitations. It is in this context that nanotechnology has the potential to improve treatment of ocular infection and diseases. The benefits of nanopharmaceuticals in ocular treatment are that they can significantly increase the bioavailability of the drug in the ocular tissue and provide a controlled release profile in order to prevent overdosing and allow for long-term treatment.

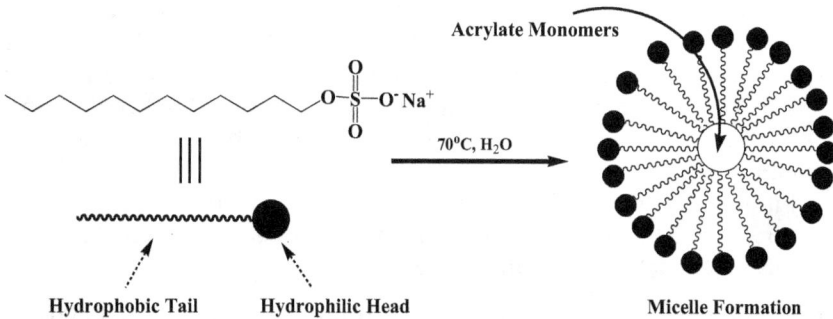

Fig. 4. Schematic of initial micelle formation during an emulsion polymerization. Image depicts how polyacrylate nanoparticles can be formed through an emulsion polymerization process. Once micelle formation occurs, radicals are used to initiate polymerization within the micelle, forming the polyacrylate nanoparticle. Published with permission from Kerriann Greenhalgh (2008).

Dermal Wound Application

One of the ultimate goals in the treatment of dermal wounds is to minimize infection. Seventy-five percent of all deaths from burn wounds occurring over 40% of the body are the direct result of infection (Church et al. 2006).

With the increasing capacity of microbes to develop drug resistance, the need to find an effective means of topical drug delivery that is capable of reaching the site of infection is paramount. The majority of antibiotics are water soluble, which is problematic when treating infections in the grossly hydrophobic dermal layers. Nanopharmaceuticals can provide an immense benefit here. The nanoparticle drug carrier permits penetration of the drug to the site of infection that would otherwise not occur. Encapsulating the antibiotic in a mostly hydrophobic nanoparticle would enhance penetration further and provide the ultimate topical dermal wound care product.

Key Facts

- Nanoparticles are particles that are 1 to 100 nanometers (nm) in diameter and can be made up of many different organic and inorganic materials. Nanoparticles are most often polymer or metallic in composition. They were first used in therapeutics in 1976 by Kreuter and Speiser as a flu vaccine. The most common nanoparticles are metallic nanoparticles because of their straightforward preparation and natural occurrence. Silver-based nanoparticles, poly(lactic acid) (PLA) and poly(glycolic acid) (PGA) nanoparticles are the current leaders in pharmaceutical applications, including topical application.
- Silver-based nanoparticles have played a key role in the development of nanoparticles in medicine, especially in relation to topical applications. Silver is a well-known antimicrobial agent, or substance that kills bacteria. Nanocrystalline silver can be found in at least two commercially available wound care products, Acticoat™ and Silverlon®. These products have demonstrated enhanced antimicrobial efficacy over more traditional silver-containing wound care products. Pitfalls of these products are that excess silver deposits in the skin can leave a bluish color and cellular toxicity has also been observed, ultimately delaying wound healing.

- It has been reported by the American Burn Association that 500,000 Americans seek medical attention, 40,000 are hospitalized, and 4,000 die every year because of burn injury (Burn Incidence Fact Sheet 2005). Mortality rates due to burn wounds are staggering when the total body surface area is greater than 50%, and even greater when the majority of the wounds are third degree, compromising all of the patient's skin. Infection in these wounds can slow healing, damage the remaining healthy tissue surrounding the wound, and lead to sepsis and toxic shock syndrome. All of these complications can be life threatening if left untreated. The key for successful treatment of burn wounds is immediate antimicrobial treatment combined with debridement, removal of the dead tissue, and constant moisture.

Definitions

Acute: Something pertaining to a short period of time. In terms of wounds, an acute wound is one that is brought on by a sudden disturbance to the skin, such as a burn or laceration, and is expected to heal within a short period of time (less than 30 days).

Bactericidal: An agent capable of killing bacteria. Antibiotics, antiseptics, and disinfectants can be bactericidal. This can also refer to the level at which the agent is capable of killing bacteria, as opposed to retarding its growth.

Bacteriostatic: An agent capable of inhibiting the growth or reproduction of bacteria. Different from bactericidal (capable of killing bacteria outright).

Chronic: Something pertaining to an extended period of time. In terms of wounds, a chronic wound is one that has developed over a long period of time and is persistent; difficult to treat.

Hydrophilicity: The physical property of a molecule that attracts water. Hydrophilic molecules tend to be polar and mix easily with water. Most salts are hydrophilic and dissolve readily in water. One example of a common hydrophilic molecule is table salt.

Hydrophobicity: The physical property of a molecule that repels water. Hydrophobic molecules tend to be non-polar and do not dissolve or mix well with water. Most oils and waxes are hydrophobic. One example of a common hydrophobic molecule is olive oil.

In vivo: "Within the living." The term refers to experimentation using a whole, living organism as opposed to a partial or dead organism (*in vitro*) in a controlled environment.

Nanoparticle: An ultrafine particle on the order of 1 to 100 nanometers in diameter.

Nanopharmaceutical: A medical drug produced using nanotechnology. The nanopharmaceutical typically comprises a therapeutic agent coupled with a nano-sized carrier or delivery agent, such as a particle, tube, colloid or micelle.

Nanotechnology: The study of the controlling of matter on an atomic and molecular scale.

Summary Points

- Nanopharmaceuticals have become widely used in areas of transdermal, ocular, and topical application because of their ability to increase transport in lipophilic tissues and across epithelium and improve intracellular penetration. Stability of the drug, retention of drug activity and sustained release profiles are all key factors when designing a nanopharmaceutical and each factor must be tailored to the specific topical application desired for maximum effectiveness.
- Nanoparticles have played a critical role in improving the treatment of infection, especially in topical applications, by allowing antimicrobials with previously limited bioavailability to reach deep-rooted infections in the skin and eye.
- Nanocarriers have played a critical role in improving drug stability and bioavailability *in vivo*. Nanocarriers have provided these drugs the protection they need to survive *in vivo* for longer periods of time. They have also allowed them to travel to places within the body that their hydrophobic-hydrophilic interactions previously prevented.
- Nanocrystalline silver can be found in at least two commercially available wound care products, ActicoatTM and Silverlon®. It has shown enhanced antimicrobial efficacy over more traditional silver-containing wound care products. However, silver-based antimicrobials have disadvantages and the wound care market is trending away from them.
- Chitosan nanoparticles have played a critical role in improving delivery of many pharmaceuticals by providing increased bioavailability, drug stability, targeted delivery, and improved cytostatic activity over longer periods of time (Lance et al. 1995).

Abbreviations

MRSA	:	methicillin-resistant Staphylococcus aureus
PEG	:	poly(ethylene glycol)
PGA	:	poly(glycolic acid)
PLA	:	poly(lactic acid)

References

Allemann, E., R. Gurny, and E. Doelker. 1993. Drug-loaded nanoparticles—preparation methods and drug targeting issues. Euro. J. Pharm. Biopharma. 39: 173–191.

Atiyeh, B.S., et al. 2007. Effect of silver on burn wound infection control and healing: Review of the literature. Burns 33: 139–148.

Atiyeh, B.S., S.N. Hayek, and S.W. Gunn. 2005. New technologies for burn wound closure and healing—Review of the literature. Burns 31: 944–956.

Burn Incidence Fact Sheet. 2005. American Burn Association. http://www.ameriburn.org/resources_factsheet.php.

Cavallaro, G., et al. 1994. Entrapment of beta-lactams antibiotics in polyethylcyanoacrylate nanoparticles—studies on the possible *in-vivo* application of this colloidal delivery system. Int. J. Pharma. 111: 31–41.

Chiu, T., et al. 2004. Porcine skin: friend or foe? Burns 30: 739–741.

Church, D., S. Elsayed, O. Reid, B. Winston and R. Lindsay. 2006. Burn wound infections. Clin. Micro. Rev. 19: 403–434.

Couvreur, P., C. Dubernet, and F. Puisieux. 1995. Controlled drug-delivery with nanoparticles—current possibilities and future trends. Euro. J. Pharma. Biopharma. 41: 2–13.

Couvreur, P., et al. 1979. Polycyanoacrylate nanocapsules as potential lysosomotropic carriers—preparation, morphological and sorptive properties. J. Pharm. Pharmacol. 31: 331–332.

Couvreur, P., et al. 1986. Polymeric Nanoparticles and Microspheres, Vol. 27. CRC Press, Boca Raton, Florida.

De Campos, A.M., A. Sanchez, and M.J. Alonso. 2001. Chitosan nanoparticles: a new vehicle for the improvement of the delivery of drugs to the ocular surface. Application to cyclosporin A. Int. J. Pharma. 224: 159–168.

Duncanson, W.J., et al. 2007. Targeted binding of PLA microparticles with lipid-PEG-tethered ligands. Biomaterials 28: 4991–4999.

Greenhalgh, K., and E. Turos. 2006. *In vivo* studies of polyacrylate nanoparticle emulsions for topical and systemic applications. Nanomed. Nanotech. Bio. Med. 5: 46–54.

Gu, H.W., et al. 2003. Presenting vancomycin on nanoparticles to enhance antimicrobial activities. Nano Lett. 3: 1261–1263.

Holder, I.A., et al. 2003. Assessment of a silver-coated barrier dressing for potential use with skin grafts on excised burns. Burns 29: 445–448.

Kim, K., et al. 2006. Cell-permeable and biocompatible polymeric nanoparticles for apoptosis imaging. J. Amer. Chem. Soc. 128: 3490–3491.

Kreuter, J., and P.P. Speiser. 1976. New adjuvants on a polymethylmethacrylate base. Infect. Immun. 13: 204–210.

Kumar, M., et al. 2003. Genetic Vaccines and Therapy 1: 3.

Lamb, K.A., et al. 1991. Toxicity of amphotericin-B emulsion to cultured canine kidney cell monolayers. J. Pharm. Pharmacol. 43: 522–524.

Lance, M.R., C. Washington, and S.S. Davis. 1995. Structure and toxicity of amphotericin B/triglyceride emulsion formulations. J. Antimicrob. Chemother. 36: 119–128.

Leroux, J.-C., E. Allemann, F. De Jaeghere, E. Doelker, and R. Gurny. 1996. Biodegradable nanoparticles from sustained release formulations to improved site specific drug delivery. J. Control Rel. 39: 339–350.

Maincent, P., et al. 1986. Disposition kinetics and oral bioavailability of vincamine-loaded polyalkyl cyanoacrylate nanoparticles. J. Pharma. Sci. 75: 955–958.

Morones, J.R., et al. 2005. The bactericidal effect of silver nanoparticles. Nanotechnology 16: 2346–2353.

Pankhurst, Q.A., et al. 2003. Applications of magnetic nanoparticles in biomedicine. J. Phys. D-App. Phys. 36: R167–R181.

Polymeric Drug Delivery: A Brief Review. 2004–2005. Drugdel.com. http://www.drugdel.com/polymer.htm

Rustogi, R., et al. 2005. The use of Acticoat (TM) in neonatal burns. Burns 31: 878–882.

Silver, S., L.T. Phung, and G. Silver. 2006. Silver as biocides in burn and wound dressings and bacterial resistance to silver compounds. J. Indus. Micro. Biotech. 33: 627–634.

Sorkine, P., et al. 1996. Administration of amphotericin B in lipid emulsion decreases nephrotoxicity: Results of a prospective, randomized, controlled study in critically ill patients. Crit. Care Med. 24: 1311–1315.

Souza, L.C., et al. 1993. *In-Vitro* and *in-vivo* studies of the decrease of amphotericin-B toxicity upon association with a triglyceride-rich emulsion. J. Antimicrob. Chemother. 32: 123–132.

Steinstraesser, L., et al. 2002. Activity of novispirin G10 against Pseudomonas aeruginosa *in vitro* and in infected burns. Antimicrob. Agents Chemother. 46: 1837–1844.

Tian, J., et al. 2007. Topical delivery of silver nanoparticles promotes wound healing. Chem. Med. Chem. 2: 129–136.

Tom, R.T., et al. 2004 Ciprofloxacin-protected gold nanoparticles. Langmuir 20: 1909–1914.

Turner, J.L., and K.L. Wooley. 2004. Nanoscale cage-like structures derived from polyisoprene-containing shell cross-linked nanoparticle templates. Nano Lett. 4: 683–688.

Turos, E., et al. 2007a. Antibiotic-conjugated polyacrylate nanoparticles: New opportunities for development of anti-MRSA agents. Bioorg. Med. Chem. Lett. 17: 53–56.

Turos, E., et al. 2007b. Penicillin-bound polyacrylate nanoparticles: Restoring the activity of beta-lactam antibiotics against MRSA. Bioorg. Med. Chem. Lett. 17: 3468–3472.

Wang, Y. 2006. PhD Dissertation. University of South Florida.

Washington, C., M. Lance and S.S. Davis. 1993. Toxicity of amphotericin-B emulsion formulations. J. Antimicrob. Chemother. 31: 806–808.

Yarkoni, E., and H.J. Rapp. 1978. Toxicity of emulsified trehalose-6,6'-dimycolate (cord factor) in mice depends on size distribution of mineral-oil droplets. Infect. Immun. 20: 856–860.

5

Nanocrystal Formulations for Improved Delivery of Poorly Soluble Drugs

Jan Möschwitzer[1],* and Rainer H. Müller[1]

ABSTRACT

Drug delivery technologies based on nanosizing approaches are now viable and well-established ways to formulate poorly soluble drug molecules. Over the last 30 years many researchers have laid the foundation for the success of this formulation approach. Various techniques were explored; some of them proved to be very successful and some of them too complex to get the attention of the pharmaceutical industry. In the meantime, some very successful drug products containing nanocrystals have reached the market. Although nanosizing can be regarded as a well-established approach it is still an active area in academic and industrial research. Advances in production methods open up new application areas, such as intracellular and targeted drug delivery. In this chapter we discuss the unique and specific properties that make drug nanocrystals such a useful tool for the formulator of patient-friendly drug delivery systems. We give an

[1]Department of Pharmaceutics, Biopharmaceutics & NutriCosmetics Freie Universität Berlin, Freie Universität Berlin, Kelchstr. 31, 12169 Berlin, Germany.
*Corresponding author: Email: info@janmoeschwitzer.de

List of abbreviations after the text.

overview of industrial applicable production techniques for drug nanocrystals including the most recent developments and indicate the pros and cons of various methods. By discussing the different products on the market, we show how versatile the nanosizing approach is. We explain the link between the special properties of the drug nanocrystal formulations and the specific requirements of the various administration routes.

INTRODUCTION

Poor water solubility of new chemical entities has reached a high level of importance in the pharmaceutical world. The formulation of these challenging drug candidates required the implementation of novel drug delivery technologies over the last 30 years. The need for new technological approaches has arisen from the applications of novel drug discovery methods, such as high throughput screening and computer-aided drug design. However, the large number of poorly soluble compounds is also a result of novel diseases with novel molecular targets, which require often more lipophilic structures for a better receptor fitting and membrane permeability (Rabinow 2004).

Although the general formulation philosophy can vary in the different pharmaceutical companies, particle size reduction would be generally not the first attempt on the way to a drug product. The standard approach is normally to improve the oral bioavailability of these problematic compounds by identification of a pharmaceutically acceptable and feasible salt form. This approach can sometimes significantly increase the solubility and dissolution rate, leading to an increased oral bioavailability. An alternative, but more difficult approach would be the identification of a suitable pro-drug structure (Merisko-Liversidge and Liversidge 2008). Depending on the physico-chemical properties of the new chemical entity (NCE), other enabling technologies, such as high-energy complexes or lipid-based systems, are often applied. However, there are more and more situations in which all these alternative approaches fail and then, very often, particle size reduction is the last resort before the molecule is abandoned. With the increasing number of successful examples of drug products using nanosized active pharmaceutical ingredients (APIs) on the market, particle size reduction became an established formulation technology that is today often applied as a standard approach. The technological examples range from formulation support for early pharmacokinetic studies in using nanosuspensions to large-scale production of blockbuster products, such as Abbott's Tricor®. The biological advantages of drug nanoparticles have been summarized by Liversidge, the inventor of élan's NanoCrystal®

technology, as follows: formulation performance improvement related to enhanced dissolution, safer and more patient-compliant dosage forms as well as better potential for dose escalation for improved efficacy (Merisko-Liversidge and Liversidge 2008). In this chapter we discuss the special features of drug nanocrystals, give an overview of industrially applicable production techniques for large-scale production, and discuss the use of nanosized products for oral and parenteral administration (Table 1).

Table 1. Key features of drug delivery systems.

1. DDS are used to administer an API to achieve the desired therapeutic effect in humans or animals in a controlled manner.
2. DDS are very often patent-protected formulation technology platforms.
3. Poorly soluble APIs require tailor-made drug delivery technologies to formulate patient-friendly drug presentations with increased bioavailability, reduce food effects, decrease patient-to-patient variability.
4. Drug delivery platforms are used to reformulate drugs to improve performance.
5. Over the last 20 years, various drug nanocrystal delivery systems were developed by several specialized drug delivery companies.
6. Novel DDS currently under development aim to achieve a better drug targeting including systems for intracellular delivery and brain.

DRUG NANOCRYSTALS—DEFINITIONS AND SPECIAL FEATURES

Drug nanoparticles, often also referred to as drug nanocrystals, can be defined as drug particles consisting of pure API with mean particle size in the nanometer range. The typical mean particle size ranges from 100 nm to 1000 nm; for some newer techniques the lower particle size limit may be 50 nm or less. In order to obtain long-term stability the particles are often stabilized with surfactants and other stabilizers, e.g., polymers. Drug nanocrystals suspended in a dispersion medium that can contain surfactants or stabilizers are referred to as nanosuspensions. The API of drug nanoparticles can exist in different solid states, from amorphous or partially amorphous to fully crystalline.

For a better understanding they have to be distinguished from polymeric nanoparticles, which exist in the same particle size range. In contrast to drug nanocrystals, polymeric nanoparticles consist of polymeric matrix material that is used as a template structure to hold API, which is distributed in the matrix. One significant difference between the two systems is the payload. Since a drug nanocrystal is mainly composed of pure API the payload is much higher than for polymeric nanoparticles; the payload is practically 100%.

PHARMACEUTICAL-RELEVANT PHYSICAL PROPERTIES

At present particle size reduction techniques, e.g., drug nanocrystals, are primarily employed for drug molecules showing oral bioavailability dependent on dissolution rate. The rate of absorption of these compounds can be described as a function of the particle size and the solid state of the particles. Many published examples show this trend; the smaller the particle size, the better the drug absorption (Jinno et al. 2006). The improved performance of drug products with nanosized API can be explained by using the Noyes-Whitney equation (Eq. 1). The dissolution rate can be accelerated by increasing the relevant surface area (A). Because of the distinctly smaller particle size of typical drug nanocrystals, such as 200 nm, the surface area is increased by a factor of 50 compared to standard micronized API of 10 μm. Another positive aspect is the decrease in the diffusional distance h. According to the Prandtl equation it can be stated that the diffusional distance h is reduced with increasing the curvature of ultrafine particles. Consequently, the dissolution rate is increased for smaller particles. The concentration gradient (C_s-C_x) also plays an important role. It is reported that drug nanocrystals often show an increased saturation solubility C_s. This can be explained by the Ostwald-Freundlich equation (Kipp 2004) and by the Kelvin equation (Müller 1998a). The extent of increased saturation solubility for nanosized APIs ranges from 10% to several fold (Dai et al. 2007; Hecq et al. 2005; Müller 1998b). It is still under investigation to what extent the increased saturation solubility can be explained solely by the reduced particle size. It is very likely that changes in the solid state properties of the drug nanocrystals during processing also contribute to the increased saturation solubility. However, it is indisputable that drug nanocrystals show increased dissolution velocity, which is one of the most important features for its use in oral pharmaceutical dosage forms.

$$\frac{dc_x}{d_t} = \frac{D \cdot A}{h}\left(c_s - c_x\right)$$

Equation 1

where:

$\dfrac{dc_x}{d_t}$ = dissolution velocity

D = diffusion coefficient
A = surface of drug particle
h = thickness of diffusional layer
c_s = saturation solubility of the drug
c_x = concentration in surrounding liquid at time x

PRODUCTION TECHNIQUES FOR DRUG NANOCRYSTALS

There are four different ways to produce drug nanoparticles (Fig. 1).

The first more theoretical way leads via a chemical reaction step directly to nanoparticles (chemical reaction approach). This approach is mainly used for the production of pigments and coating dispersions and not yet widely applied for the production of drug nanoparticles. The other three ways, i.e., bottom-up, top-down and combination technologies, are the most important principles for the production of drug nanocrystals (Möschwitzer 2010a).

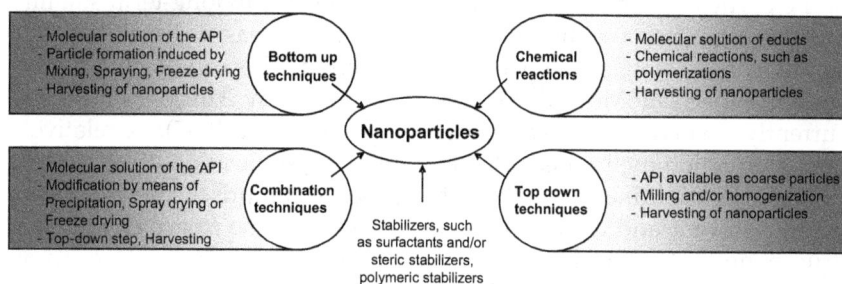

Fig. 1. Overview of different nanosizing principles including main process steps (modified with permission from Möschwitzer 2010b).

Bottom-up Technologies

Bottom-up approaches are the first and, from the technological requirements point of view, the easiest way to produce drug nanocrystals. One prerequisite to use bottom-up approaches is that the API has to have sufficient solubility in a pharmaceutically acceptable solvent, since the process starts with drug molecules in solution. The creation of nanoparticles via precipitation is induced by changing the conditions of the system. Classical precipitation processes, often referred to as solvent/antisolvent approaches, use a water-miscible organic solvent, such as ethanol or isopropanol, in which the poorly soluble API is dissolved. Mixing of the organic drug solution with an aqueous phase leads to a change in solvent power of the system resulting in the formation of drug nanoparticles. This approach is one of the easiest ways to produce drug nanoparticles. It is known as "hydrosol process" and was already developed in the 1980s by Sucker and colleagues (List 1991; Gassmann 1994). The difficulty of this process lies in avoiding further particle growth and preserving the small particle size. Owing to the presence of the organic solvent in the system after the precipitation process, the solubility of the API can be increased. This leads to a higher molecular mobility that can result in significant crystal growth. For that reason, hydrosols have to be lyophilized directly after production.

The improvement of the standard precipitation principle has been the focus of many research groups from academia and industry. One group suggests the use of water-soluble polymeric stabilizers, such as hydroxypropyl methylcellulose or polyvinylalcohol, to preserve the size of the precipitated nanocrystals. The stabilizers lead to an increased viscosity of the aqueous phase, which reduces the particle growth. The resulting suspension is subsequently spray-dried to obtain a dry powder with a relatively high drug loading (Rasenack and Muller 2004). The solid state of the particles is an important aspect, with regard to long-term stability but also regarding the dissolution rate. For an increased dissolution rate an amorphous state of the particles is preferred. This can be obtained by applying the NanoMorph® technology, invented by Auweter et al., and currently owned by the company Soliqs (Auweter 1997). A relatively new production method is called controlled crystallization during freeze drying (de Waard et al. 2008). This process comprises the use of an organic solvent and a sugar as stabilizer. During the freeze drying step drug nanocrystals are formed in a very controlled manner. Depending on the shelf temperature during the freeze drying process smaller or larger particles are formed. During the last ten or twenty years the field of bottom-up technologies has evolved further. Many particle engineering technologies have been developed. The first approach to mention is known as Evaporative Precipitation into Aqueous Solution (Chen et al. 2002). The first step is again the preparation of an organic solution of the API; in this case the organic solvent is not miscible with water. This organic drug solution is sprayed into heated water, which results in an immediate evaporation of the organic solvent. As a result, drug nanoparticles are formed and the organic solvent is removed from the system. The same research group has developed two other particle engineering techniques, known as spray-freezing into liquid and ultra-rapid freezing (Hu et al. 2003; Rogers et al. 2002).

Besides the use of cryogenic techniques, supercritical fluid technologies have gained some importance (Jung 2001). Most processes use carbon dioxide as supercritical fluid. The easily available carbon dioxide shows dual behavior when passing the supercritical point; it has then a low density but can also act as solvent. In general one can distinguish two fundamentally different principles: either the supercritical fluid is used as solvent for the API or the supercritical fluid is used as antisolvent. For compounds possessing a good solubility in supercritical carbon dioxide, the RESS (Rapid Expansion of Supercritical Solutions) can be used (Maston 1987). The API is dissolved in supercritical carbon dioxide and pumped via a nozzle system into an expansion chamber. This results in an extremely fast phase change of the carbon dioxide from the supercritical to the gas-like state. The API precipitates because of the loss of solvent power of the system. The process conditions lead normally to an amorphous product with a highly porous structure. The RESS technology can be regarded as solvent-free process since the carbon dioxide is ubiquitously available. Supercritical fluids can be also used as antisolvents in case the API possesses insufficient solubility for using the RESS technology. In this case the API is dissolved in an organic solvent. This solution is mixed with the supercritical antisolvent in a special nozzle type. The organic solvent dissolves in the supercritical antisolvent and the system loses the solvent power for the API. This results in a precipitation of small API nanoparticles, which are collected as a fine powder either directly in the expansion chamber or in a special cyclone system. This process is often referred to as the supercritical antisolvent process (Kim et al. 2008). Many drug delivery companies offer particle engineering processes using supercritical fluids. The companies have developed special proprietary processes and nozzle systems to increase the production efficiency and yield. At present, production capacity from lab scale up to pilot scale is available.

Top-down Technologies

In the top-down approach, one starts with a coarse API particle. The particles are broken down to drug nanocrystals by applying attrition forces using different particle size reduction equipment. Since API can be processed without using further solvents these techniques are relatively universal. Almost any API can be theoretically processed into drug nanocrystals. The only prerequisite is that pharmaceutically acceptable dispersion medium exists in which the API is poorly or not soluble. This broad applicability has contributed to the commercial success of top-down approaches for pharmaceutical applications.

A first, very important technology is based on wet ball milling (WBM). This very common approach is known as NanoCrystal® technology and was developed by Liversidge et al. The company élan uses this approach for the production of surface-modified drug nanoparticles (Liversidge 1992). The process steps are relatively straightforward. A milling chamber is charged with milling media (e.g., zirconium dioxide beads, silicium nitride beads, and polystyrene beads), aqueous stabilizer/surfactant solution and micronized API. Various mills from ultra-small lab scale up to large production scale are available. In general one can distinguish low-energy bead mills with a slowly moving milling media and high-energy ball mills. In most cases an agitated high-energy ball mill is used as milling equipment. The moving agitator moves the milling beads; shear forces are thereby induced, leading to the attrition of the drug particles. In the course of the process the milling media is exposed to a very high-energy input. This can lead to various amounts of abrasion from the beads depending on the resistance of the material of the milling beads. One way to reduce the quantity of impurities is the use of more durable milling media consisting of highly cross-linked polystyrene resin (Bruno 1992; Merisko-Liversidge and Liversidge 2008). Large production-scale mills are operated in circulation mode, which means that the suspension is continuously pumped through the milling chamber until the desired particle size of the drug nanocrystals is obtained. The drug nanoparticles are separated from the milling media either by a separating gap or a filter cartridge. The WBM technology is by far the most important particle size reduction method at present. There are five products on the market using this technology (Table 2); many others are still being developed.

Different types of high pressure homogenizers can also be used for particle size reduction. High pressure homogenization (HPH) can be performed with either piston-gap homogenizers or jet stream homogenizers.

Two methods exist for piston-gap homogenizers. The first technology is known as the Dissocubes® process, which can be described as homogenization performed in aqueous media at room temperature (Keck and Muller 2005). For the homogenization process one starts with the preparation of a coarse suspension containing the API, water as dispersion medium and stabilizers or surfactants. This suspension is forced by a piston through a very tiny homogenization gap. The size of the API particles is mainly reduced by cavitation forces, shear forces and particle collision. The same piston-gap homogenization can also be performed using water-reduced or even completely water-free dispersion media. This process is called Nanopure® technology and currently owned by Abbott (Mueller 2000). Drug nanoparticles can also be generated by using jet stream homogenizers. This process is known as IDD-P® technology, a technology of SkyePharma. A suspension or an emulsion, in case of molten API, is pumped under high pressure of up to 1,700 bar through the system. When the fluid streams pass the collision chamber of the Z-type or the Y-type it comes to particle collision, shear forces and also cavitation forces. The particle size is preserved by using phospholipids or other surfactants and stabilizers. Because of the relatively low power density of the standard equipment, 50 or more passes of the suspension are necessary for a sufficient particle size reduction (Mishra 2003). The IDD-P® technology has proven its commercial applicability with one product on the market.

Table 2. Marketed products using various nanosizing approaches.

Trade name and API	FDA approval	Nanosizing approach (DDS)	Company	Administration route	Reason for using nanosized API
Rapamune® Sirolimus	2000	NanoCrystal®	Wyeth (now Pfizer)	Oral	Reformulation, Patient-friendly tablet instead of solution
Emend® Aprepitant	2003	NanoCrystal®	Merck	Oral	NCE, High bioavailability, no food effects
Tricor®Lyphantyl® Fenofibrate	2004	NanoCrystal®	Fournier Pharma, Abbott Laboratories	Oral	Reformulation, No food effects
Triglide® Fenofibrate	2005	IDD-P®	Sciele, Shionogi Pharma Inc.	Oral	Reformulation, No food effects
Megace® ES Megestrol acetate	2005	NanoCrystal®	PAR Pharmaceuticals	Oral	Reformulation, No food effects, more patient-friendly
Invega® Sustenna®	2009	NanoCrystal®	Janssen	Parenteral, Intramuscular	Reformulation

Table shows all marketed products using a specialized drug delivery technology based on drug nanocrystals, information available in public domain.

Combination Technologies

Standard top-down approaches normally require the use of micronized starting material in order to avoid a clogging of the machine, e.g., media separator in a ball mill or the homogenization gap in a high pressure homogenizer, respectively. In general, the API is therefore micronized by jet-milling before it can be processed with a ball mill or a high pressure homogenizer. This is relatively costly and leads to API losses. Furthermore, standard top-down techniques have limitations regarding the efficiency of the process. Relatively long milling times or a high number of homogenization cycles are required to obtain drug nanocrystals in the desired size range. These two common disadvantages have triggered the development of a completely new way to produce drug nanocrystals— the combination approaches. In a combination approach, two particle size reduction principles are coupled to obtain better efficiency. Normally, bottom-up technologies are coupled with top-down approaches.

The first combination approach was developed by the company Baxter and is known as Nanoedge™ technology. This approach combines microprecipitation as bottom-up step with HPH (Kipp 2001). The API is at first dissolved in an organic solvent suitable for pharmaceutical applications. In some cases the API can also be dissolved in the pure surfactant phase. This organic phase is then mixed with an aqueous phase acting as antisolvent, resulting in a precipitation of preferably friable crystals. The next step breaks these particles and causes an annealing of the particles, leading to a more stable drug product. The disadvantage of this process is that the organic solvent comes still in contact with the aqueous antisolvent. Therefore, it has to be carefully removed from the resulting nanosuspension, otherwise the organic solvent can act as cosolvent and increase the solubility of the API in the aqueous dispersion medium. Eventually, this could result in increased particle growth upon storage. A further development is known as Cavi-precipitation technology. Cavi-precipitation means that the precipitation is performed in the cavitation zone of the high pressure homogenizer. This reduces the time from the formation of the first nuclei until the energy input. Consequently, the particle size of the obtained particles can be much smaller than in the Nanoedge™ process. In some cases it was shown that particle size down to 22 nm could be obtained (Müller 2005). The technology was developed under the codename H69 and is part of the smartCrystals® patent portfolio, which is currently owned by the company Abbott.

The understanding of the potential disadvantages of the Nanoedge™ process as well as the H69 process led to a further development of the combination technologies—the separation of the bottom-up step and the subsequent top-down step. The first approach was developed under the codename H42 and is also part of the smartCrystals® technology platform. It can be described as a combination of non-aqueous spray drying as bottom-up step followed by HPH (Möschwitzer 2005a). The modification of the starting material via spray drying results in API powder, which may even already contain the surfactants or other stabilizers. The modification step changes the hardness of the API, therefore the subsequent HPH is much more effective than the conventional HPH process using micronized API as starting material. It was shown that with only a few homogenization cycles a nanosuspension with a very small mean particle size and a narrow particle size distribution can be obtained (Fig. 3). For the model compound hydrocortisone acetate it could be also shown that the modification by spray drying does not necessarily lead to the formation of amorphous API (Möschwitzer and Muller 2006).

The most effective combination technology in terms of particle size reduction was developed under the codename H96 and is also part of the smartCrystals® technology platform. It is a combination of a non-aqueous freeze-drying process (bottom-up) with HPH (top-down) (Möschwitzer 2005b). The major advantage of this technology is its superior particle size reduction effectiveness in conjunction with a very controlled way of API modification. Processing of the modified API powder with HPH leads to drug nanoparticles significantly smaller than 100 nm after only a few homogenization cycles (Fig. 2) (Salazar 2010). The small particle size opens completely new applications for drug nanocrystals, including intracellular drug delivery with a very high payload of the nanoparticles (Staedtke et al. 2010). The disadvantage of this process is the time-consuming API modification by means of freeze drying. It is a technological challenge to perform non-aqueous freeze drying in an industrial environment of good

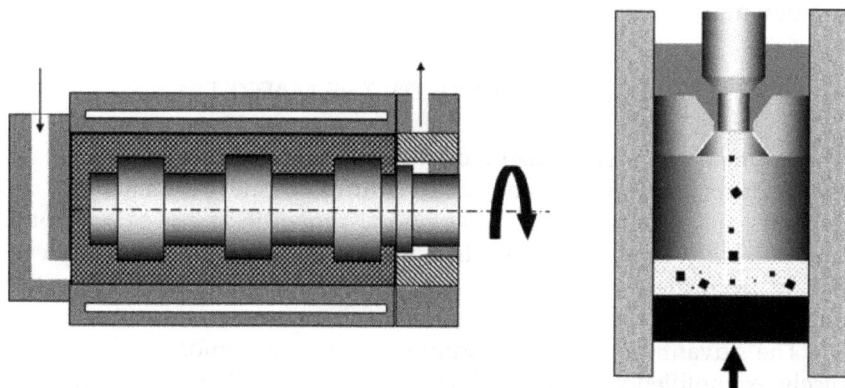

Fig. 2. Schematic drawing of top-down equipment. Left: cross-section of a high-energy agitated ball mill. Right: cross-section of a piston-gap high pressure homogenizer (with permission from Möschwitzer 2005c).

manufacturing practices. However, pre-clinical and clinical studies using this technology have proven its industrial applicability.

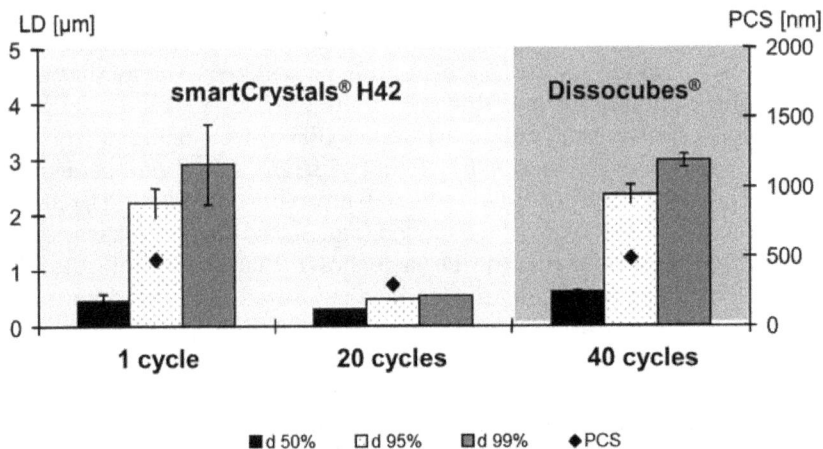

Fig. 3. Comparison of the particle size reduction effectiveness for the H42 process of the smartCrystals® technology platform versus the Dissocubes® technology, API was hydrocortisone acetate, results show mean of three measurements ± SD (with permission from Möschwitzer 2005c).

The smartCrystals® formulation platform comprises another combination process, known under the code CT, which stands for combined technology. This technology is not a classical combination process, but rather an improved version of a top-down process. The API is pre-treated in a milling procedure and subsequently processed with high pressure homogenization. Since it represents a combination of two top-down processes and includes no bottom-up step, it falls in the category of top-down processes (Petersen 2006).

ORAL PRODUCTS ON THE MARKET

The approach of particle size reduction was first employed to increase the performance of drug molecules with low oral bioavailability. In general, compounds showing suboptimal properties, such as poor oral bioavailability, strong food effects, and lack of dose proportionality can benefit from the formulation as drug nanocrystals (Merisko-Liversidge and Liversidge 2008).

The advantages of orally administered drug nanocrystals can be nicely exemplified with the products on the market. Rapamune®, Whyth' sirolimus formulation, was introduced to the market as an oral solution. Because of stability issues, the formulation had to be stored in a refrigerator and reconstituted prior to use in a complicated procedure. The systemic

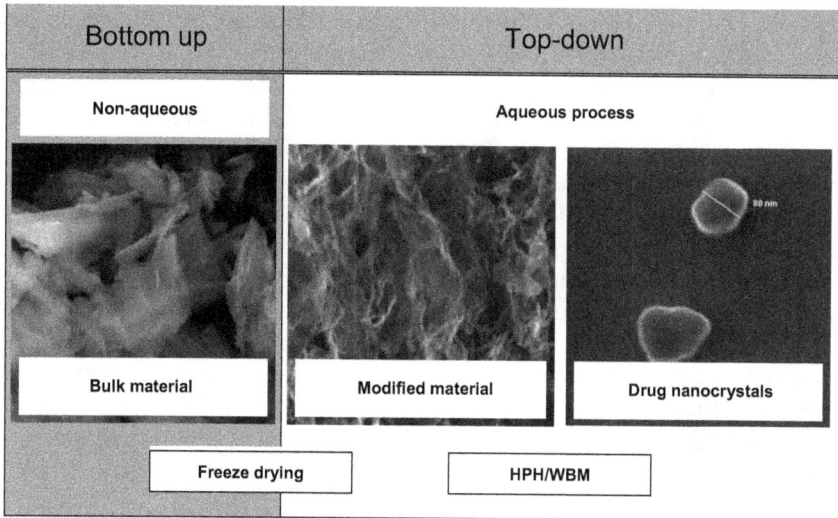

Bottom up	Top-down	
Non-aqueous	Aqueous process	
Bulk material	Modified material	Drug nanocrystals
Freeze drying	HPH/WBM	

Fig. 4. Schematic description of the combinative process H 96, part of the smartCrystals® technology portfolio. Freeze drying as bottom-up step is coupled with high pressure homogenization (HPH) or wet ball milling (WBM) (modified with permission from Salazar 2010).

availability of this solution was only 14%. The objective of the reformulation was the development of a user-friendly dosage form. élan has developed a tablet formulation based on its proprietary NanoCrystal® technology platform. In 2000, the improved formulation was approved by the FDA as the first product on the market using élan's NanoCrystal® technology. The more user-friendly tablet formulation can now be stored at room temperature. In addition to that, the bioavailability after administration as a tablet containing Sirolimus NanoCrystals® was 27% higher relative to the solution. Clinical data for the tablet formulation show that the food effect was different between the solution and the tablet, but still present in both formulations. Therefore, Rapamune® in the form of a solution or tablet formulation has to be taken consistently with or without food. Emend® from Merck is the second product employing the NanoCrystal® technology, approved by the FDA and introduced to the market in 2003. In this case the nanosized product was not a reformulation, but particle size reduction was chosen as approach for the market entry of the NCE aprepitant. Emend® capsules contain spray-coated pellets with the nanocrystalline aprepitant, sucrose, microcrystalline cellulose, hyprolose and sodium dodecylsulfate (Möschwitzer 2005c). The particle size reduction led to an increase in oral bioavailability of about 600%. In addition, the positive food effect in feed state could be significantly reduced, enabling administration independently from food intake. In 2004 Tricor®, a reformulated fenofibrate formulation was placed onto the market by the companies Fournier and Abbott as the third product using the NanoCrystal® technology. In this case the particle size reduction led to a slightly improved bioavailability and significantly reduced food effects. Therefore, the dose could be reduced from 200 mg to 145 mg, achieving a bioequivalence. The most important advantage of the new formulation is the reduced food effect, which allows a convenient administration independently from meals. The first nanoproducts were developed as conventional solid dosage forms, which is obviously the most preferred way of administration. The proper development of oral solid dosage forms reduces also the risk of particle growth upon storage. The major challenge for the development of the most oral solid immediate release dosage form is to achieve a quick and complete redispersion of the nanocrystalline API particles. Besides conventional solid dosage forms, nanosized APIs can be also administered in form of a nanosuspension. Although nanosuspensions bear the same risk of particle growth and physical instability as any other highly dispersed system, it is possible to develop pharmaceutical products with sufficient shelf life. Megace® ES was approved in 2005 by the FDA as improved, nanocrystalline formulation of megestrole acetate for the treatment of anorexia or cachexia in patients diagnosed with AIDS. The formulation as nanosuspension led to significant patient benefits, such as a 16-fold reduced viscosity and a 75% reduced administration volume; instead of 20 ml suspension, only 5 ml nanosuspension was needed. Furthermore, food effects of the original suspension could be eliminated by the formulation as nanosuspension. Therefore, the drug product can now be taken with or without food, which is a tremendous improvement in terms of acceptance and patient compliance. This example shows also that nanosuspensions can have sufficient long-term stability. This example has certainly contributed to a broader acceptance of the formulation principle in general.

Although the market for nanosized APIs is currently dominated by products based on media milling using élan's NanoCrystal® technology, production techniques are available for commercial-scale manufacturing. One example is Triglide, a nanosized fenofibrate formulation produced by means of HPH. For this drug product, SkyePharma's IDD-P™ technology was applied to produce nanosized fenofibrate. The product was approved by the FDA in 2005 and marketed in the US by Shionogi Pharma Inc. The 160 mg dose is bioequivalent to the product containing 200 mg micronized fenofibrate and shows no food effects, like Tricor®.

INJECTABLE FORMULATIONS

The development of injectable formulations for poorly soluble compounds can be extremely challenging. The standard approach is the use of co-solvents, solubilization with surfactants or lipids as well as complexation with different types of cyclodextrines (Strickley 2004). However, the administration of these excipients can cause undesired effects, such as pain at the injection site (Irizarry et al. 2009). Sometimes relatively large injection volumes are needed to administer the required dose levels. The use of lipidic systems is often also associated with pain upon injection due to an increased viscosity of the system. The disadvantages of the standard systems along with the success of the first nanocrystalline drug products on the market have triggered the exploration of nanosuspensions for injectable dosage forms.

Animal studies have revealed that nanosuspensions lead to a different pharmacokinetic profile compared to solutions. Studies done with itraconazole cyclodextrine solutions (Sporanox®) and nanosuspensions of the same compound have shown lower C_{max} levels and significantly prolonged mean residence times for the nanosuspensions. This led to a significantly improved tolerability of the API, with no mortality up to 320 mg/kg dose for the nanosuspension compared to the death of some animals already at 30 mg/kg dose levels in the Sporanox® group (Rabinow et al. 2007).

Drug nanocrystals injected intravenously can be used to target different organs. A suitable application is targeting of the monophasic phagocytic system. Because of the small particle size, drug nanocrystals are not filtered out by the lung, but will be deposited in the liver, spleen and bone marrow. This effect was shown for clofazime, a drug for the treatment of *Mycobacterium avium* infections. After administration of a nanosuspension, much higher drug concentrations were found in the spleen and in the liver than in the lung (Peters et al. 2000). Depending on the particle size, nanosuspensions can possess the pharmacokinetic profile of a solution. Therefore, the nanocrystals need to be much smaller than 100 nm. The different organ distribution of drug nanoparticles depending on their particle size was demonstrated with the API oridonin, by comparing the pharmacokinetics of two different sizes, 103 nm versus 897 nm, respectively. It was shown that the small particles have a plasma concentration profile and organ distribution similar to a solution. In contrast, the larger particles accumulated in the organs of the reticoluendothelial system (Gao et al. 2008).

For direct brain targeting, drug nanocrystals need to be modified by surface modification using the concept of "differential protein adsorption". Surface modification of drug nanocrystals with the surfactant Tween 80 leads to a preferential adsorption of apolipoprotein E (Kreuter 1997). This protein adsorption enables a targeted delivery of drug nanocrystals to the brain. Atovaquone drug nanocrystals modified with Tween 80 have shown excellent efficacy in the treatment of Toxoplasmosis (Schöler et al. 2001). A different strategy is indirect brain targeting. For the anti-viral compound indinavir a passive brain targeting via macrophages could be shown (Dou et al. 2007). Another approach is to load ultrasmall drug nanocrystals of a drug into red blood cells as a kind of living container in an incubation procedure. The API is released slowly after re-administration of the loaded red blood cells back into the patient. The feasibility of this system could be shown *in vitro* for the antifungal compound amphotericin B (Staedtke et al. 2010).

In 2009, Invega® Sutenna® was approved by the FDA as first injectable product using the NanoCrystal® technology. The product is a once-a-month, extended release formulation of the poorly soluble API paliperidone palmitate. The aqueous-based nanosuspension can be administered intramuscularly using a small bore needle and a small volume without the need for a power injector. This can contribute to improved compliance of schizophrenic patients. This exemplifies that long-term, stable, injectable formulations can be developed on the basis of nanosizing approaches.

DERMAL DELIVERY OF DRUG NANOCRYSTALS

Nanoparticles in general possess many advantageous properties, such as increased bioadhesion, membrane penetration and permeation potential. These properties make them also interesting for mucosal and in particular dermal drug delivery systems.

The use of nanoparticles for dermal applications started first with cosmetic applications, using inorganic nanoparticulates, such as titanium dioxide, as pigment in sunscreens. Due to lower regulatory burden also the first nanosized active compounds were developed for the cosmetic market. Antioxidants such as rutin, apigenin and hesperidin were formulated as nanosuspensions and added to the final products by simply being admixed to the aqueous phase (Al 2010; Mauludin 2008). The first products containing nanosized rutin were placed on the market by the company Juvena in the series Juvedical® in 2007. In 2009 another product, platinium rare, was marketed that contains nanosized hesperidin. The use of nanosized rutin results in a significantly increased sun protection factor than the use of the more water-soluble rutin derivative. Two factors contribute to the better performance of the nanosized rutin. Becaue of the decreased particle size, the dissolution rate and the solubility of the poorly soluble rutin are increased, which leads to a higher concentration of the active ingredient between the formulation and the skin. This results in an increased penetration. Additionally, the original rutin is more lipophilic than the rutin derivative, which also leads to better penetration. Because of the higher dissolution rate of small nanocrystals, the penetrated fraction is rapidly replaced as long as undissolved particles are still present (Shegokar and Muller 2010). The same effect was also shown for the API diclofenac administered as nanosuspension. In Yucatan micropigs, a 3.8-fold increase in drug flux was measured compared to the control group (Piao et al. 2008).

CONCLUSION

Nanosizing has evolved from a rescue technology to a major, well-established and accepted formulation approach for poorly soluble compounds after many efforts by pioneers in the field. The technological portfolio is so broad that almost any compound can be theoretically nanosized. However, nanotechnology is a multidisciplinary field that requires a very broad knowledge and understanding of many underlying principles from different disciplines. This understanding of the fundamentals in colloidal sciences is just as important as knowledge about biopharmaceutical aspects, pharmacokinetics and processing to final dosage forms. It needs more than just the right equipment and excipients to develop and produce reproducible high quality products. Although there might still be room for new technological ways to come to nanoparticles, there is nowadays also a clear trend to more systematic investigations of the existing systems. The unique properties of nanosized particles justify also the continuous efforts to find new applications in order to extend the area of use further. With the development of techniques for the production of drug nanoparticles smaller than 100 nm, new areas such as intracellular drug uptake can also be explored. The short time between the first patents filed in this area and the market entry of the first products emphasizes the urgent need for such smart formulation technologies. Nanosizing is a formulation approach that can really contribute to more patient-friendly formulations.

Definitions

Active pharmaceutical ingredient: The pharmaceutical active moiety in a drug formulation.

Bioavailability: The fraction of an administered dose of unchanged API that reaches the systemic circulation.

New chemical entity: A new chemical substance or an active moiety that emerges from the drug development process of pharmaceutical innovator companies. Before it can be used in marketed drug products it has to be approved by regulatory agencies, such as the FDA or the European Medicines Agency.

Ostwald ripening: A typical physical instability of dispersed systems, e.g., suspensions, with an non-homogeneous particle size distribution. The particles of the smaller fraction dissolve and the dissolved species redeposits on the surface of larger particles. This leads eventually to an increase in the mean particle size.

Particle size distribution: General measure for the quality of a suspension, the broader the particle size distribution the more non-homogeneous the individual particle sizes of the drug nanocrystals in the sample are. A broad particle size distribution can lead to Ostwald ripening.

Poorly soluble drug molecules: Drugs showing a solubility of less than 0.1 mg/ml in aqueous or biorelevant media. Independently of the absolute solubility number, which may vary between the companies, it is in general very challenging to formulate poorly soluble compounds to achieve sufficiently high *in vivo* concentrations for a pharmacodynamic effect.

Solid state properties: The solid state of an API is determined by the molecular arrangement within the solid drug particle. In general one can distinguish between amorphous and crystalline material. The solid state properties, such as melting point or solubility, depend also on the form of the API, i.e., whether the API exists as free form or as salt form, solvate, hydrate or cocrystal. Solid state properties can have a significant influence on the processability of the API, but also on the bioavailability *in vivo*.

Summary Points

- An increasing number of poorly soluble compounds in development pipelines of pharmaceutical companies led to an increased demand for novel drug delivery technologies.
- Nanosizing approaches are a very useful tool to formulate poorly soluble compounds, especially for those showing oral bioavailability dependent on dissolution rate.
- Drug nanocrystals can be produced via bottom-up or top-down techniques, or a combination of the two.

- For oral drug delivery, drug nanocrystals are the method of choice for high dose compounds to increase bioavailability, reduce food effects, decrease patient-to-patient variability as well as incomplete or erratic absorption, and accelerate the onset of action, as demonstrated by the several products on the market and in development.
- Drug nanocrystals can be also used in injectable dosage forms to avoid the use of solubilizing excipients, such as co-solvents, cyclodextrines or lipids, which can cause undesired side effects, like pain on the injection site or difficult administration procedures due to large volumes or high viscosity.
- Besides the use in standard applications, drug nanocrystals can be also used to achieve better drug targeting.
- In dermal applications, nanosized APIs can lead to a better drug flux, resulting in better effects.

Abbreviations

API	:	active pharmaceutical ingredient
DDS	:	drug delivery system
FDA US	:	Food and Drug Administration (a regulatory agency of the US Department of Health and Human Services)
HPH	:	high pressure homogenization
IDD®	:	P-insoluble drug delivery system
LD	:	laser diffractometry
NCE	:	new chemical entity
PCS	:	photon correlation spectroscopy
SD	:	standard deviation
WBM	:	wet ball milling

References

Al, S.L., R. Shegokar, and R.H. Müller. 2010. Apigenin smartCrstals for novel UV skin protection formulations. 8th European Workshop on Particulate Systems, Paris.

Auweter, H.B., H. Haberkorn, D. Horn, E. Luddecke, and V. Rauschenberger. 1997. Production of carotenoid preparations in the form of coldwater-dispersible powders, and the use of the novel carotenoid preparations. US patent application.

Bruno, J.A.D., D. Brian, E. Gustow, K.J. Illig, and N.S. Rajagopalan. 1992. Method of grinding pharmaceutical substances. US patent application. May 21, 1996.

Chen, X., T.J. Young, M. Sarkari, R.O. Williams, 3rd, and K.P. Johnston. 2002. Preparation of cyclosporine A nanoparticles by evaporative precipitation into aqueous solution. Int. J. Pharm. 242: 3–14.

Dai, W.G., L.C. Dong, and Y.Q. Song. 2007. Nanosizing of a drug/carrageenan complex to increase solubility and dissolution rate. Int. J. Pharm. 342: 201–207.

De Waard, H., W.L. Hinrichs, and H.W. Frijlink. 2008. A novel bottom-up process to produce drug nanocrystals: controlled crystallization during freeze-drying. J. Control Release 128: 179–183.

Dou, H., J. Morehead, C.J. Destache, J.D. Kingsley, L. Shlyakhtenko, Y. Zhou, M. Chaubal, J. Werling, J. Kipp, B.E. Rabinow, and H.E. Gendelman. 2007. Laboratory investigations for the morphologic, pharmacokinetic, and anti-retroviral properties of indinavir nanoparticles in human monocyte-derived macrophages. Virology 358: 148–158.

Gao, L., D. Zhang, M. Chen, C. Duan, W. Dai, L. Jia, and W. Zhao. 2008. Studies on pharmacokinetics and tissue distribution of oridonin nanosuspensions. Int. J. Pharm. 355: 321–327.

Gassmann, P., M. List, A. Schweitzer, and H. Sucker. 1994. Hydrosols—Alternatives for the Parenteral Application of Poorly Water Soluble Drugs. Eur. J. Pharm. Biopharm. 40: 64–72.

Hecq, J., M. Deleers, D. Fanara, H. Vranckx, and K. Amighi. 2005. Preparation and characterization of nanocrystals for solubility and dissolution rate enhancement of nifedipine. Int. J. Pharm. 299: 167–177.

Hu, J., K.P. Johnston, and R.O. Williams, 3rd 2003. Spray freezing into liquid (SFL) particle engineering technology to enhance dissolution of poorly water soluble drugs: organic solvent versus organic/aqueous co-solvent systems. Eur. J. Pharm. Sci. 20: 295–303.

Irizarry, M.C., D.J. Webb, Z. Ali, B.A. Chizh, M. Gold, F.J. Kinrade, P.D. Meisner, D. Blum, M.T. Silver, and J.G. Weil. 2009. Predictors of placebo response in pooled lamotrigine neuropathic pain clinical trials. Clin. J. Pain 25: 469–476.

Jinno, J., N. Kamada, M. Miyake, K. Yamada, T. Mukai, M. Odomi, H. Toguchi, G.G. Liversidge, K. Higaki, and T. Kimura. 2006. Effect of particle size reduction on dissolution and oral absorption of a poorly water-soluble drug, cilostazol, in beagle dogs. J. Control Release 111: 56–64.

Jung, J., and M. Perrut. 2001. Particle design using supercritical fluids: Literature and patent survey. J. Supercritical Fluids 20: 179–219.

Keck, C.M., and R.H. Muller. 2005. Drug nanocrystals of poorly soluble drugs produced by high pressure homogenisation. Eur. J. Pharm. Biopharm.

Kim, M.S., S.J. Jin, J.S. Kim, H.J. Park, H.S. Song, R.H. Neubert and S.J. Hwang. 2008. Preparation, characterization and in vivo evaluation of amorphous atorvastatin calcium nanoparticles using supercritical antisolvent (SAS) process. Eur. J. Pharm. Biopharm. 69: 454–465.

Kipp, J.E. 2004. The role of solid nanoparticle technology in the parenteral delivery of poorly water-soluble drugs. Int. J. Pharm. 284: 109–122.

Kipp, J.E.W., Joseph Chung Tak, M.J. Doty, and C.L. Rebbeck. 2001. Microprecipitation method for preparing submicron suspensions. US patent application.

Kreuter, J.A., D.A. Karkevich and B.A. Sabel. 1997. Drug targeting to the nervous system by nanoparticles. USA patent application.

List, M., and H. Sucker. 1991. Hydrosols of pharmacologically active agents and their pharmaceutical compositions comprising them. US patent application.

Liversidge, G.G., K.C. Cundy, J.F. Bishop, and D.A. Czekai. 1992. Surface modified drug nanoparticles. USA patent application.

Maston, D.W., J.L. Fulton, R.C. Petersen, and R.D. Smith. 1987. Rapid Expansion of Supercritical Solutions: Solute Formation of Powders, Thin Film, and Fibers. Ind. Eng. Chem. Res. 26: 2298–2306.

Mauludin, R. 2008. Nanosuspensions of poorly soluble drugs for oral administration. Freie Universität, Berlin.

Merisko-Liversidge, E.M., and G.G. Liversidge. 2008. Drug nanoparticles: formulating poorly water-soluble compounds. Toxicol. Pathol. 36: 43–48.

Mishra, A.K., M.G. Vachon, P. Guivarch, R.S. Snow, and G.W. Pace. 2003. IDD Technology: Oral Delivery of Water-Insoluble Drugs Using Phospholipid-Stabilized Microparticulate IDD Formulations. In: M.J. Rathbone, J. Hadgraft and M.S. Roberts (eds.). Modified-Release Drug Delivery Technology. Marcel Dekker, New York.

Möschwitzer, J. 2005a. Method for Producing Ultrafine Submicronic Suspensions. Germany patent application DE 10 2005 011 786.4.

Möschwitzer, J. 2010a. Particle Size Reduction Technologies in the Pharmaceutical Development Process. Amer. Pharm. Rev. 54–59.

Möschwitzer, J. 2010b. Special aspects of nanomedicines. European Medicines Agency: 1st International Workshop on Nanomedicines, London.

Möschwitzer, J., and A. Lemke. 2005b. Method for Carefully Producing Ultrafine Particle Suspensions and Ultrafine Particles and Use Thereof.

Moschwitzer, J. and R.H. Muller. 2006. New method for the effective production of ultrafine drug nanocrystals. J. Nanosci. Nanotechnol. 6: 3145–3153.

Möschwitzer, J.P. 2005c. Drug nanocrystals prepared by high pressure homogenization—the universal formulation approach for poorly soluble drugs. Ph.D., Free University of Berlin.

Mueller, R.H., K. Krause, and K. Maeder. 2000. Method for controlled production of ultrafine microparticles and nanoparticles. EP patent application 00947962. 10.04.2002.

Müller, R.H., and B.H.L. Böhm. 1998a. Nanosuspensions. In: R.H. Müller, S. Benita and B. Böhm (eds.). Emulsions and Nanosuspensions for the Formulation of Poorly Soluble Drugs. Medpharm, Stuttgart.

Müller, R.H., and J. Möschwitzer. 2005. Method and Device for Producing Very Fine Particles and Coating Such Particles. Germany patent application DE 10 2005 053 862.2.

Müller, R.H., and K. Peters. 1998b. Nanosuspensions for the formulation of poorly soluble drugs. I. Preparation by a size-reduction technique. Int. J. Pharm. 160: 229–237.

Peters, K., S. Leitzke, J.E. Diederichs, K. Borner, H. Hahn, R.H. Muller, and S. Ehlers. 2000. Preparation of a clofazimine nanosuspension for intravenous use and evaluation of its therapeutic efficacy in murine Mycobacterium avium infection. J. Antimicrob. Chemother. 45: 77–83.

Petersen, R. 2006. Nanocrystals for Use In Topical Cosmetic Formulations and Method of Production Thereof US patent application US 60/866,233.

Piao, H., N. Kamiya, A. Hirata, T. Fujii and M. Goto. 2008. A novel solid-in-oil nanosuspension for transdermal delivery of diclofenac sodium. Pharm. Res. 25: 896–901.

Rabinow, B.E. 2004. Nanosuspensions in drug delivery. Nat Rev Drug Discov, 3: 785–96.

Rabinow, B., J. Kipp, P. Papadopoulos, J. Wong, J. Glosson, J. Gass, C.S. Sun, T. Wielgos, R. White, C. Cook, K. Barker, and K. Wood. 2007. Itraconazole IV nanosuspension enhances efficacy through altered pharmacokinetics in the rat. Int. J. Pharm. 339: 251–260.

Rasenack, N., and B.W. Muller. 2004. Micron-size drug particles: common and novel micronization techniques. Pharm. Dev. Technol. 9: 1–13.

Rogers, T.L., A.C. Nelsen, J. Hu, J.N. Brown, M. Sarkari, T.J. Young, K.P. Johnston, and R.O. Williams, 3rd 2002. A novel particle engineering technology to enhance dissolution of poorly water soluble drugs: spray-freezing into liquid. Eur. J. Pharm. Biopharm. 54: 271–280.

Salazar, J., O. Heinzerling, R.H. Müller, and J. Möschwitzer. 2010. Optimization of a Novel Combinatory Production Method for Nanosuspensions. 8th European Workshop on Particulate Systems, Paris.

Schöler, N., K. Krause, O. Kayser, R.H. Muller, K. Borner, H. Hahn, and O. Liesenfeld. 2001. Atovaquone nanosuspensions show excellent therapeutic effect in a new murine model of reactivated toxoplasmosis. Antimicrobial Agents and Chemotherapy 45: 1771–1779.

Shegokar, R., and R.H. Muller. 2010. Nanocrystals: industrially feasible multifunctional formulation technology for poorly soluble actives. Int. J. Pharm. 399: 129–139.

Staedtke, V., M. Brahler, A. Muller, R. Georgieva, S. Bauer, N. Sternberg, A. Voigt, A. Lemke, C. Keck, J. Moschwitzer, and H. Baumler. 2010. In vitro inhibition of fungal activity by macrophage-mediated sequestration and release of encapsulated amphotericin B nanosupension in red blood cells. Small 6: 96–103.

Strickley, R.G. 2004. Solubilizing excipients in oral and injectable formulations. Pharm. Res. 21: 201–230.

Antioxidant Nanoparticles

Beverly A. Rzigalinski,[1,]* Kathleen Meehan,[2]
Mark D. Whiting,[3] Candace E. Dillon,[1] Kevin Hockey[1] and
Michael Brewer[1]

ABSTRACT

The field of nanomedicine implies cross-disciplinary interactions between biomedical sciences, physics and engineering to provide cutting edge technology for maintenance and improvement of public health. The emergence of nanoparticle antioxidants as potential therapeutics is one important result of these cross-disciplinary endeavors. Nanoparticle antioxidants of gold, platinum, fullerene derivatives, and cerium oxide are potent free radical scavengers that have potential in treatment of disorders associated with oxidative stress including neurodegenerative disorders, cardiovascular disease, inflammatory disorders, and cancer. The biological activity of these nanoparticles and their mechanism of action is the focus of this chapter, and the critical knowledge gaps in need of future study are also highlighted. Particular focus is given to cerium oxide nanoparticles, which

[1]Virginia College of Osteopathic Medicine, NanoNeuroLab, 1861 Pratt Drive, Blacksburg, VA, USA 24141.

[2]Virginia Polytechnic and State University, Bradley Dept. of Computer & Electrical Engineering, 302 Whittemore Hall, Blacksburg, VA 24060, 540-231-4442.

[3]Radford University, Dept. of Psychology, Radford, Virginia 24141, 540-831-5849.

*Corresponding author; Email: brzigali@vcom.vt.edu

List of abbreviations after the text.

appear to be the most potent with the least potential deleterious effects of the group. Through future interdisciplinary collaboration, the superior antioxidant effects of these nanoparticles may be applied to prevention of diseases associated with oxidative stress and improvement in their pharmacological treatment.

INTRODUCTION

Oxidative stress is a component of all human disease, including inflammatory disorders, neurodegenerative disorders, cardiovascular disease, and musculoskeletal diseases. Aging itself is hypothesized to progress, in part, via increased cellular and tissue damage associated with cumulative exposure to oxidative stress during a lifetime. Although oxidative stress is a major element of disease, medicine is limited in its capability to counteract the deleterious consequences of oxidative stress in an efficacious manner. The emerging field of nanomedicine has provided us with immense potential for counteracting oxidative stress in the form of antioxidant nanoparticles, which may dramatically improve outcome in disorders associated with oxidative stress.

APPLICATION TO AREAS OF HEALTH AND DISEASE

Antioxidant nanoparticles have relevance to all areas of health and disease. It is well-recognized that optimal health can be preserved with minimization of oxidative stress and free radial production. Given that excess free radicals and oxidative stress are associated, in some form, with all disease pathologies, efficient antioxidants have the potential to improve outcome in all disease states. Despite our efforts to produce effective antioxidants to date, our attempts have provided only modest benefit to public health. Hence, the applicability of potent and regenerative antioxidant nanoparticles to preservation of public health is vast.

OXIDATIVE STRESS AND FREE RADICALS

Free radicals are highly unstable molecules, which lack the full complement of electrons necessary for atomic stability. Hence, they are highly reactive and interact with molecules in the reduced intracellular environment, resulting in destruction of lipids, proteins, and DNA. A primary component of free radical species is oxygen, hence free radicals are often termed "reactive oxygen species" or ROS. Among the ROS encountered within the cellular environs are superoxide (O_2^-), the hydroxyl radical ($OH\bullet$), nitric oxide (NO), peroxynitrite (ONOO-), and lipid hydroperoxides

(Halliwell 2006). Such highly reactive ROS impart considerable damage to the molecular structure of biological organisms, thereby contributing to the pathologies of numerous disease states.

Cells and organisms normally function within a limited range of free radical production. In fact, free radical generation is a necessary component of several normal signal transduction pathways, such as control of blood pressure by NO. Therefore, a basal level of free radical production is critical to normal cell function. When excess ROS production occurs, such as in cell damage or disease, cells have innate protective measures to counteract damaging ROS. These endogenous protectants include enzymes that degrade radicals such as superoxide dismutase, catalase, and glutathione peroxidase; and endogenous chemical reductants such as *n*-acetyl cysteine, vitamins C and E, and carotenes. Disease states arise when the production of free radicals exceeds the capacity of these innate defense mechanisms or when innate defenses are compromised or damaged by aging or disease. In this case, aberrant free radical production induces considerable tissue damage, resulting in oxidative stress and cell/tissue dysfunction. In summary, within the cell there is a balance between ROS production for signaling, excess ROS production, and endogenous antioxidants. When this balance is disturbed, disease ensues.

Free radical scavengers or antioxidants have been the subject of much pharmaceutical investigation. If one could counteract the demise associated with excessive ROS production, disease could be prevented or more successful treatments developed. However, to date, our development of pharmaceutical antioxidants has met with very limited success for several reasons. One confounding factor is the subject of localization. Once generated within the cell, ROS are highly reactive and are unlikely to diffuse great distances within the reduced intracellular milieu, before damaging a cellular macromolecule. To date, delivery of antioxidants has been through the bulk extracellular environment (such as oral vitamin C or E) and relies on absorption and distribution to carry the antioxidant to the precise site where it is needed. The amount of antioxidant that actually reaches the site of ROS generation is likely to be very low. In cases such as the brain, where the blood-brain barrier constrains substances from crossing it, penetration of many antioxidants is not possible. Therefore, localization of antioxidants to the "right place at the right time" is critical to disease intervention. Additionally, the antioxidants used to date scavenge only a single free radical and are destroyed themselves by the interaction with a single ROS. Yet sites of oxidative stress in disease produce high, sustained levels of ROS, requiring a high, sustained level of antioxidants that we have not yet been able to achieve.

The ideal antioxidant would have several salient characteristics. First, it would be readily dispersed to sites that are high in oxidative stress, particularly brain, heart, and lung. Second, it would be effective at low concentrations. Third, it would be durable, producing long-lasting and potent effects at sites of ROS production. Last, it would be non-toxic and well tolerated by cells, tissues, and organisms. As yet, we have no such pharmacological agent. However, with the advent of nanomedicine, the construction of such a nanopharmaceutical appears within reach.

A RADICAL APPROACH TO OXIDATIVE STRESS

Nanotechnology and materials science have provided highly efficient technology to reduce oxidative stress in industrial applications. Nanoparticles with antioxidant properties are routinely used to improve combustion, remove environmental contaminants from engine exhausts, and reduce oxidation on various material surfaces. The reactions involved in materials applications are strikingly similar to endogenous antioxidant activity within the cell. These similarities have led biomedical science into promising investigations on the use of nanoparticle antioxidants in disease prevention. A summary of antioxidant nanoparticles and their biological effects is presented in Table 1. The most active antioxidant nanoparticles include the metals gold and platinum, the fullerenes and their hydroxylated derivatives, and nanoparticles composed of the rare earth oxide, cerium oxide. As evidenced by the information displayed in Table 1, application of nanoparticle antioxidants to disease is far reaching, encompassing longevity extension, neurodegenerative diseases such as Alzheimer's Disease or Parkinson's Disease, traumatic brain injury, retinal degeneration, cardiovascular disease, radioprotection, inflammation, diabetes, and cancer therapy.

What accounts for the broad applicability of this wide range of nanoparticles to so many diseases? What is their mechanism of action? Could they possibly be the "magic bullet" for disease in the future? These questions are addressed in the remainder of this chapter. For the sake of brevity, we focus primarily on cerium oxide nanoparticles (CeONP), which appear to be the most highly investigated and exhibit the least potential negative effects of the group.

CERIUM OXIDE NANOPARTICLES: BIOLOGICAL EFFECTS

For cerium, as well as the other nanoparticles in Table 1, the basic mechanism is one of an antioxidant, which chemically deactivates and detoxifies ROS produced within the cell. Cerium is a rare earth element of the lanthanide series, with multiple valence states, either +3 (fully reduced)

Table 1. Nanoparticle antioxidants and their biological actions.

Nanoparticle	Biological action in disease	References
Platinum		
	• Increase lifespan of *C. elegans*	Kim et al. 2008
	• Inhibition of oxidative damage from cigarette smoke	Onizawa et al. 2009
	• Inhibition of cancer cell growth	Saitoh et al. 2009
Gold	• Antidiabetic agent in mice, reduces oxidative stress and organ damage in hyperglycemia	BarathManiKanth et al. 2010
	• Antioxidant for reduction of osteoporosis	Ok-Joo et al. 2010
Fullerene and derivatives		
	• Antioxidant properties, neuroprotectant, radioprotectant, protection against ischemia-reperfusion injury	Reviewed in Markovic and Trajkovic 2008
	• Protection against doxorubicin toxicity	Injac et al. 2008
	• Protection of endothelial cells from NO-induced damage	Lao et al. 2009
Cerium oxide	• Extends neuronal lifespan, extends *Drosophila* lifespan	Rzigalinski et al. 2006, 2009
	• Neuroprotectant in traumatic brain injury	Rzigalinski et al. 2009a; Whiting & et al. 2009
	• Neuroprotectant in Alzheimer's and Parkinson's diseases	Rzigalinski et al. 2009b; Singh et al. 2007
	• *In vitro* neuroprotection	Schubert et al. 2006
	• Radioprotectant *in vitro* and *in vivo*	Rzigalinski 2005; Colon et al. 2009
	• Mitochondrial protectant	Rzigalinski et al. 2009b
	• Cardioprotectant, reduces cardiovascular disease	Niu et al. 2007
	• Protection of retinal cells from light-induced oxidative stress	Chen et al. 2006
	• Anti-inflammatory actions	Rzigalinski et al. 2006; Rzigalinski and Clark 2005; Hirst et al. 2008

or +4 (fully oxidized), and numerous excited sub-states. In the oxide form, the lattice structure affords even more antioxidant capacity, because of electron "holes" or oxygen vacancies within the lattice structure. Creation and annihilation of oxygen vacancies and alterations in cerium valence impart exceptional redox activity to CeONP, which is further enhanced by the increased surface area and quantum lattice alterations that occur at the nano-scale. Further, CeONP are unique in that there is a high hydrogen- and oxygen-absorbing capacity on the surface, providing for ease of reaction with H_2, O_2, or H_2O and their associated radical species. In the materials industry, CeONP are used to prevent oxidation, remove carbon monoxide, and reduce nitrogen oxide emissions. The reactions of CeONP as used in materials applications are strikingly similar to the actions of antioxidants in the biological context.

The basic antioxidant functions of CeONP *in vitro* are demonstrated using electron paramagnetic resonance spectroscopy (EPR) as shown in Fig. 1. In panel A, superoxide radicals were generated in the first (uppermost) trace by irradiating a solution of riboflavin (0.53 mM) and the free radical spin trap DMPO (0.1 M) in PBS, pH 7.4. In the lower traces, 5 µM cerium oxide nanoparticles of varying sizes were added to

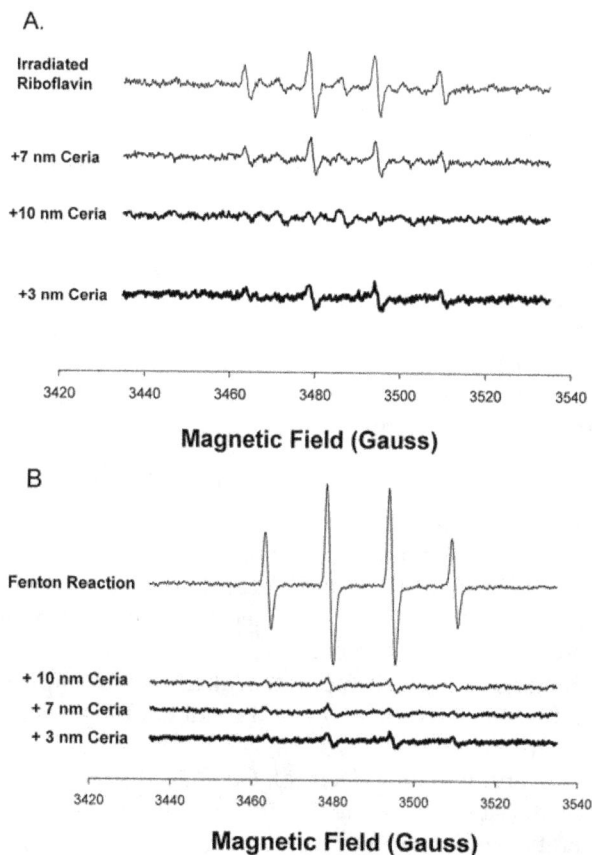

Fig. 1. Antioxidant activity of cerium oxide nanoparticles. The superoxide (A) and hydroxyl (B) radical scavenging activity of different sizes of cerium oxide nanoparticles is demonstrated using electron paramagnetic resonance (Rzigalinski et al., unpublished).

the mixture. Notice that cerium oxide nanoparticles of the average size of 10 nm completely scavenged all superoxide radicals generated. CeONP with average sizes of 3 and 7 nm were less effective at scavenging the superoxide radical. In panel B, hydroxyl radicals were generated by the Fenton reaction (upper trace), with CeONP of varying sizes added as described above. Note, once again, that CeONP of all the size ranges effectively scavenged hydroxyl radicals. These results demonstrate the role of CeONP as a free radical scavenger, or antioxidant. But what of the size differences? As shown in panel A, different-sized CeONP appear to differentially scavenge superoxide radicals, suggesting that size of the nanoparticle affects its activity. This has also been noted in cell culture and *Drosophila* studies (Rzigalinski and Clark 2005; Rzigalinski et al. 2006; Singh et al. 2007), where decreasing particle size below 10 nm decreases neuroprotective and antioxidant effects, and increasing size to 50 nm also abrogates protective effects. Hence, there appears to be an optimal size range at which CeONP (and possibly other nanoparticles) exert their beneficial biological effects.

At the biological level in cells and organisms, our current knowledge of CeONP has been reviewed by our group (Rzigalinski et al. 2006). Briefly, CeONP dramatically preserve the lifespan of mixed organotypic cultures of brain cells and pure neuronal cultures, while preserving normal calcium signaling during the extended lifespan of the cultures, demonstrating a durable and possibly regenerative action for a single dose. Further, CeONP protect neurons and other cell types from free radical challenge, as reported by our group and others (Rzigalinski et al. 2006, 2007; Schubert et al. 2006; Colon et al. 2009; Chen et al. 2006). Additionally, CeONP protected traumatically injured neurons from calcium dysregulation and cell death, and decreased inflammatory functions in microglia and other immune cells (Rzigalinski and Clark 2005). Notably, the antioxidant effects of CeONP were produced by a single 10 nM dose (avg particle size 10 nm) delivered once, on day 10, *in vitro*. CeONP were incorporated into the cytoplasm and mitochondria and were retained within the cells for the culture lifespans. *In vivo*, we found that CeONP induce dramatic extension of median and maximal lifespan in *Drosophila* and preserve motor function with aging (Rzigalinki et al. 2006; Singh et al. 2007) and improve learning, memory, and functional recovery in a rat model of traumatic brain injury (Whiting et al. 2009).

The potent antioxidant effects of CeONP show great promise in treatment of diseases associated with oxidative stress, particularly neurodegenerative and inflammatory disorders. For example, Alzheimer's Disease is associated with high levels of oxidative stress, some of which may be induced by free radicals produced during amyloid beta (1–42) ($A\beta$(1–42)) aggregation (Yatin et al. 1999). By virtue of its antioxidant properties, CeONP may be able to counteract the damaging effects of ROS in Alzheimer's Disease. In Fig. 2, EPR spectra were obtained during Aβ(1-42) (1 mg/ml) aggregation with the spin trap PBN (50 mM) incubated at 37°C in chelexed PBS according to procedures described by Yatin et al. (1999). In the upper trace, Aβ(1–42) yielded a characteristic 6 line spectrum indicative of ROS formation. In the second trace, Aβ(1–42) was incubated with CeONP (58 µM). As shown, no radial species were produced, suggesting that CeONP either blocks the production of free radicals or scavenges the free radicals produced by aggregation of Aβ. The third trace displays an albumin control, and the fourth trace represents CeONP incubated in presence of the spin trap PBN alone, indicating that CeONP does not generate free radicals under these conditions. These results suggest that CeONP may be effective at reducing neuronal damage associated with oxidative stress during Aβ(1–42) aggregation.

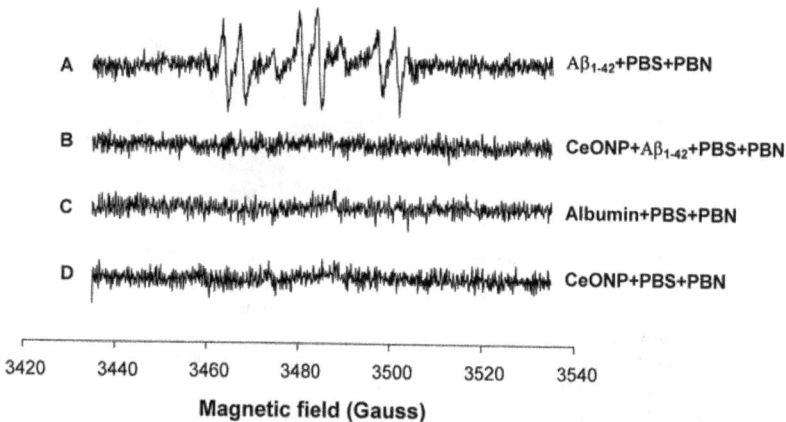

Fig. 2. Cerium oxide nanoparticles inhibit amyloid-beta–induced free radical formation *in vitro*. Trace A shows free radicals generated by amyloid beta, measured by electron paramagnetic resonance. Trace B demonstrates the ability of cerium oxide nanoparticles to scavenge amyloid-beta–generated free radicals. Trace C is an albumin control, and trace D demonstrates that cerium oxide nanoparticles alone do not generate free radicals. For detailed explanation, see text (Rzigalinski et al., previously unpublished).

This hypothesis was tested in a tissue culture model of aggregated Aβ(1-42)-induced neuronal damage, as shown in Fig. 3. Pure cultures of cortical neurons were treated with or without CeONP on day 10 *in vitro*, followed by exposure to aggregated Aβ(1–42) (10 µM) on day 12. Neuronal damage was assessed by light microscopy 24 h later. Note that CeONP at doses of 10 and 100 nM protected cortical neurons from Aβ(1–42)-induced cell damage. Importantly, low doses were delivered only once, prior to Aβ(1–42) challenge, highlighting the efficacy of CeONP as an antioxidant.

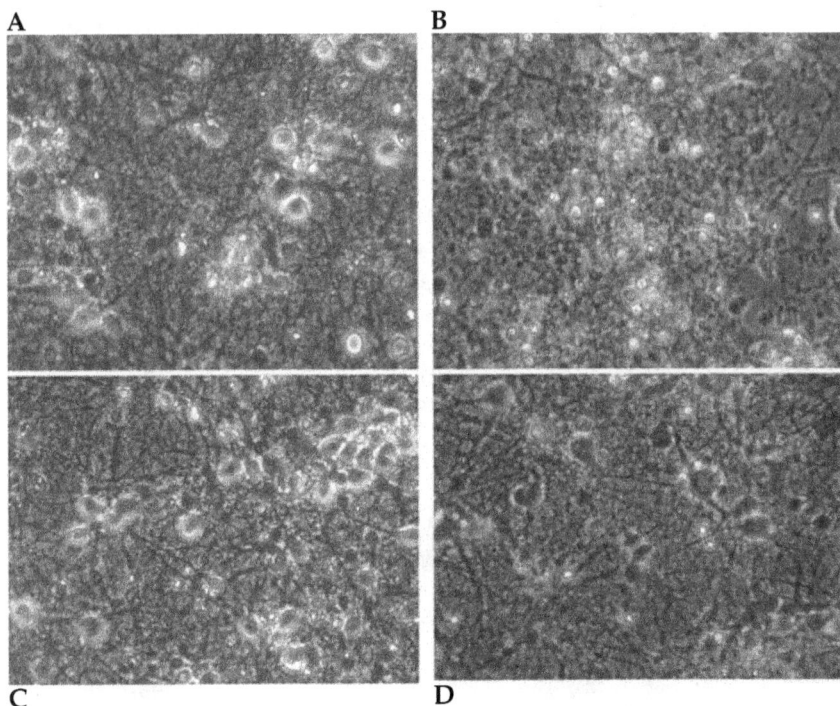

Fig. 3. Cerium oxide nanoparticles prevent neuronal damage associated with exposure to Aβ(1–42). Panel A shows a healthy culture of rat cortical neurons. Panel B shows a similar culture, treated with Aβ(1–42) for 24 h. Note fragmentation of neuronal processes and overall degradation of culture heath. In panels C and D, cultures were pretreated with 10 and 100 nM CeONP respectively, followed by Aβ(1–42) exposure. Note protective effects of cerium oxide nanoparticles (previously unpublished data).

Color image of this figure appears in the color plate section at the end of the book.

Similar observations were noted in a *Drosophila* model of Parkinson's Disease, based on exposure to the superoxide generator paraquat. For these experiments, cohorts of 100 flies were fed standard fly food with or without 10 and 100 μM CeONP for 21 d (approximately middle age). At this time, flies were starved for 3 h and exposed to filter paper saturated with 5% sucrose containing the LD50 for paraquat (10 mM) for 3 h. As shown in Fig. 4, paraquat rapidly led to the demise of 80% of control flies by 24 h. However, flies fed CeONP experienced significant protection against paraquat toxicity. Paraquat also induces a Parkinson's-like decrease in motor function in *Drosophila*, and this parameter was also examined using a Trikinetics Activity monitor as previously described (Rzigalinski et al. 2006, 2009). Flies are placed in a vial and activity is monitored when a fly crosses an infra-red beam at different levels in the vial. The number of beam crosses per fly during a 5 min time frame is expressed as an activity

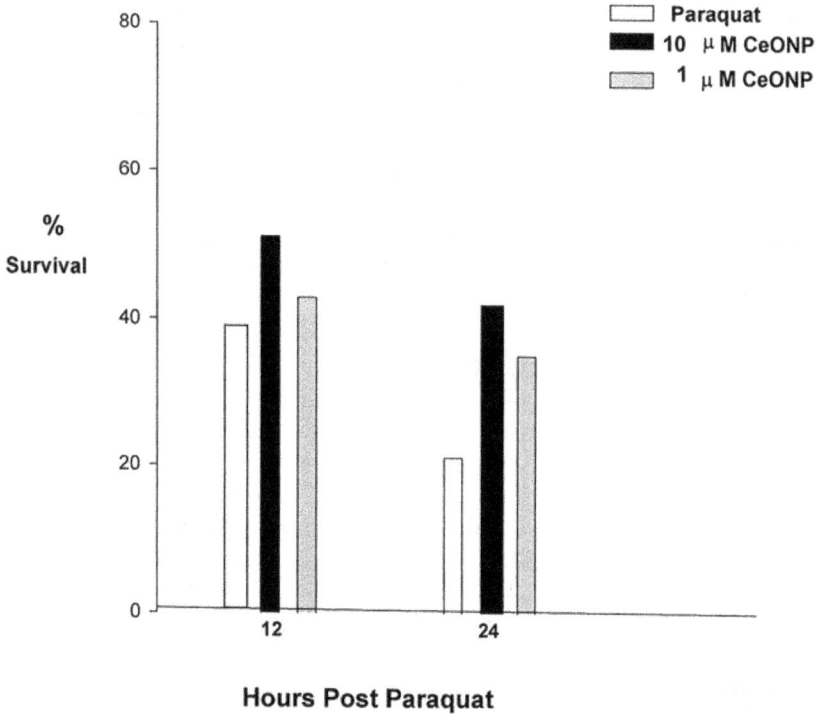

Fig. 4. Cerium oxide nanoparticles protect *Drosophila* from paraquat toxicity. A cohort of 100 male flies were fed the indicated concentrations of CeONP for 21 d, followed by exposure to paraquat for 3 h. Survival was measured every 3 h. Results are shown from a previously unpublished cohort of flies (Rzigalinski et al., previously unpublished data).

score, as shown on the y axis of Fig. 4. Activity was dramatically reduced 24 h post-paraquat. However, flies fed CeONP prior to exposure maintained significantly higher activity, suggesting improved motor function. These results suggest that CeONP may effectively prevent motor dysfunction induced by the paraquat model of Parkinson's Disease in *Drosophila*.

These examples demonstrate the efficacy and utility of CeONP in abrogating diseases associated with oxidative stress. In the examples given above, pretreatment with CeONP prior to oxidant challenge was used. At the time of this writing, delivery of CeONP after oxidative insult is presently underway, with equally dramatic results.

MECHANISM OF ACTION

It appears obvious and well founded in the literature that cerium oxide and other nanoparticles listed in Table 1 act via antioxidant effects. But how are these effects produced and what makes them superior to our traditional antioxidants such as vitamins E or C? What properties account for their apparent durable and possibly regenerative actions?

For CeONP, we can identify three primary physicochemical characteristics: size, cerium valence state and oxygen vacancies. Cerium atoms within the nanoparticle lattice exist in either the +3 (fully reduced) or +4 (fully oxidized) state, and flip-flop between the two during a radical scavenging electron transfer event. CeONP also have numerous oxygen vacancies present throughout the nanoparticle. These oxygen vacancies are a point in the lattice where an electron is missing from the oxygen shell. Such oxygen vacancies also participate in the antioxidation process, by reacting with free electrons of free radicals. Therefore, it would seem logical to conclude that valence state and oxygen vacancies account for the superb antioxidant properties of CeONP. Regarding size, as the size of the nanoparticle decreases, surface area increases, resulting in an increase in available area for ROS scavenging through increased exposure of cerium atoms and associated oxygen vacancies. Taken together, the effectiveness of a nanoparticle antioxidant such as cerium should depend on the amount of cerium in the 3+ state, the amount of oxygen vacancies, and the size.

To test these concepts, we assessed characteristics of several different CeONP of varying size and physicochemical structure, as shown in Table 2. Samples 1 and 2 were commercially prepared pharmaceutical grade, of two different sizes. Samples 3 and 4 were commercially prepared particles of the same average size (7 nm), from two different suppliers.

Table 2. Physicochemical properties of nanoparticles vs. biological effects.

Sample	Avg. size (nm)	XPS [Ce^{3+}] (%)	γ	Reactivity constant	PrI uptake (injured cells/mg protein x 10^6)	Protection from H$_2$O$_2$ challenge (% increase in survival over control)	Protection from O$_2$. challenge (% increase in survival over control)
1.	50	27.9	0.140	0.0137	0.14	0	2
2	10	31.2	0.160	0.012	0	78	83
3	7	33.7	0.169	0.0374	0	52	58
4	7	34.5	0.173		0.09	8	10
5	5	28.2	0.141	0.459	0.21	12	21

Size, amount of Ce3+, and oxygen vacancies (γ) are compared to biological parameters of toxicity (propidium iodide, PrI, uptake), and antioxidant activity. See text for complete explanation.

Sample 5 was custom synthesized, with the smallest average size (5 nm). The amount of cerium in the +3 state is hypothesized to be a key regulator of antioxidant capacity and thought to increase the number of oxygen vacancies (Pirmohamed et al. 2010). Therefore, we assessed the percentage of Ce3+ in the different nanoparticles, by X-ray photoelectron spectroscopy

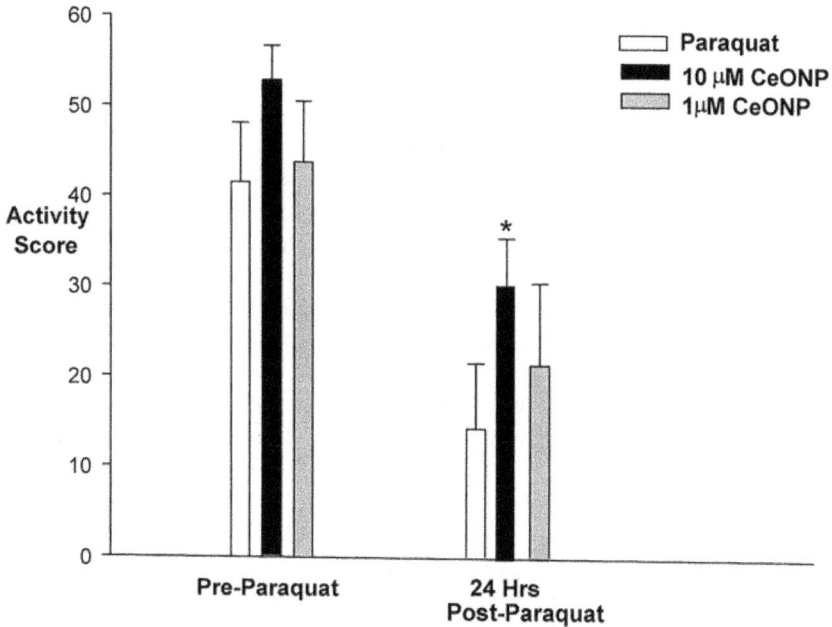

Fig. 5. Cerium oxide nanoparticles preserve motor function in *Drosophila* challenged with paraquat. Flies were exposed to paraquat as in Fig. 4. Activity was measured in a Trikinetics activity monitor. Results are mean +/- SEM. Results are representative of three separate experiments (Rzigalinski et al., unpublished data).

(XPS). As an additional measure of oxygen vacancy content, the "γ" figure was calculated, which represents the departure from stoichiometry (or CeO_2-x content). A "reactivity constant" was also measured, which is the kinetic rate constant for the dissociation of H_2O_2 in the presence of CeONP, a measure of chemical redox activity. For biological measures, we assessed cell damage by uptake of propidium iodide (Singh et al. 2007), an index of toxicity. Protection from H_2O_2 challenge, and protection from superoxide challenge generated by the xanthine/xanthine oxidase reaction, were also measured. Biological parameters were tested in mixed organotypic rat cortical brain cell cultures. All particles were delivered at a dose of 100 nM, and biological assessments were made 24 h later.

Surprisingly, beneficial biological effects were not solely related to particle size. As expected, 10 nm average size particles exhibited no toxicity at the 100 nM dose. One commercial sample of 7 nm size also had no toxicity. All other sizes, including larger and smaller, exhibited a small toxic effect at the 100 nM concentration, as evidenced by increased uptake of propidium iodide. We speculate that these small but significant deleterious effects may be due to alterations in physicochemical properties that promote radical generation, rather than radical scavenging.

Beneficial biological effects of CeONP in hydroxyl or superoxide radical scavenging were also not clearly related to size or Ce3+ content, as shown. However, there appeared to be a range of Ce3+ content, from 31 to 33%, which produced the best biological antioxidant activity. Likewise, a size between 7 and 10 nm also provided the most antioxidant activity within the group, but clear correlations could not be made. The upshot here is that particle size, Ce3+ content, and non-stoichiometry could not be directly coupled to biological effects. However, there did appear to be a range at which beneficial effects were observed, and a decline in antioxidant effects were noted with a decrease in particle size. We would note that the 10 nm average size particles are those routinely used in our prior experiments.

It is obvious, from Table 2, that size, cerium valence, and oxygen vacancies contribute to CeONP antioxidant activity. But the question remains as to how the potent, regenerative radical scavenging action of these nanoparticles is afforded. And what of the other nanoparticle antioxidants in Table 1, which do not have an oxide lattice? Could there be a unifying principle behind nanoparticle antioxidant activity? To answer these questions, we must consider first that actions at the nano-scale are highly dependent on quantum effects at the electronic level. Next, we must visit some elegant work by Andrievsky and colleagues (2009) and Zheng et al. (2006) with respect to a hydration shell. It has been shown that antioxidant activity of nanoparticles is more robust in the presence of water, as opposed to other solvents. Andrievsky et al. (2009) demonstrate that the antioxidant activity of derivitized fullerenes is intimately connected to an ordered water hydration shell, containing hydroxyl, hydronium ions, and water, which forms around the nanoparticle in aqueous media. In this model, the ionic components of ordered water directly participate in free radical annihilation by the nanoparticle. This makes logical sense for the antioxidant activity of the nanoparticles described in Table 1 as a group, including cerium oxide. Further, CeONP have been shown to adsorb water and its associated ionic species onto the surface of the nanoparticle matrix. Thus, the nanoparticle initiates the radical scavenging event and electron transfer, with final acceptance of the charge to ionized species in water at neutral pH.

Let us take cerium oxide as an example. Prior studies have shown that durable, persistent antioxidant activity is afforded from a single dose of CeONP. It is further hypothesized that the antioxidant actions of CeONP are regenerative (Rzigalinski et al. 2006). How might this be so? If a single radical interacted with an oxygen vacancy in a CeONP nanoparticle, which in turn altered the cerium valence state, the radical would be scavenged. But in order to regenerate the original antioxidant matrix, the electron has to go somewhere! Andrievsky and colleagues supply the potential answer in that it is transferred to ion species in the hydrated water shell surrounding the nanoparticle.

The hypothesized mechanism of action is summarized in Fig. 6. A free radical is generated by a disease state, metabolic activity, or toxin. This radical has a free ion associated with it that initially reacts with the oxygen vacancies and cerium valence changes occurring in CeONP. The regenerative scavenging activity of the nanoparticle is accomplished by eventual dispersion of the electron or proton in the ordered hydration shell around the nanoparticle. This model would also explain why nanoparticle antioxidant activity appears to depend on a neutral pH, in the aqueous environment, and why CeONP and some fullerene derivatives afford radioprotection to normal cells, while affording no protection to tumor cells, which have a slightly more acidic intracellular pH. The precise contributions of all the elements of the model remain to be further investigated and will likely be elaborated upon in future studies.

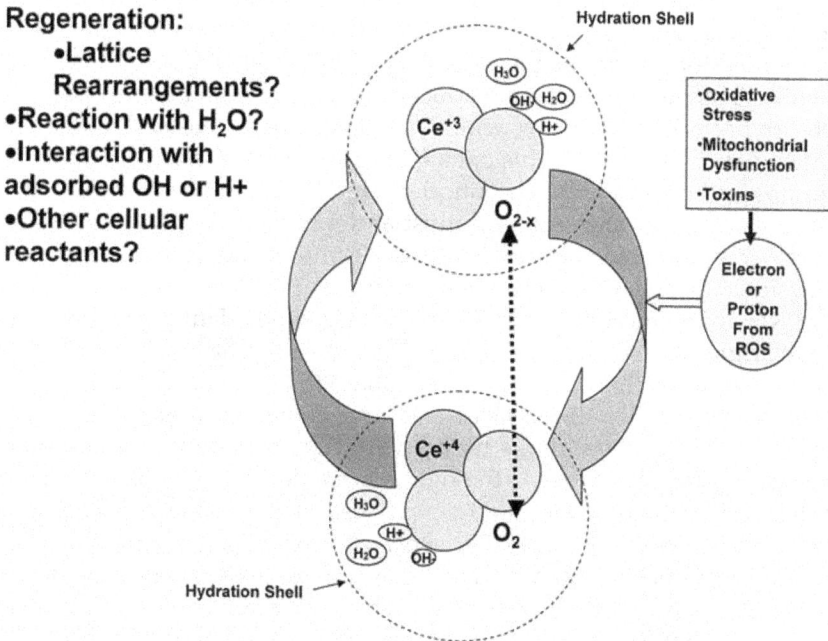

Fig. 6. Hypothesized mechanism of action of antioxidant cerium oxide nanoparticles. A free radical is generated by a disease state, metabolic activity, or a toxin. The free radical initially reacts with the oxygen vacancies and cerium valence in CeONP. The regenerative scavenging activity of the nanoparticle is accomplished by eventual dispersion of the electron or proton in the ordered hydration shell around the nanoparticle.

PHARMACOLOGICAL AND TOXICOLOGICAL PARAMETERS

To move the use of nanoparticle antioxidants from the bench to the bedside, nanomedicine must tackle the pharmacological concepts of absorption, distribution, metabolism, excretion, and toxicology. Since nanomedicine incorporates a new element of physics into drug design, we will no doubt have to change our modes of thinking. First, the traditional drugs and antioxidants we have used for centuries all have one thing in common: chemical structure. With the advent of nanomedicine, we have not only drugs with a chemical structure, but also drugs in which physical principles, size, shape, electronic lattice structure, and quantum effects are defining properties. Too often we hear the words platinum nanoparticles or cerium oxide nanoparticles thrown about loosely. What is the size? The shape? The lattice structure? We have now moved into an era where the physics of the particles take on defining properties and we must be insistent upon proper and complete particle characterization of the nanoparticles with which we experiment. Synthesis methods of antioxidant nanoparticles are also critical, since any tailing solvents need to be removed or otherwise accounted for, as they may impart toxic or biological effects of their own. Synthesis environment is also critical, since nanoparticles are known to absorb bacterial endotoxins, which can also damage cells and initiate the inflammatory response. This entails the adherence of many engineering groups to the term "pharmaceutical grade", which implies measurement of contaminating impurities—a must at the nano-scale. Likewise, it also entails the movement of biomedical science to incorporate principles of physics in drug design and development.

Key to the successful use of nanoparticle antioxidants is getting them to the right place at the right time in the right concentration. For CeONP, we have shown that they readily distribute to the most highly oxidative organs in the body, brain, heart and lung (Rzigalinski et al. 2006) and appear to be retained by the tissues with little metabolism or excretion. Hence, they appear to get to the right place in a concentration that, when combined with their regenerative actions, enables potent and durable antioxidant activity. Figures 7–9 show electron micrographs of CeONP within the intracellular environment. In Fig. 7, we see a nanoparticle cluster in the cytoplasm. Note that there appears to be a matrix, possibly protein or cytoskeletal, surrounding the nanoparticle. This observation begs the question, what proteins associate with nanoparticles in the cell that may contribute to their antioxidant capacity and further interact with the hydration shell in deactivating free radicals? In Fig. 8, we see nanoparticles within the cell and at cellular edges of neuronal cultures. Finally, in Fig. 9, we observe nanoparticles in the middle of the mitochondria, a site of high ROS production. These micrographs demonstrate that CeONP readily enter cells and appear to associate with cellular components.

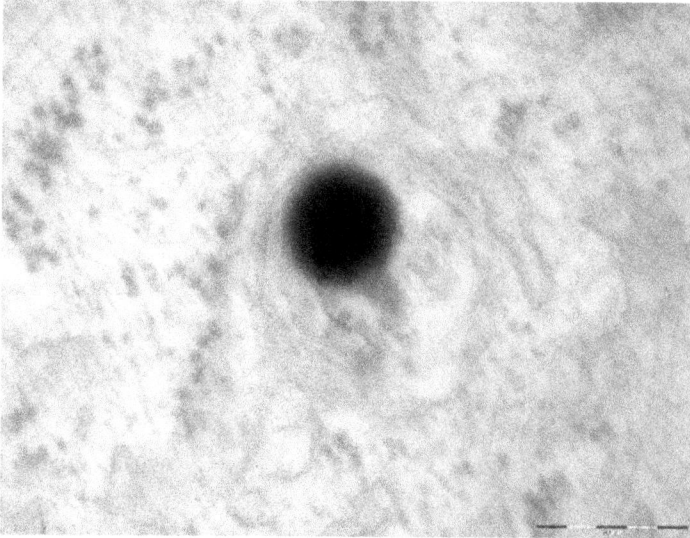

Fig. 7. Intracellular distribution of cerium oxide nanoparticles. Electron micrograph of a cerium oxide nanoparticle in the cytoplasm of a rat cortical neuron. Note fibrous elements surrounding the nanoparticle (Rzigalinski et al., previously unpublished).

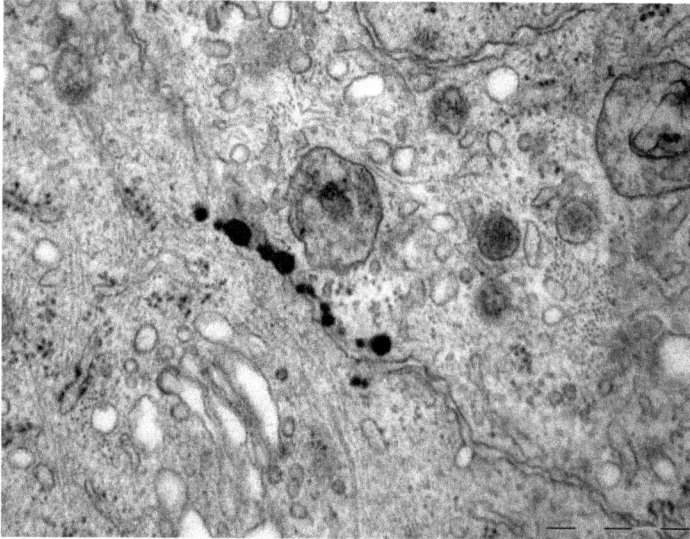

Fig. 8. Cerium oxide nanoparticles in mixed organotypic brain cell cultures. Shown are cerium oxide nanoparticles near cell edges in a mixed rat cortical cell culture (previously unpublished).

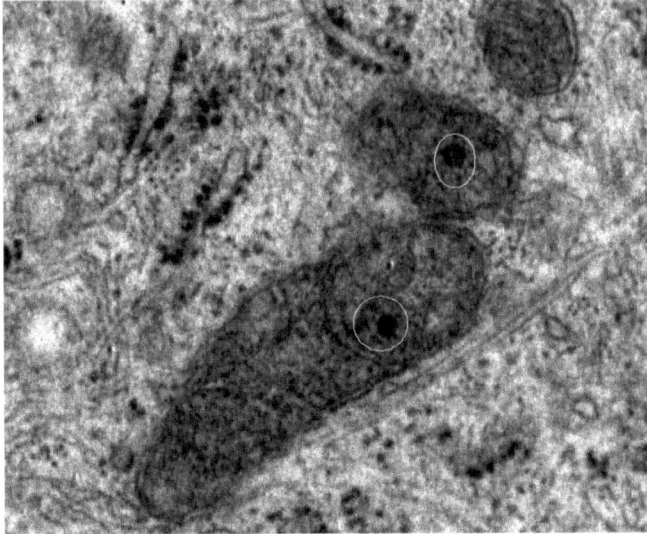

Fig. 9. Cerium oxide nanoparticles enter mitochondria. The electron micrograph above shows cerium oxide nanoparticles inside the mitochondria of mixed organotypic rat brain cell cultures (previously unpublished data).

Along with the superior antioxidant activity of CeONP and other antioxidant nanoparticles, we must also consider potential toxicity, a subject about which we know very little. For CeONP, there is little to no metabolism of nanoparticles. Once within tissues, CeONP remain for up to 6 mon. Thus far, doses up to 50 µg/g showed no toxic effects in mice (Rzigalinski et al., unpublished data). Nonetheless, studies involving the effects of nanoparticle persistence on biological function and behavior need to be performed.

Combining their persistence with the regenerative, durable antioxidant activity, CeONP hold potential as long-term, low-dose antioxidants that afford long-term protection against oxidative stress. But what are the caveats of too high a dose? We must remember that organisms use ROS for normal signaling, which can be disturbed by excessive antioxidant activity. In our work with CeONP to date, most responses produce a bell curve, with an optimal range above which antioxidant protection is not observed. Although deleterious effects have not yet been identified for the concentrations used as antioxidants, further investigations here are warranted. For example, at what cumulative dose are normal signaling pathways disturbed by radical scavenging? What physiological parameters could be affected by too high a cumulative dose? Let us consider vascular tone as an example. Vascular tone is regulated, in part, by the free radical NO. CeONP and other nanoparticle antioxidants scavenge NO, as well as other free radicals. However, we do not know the effects of CeONP on vascular function. Although our group has observed no deleterious effects after injections of low concentrations in hundreds of animals, long-term effects on signaling pathways such as those associated with vascular tone remain to be examined before these antioxidants are used in human pharmacotherapy.

Regarding our ideal antioxidant described in the beginning of this chapter, how do our nanoparticle antioxidants stack up? First, let us consider getting to the right place and the right time. CeONP distribute primarily to the most oxidative organs in the body: brain, heart and lung. Therefore, they get to the sites of highest oxidative stress, particularly the brain. Biodistribution of other nanoparticle antioxidants have not been well studied, but effects appear to support wide biodistribution. Next, our ideal antioxidant would be highly potent and effective in low concentrations. Our nanoparticle antioxidants have precisely that quality, in that they are effective in low doses and are more potent and efficacious than our traditional pharmacological antioxidants. Additionally, the ideal antioxidant would have long-lasting effects, enabling extended ROS-scavenging activity and increasing efficacy. With CeONP, we have a very long-lasting, regenerative, antioxidant activity, consistent with this need for the ideal antioxidant. Lastly, the antioxidants should be well tolerated, non-toxic, and without adverse effects. To date, at the doses used for antioxidant activity, deleterious effects have not been noted. However, further toxicological studies are necessary here, at the relevant doses, and for extended time periods due to the duration of nanoparticle antioxidants in biological systems. Ideally, nanotechnology will provide us with the ability to construct "smart antioxidants" whose activity is turned on when oxidative stress reaches a critical concentration in a tissue, and turned off at low levels of oxidative stress. Recent developments in microgel matrices (Ballauff and Lu 2007) that regulate catalytic activity and construction of passivation coatings (Rzigalinski et al. 2006) make this a realistic goal. Are nanoparticle antioxidants the magic bullet we have long awaited? Perhaps, but only more advanced studies will answer that question for certain.

CeONP and other antioxidant nanoparticles represent a new field of nanomedicine that has the potential to impact a wide range of diseases. To date, they are some of the most potent antioxidants yet identified. *In vivo* studies in progress suggest the capacity to improve treatment of many neurodegenerative, cardiovascular, and inflammatory disorders. Although much further work is required to move this field into the future, nanoparticle antioxidants hold the potential to revolutionize disease treatment with nanopharmaceuticals.

Summary Points

- Nanoparticle antioxidants have the potential to revolutionize treatment of disease associated with oxidative stress. Nanoparticle antioxidants include constructs of gold, platinum, fullerene derivatives and particularly cerium oxide.
- Biological activities include treatment of neurotoxicity and neurodegeneration, anti-inflammatory actions, radioprotection, treatment of cardiovascular disease; treatment of diabetes and complications, mitochondrial protection, neuroprotection in traumatic brain injury, and inhibition of tumor cell growth.
- Cerium oxide nanoparticles are some of the most highly potent nanoparticle antioxidants and regenerate their radical-scavenging activity. The mechanism of action of cerium oxide nanoparticles is based on the atomic valence of cerium, oxygen defect structure in the nanoparticle lattice, size, and interactions with water through the hydration shell.
- Cerium and other nanoparticle antioxidants are not subject to traditional pharmaceutical parameters of absorption, distribution, metabolism, and excretion. Many nanoparticle antioxidants are not metabolized or excreted and persist in the system. This makes low, single dosing possible, but poses challenges to traditional pharmacology.
- Owing to the lack of metabolism and excretion of nanoparticle antioxidants, further study is necessary to assess long-term effects such as toxicity. However, experiments to date, for cerium, show few toxicological difficulties.

Key Facts

- Free radicals are molecules with unstable electrons. When generated within a cell, they interact with endogenous molecules and render them dysfunctional. Free radicals are generated during normal metabolism, aging, and disease states. Cells have defense mechanisms, or antioxidants, to chemically counteract free radicals, to a limited degree. When the production of free radicals exceeds a cell or organism's natural defense mechanisms, oxidative stress ensues. Oxidative stress is a component of almost all human diseases.
- Endogenous or "natural" antioxidants, such as vitamins C and E, scavenge a single free radical and are destroyed during antioxidant activity. Cellular enzymes such as superoxide dismutase and catalase also act as antioxidants and chemically inactivate free radicals. Both are ineffective during high levels of oxidative stress.
- The ideal antioxidant would have several salient characteristics. First, it would be readily dispersed to sites that are high in oxidative stress.

Second, it would be effective at low concentrations. Third, it would be durable, producing long-lasting and potent effects at sites of free radical production. Last, it would be non-toxic and well tolerated by cells, tissues, and organisms. As yet, we have no such pharmacological agent. However, with the advent of nanomedicine, the construction of such a nanopharmaceutical appears within reach.

- Nanoparticle antioxidants provide a new, more effective, class of compounds with greater efficacy in treatment of oxidative stress. At the nano-scale, increased surface area and quantum effects increase the antioxidant capacity of nanoparticle antioxidants. Size, atomic valence states, electronic structure, and quantum effects are critical to nanoparticle antioxidant activity. Nanoparticle antioxidants include gold, platinum, fullerene derivatives, and cerium oxide.
- Nanoparticle antioxidants show promise in treatment of a wide range of disease states including neurodegenerative, inflammatory, and cardiovascular disease, as well as cancer therapy and diabetes.
- Cerium oxide nanoparticles are particularly efficacious and potent antioxidants, with little deleterious effects identified. The mechanism of action of cerium oxide nanoparticles, and possibly other nanoparticles, involves transmission of free radical electrons through the nanoparticle matrix and, in some cases, into the hydration shell of the nanoparticle.
- Progress in evaluation of nanoparticle antioxidants involves interdisciplinary thinking and development of new paradigms based on electronic and quantum effects.

Definitions

Aβ(1-42): Amyloid Beta fragment 1–42, the fragment of amyloid beta responsible for formation of amyloid beta plaques and neuronal death associated with Alzheimer's Disease.

Antioxidants: Substances that chemically react with, deactivate, and detoxify free radicals.

Efficacy: The maximal response possible for a given drug.

Free radical: A highly unstable molecule lacking the full complement of electrons to provide atomic stability.

Oxidative stress: A condition arising in a cell or tissue where free radical production exceeds the body's innate endogenous protective systems.

Oxygen vacancy: An electron "hole" in a nanoparticle lattice. The oxygen vacancy represents an electron missing from the outer shell of the element oxygen.

Potency: The concentration of a drug required to produce a 50% response.

Spin trap: A reagent used in EPR studies that reacts with a free radical to give an EPR signal.

Valence state: The amount of electrons in the outer orbital of an atom.

Abbreviations

CeONP	:	Cerium oxide nanoparticles
EPR	:	Electron paramagnetic resonance
ROS	:	Reactive oxygen species
XPS	:	X-ray photoelectron spectroscopy

This work was supported, in part, by a grant from the United Mitochondrial Disease Foundation

References

Ballauff, M., and Y. Lu. 2007. "Smart" nanoparticles: Preparation, characterization, and applications. Polymer 48: 1815–1823.

BarathManiKanth, S., K. Kalishwaralal, M. Sriram, S.K. Pandian, H.-S. Youn and S. Gurunathan. 2010. Anti-oxidant effect of gold nanoparticles restrains hyperglycemic conditions in diabetic mice. J. Nanobiotech. 8: 16–30.

Chen, J., P. Patil, S. Seal, and J.F. McGinnis. 2006. Rare earth nanoparticles prevent retinal degeneration induced by intracellular peroxides. Nature Nanotechnol. 1: 142–150.

Colon, J., L. Herrera, J. Smith, S. Patil, C. Komanski, P. Kupelian, S. Seal, D.W. Jenkins, and C.H. Baker. 2009. Protection from radiation-induced pneumonitis using cerium oxide nanoparticles. Nanomed.: Nanotech. Biol. Med. 5: 225–231.

Halliwell, B. 2006. Oxidative stress in neurodegeneration: where are we now? J. Neurochem. 97: 1634–1658.

Hirst, S.M., A.S. Karakoti, R.D. Tyler, N. Sriranganathan, S. Seal and C.M. Reilly. 2009. Anti-inflammatory properties of cerium oxide nanoparticles. Small 5: 2848–2856.

Injac, R., and B. Strukelj. 2008. Recent advances in protection against doxorubicin-induced toxiocity. Technol. Cancer Res. Treat. 7: 497–516.

Kim, J., M. Takahashi, T. Shimizu, T. Shirasawa, M. Kajita, A. Kanayama, and Y. Miyamoto. 2008. Effects of a potent antioxidant, platinum nanoparticle, on the lifespan of Caenorhabditis elegans. Mech. Ageing Dev. 129: 322–331.T.

Lao, F., W. Li, D. Han, Y. Qu, Y. Liu, Y. Zhao, and C. Chen. 2009. Fullerene derivatives protect endothelial cells against NO-induced damage. Nanotechnol. 20: 225103–225112.

Markovic, Z., and V. Trajkovic. 2008. Biomedical potential of the ractive oxygen species generation and quenching by fullerenes (C60). Biomaterials 29: 3561–3573.

Niu, J., A. Azfer, L.M. Rogers, X. Wang, and P.E. Kolattukudy. 2007. Cardioprotective effects of cerium oxide nanoparticles in a transgenic murie model of cardiomyopathy. Cardiovasc. Res. 73: 549–559.

Ok-Joo, S., K. Jin-chun, K. T.-W. Kyung, H.-J. Kim, Y.-Y. Kim, S.-H. Kim, J.-S. Kim, and H.-S. Choi. 2010. Gold nanoparticles inhibited the receptor activator of nuclear factor-kB ligand (RANKL)-induced osteoclast formation by acting as an antioxidant. Biosci. Biotechnol. Biochem. 74: 100375-1-5. n

Onizawa, S., K. Aoshiba, M. Kajita, Y. Miyamoto, and A. Nagai. 2009, Platinum nanoparticle antioxidants inhibit pulmonary inflammation in mice exposed to cigarette smoke. Pulm. Pharm. Ther. 22: 340–349.

Pirmohamed, T., J.M. Dowding, S. Singh, B. Wasserman, E. Heckert, A.S. Karakoti, J.E.S. King, S. Seal, and W.T. Self. 2010. Nanoceria exhibit redox state-dependent catalase mimetic activity. Chem. Comm. 46: 2736–2738.

Rzigalinski, B.A. 2005. Nanoparticles and cell longevity. Technol. Cancer Res. Treat. 4: 651–659.

Rzigalinski, B.A., and A. Clark. 2005. Anti-inflammatory, radioprotective, and longevity enhancing capabilities of cerium oxide nanoparticles. Patent pending WO60693930.

Rzigalinski, B.A., C.A. Cohen, and N. Singh. 2009. Cerium oxide nanoparticles for treatment and prevention of Alzheimer's Disease, Parkinson's Disease, and disorders associated with free radical production and/or mitochondrial dysfunction. Patent pending WO2009/052295.

Rzigalinski, B.A., D. Bailey, S.E. Merchant, S.C. Kuiry, and S. Seal. Cerium oxide nanoparticles and use in enhancing cell survivability. US Patent #7534453.

Rzigalinski, B.A., K. Meehan, R.M. Davis, Y. Xu, W.C. Miles, and C.A. Cohen. 2006. Radical nanomedicine. Nanomedicine 1: 399–412.

Saitoh, Y., Y. Yoshimura, K. Nakano, and N. Miwa. 2009. Platinum nanocolloid-supplemented hydrogen dissolved water inhibits growth of human tongue carcinoma cells preferentially over normal cells. Exp. Oncol. 31: 156–162.

Schubert, D., R. Dargusch, J. Raitano, and S. Chan. 2006. Cerium and yytrium oxide nanoparticles are neuroprotective. Biochem. Biophys. Res. Comm. 342: 86–91.

Shi, Y., S. Yadav, F. Wang, and H. Wang. 2010. Endotoxin promotes adverse effects of amorphous silica nanoparticles on lung epithelial cells in vitro. J. Toxicol. Env. Health A. 73: 748–756.

Singh, N., C.A. Cohen, and B.A. Rzigalinski. 2007. Treatment of neurodegenerative disorders with radical nanomedicine. Ann. NY Acad. Sci. 1122: 219–230.

Watanabe, A., K. Masashi, J. Kim, A. Kanayama, K. Takahashi, T. Mashino, and Y. Miyamoto. 2009. In vitro free radical scavenging activity of platinum nanoparticles. Nanotechnology 20: 455105–455114.

Whiting, M.D., B.A. Rzigalinski, and J.S. Ross. 2009. Cerium oxide nanoparticles improve neuropathological and functional outcome following traumatic brain injury. J. Neurotrauma 26: 101.

Yatin, S.M., S. Varadarajan, C.D. Link, and D.A. Butterfield. 1999. In vitro and in vivo oxidative stress associated with Alzheimer's amylid beta-peptide (1–42). Neurbiol. Aging 20: 325–330.

7

Lipid Nanocapsules in Nanomedicine

Marie Morille,[1] Patrick Saulnier,[1] Jean-Pierre Benoit[1] and Catherine Passirani[1,]*

ABSTRACT

Applications of nanotechnology for treatment, diagnosis, and monitoring have recently led to the expansion of a new investigatory field called "nanomedicine". Indeed, research now focuses on personalized treatment, especially in anti-cancer therapy, that could require the intervention of smart nanocarriers able to minimize side effects and to target the "sick" cells. In this way, lipid nanocapsules (LNCs) have attracted increasing attention during recent years. These lipid-based colloidal carriers, initially designed in the early 2000s as a pharmaceutical alternative to liposomes and polymer nanoparticles, are composed of an oily core surrounded by a rigid tension-active shell. Formulated with a solvent-free process based on a phase-inversion temperature method, the control of the formulation parameters enables adjustments of the particles' mean diameter within the range of 20–100 nm. The particles have a monodispersed size distribution,

[1]Inserm U646, Université d'Angers, Institut de Biologie en Santé, 4 rue Larrey, F-49333 Angers cedex 9.
*Corresponding author; Email: catherine.passirani@univ-angers.fr
List of abbreviations after the text.

allowing them to be used in different applications. These nanocarriers have been used for the encapsulation of different drugs, including anti-cancer ones, and also for nucleic acids. This chapter highlights the formulation processes involved in LNC production, and also the interaction of LNCs with their environment, at the extracellular as well as at the intracellular level. The use of LNCs for relevant applications in nanomedicine is then discussed.

INTRODUCTION

At present, modalities of diagnosis and treatment of various diseases, especially cancer, present major limitations, such as poor sensitivity or specificity, which causes drug toxicity. Personalized medicine has become a hopeful device for biomedical research, with pharmaceutical agents tailored for specific diseases and patients. The use of nanotechnology for the development of nano-therapeutics that can target specific cells and tissue types provides great promise for personalized medicine applications.

Nanocarriers can improve the kinetic profile of drug release, providing improved sustained release of drugs and a reduced requirement for frequent dosing. The *in vivo* fate of the encapsulated drug is determined by the properties of the vector; this can allow for a controlled and localized release of the active drug according to the specific needs of the therapy. As an example, tumour tissues display several distinct characteristics such as hyper-vascularization, defective vascular architecture, and a deficient lymphatic drainage system. This means that long-circulating carriers (or stealth carriers) are able to accumulate preferentially, and to be retained better, in tumour tissue than in normal tissue. This concept is called the enhanced permeability and retention effect (EPR effect) (Fig. 1) (Maeda et al. 2000). The modification of vectors following this concept, allowing an extended circulation time and accumulation in tumours, is called "passive targeting". Additionally, vectors can be tailored by ligand attachment to allow specific cell types to be targeted by using interactions between cell surface receptors and these ligands present on their surface. This strategy is called "active targeting".

A principal challenge in the formulation of vectors is to adapt the choice of their structure to the final aims of drug delivery, such as the biocompatibility of the used compounds, the physicochemical properties of the drug, and the therapeutic goals. In this way, lipid nanocapsules (LNCs) can be prepared by a solvent-free, soft-energy procedure (Heurtault et al. 2002). The capsules are composed of a liquid core surrounded by a cohesive interface and represent an alternative system

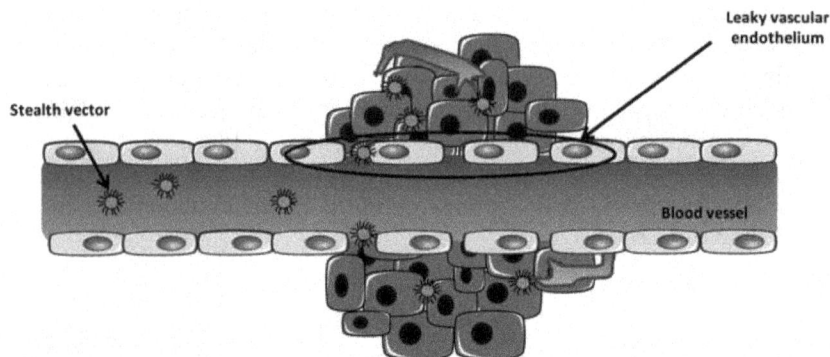

Fig. 1. Schematic representation of the EPR effect. Following the EPR effect, a stealth vector penetrates the tumour thanks to the leaky vasculature of the neoformed tumoral blood vessels, and is retained in this area by weak lymphatic drainage. These parameters lead to the preferential accumulation of vectors in tumours.

to polymer nanoparticles, emulsions, liposomes, and lipid nanoparticles. Indeed, LNCs have the attractive advantage of improving solubility of hydrophobic compounds in an aqueous medium to render them suitable for parenteral administration. Control of the formulation parameters allows the particle mean diameter to be adjusted within the range of 20–100 nm with a monodispersed size distribution. Their size range and their concentration in the related suspension allows them to be injected without occluding needles and capillaries, and is ideal for targeted drug delivery in light of the physiopathology of certain disorders such as cancer and inflammation.

This chapter reviews the state of the art in drug delivery with LNCs. It describes the design of these carriers, as well as the interaction of LNCs with biological systems (cellular and extracellular interaction). Finally, the potential perspectives in nanomedicine applications are highlighted.

LNC FORMULATION, CHARACTERIZATION AND SURFACE MODIFICATION

Formulation

Composition

LNC formulation is based on three principal components: an oily phase, an aqueous phase and a non-ionic surfactant. The preparation of optimal LNCs (i.e., nanometre range, monodispersity, and stability upon dilution) is strongly dependent on the proportions of the compounds (Heurtault et al. 2003). According to the proportions of the components leading to nanocapsule formation, a feasibility domain has been defined on a ternary diagram representing the range of proportions of the three major LNC constituents (Fig. 2).

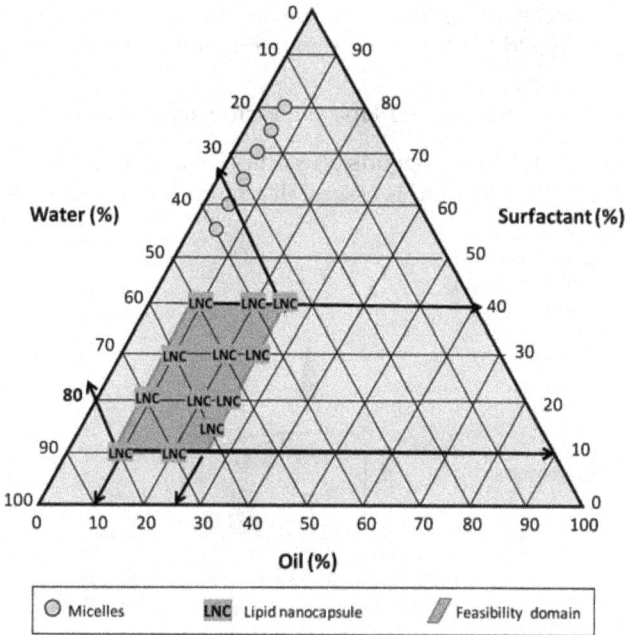

Fig. 2. Influence of the component proportions on the formation of lipid nanocapsules. Amounts of between 10% and 40% (w/w) of hydrophilic surfactant, 50% and 80% (w/w) of water, and 10% and 25% (w/w) of oil lead to particle formation.

The oily phase is essentially constituted of triglycerides of capric and caprylic acids known under the commercial name of Labrafac® WR 1349. The hydrophilic surfactant, Solutol® HS 15, is derived from polyethyleneglycol (PEG) and is a mixture of free PEG 660 and PEG 660 hydroxystearate. The aqueous phase consists of deionized water plus sodium chloride salt. Furthermore, another surfactant, Lipoid®S75-3, composed of 69% phosphatidylcholine soya bean lecithin, is used in small proportions to significantly increase LNC stability. All these components are approved by the Food and Drug Administration (FDA) for oral, topical and parenteral administration. Additionally, preliminary studies have been performed to check the genetically modified organism status of the LNC components (Hureaux et al. 2009). It is, however, important to note that some of these components, especially the oily phase, can be adapted as a function of the encapsulated drugs (Hureaux et al. 2009; Morille et al. 2009; Vonarbourg et al. 2009; Morille et al. 2010; Roger E et al. 2009). The composition of LNCs is therefore adjustable with regard to the properties of the compound to be encapsulated. Even hydrophilic compounds can be encapsulated by means of a recently patented process based on the use of a reverse, micelles-in-oil-in-water structure (Anton et al. 2010). Aqueous-core lipid nanocapsules, able to encapsulate hydrophilic species with relatively good yields, have also been formulated (Anton et al. 2009).

Preparation Processes by the Phase-Inversion Temperature Method

The formulation of LNCs depends on a phase-inversion temperature (PIT) method (Fig. 3). This method is essentially based on the solubility changes

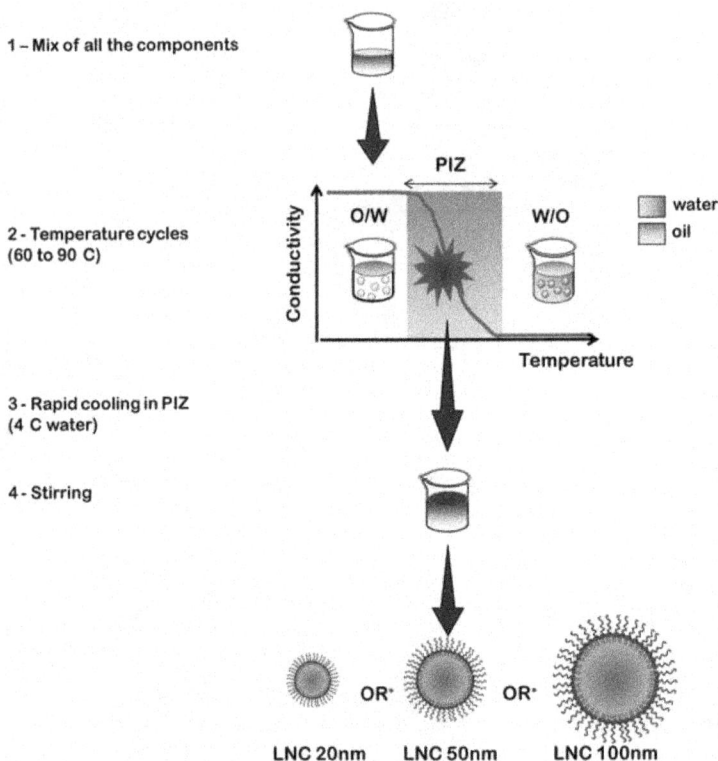

Fig. 3. Formulation of LNCs by the phase-inversion temperature method. Firstly, all the components are mixed. The suspension is then heated and cooled between 60°C and 90°C to obtain conductivity variation, and therefore phase inversion between the oil and water phase. After several temperature cycles, and in the phase-inversion zone, where the conductivity is much lower, a rapid dilution with cold water is performed to fix the suspension. The LNCs obtained are then stirred for 5 min. *The size of the obtained particle depends on the composition of non-ionic surfactants and oil.

Color image of this figure appears in the color plate section at the end of the book.

of PEG-type, non-ionic surfactants according to the temperature. These types of surfactant become less hydrophilic with increasing temperature as a consequence of the dehydration of PEG chains due to the breakdown of hydrogen bonds with water molecules. This results in a phase-inversion process from an oil-in-water (O/W) to a water-in-oil (W/O) emulsion, which takes place in the phase-inversion zone (PIZ).

The preparation process of lipid nanocapsules is based on two steps. Step I consists of mixing all the components (whose proportions vary according to the study) under magnetic stirring and in applying several temperature cycles between 85 and 60°C to cross the PIZ (Heurtault et al. 2002; Saulnier et al. 2008). Step II is an irreversible shock, induced by sudden dilution with cold water added to the mixture. This is done to break the bicontinuous system obtained in the PIZ and leads to the formation of lipidic nanocapsules that are composed of a liquid core surrounded by a cohesive interface and that are dispersed in an aqueous medium (Fig. 4). The formed nanoparticles exhibit very narrow dispersal levels (0.05) under 0.3, and a good level of stability (> 18 mon) under drastic *in vitro* conditions (change of pH, temperature, dilution and stirring) (Heurtault et al. 2002).

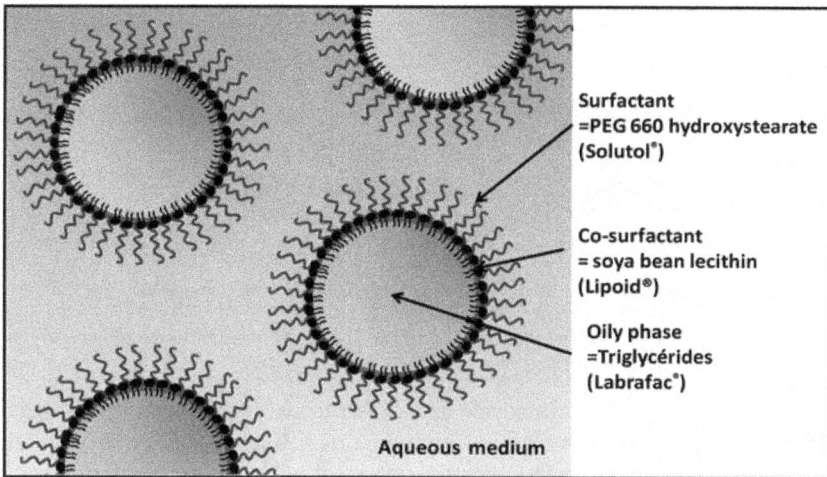

Fig. 4. Schematic representation of LNC structure. LNCs, dispersed in an aqueous medium, are composed of an oily core of triglycerides and surrounded by a shell of surfactants (PEG 660 hydroxystearate and soya bean lecithin composed of 37% of phosphatidylcholine).

Moreover, the formulation temperature range can be modified to allow the encapsulation of thermolabile drugs, such as DNA, with an increased proportion of salt, and the addition of polyglyceryl-6 dioleate (Oleic Plurol®) substituting a portion of Labrafac® ((Morille et al. 2009, 2010; Vonarbourg et al. 2009).

Physicochemical Characterization

LNC size is measured by dynamic light scattering (DLS). This method measures the fluctuations of the intensity of the scattered light caused by particle movement. Additional techniques such as microscopy observation can also be useful. According to these methods, the average volume size of LNCs ranges from 20 to 100 nm, with a very narrow range of polydispersity (pdl < 0.3) (Heurtault et al. 2002). Moreover, LNC shapes, observed by cryoTEM (Fig. 5) and AFM, show regular, spherical shapes.

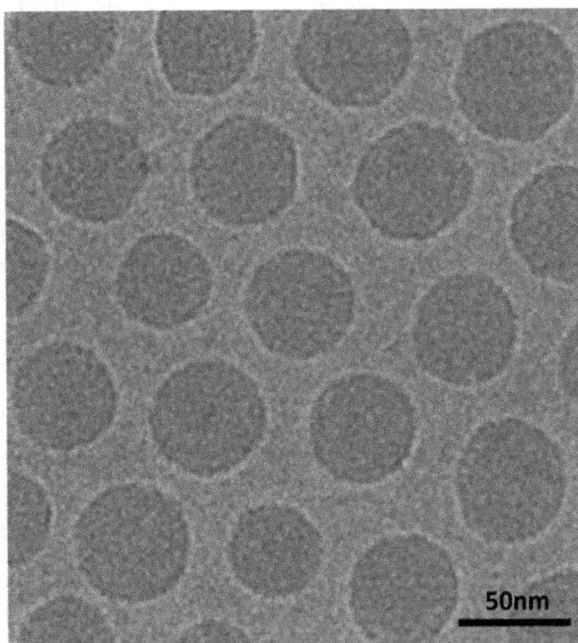

Fig. 5. Observation of LNCs by cryoTEM. CryoTEM visualization of nanocapsules allows the monodispersity of such objects to be seen.

Another way to characterize LNCs is the study of the zeta potential (Fig. 6). Almost all particles in contact with a liquid acquire an electric charge on their surface. The electric potential at the shear plane is called the zeta potential. The shear plane is an imaginary surface separating the thin layer of liquid (a liquid layer constituted of counter-ions) bound to the solid surface in motion. Zeta potential is an important and useful indicator of the particle surface charge, and this can be used to predict and control the stability of colloidal suspensions or emulsions. Indeed, particle aggregation is less likely to occur for charged particles (high zeta potential) due to electric repulsion. However, this rule cannot strictly be applied for systems that contain steric stabilizers, such as PEG chains for

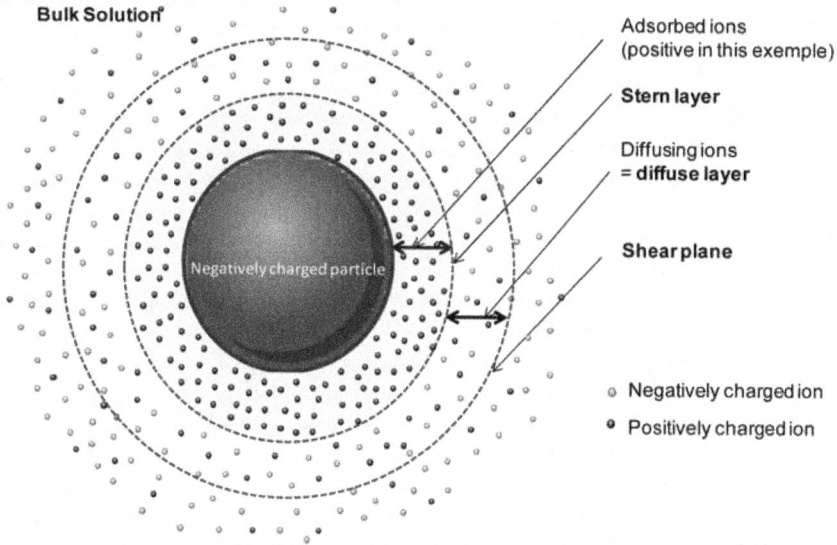

Fig. 6. Schematic representation of surrounding ions associated to a colloidal particle in a bulk solution. The layer closest to the surface of the charged particle is composed of absorbed counter ions (positive when the particle is negatively charged). The outer boundary of this layer is defined as the Stern layer. The second layer is composed of ions that diffuse around the charged particle. The boundary of this layer is defined as the shear plane, and it is the summation of charges within the slipping boundary that defines the strength of the electrostatic interaction. The charge or electrostatic potential at the shear-plane is defined as the zeta potential.

LNCs. In general, empty LNCs have a negative zeta potential (around -10 mV) due to the negative contribution of phospholipid molecules and the presence of dipoles in the PEGylated chains at the surface of the shell (Vonarbourg et al. 2005).

Surface Modification

To increase stability and avoid non-specific interactions with blood components, the modification of the LNC surface can be accomplished by "shielding" the vector surface with a hydrophilic and flexible polymer such as PEG (Vonarbourg et al. 2009; Perrier et al. 2010) (Fig. 7).

Post Insertion

The post-insertion method consists of a co-incubation step of preformed LNCs with amphiphilic PEG derivates, followed by a rapid cooling step that stabilizes the system (Hoarau et al. 2004; Perrier et al. 2010). The standard condition chosen for the post insertion of LNCs is incubation for

1h30 at 60°C. The suspension is vortexed every 15 min and then quenched in an ice bath for 1 min. It is important to note that the co-incubation step can be carried out at a lower temperature with a longer incubation time, this being necessary with the use of thermo-sensitive molecules (Hoarau et al. 2004; Morille et al. 2010).

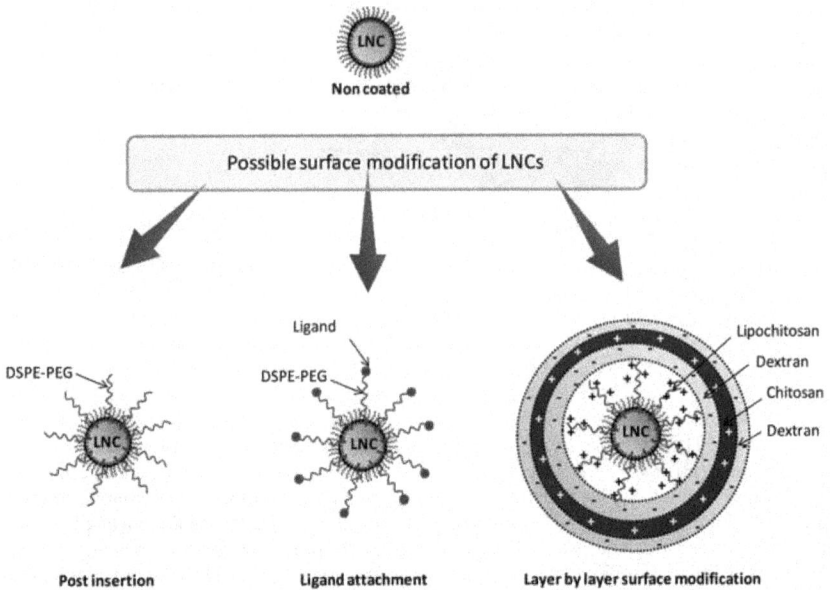

Fig. 7. Schematic representation of various LNC surface modifications. Modification of the LNC surface can be achieved to allow their application as a function of the administration pathway, as well as the envisaged therapeutic strategy.

Layer-by-layer Surface Modification

In parallel to the post-insertion process, LNCs have recently been modified by a layer-by-layer (LBL) approach based on electrostatic interactions between polyelectrolytes (PEs) and the LNC surface (Hirsjarvi et al. 2010). Initially, LNCs are post-inserted with a cationic surfactant, lipochitosan (LC) coupled to a lipid anchor (stearic anhydride). The negatively charged (dextran sulphate, DS) and positively charged (chitosan, CS) are then adsorbed on the LNC surface as can be verified by the alternating zeta potentials and increase in size. As an example of therapeutic application, fondaparinux sodium (FP), a heparin-like synthetic pentasaccharide, has been introduced on LNC surfaces instead of DS, thus validating the layer-by-layer process of this model therapeutic polyelectrolyte.

Ligand Attachment

Immuno-nanocapsules can be synthesized by conjugating whole OX26 monoclonal antibodies (OX26 mAb) to LNCs; these are directed against the transferrin receptor (TfR). The TfR is over-expressed on the cerebral endothelium and mediates the transcytosis mechanism. To allow this association, DSPE-PEG2000 is functionalized with reactive-sulphhydryl maleimide groups (DSPE–PEG2000–maleimide). The mAbs are then thiolated to react with maleimide functions, whereas thiol residues were already present on Fab' fragments (Beduneau et al. 2007). This method of association can also be used for the coupling of other mAbs, and studies are currently under consideration.

In parallel to antibodies, other molecules can be associated with LNCs. As an example, galactose moieties (gal) can be added covalently at the distal end of long PEG chains, in order to provide active targeting of the asialoglycoprotein-receptor (ASGP-R) present on hepatocytes. Two kinds of PEG polymers can be used: F108, a block copolymer (ethylene oxide 132– propylene oxide 50– ethylene oxyde 132) and the already described DSPE-PEG2000. The synthesis of galactosylated F108 is performed by enzymatic galactosylation, whereas the synthesis of DSPE-PEG-gal is performed by chemical galactosylation via a reductive amination that requires the use of lactose (Morille et al. 2009).

LNC INTERACTIONS WITH THEIR ENVIRONMENT

LNC Pharmacokinetic

Clearance/Interaction with Blood Components

Particle size, shape, and charge are key factors for intravenous injection; a small and weakly charged carrier seems to be better adapted to this pathway. Controlling the size of LNCs is a therefore a great advantage, because this parameter is known to alter the biological distribution profile of a carrier. Between 1 and 20 nm, nanocarriers can be rapidly eliminated by renal clearance. A size of 30 to 100 nm seems to be optimal to avoid phagocytic clearance, in contrast to larger particles that can rapidly be cleared by the MPS.

In parallel to the influence of size, surface characteristics (e.g., hydrophilic profile, charge) can also strongly influence the behaviour of the vector (Vonarbourg et al. 2006). As already mentioned (in the section on Surface Modification), the use of hydrophilic PEGylation can enhance the mean residence time of the vectors in the vascular compartment. One of the major landmarks of LNCs is their PEG corona, which provides them stealth properties, i.e., it renders them almost invisible with respect to the immune system. Indeed, the PEG corona, which forms dense conformational clouds, creates a steric barrier, preventing interaction with immune cells and plasma proteins. The presence of short and dense PEG chains at the surface of LNCs leads to a weak level of uptake by macrophages, and a low level of opsonization by the complement protein (tested by the CH50 method) (Vonarbourg et al. 2006). This study also demonstrated a slight size effect (LNC 20 nm vs. 50 nm vs. 100 nm) on complement activation; this was attributed to differences in the density and flexibility of PEG chains related to LNC curvature radius. Indeed, PEG molecular weight, density, conformation and the flexibility of the chain are known to be important parameters that can compensate and limit interaction with immune cells (for review see Vonarbourg et al. 2006).

LNC Residence Time in Blood

In a 2006 study, following intravenous injection, ^{188}Re/^{99m}Tc-labelled LNCs exhibited a blood half-life of 21 ± 1 min for ^{99m}Tc and 22 ± 2 min for ^{188}Re in rats (Ballot et al. 2006). This half-life is nevertheless insufficient for several applications. The low PEG chain length (660 Da) can explain the rapid elimination of conventional LNCs from the blood circulation, despite their very weak complement activation. Indeed, a dense coating can prevent opsonization by high molecular weight proteins but cannot systemically improve the circulation profile (Vonarbourg et al. 2006).

To enhance LNC blood behaviour, longer PEG chains were post-inserted on preformed LNCs (Hoarau et al. 2004; Beduneau et al. 2006). This surface modification provided half-lives of up to 6 h after intravenous administration in rat tail veins. We can therefore see that post-inserted nanocapsules can be considered as potential carriers for drug delivery, particularly into solid tumours due to the previously described EPR effect. As expected, docetaxel-loaded LNCs coated with DSPE-PEG 2000 significantly and substantially accumulate in C26 colon adenocarcinoma subcutaneous tumours in comparison to uncoated LNCs as assessed by radioactive tracking of both docetaxel ([^{14}C]DTX), and LNCs ([^{3}H] CHE) (Khalid et al. 2006). This study showed that, following intravenous injection of coated LNCs, tumoural docetaxel concentrations increased over a 12 h sampling period and were much higher than those of a control docetaxel formulation (Taxotere®). Additionally, DNA LNCs covered with DSPE-PEG2000 exhibited strong tumour accumulation thanks to their long circulation properties (5-fold improvement of half-life in mice, from 1.4 h to 7.1 h) (Morille et al. 2010). This accumulation provides a specific transfection of tumour tissues thanks to passive targeting (data submitted for publication).

Cellular Interactions

Internalization

Garcion et al. (2006) showed that LNCs are able to penetrate intracellular compartments of glioma cells by interacting with cholesterol-rich microdomains. Recent studies carried by Roger et al. (2009) and Paillard et al. (2010) on polarized Caco2 cells and F98 glioma cells, respectively, have provided us with a broad knowledge of LNC cellular behaviour. These studies showed that LNC cell internalization was a rapid, active and saturable process, predominantly mediated through a cholesterol-dependent pathway (Fig. 8). Remarkably, size appeared to be a major feature that influences LNC endocytosis. Indeed, large, 100 nm LNCs are endocytosed by a cholesterol-dependent pathway, but also by clathrin/caveolae-dependent pathway (Paillard et al. 2010). This particularity seems to be explained by a difference in curvature index of LNCs modifying the area of contact with cell membranes. In parallel, LNC composition (different as a function of the tailored size) also seems to govern this internalization.

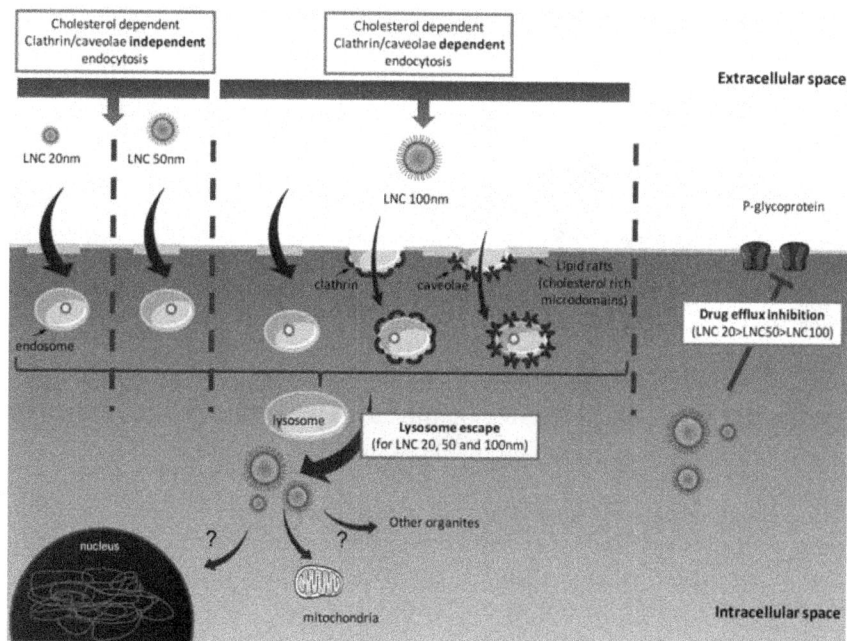

Fig. 8. Schematic representation of known LNC interactions with cells. 20 nm and 50 nm LNCs seem to be exclusively endocytosed by the cholesterol-dependent pathway. 100nm LNCs are also internalized by this pathway, as well as by clathrin-dependent mechanisms. LNCs (20, 50, and 100 nm) are then supposed to escape from lysosomal degradation. LNCs are also known to inhibit drug efflux pumps, such as P-glycoprotein.

Color image of this figure appears in the color plate section at the end of the book.

Lysosomal Escape

Interestingly, following internalization, LNCs do not accumulate within late endosomes or lysosomes. This shows that LNCs display specific skills for endosome or lysosome escape, probably due to the presence of Solutol®. Indeed, one of the specificities of LNCs is their capacity, thanks to the presence of HS-PEG, to inhibit activation of the plasma membrane multi-drug resistance pump (such as glycoprotein P (P-gp), allowing a decrease of drug resistance (Garcion et al. 2006)). In this way, HS-PEG might also affect the biological activity of lysosomal pumps, such as the V-ATPases, responsible for the low pH of this organelle. After escaping these organelles, LNCs might colonize other subcellular compartments, such as cytosol or other organites, and provide sufficient intracellular drug bioavailability. Proof of this bioavailability is the transfection efficiency of DNA LNC in various cell lines, demonstrating nucleus entry (Morille et al. 2009).

LNCS FOR DRUG DELIVERY

LNC formulation and characterization studies have been carried out as a function of the injection pathway and the encapsulated drug (Table 1). All these kinds of application to nanomedicine, peculiar to each application, are dealt with in this section.

LNCs as Drug Delivery Systems for Ibuprofen in Pain Treatment

In order to improve ibuprofen bioavailability, ibuprofen-loaded LNCs were developed in a size range of around 50 nm, with a high level of drug incorporation (94–98%). They exposed an *in vitro* drug release of 24 h in phosphate buffer (Lamprecht et al. 2004). Following oral administration, the antinociceptive effect, tested by the tail flick test, was prolonged up to 4 h in the LNC group. The pain relief after intravenous administration was prolonged for at least 2 h after the LNC formulation was administered. This study thus showed the propensity of LNCs for intravenous administration of ibuprofen, which exhibits sustained-release properties and may be interesting in the treatment of post-operative pain. Moreover, the most interesting finding consisted in the design of an injectable carrier of ibuprofen that offers an alternative to other formulations based on co-solvents and less biocompatible surfactants.

Oral Administration of Paclitaxel

The neoplastic agent paclitaxel is known to be pumped out from enterocytes by P-gp, and this strongly limits its oral uptake (Roger E. et al. 2009 and

Table 1. Various strategies applied for LNC formulation and their applications in health and disease

	Administration pathway	Coating	Ligand	Encapsulated compound	Application	Ref.
Ibuprofen-LNC	Oral Intravenous	- - -	- - -	Ibuprofen	Pain treatment	Lamprecht *et al.,* 2004.
PTX-LNC	Oral	- - -	- - -	Paclitaxel	Oral administration of anti-cancer drugs	Peltier *et al.,* 2006. Roger *et al.,* 2010.
PTX-LNC	Aerosol	- - -	- - -	Paclitaxel	Lung tumour	Hureaux *et al.,* 2009b
OX26 Immuno LNC	Intravenous	DSPE-PEG_{2000}	OX26mAb	^{188}Re lipophilic complex	Brain tumour	Beduneau *et al.,* 2007. Beduneau *et al.,* 2008.
^{188}Re-SSS Complex - LNC	Intratumoral CED	- - -	- - -	^{188}Re-SSS complex	Brain tumour	Allard *et al.,* 2008a.
Ferrociphenol-LNC	Intratumoral CED	- - -	- - -	Ferrociphenol (Fc-diOH)	Brain tumour	Allard *et al.,* 2010.
Docetaxel-LNC	Intravenous	DSPE-PEG_{2000}	- - -	Docetaxel	Passive tumour targeting	Khalid *et al.,* 2006
PEGylated DNA LNC	Intravenous	DSPE-PEG_{2000}	- - -	Lipoplexes (DOTAP/DOPE-DNA complex)	Passive tumour targeting	Morille *et al.,* 2010
Galactosylated DNA LNC	Intravenous	DSPE-PEG_{2000}	Galactose	Lipoplexes (DOTAP/DOPE-DNA complex)	Active liver targeting	Morille *et al.,* 2009

The attractive characteristics of LNCs allow their use in many therapeutic applications, not only for drug delivery, cancer diagnosis and therapy, but also for gene and cell therapy.

Roger E. et al. 2010) . This problem can be solved by the entrapment of drug molecules in the LNCs (Peltier et al. 2006), by using their potential to inhibit drug efflux. By loading paclitaxel into the oily core of LNCs, and after oral administration, the plasmatic concentration of paclitaxel has been shown to be three times that of the conventional formulation (Taxol®). Further studies have been carried out to elucidate the cellular mechanism involved in the improvement of oral bioavailability (Roger et al. 2009, 2010). Transport and uptake experiments of LNCs have been performed using a Caco-2 cell monolayer, a well-established *in vitro* model for drug transport studies. It has been shown that LNCs increase PTX transport across intestinal Caco-2 epithelial cells using vesicle-mediated transcytosis. Interestingly, another unsuspected negative effect of the P-gp on the uptake of LNCs has been shown, indicating a role for P-gp in endocytosis regulation and not only in drug efflux.

LNC Aerosols to Treat Lung Cancer

The encapsulation of paclitaxel in LNCs was envisaged to provide pulmonary drug delivery by nebulization with the goal of treating other bronchial and pulmonary diseases such as lung tumours (Hureaux et al. 2009). LNC dispersions have been made into aerosols with commercial nebulizers. The results have shown that LNC dispersions can be formulated into aerosols by using mesh nebulizers without altering the LNC structure (in terms of drug payload, structure and cytotoxicity). Only eFlow® rapid-produced aerosols were compatible with human use and no modifications of drug payload or cytotoxicity effects of paclitaxel-loaded LNCs (PTX–LNCs) have been observed. In order to carry out preclinical studies, a scaled-up LNC formulation protocol has been carried out. Chemical parameters, such as acidity and osmolarity, were optimized, and a storage procedure for PTX–LNC batches was set up. This study demonstrates the feasibility of an LNC paclitaxel aerosol. Animal studies are now needed to determine the tolerance and therapeutic potential of LNC dispersion aerosols.

LNC as a Vector in Glioblastoma Treatment

Malignant gliomas are still associated with a poor prognosis, despite advances in neurosurgery, radiotherapy and chemotherapy. The treatment of brain cancer is one of the most difficult challenges in oncology. Indeed, the delivery of drugs to brain tumours is limited by the presence of the blood-brain barrier (BBB) separating the blood from the cerebral parenchyma, as well as the significant infiltrative capacity of these tumour cells.

Active Targeting by Intravenous Injection

The concept of active targeting involving the attachment of a targeting moiety to LNCs (such as monoclonal antibodies, peptides, polysaccharides) was envisaged in the treatment of glioblastomas after intravenous injection. As previously mentioned, LNCs were conjugated to OX26 monoclonal antibodies (OX26 mAb). The OX26 mAb is directed to the transferrin receptor (TfR), this being highly expressed on the cerebral endothelium. The obtained immuno-nanocapsule size was around 150 nm in diameter, with a coupling efficiency of between 20 and 29%. A size increase after ligand conjugation showed the location of OX26 mAb to be outside the PEG brush, which facilitated cell association. The specific association of immuno-nanocapsules with over-expressing TfR cells was demonstrated *in vitro* (Beduneau et al. 2007). The biodistribution of immuno-nanocapsules, labeled with a ^{188}Re lipophilic complex, was determined in healthy rats. This study demonstrated a 2-fold higher accumulation of immuno-nanocapsules in the brain compared to non-targeted LNCs (Beduneau et al. 2008). Such immuno-nanocapsules represent promising nanocarriers for the active targeting of drug delivery to brain tumours since they are able to cross the BBB on physiological animal models.

Local Treatment and Convection-enhanced Delivery

In an attempt to overcome the limitations of systemic delivery of anti-cancer drugs aimed at brain targeting, due to the presence of the BBB, several methods of regional delivery have been developed. Among the different injection strategies by stereotaxy, convection-enhanced delivery (CED), using an external pressure gradient inducing fluid convection in the brain via a surgically implanted catheter, allows a greater volume to be distributed than diffusion alone (Allard et al. 2009). Local treatment with radio-pharmaceutical drug- (^{188}Re-SSS complex) loaded LNCs by means of CED on a 9L intra-cranial xenograft model of a rat brain tumour has been investigated (Allard et al. 2008). Interestingly, the ^{188}Re-SSS LNC-treated group showed a significant improvement in median survival time as compared to a control group and a blank LNC-treated group. The increase in the median survival time was about 80% compared to the control group, and 33% of the animals were long-term survivors (over 100 d). In addition, this formulation of LNCs was eliminated more slowly than the classical solution of ^{188}Re (perrhenate ^{188}ReO$_4^-$), which is recovered very quickly in the urine. LNCs ensure a prolonged therapeutic effect and can be considered as a promising radio-pharmaceutical carrier for internal radiotherapy of brain tumours. In parallel, the CED method has recently been used for the local delivery of LNCs loaded with organometallic molecules based on diphenol ferrocifen (Fc-diOH) followed by external beam irradiation, on a rat brain tumour model (Allard et al. 2010). The results showed a synergistic anti-proliferative effect between Fc-diOH–LNCs and radiotherapy on *in vitro* 9L glioma cells as well as in an orthotopic 9L glioma model. Moreover, in the same study, an antitumour effect was also obtained after a single intra-tumoural injection at Day 6 after subcutaneous 9L injection on Fischer F344 rats where Fc-diOH–LNC treatment dramatically reduced both tumour mass and tumour volume without any irradiation.

Use of Mesenchymal Stem Cells as Vectors for LNCs

Another alternative to enhance NP delivery in brain tumours is the use of cellular vectors that have endogenous tumour-homing activity and can thereby chaperone NP delivery *in vivo*. In this way, mesenchymal stem cells, also called multipotent mesenchymal stromal cells (MSCs), have a tendency to distribute at the site of tumours and could be potential candidates to encapsulate LNCs. A recent study has demonstrated that MIAMI cells can serve as cellular carriers for LNCs to brain tumours (Roger M. et al. 2010). Results showed MIAMI cells distributed around the tumour after their direct tumoral injection. This proof of concept holds great promise for the therapy of gliomas. Indeed, the combination of MIAMI cells carrying radiosensitizer-loaded LNCs such as paclitaxel or ferrocifen could be interesting. Due to the distribution of MIAMI cells at the border between tumour and parenchyma, another alternative could be the encapsulation of therapeutic agents acting as antiangiogenic factors targeting the tumour microenvironment.

LNC in Hearing Loss Treatment

Hearing loss is a major public health problem, and its treatment with traditional therapy strategies is often unsuccessful because of limited drug access deep in the temporal bone. In this context, the ability of LNCs to pass through the round window membrane and reach inner ear targets has been evaluated (Zou et al. 2008): FITC was incorporated as a tag for the LNCs and Nile Red was encapsulated inside the oily core to assess the integrity of the LNCs. The capability of LNCs to pass through the round window membrane and the distribution of the LNCs inside the inner ear were evaluated in rats via confocal microscopy in combination with image analysis using ImageJ. After round window membrane administration, LNCs reached the spiral ganglion cells, nerve fibres, and spiral ligament fibrocytes within 30 min. The paracellular pathway was the main approach for LNC penetration of the round window membrane. LNCs could also reach the vestibule, the middle ear mucosa, and the adjacent artery. Nuclear localization was detected in the spiral ganglion, though infrequently. These results suggest that LNCs are potential vectors for drug delivery into the spiral ganglion cells, nerve fibres, hair cells, and spiral ligament (Zou et al. 2008).

LNC as Vectors for Gene Delivery

DNA, complexed with cationic lipids, i.e., DOTAP/DOPE, has been encapsulated into LNCs, leading to the formation of stable nanocarriers (DNA LNCs) with a size inferior to 130 nm (Vonarbourg et al. 2009). Amphiphilic and flexible PEG polymer coatings [PEG lipid derivative (DSPE PEG2000) or F108 Poloxamer®] at different concentrations were selected to provide DNA LNCs with stealth capability (Morille et al. 2010). After intravenous injection, a significant increase of *in vivo* circulation time in mice, especially for DSPE PEG2000 10 mM, and an early half-life time ($t_{1/2}$ of distribution) was shown to be 5 times that of non-coated DNA LNCs (7.1 h vs 1.4 h). Tumour accumulation, assessed by an *in vivo* fluorescence imaging system, was evidenced for these coated LNCs as passive targeting without causing any hepatic damage. This DNA delivery system has been shown to provide an efficient *in vivo* transfection in tumours by the enhanced and permeability retention (EPR) effect (results submitted for publication). Nevertheless, this efficient transfection could probably be improved by the use of tumour-targeting moieties, such as folate, or RGD peptides.

In parallel to tumour transfection, another strategy of gene therapy has been developed for the transfection of the liver (Morille et al. 2009). In order to overcome internalization difficulties encountered with the PEG shield, a specific ligand (galactose) can be covalently added at the distal end of PEG chains, in order to provide active targeting of the asialoglycoprotein-receptor present on hepatocytes. A study has shown that DNA LNCs are as efficient as positively charged DOTAP/DOPE lipoplexes for transfection. In primary hepatocytes, when non-galactosylated, the two polymers significantly decrease transfection, probably by creating a barrier around the DNA LNCs. Interestingly, in this study, galactosylated F108 coated DNA LNCs led to an 18-fold increase in luciferase expression compared to non-galactosylated ones. This demonstrates a specific interaction between LNCs and the ASPG-R of the targeted cells (Morille et al. 2009).

APPLICATION TO AREAS OF HEALTH AND DISEASE

Lipid nanocapsules provide a new tool that contributes to nanomedicine development. Using FDA-approved constituents, LNCs are prepared by a solvent-free process to obtain particles of less than 100 nm with a low level of dispersity. These LNCs provide considerable drug encapsulation capacity and also exhibit sustained-release functions at the site of action. Moreover, thanks to the PEG surfactant surface, LNCs display a P-gp inhibitory effect, harmonized with a stealth effect, which acts against the complement system as well as MPS uptake; they also can be grafted with ligands for the purpose of actively targeting drug delivery.

The interesting characteristics of LNCs allow them to be used in many therapeutic applications, not only for drug delivery, cancer diagnosis and therapy, but also for gene and cell therapy. Moreover, issues related to scaling up and the manufacturing of LNCs should not pose a problem since long-term storage and low costs of formulation could allow the use of LNCs for clinical application. Nevertheless, long-term toxicity studies should be carried out to address the challenge of safe use of LNCs in nanomedicine.

Summary Points

- Lipid nanocapsules are formulated with biocompatible compounds and without organic solvents.
- Lipid nanocapsule size can be controlled from 20 nm to 100 nm.
- Lipid nanocapsules can be simply formulated allowing for possible large-scale production.
- Lipid nanocapsules efficiently inhibit the drug efflux caused by the glycoprotein P, thanks to the presence of Solutol®.
- The lipid nanocapsule poly(ethylene glycol) corona allows a weak interaction with cells of the mononuclear phagocyte system, and this can be enhanced by a PEGylation with longer poly(ethylene glycol) chains.
- The use of ligand (mAb, polysaccharides, ...) can permit the specific active targeting of wanted cells as a function of the therapeutic strategy.
- Lipid nanocapsules are internalized in cells by means of a rapid and saturable process, mediated by a cholesterol-dependent and clathrin-caveolae-independent pathway.
- Lipid nanocapsules display specific tools for avoiding lysosome degradation.
- Lipid nanocapsules are involved in the treatment of cancer (e.g., glioma, lung cancer), hearing loss and other conditions, and in pain treatment. This is possible by means of the encapsulation of various molecules such as anticancer drugs as well as nucleic acids.

Definitions

Convection-enhanced delivery (CED): Continuous injection under positive pressure of a fluid containing a therapeutic agent.

Lipid nanocapsules: Lipid-based colloidal carriers composed of an oily core surrounded by a rigid, tension-active shell.

Mononuclear phagocyte system (MPS): Represented by Kupffer cells in the liver, as well as macrophages in the spleen and bone marrow.

Multidrug resistance: Resistance to therapy that has been correlated to the presence of at least two molecular "pumps" in tumour-cell membranes that actively expel chemotherapy drugs from the interior of the cells (P-glycoprotein and the so-called multidrug resistance–associated protein (MRP)). This allows tumour cells to avoid toxic effects.

P- Glycoprotein (P-gp): An ATP-dependent efflux pump that exports drugs and endogenous metabolites out of a cell.

PEGylation: The action of modifying a vector surface with PEG in order to obtain enhanced blood circulation.

Stealth vector: A vector able to avoid recognition by the immune system.

Key Facts

- Most therapeutic drugs are distributed to the whole body, which results in general toxicity and poor acceptance of the treatment by patients.
- The concept of vectorization (or drug delivery) was created to modulate, and if possible to control, the distribution of a drug by means of an association with an appropriate carrier, called a vector, or drug-delivery system.
- First-generation vectors (e.g., liposomes, micelles, nanoparticles) are formulated to provide controlled drug release and are mainly used for *in situ* administration. They allow passive targeting, i.e., a passive accumulation in tumours by means of the EPR effect.
- Second-generation vectors, also called stealth vectors, are synthesized to avoid both opsonization and recognition by the MPS; they are often covered by hydrophilic and flexible polymers such as poly(ethylene glycol).
- Third-generation vectors are covered with targeting moieties such as antibodies, peptides, polysaccharides or vitamins. They allow, by binding to over-expressed receptors in the targeted-cell membranes, active targeting at the cellular level.
- Passive and active targeting can be combined to allow both accumulation in tumour zone and cell penetration.

Abbreviations

AFM	:	atomic-force microscopy
ASGP-R	:	asialoglycoprotein receptor
BBB	:	blood-brain barrier
CED	:	convection-enhanced delivery
CS	:	chitosan
DLS	:	dynamic light scattering
DNA	:	deoxyribonucleic acid
DOPE	:	1,2-DiOleylsn- glycero-3-PhosphoEthanolamine
DOTAP	:	1,2-DiOleoyl-3-TrimethylAmmonium-Propane
DS	:	dextran sulphate
DSPE	:	distearoylphosphatidylethanolamine
EPR	:	enhanced permeability and retention
FcdiOH	:	ferrociphenol
FDA	:	Food and Drug Administration
FITC	:	fluorescein isothiocyanate
Gal	:	galactose
HS-PEG660	:	PEG 660 hydroxystearate
LBL	:	layer-by-layer
LNC	:	lipid nanocapsules
mAb	:	monoclonal antibody
MPS	:	mononuclear phagocyte system
MSC	:	mesenchymal stromal cells
O/W	:	oil-in-water
PE	:	polyelectrolyte
PEG	:	poly(ethylene glycol)
P-gp	:	glycoprotein P
PIT	:	phase-inversion temperature
PIZ	:	phase-inversion zone
PTX	:	paclitaxel
Re	:	rhenium
Tc	:	technecium
TfR	:	transferrin receptor
W/O	:	water-in-oil

References

Allard, E., C. Passirani, and J.P. Benoit. 2009. Convection-enhanced delivery of nanocarriers for the treatment of brain tumors. Biomaterials 30: 2302–2318.

Allard, E., F. Hindre, C. Passirani, L. Lemaire, N. Lepareur, N. Noiret, P. Menei, and J.P. Benoit. 2008. ^{188}Re-loaded lipid nanocapsules as a promising radiopharmaceutical carrier for internal radiotherapy of malignant gliomas. Eur. J. Nucl. Med. Mol. Imag. 35: 1838–1846.

Allard, E., D. Jarnet, A. Vessieres, S. Vinchon-Petit, G. Jaouen, J.P. Benoit, and C. Passirani. 2010. Local delivery of ferrociphenol lipid nanocapsules followed by external radiotherapy as a synergistic treatment against intracranial 9L glioma xenograft. Pharm. Res. 27: 56–64.

Anton, N., H. Mojzisova, E. Porcher, J.P. Benoit, and P. Saulnier. 2010. Reverse micelle-loaded lipid nano-emulsions: New technology for nano-encapsulation of hydrophilic materials. Int. J. Pharm. 398: 204–209.

Anton, N., P. Saulnier, C. Gaillard, E. Porcher, S. Vrignaud, and J.P. Benoit. 2009. Aqueous-core lipid nanocapsules for encapsulating fragile hydrophilic and/or lipophilic molecules. Langmuir 25: 11413–11419.

Ballot, S., N. Noiret, F. Hindre, B. Denizot, E. Garin, H. Rajerison, and J.P. Benoit. 2006. 99mTc/188Re-labelled lipid nanocapsules as promising radiotracers for imaging and therapy: formulation and biodistribution. Eur. J. Nucl. Med. Mol. Imaging 33: 602–607.

Beduneau, A., F. Hindre, A. Clavreul, J.C. Leroux, P. Saulnier, and J.P. Benoit. 2008. Brain targeting using novel lipid nanovectors. J. Control. Release 126: 44–49.

Beduneau, A., P. Saulnier, F. Hindre, A. Clavreul, J.C. Leroux, and J.P. Benoit. 2007. Design of targeted lipid nanocapsules by conjugation of whole antibodies and antibody Fab' fragments. Biomaterials 28: 4978–4990.

Beduneau, A., P. Saulnier, N. Anton, F. Hindre, C. Passirani, H. Rajerison, N. Noiret, and J.P. Benoit. 2006. Pegylated nanocapsules produced by an organic solvent-free method: Evaluation of their stealth properties. Pharm. Res. 23: 2190–2199.

Garcion, E., A. Lamprecht, B. Heurtault, A. Paillard, A. Aubert-Pouessel, B. Denizot, P. Menei, and J.P. Benoit. 2006. A new generation of anticancer, drug-loaded, colloidal vectors reverses multidrug resistance in glioma and reduces tumor progression in rats. Mol. Cancer Ther. 5: 1710–1722.

Heurtault, B., P. Saulnier, B. Pech, J.E. Proust, and J.P. Benoit. 2002. A novel phase inversion-based process for the preparation of lipid nanocarriers. Pharm. Res. 19: 875–880.

Heurtault, B., P. Saulnier, B. Pech, M.C. Venier-Julienne, J.E. Proust, R. Phan-Tan-Luu, and J.P. Benoit. 2003. The influence of lipid nanocapsule composition on their size distribution. Eur. J. Pharm. Sci. 18: 55–61.

Hirsjarvi, S., Y. Qiao, A. Royere, J. Bibette, and J.P. Benoit. 2010. Layer-by-layer surface modification of lipid nanocapsules. Eur. J. Pharm. Biopharm.

Hoarau, D., P. Delmas, S. David, E. Roux, and J.C. Leroux. 2004. Novel long-circulating lipid nanocapsules. Pharm. Res. 21: 1783–1789.

Hureaux, J., F. Lagarce, F. Gagnadoux, A. Clavreul, J.P. Benoit, and T. Urban. 2009. The adaptation of lipid nanocapsule formulations for blood administration in animals. Int. J. Pharm. 379: 266–269.

Hureaux, J., F. Lagarce, F. Gagnadoux, L. Vecellio, A. Clavreul, E. Roger, M. Kempf, J.L. Racineux, P. Diot, J.P. Benoit, and T. Urban. 2009. Lipid nanocapsules: Ready-to-use nanovectors for the aerosol delivery of paclitaxel. Eur. J. Pharm. Biopharm. 73: 239–246.

Khalid, M.N., P. Simard, D. Hoarau, A. Dragomir, and J.C. Leroux. 2006. Long circulating poly(ethylene glycol)-decorated lipid nanocapsules deliver docetaxel to solid tumors. Pharm. Res. 23: 752–758.

Lamprecht, A., J.L. Saumet, J. Roux, and J.P. Benoit. 2004. Lipid nanocarriers as drug delivery system for ibuprofen in pain treatment. Int. J. Pharm. 278: 407–414.

Maeda, H., J Wu, T. Sawa, Y. Matsumura, and K. Hori. 2000. Tumor vascular permeability and the EPR effect in macromolecular therapeutics: a review. J. Control. Release 65: 271–284.

Morille, M., C. Passirani, E. Letrou-Bonneval, J.P. Benoit, and B. Pitard. 2009. Galactosylated DNA lipid nanocapsules for efficient hepatocyte targeting. Int. J. Pharm.

Morille, M., T. Montier, P. Legras, N. Carmoy, P. Brodin, B. Pitard, J.P. Benoit, and C. Passirani. 2010. Long-circulating DNA lipid nanocapsules as new vector for passive tumor targeting. Biomaterials 31: 321–329.

Paillard, A., F. Hindre, C. Vignes-Colombeix, J.P. Benoit, and E. Garcion. 2010. The importance of endo-lysosomal escape with lipid nanocapsules for drug subcellular bioavailability. Biomaterials 31: 7542–7554.

Peltier, S., J.M. Oger, F. Lagarce, W. Couet, and J.P. Benoit. 2006. Enhanced oral paclitaxel bioavailability after administration of paclitaxel-loaded lipid nanocapsules. Pharm. Res. 23: 1243–1250.

Perrier, T., P. Saulnier, and J.P. Benoit. 2010. Methods for the Functionalisation of Nanoparticles: New Insights and Perspectives. Chemistry.

Perrier, T., P. Saulnier, F. Fouchet, N. Lautram, and J.P. Benoit. 2010. Post-insertion into Lipid NanoCapsules (LNCs): From experimental aspects to mechanisms. Int. J. Pharm. 396: 204–209.

Roger, E., F. Lagarce, E. Garcion, and J.P. Benoit. 2009. Lipid nanocarriers improve paclitaxel transport throughout human intestinal epithelial cells by using vesicle-mediated transcytosis. J. Control. Release 140: 174–181.

Roger, E., F. Lagarce, E. Garcion, and J.P. Benoit. 2010a. Biopharmaceutical parameters to consider in order to alter the fate of nanocarriers after oral delivery. Nanomedicine (Lond.) 5: 287–306.

Roger, E., F. Lagarce, E. Garcion, and J.P. Benoit. 2010b. Reciprocal competition between lipid nanocapsules and P-gp for paclitaxel transport across Caco-2 cells. Eur. J. Pharm. Sci. 40: 422–429.

Roger, M., A. Clavreul, M.C. Venier-Julienne, C. Passirani, L. Sindji, P. Schiller, C. Montero-Menei, and P. Menei. 2010. Mesenchymal stem cells as cellular vehicles for delivery of nanoparticles to brain tumors. Biomaterials.

Saulnier, P., N. Anton, B. Heurtault, and J.P. Benoit. 2008. Liquid crystals and emulsions in the formulation of drug carriers. Comptes Rendus Chimie 11: 221–228.

Vonarbourg, A., C. Passirani, L. Desigaux, E. Allard, P. Saulnier, O. Lambert, J.P. Benoit, and B. Pitard. 2009. The encapsulation of DNA molecules within biomimetic lipid nanocapsules. Biomaterials 30: 3197–3204.

Vonarbourg, A., C. Passirani, P. Saulnier, and J.P. Benoit. 2006. Parameters influencing the stealthiness of colloidal drug delivery systems. Biomaterials 27: 4356–4373.

Vonarbourg, A., C. Passirani, P. Saulnier, P. Simard, J.C. Leroux, and J.P. Benoit. 2006. Evaluation of pegylated lipid nanocapsules versus complement system activation and macrophage uptake. J. Biomed. Mater. Res. A 78: 620–628.

Vonarbourg, A., P. Saulnier, C. Passirani, and J.P. Benoit. 2005. Electrokinetic properties of noncharged lipid nanocapsules: influence of the dipolar distribution at the interface. Electrophoresis 26: 2066–2075.

Zou, J., P. Saulnier, T. Perrier, Y. Zhang, T. Manninen, E. Toppila, and I. Pyykko. 2008. Distribution of lipid nanocapsules in different cochlear cell populations after round window membrane permeation. J. Biomed. Mater. Res. B Appl. Biomater. 87: 10–18.

8

Nanotheragnostic Colloids in Disease

José L. Arias

ABSTRACT

Drug delivery approaches based on the use of nanoparticles have tried to overcome the problem of drug inefficacy and toxicity. Nanoparticles can also be engineered for selective disease imaging. Combined disease diagnosis and therapy by a nanoparticulate platform have opened up interesting possibilities in the development of "personalized" medicines. The revolutionary concept "nanotheragnosis" is expected to permit the prediction and enhancement of the efficacy of therapeutic interventions, and to assess and understand drug delivery processes. Up to now, theragnostic nanoparticles have been principally investigated for treatment of cancer and cardiovascular disease. Ideally, theragnostic nanotools would consist in a multi-modality imaging technique combined with a multi-drug nanocarrier plus supplementary treatment strategies (e.g., hyperthermia, photodynamic therapy, and/or photothermal therapy). The structure of any given theragnostic nanoparticle must contain the drug molecule(s), the imaging agent(s), the biodegradable nanoplatform, and modifications to the last element based

Department of Pharmacy and Pharmaceutical Technology, Faculty of Pharmacy, University of Granada, Campus Universitario de Cartuja s/n, 18071 Granada, Spain;
Email: jlarias@ugr.es

List of abbreviations after the text.

on passive and/or active delivery strategies. In this chapter, special attention is given to the most significant applications of theragnostic nanoparticles and image-guided drug delivery in biomedicine. The engineering aspects involving the development of such multifunctional nanoparticulate systems are further revised. Theragnosis must open the door to more effective and less toxic treatment schedules, thus substantially improving the potential of targeted therapeutic interventions.

INTRODUCTION

Pharmacotherapy is under continuous evolution towards the synthesis of more potent drug molecules against diseases. Unfortunately, the clinical efficacy of such agents is considerably hampered by a non-specific biodistribution that makes it difficult to accumulate the drug at the targeted site in therapeutic concentrations. Very frequently, poor pharmacokinetics (i.e., rapid metabolism and plasma clearance) and drug resistance mechanisms at the tissue and/or cellular level further contribute to pharmacotherapy failure (Arias 2009; Lammers et al. 2010). Poor biodistribution and lack of specificity for non-healthy tissues and organs are also very important limitations to the use of contrast agents [e.g., magnetic resonance imaging (MRI) probes, luminophores, or radionuclides] for diagnostic purposes (Lu et al. 2007; Arias 2010). The unfavorable physicochemical properties of drug and contrast agents (i.e., hydrophobicity and high electrical charge) further explain the low accumulation of these molecules at the non-healthy cellular and tissular level (Couvreur and Vauthier 2006; Arias 2010). As a result, many of these agents that have been shown to be highly effective *in vitro* not only are ineffective *in vivo* but also exhibit toxicity.

Nanomedicine, or the application of nanotechnology for pharmacological purposes, has revolutionized drug delivery to targeted sites, allowing the appearance of new treatments with improved efficacy. Nanoparticulate drug delivery systems can be exploited to control the biological fate of any given active agent, to deliver it to the targeted cells, and to overcome resistance mechanisms. Promising results have been reported in the use of these nanomedicines in treatment of cancer, cardiovascular diseases, immunological diseases, inflammation, pain, infectious diseases, and metabolic diseases (Couvreur and Vauthier 2006; Farokhzad and Langer 2009; Arias 2009, 2010). Regarding molecular imaging, investigations on the formulation of effective and specific imaging agents and probes were based on the concept of nanoparticulate drug delivery (Lu et al. 2007; Arias 2010). For example, extended plasma half-life (and, as a consequence, efficient contrast-enhanced tissue imaging) was described for MRI contrast agents loaded into properly engineered nanoparticles (Raatschen et al. 2006; Zong et al. 2006).

Nevertheless, current nanotechnologies for drug delivery and biomedical imaging present serious limitations (Arias 2010) because of (1) poor loading of the drug or imaging molecule, (2) the rapid release("burst") of the encapsulated molecule after the administration of the nanocarrier, (3) the difficulty of carrying an adequate drug amount to the intended site of action with concurrent monitoring of the targeting efficiency, and (4) the difficulty of defining easy synthesis conditions so that the formulation of the nanocarrier can be scaled up in the pharmaceutical industry. As a result, the number of marketed nanomedicines is very limited despite the rise of literature in the field.

Hence, up to now we can say that the best therapeutic and diagnostic results will only be possible if new ideas can transform the way of delivering therapeutic and imaging agents to the disease site. In this way, the development of nanotechnologies that can combine sufficient drug loading, targeted therapeutic activity, and imaging capabilities ("nanotheragnostics") is considered the "holy grail" in pharmaceutical sciences. The present contribution is focussed on the analysis of the possibilities and applications of biodegradable nanoplatforms in the formulation of theragnostic nanomedicines against disease.

THERAGNOSIS: A REVOLUTION IN BIOMEDICINE

Theragnostic nanomedicines are integrated nanoparticulate systems formulated to simultaneously diagnose, deliver targeted therapy, and monitor response to therapy (Lammers et al. 2010; Rai et al. 2010) (Fig. 1).

Potential applications of such multifunctional nanosystems are non-invasive visualization and assessment of the biodistribution, accumulation, and release of drugs within the targeted non-healthy site, optimization of strategies based on triggered drug release, and the prediction and real time monitoring of the therapeutic effect (Table 1). A better knowledge on drug delivery processes could permit engineering of better nanomedicines for the expansion of more effective and less toxic "personalized" pharmacotherapies.

Numerous non-invasive imaging techniques have been proposed for real time visualization of the biological fate (biodistribution) of theragnostic nanoparticles. For instance, optical imaging [bioluminescence and fluorescence, e.g., near-infrared fluorescence (NIRF)], MRI, radionuclide-based imaging [positron emission tomography (PET) and single photon emission computed tomography (SPECT)], fluorescence-mediated

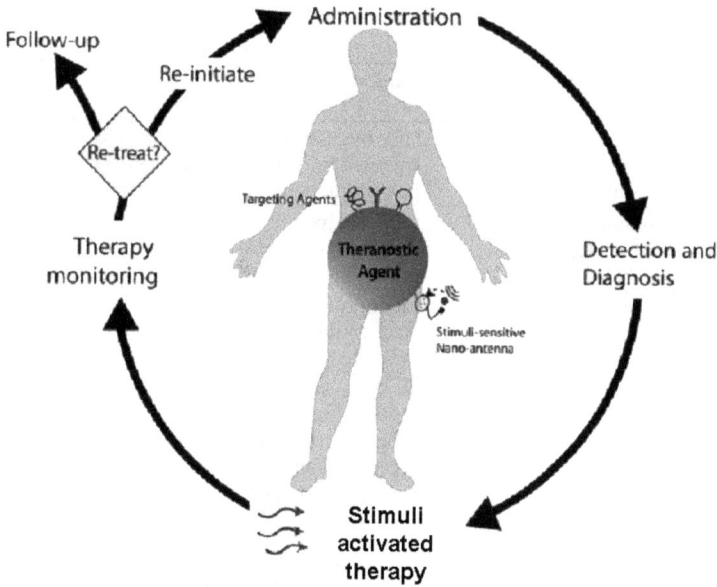

Fig. 1. Theragnostic nanoparticles in disease management. Upon administration of a single integrated theragnostic nanoparticle, a clinician could diagnose disease, detect disease location, introduce a specific stimulus (e.g., light, ultrasound, or magnetic gradient) into the disease site to activate targeted pharmacotherapy, and monitor response to therapy. Monitoring response would allow the clinician to undertake a therapy decision, whether to re-initiate treatment or establish a follow-up visit when sufficient disease regression or cure is determined. Reprinted with permission from Rai et al. (2010). Copyright Elsevier (2010).

tomography (FMT), photoacoustic tomography (PAT), and ultrasound. Limiting factors to individual imaging modalities (e.g., sensitivity, limited resolution, and anatomical information) have induced the development of multimodality imaging techniques, e.g., MRI-optical, PET-computed tomography, and MRI-PET (Fang and Zhang 2010; Janib et al. 2010). For example, PET/NIRF/MRI triple functional oleate-coated magnetite nanoparticles have been proposed for multiple disease imaging (Xie et al. 2010). In this formulation, dopamine is incorporated onto the nanoparticle surface to obtain nanoconjugates that can be easily encapsulated into a human serum albumin (HSA) matrix. The magnetic HSA-coated nanoparticles were dually labeled with ^{64}Cu-1,4,7,10-tetraazadodecane-N,N',N'',N'''-tetraacetic acid) (^{64}Cu-DOTA) and the NIRF dye cyanine 5.5 (Cy5.5). Imaging and histological examinations in subcutaneous U87MG xenograft mice revealed the good retention rate of the long-circulating nanoconjugates and a high extravasation rate into tumor interstitium.

The concept of nanotheragnosis has recently reached the preclinical arena. For example, multifunctional gold nanoparticles have demonstrated a very efficient metastatic cancer cell imaging and apoptosis. This antitumor

Table 1. Biomedical applications and clinical significance of nanotheragnosis and image-guided drug delivery.

Biomedical applications of theragnostic nanoparticles	Clinical significance of the approach
Non-invasively real time assessment (visualization) of the drug biodistribution	Easy biodistributional and pharmacokinetic analyses, and evaluation of the *in vivo* potential of targeted nanomedicines.
Real time monitoring of drug distribution into targeted sites	Deep analysis of the spatial parameters of targeted drug delivery. Better understanding and prediction of the efficacy of pharmacotherapy interventions.
Prediction of drug responses	Identification of which patients will respond to a targeted nanomedicine. Enable optimized and individualized treatment protocols.
Long-term evaluation of drug efficacy	Facilitate longitudinal experimental setups, enabling relevant efficacy analyses. Define disease progression and treatment outcome in real time. Measure how the targeted site changes in response to therapy (of great interest in cancer).
Combine disease diagnosis and therapy	Concurrent disease diagnosis and therapy.
Facilitate triggered drug release	Drug release with spatial specifity (only where high levels of nanotheragnostic agent are present in the targeted site) and temporal specificity (triggered release when optimal nanomedicine concentrations are reached into the targeted non-healthy tissue).
Monitor and quantify drug release	Drugs formulated into nanomedicines will be inactive when entrapped or conjugated in the nanocarrier.
Non-invasively assess target site accumulation	Define the potential efficacy of the therapeutic intervention.

Such clinical benefits will account for the necessity, efficacy and security of pharmacotherapy regimens (Lammers et al. 2010).

activity is supposed to be the consequence of an apoptotic event triggered by the internalized heparin moieties (Fig. 2) (Lee et al. 2010).

Ideally, theragnostic nanotools would consist in a multi-modality imaging technique combined with a multi-drug delivery nanosystem plus other treatment mechanism [e.g., hyperthermia, photodynamic therapy (PDT), photothermal therapy (PTT)] (Fang and Zhang 2010; Lammers et al. 2010). One of the most promising areas in which this concept is expected to be completely developed is cancer. In this way, preliminary results have been reported in which a theragnostic nanoplatform was able to allow the detection of a cancer cell through optical scattering, and immediately induce a mechanical cell ablation without damage to surrounding healthy tissue. Additionally, optical scattering of the multifunctional nanocarrier will confirm the destruction of the malignant cell (Wagner et al. 2010).

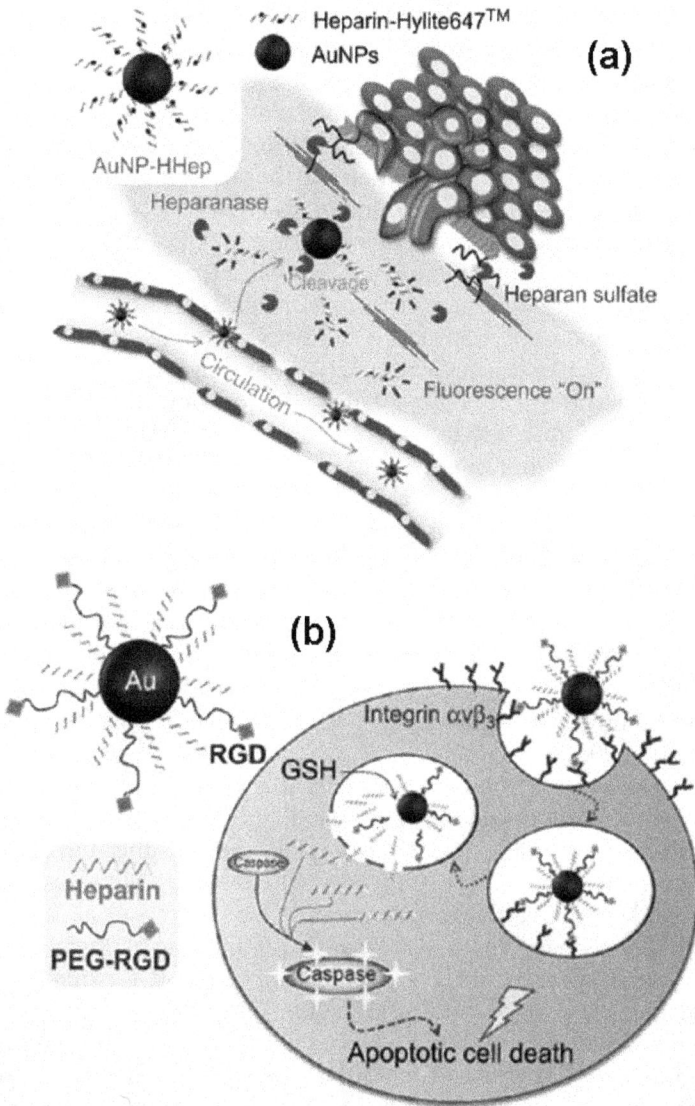

Fig. 2. Gold-based nanomedicines for metastatic cancer cell detection and apoptosis. (a) Heparin-immobilized gold nanoparticles (AuNP-HHep) were surface-modified with fluorescent dye (NIRF dye HiLyte-Fluor™ 647-amine) to detect metastatic tumor cells that overexpress heparin-degrading enzymes (i.e., heparanase). (b) Interestingly, when the nanoparticles were additionally coated with cell adhesive arginine-glycine-aspartic acid (RGD) peptides, using as spacer poly(ethylene glycol) (PEG), the nanomedicine (AuNP-Hep/PEGRGD) exhibited a great incorporation into tumor cells overexpressing RGD receptors ($\alpha_v\beta_3$ integrin) on the membrane and, as a consequence, highly specific apoptotic activities. Reprinted with permission from Lee et al. (2010). Copyright Elsevier (2010).

ENGINEERING OF BIODEGRADABLE
THERAGNOSTIC NANOPARTICLES

The development of nanoplatforms for combined medical imaging and drug delivery involves a complex combination of chemical, physical, and physicochemical factors. Only when all these parameters are properly "cooked" can an adequate *in vitro* and *in vivo* behavior be expected for the multifunctional nanomedicines. It is accepted that any given theragnostic nanoparticle must fulfil the following requisites for an efficient combined disease diagnosis and therapy: (1) very small size (below 100 nm to assure an extensive biodistribution and a uniform perfusion into the targeted tissue); (2) negligible surface electrical charge and significant hydrophilicity to delay recognition by macrophages (by opsonization) and plasma clearance; (3) the ability to transport the adequate quantities of drug and imaging agent, without overloading the body with foreign material; (4) *in vitro* and *in vivo* physical stability (null particle aggregation and/or precipitation under storage or upon administration); (5) protection of the imaging and therapeutic molecules from biodegradation by enzymatic systems; (6) null release of these molecules until the nanocarrier is completely accumulated into the non-healthy site; (7) controlled (and/or prolonged) release of the imaging and therapeutic agents into the targeted site to obtain a sustained therapeutic and diagnostic activity; (8) maximum biocompatibility, biodegradability, and minimal antigenicity (Couvreur and Vauthier 2006; Arias 2009).

It is accepted that four basic elements must be included in the structure of any given theragnostic nanoparticle (Fig. 3): the drug or therapeutic molecule, the imaging agent or signal emitter, the biodegradable nanocarrier material, and modifications to the later element based on passive and/or active delivery strategies (Jabr-Milane et al. 2008; Arias 2010; Fang and Zhang 2010; Janib et al. 2010). In these nanoplatforms, the imaging agents and the drug molecules can be either incorporated onto the nanoparticle surface and/or embedded into the particle matrix. On the other hand, modifications introduced in the biodegradable nanocarrier structure usually involve covalent attachment of targeting tools onto the nanoparticle surface and/or the use of stimuli-sensitive materials.

Only the proper design of the theragnostic agent will guarantee combined therapeutic and diagnostic activities in the form of early disease detection, the identification of disease biomarkers and signals for the choice of therapy, and an efficient (multi)drug delivery to the targeted site (Fang and Zhang 2010; Janib et al. 2010; Lammers et al. 2010).

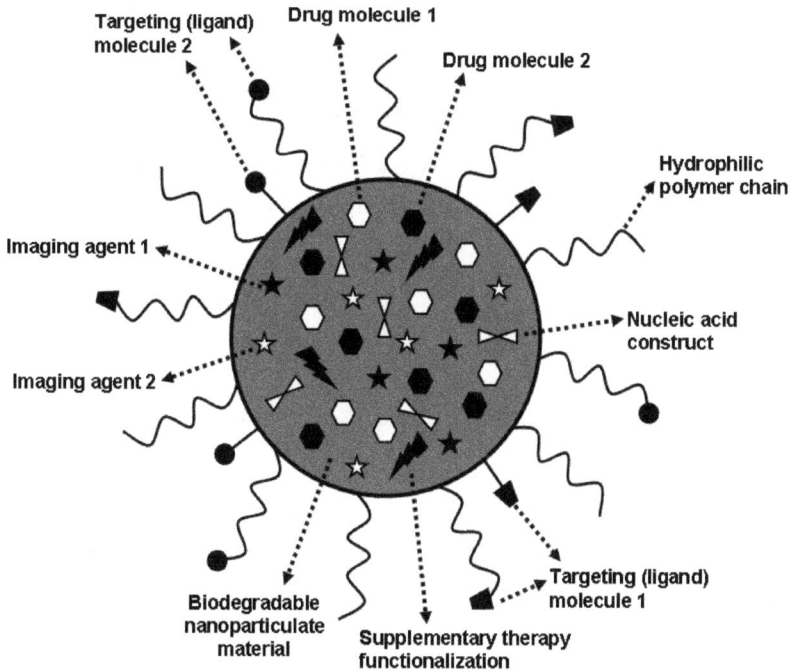

Fig. 3. Ideal structure of an efficient theragnostic nanotool for combined disease diagnosis and therapy. The multifunctional nanomedicine would be based on a biodegradable nanoparticulate system (preferably sensitive to external stimuli such as temperature, acid pHs, magnetic gradients, ultrasounds, light and enzymatic activity for active targeting to non-healthy sites) loaded with at least an imaging agent (MRI probe, luminophore, and/or radionuclide) and a drug molecule (e.g., chemotherapy agents and/or nucleic acid constructs for gene therapy or transfection), and further functionalized with hydrophilic polymer chains (i.e., PEG, poloxamers, poloxamines, or polysaccharides, for passive targeting to the disease site), ligand moieties (such as aptamers, monoclonal antibodies, folate, integrins, or transferrin, directly conjugated onto the nanoparticle surface and/or covalently bonded to the hydrophilic polymer chains for active targeting to the non-healthy site by ligand-receptor interactions), and additional engineering elements for complementary therapy (e.g., hyperthermia, photodynamic therapy, and/or photothermal therapy).

THE THERAPEUTIC MOLECULE

Theragnostic nanoparticles have been mainly investigated up to now for the treatment of cardiovascular disease (e.g., atherosclerosis, thrombosis) and cancer. However, theragnostic regimens are also finding potential applications in arthritic diseases (i.e., rheumatoid arthritis), age-related macular degeneration, atherosclerosis, neurodegenerative diseases, and psoriasis (Lanza et al. 2002; McCarthy et al. 2006, 2010; Jabr-Milane et al. 2008; Kenny et al. 2010; Kim et al. 2010; Rai et al. 2010).

To that aim, the drug or therapeutic molecule can easily be loaded to the nanoparticles through physicochemical interactions (e.g., covalent links with chemical groups of the nanocarrier, ionic interactions, hydrophobic interactions) (Fig. 3). Drug release from the nanotheragnostic platform must be controlled to occur exclusively into the targeted tissue. Preferably, an external (light excitation, alternating magnetic gradient) or environmental (pH, temperature, enzymatic degradation) stimulus will trigger the escape of the active agent from the theragnostic nanoparticle (Arias 2009; Fang and Zhang 2010).

THE SIGNAL EMITTER

The imaging agent or signal emitter must present unique optical, magnetic, or radioactive characteristics (Fig. 3). Ideally, an external stimulus (exclusively focussed on the disease site) will be responsible for physical or chemical changes in the imaging agent that will modify the amplitude or composition of the emitted signal. These changes will be detected by an external receiver and reconstructed into images. The most widely used strategies to induce changes in the emitted signals are chemical exchange saturation transfer, quenching (pairing-impairing of deactivator and fluorophores), Förster resonance energy transfer, and relaxivity changes (magnetic switch) (Fang and Zhang 2010).

Recent preclinical investigations have reported diagnostic benefits (improved signal detection) coming from the incorporation of MRI contrast agents [i.e., gadolinium (Gd^{3+}), superparamagnetic iron oxides], or luminophores into engineered biodegradable nanoparticles (Jabr-Milane et al. 2008; Arias 2010; Fang and Zhang 2010; Janib et al. 2010). If possible, the theragnostic agent should be made of more than one imaging agent. In this way, extensive efforts have been focussed on the development of multi-modality imaging agents, e.g., by conjugating superparamagnetic iron oxide nanoparticles (SPION) with a fluorescent dye (Arias 2010).

THE BIODEGRADABLE NANOCARRIER MATERIAL

The biodegradable nanocarrier could be based on an organic matrix, an inorganic component, or, more interestingly, on hybrid materials (Fig. 3) (Lu et al. 2007; Jabr-Milane et al. 2008; Fang and Zhang 2010; Lammers et al. 2010; McCarthy 2010). Inorganic platforms are generally made of quantum dots (a semiconductor nanocrystal typically composed of cadmium selenide that can act as a fluorescent label for medical imaging), metal nanoshells (usually constructed by coating a dielectric core of silica with an ultrathin gold layer, which are used as contrast agents in iridotomy, PAT, near-infrared tomography, optical coherence tomography, confocal imaging, and photothermal coagulation), SPION [e.g., magnetite or maghemite, used as MRI contrast agents, as they induce a shorter T_2 relaxation (transverse or spin-spin relaxation) which decreased signal intensity on a T_2-weighted image], and carbon nanostructures (carbon nanotube fluorescences, gadofullerenes, and gadonanotubes, which are used for fluorescence imaging). Nanotheragnostic agents could be further formulated by using organic materials: biodegradable polymers [e.g., chitosan, poly(ε-caprolactone), poly(D,L-lactide-*co*-glycolide), poly(alkylcyanoacrylate), or copolymers], lipid-based nanoparticles (liposomes, niosomes, or solid lipid nanoparticles), or carbon nanotubes. However, organic-inorganic nanohybrids are of greater interest because of the possibility of combining multi-modality imaging agents to different drug molecules, plus additional treatment possibilities (i.e., hyperthermia, PDT, PTT). As well, these nanocomposites will notably take advantage on passive and active targeting strategies (Arias 2009).

MODIFICATIONS BASED ON PASSIVE AND/OR ACTIVE DRUG DELIVERY STRATEGIES

Particular attention is given to recent advances in nanoparticle engineering to control the *in vivo* behavior of theragnostic medicines. The biological fate (and efficacy) of nanotheragnostic agents must be controlled and optimized taking advantage of well-known passive and active drug targeting strategies (Fig. 3) (Couvreur and Vauthier 2006; Arias 2009; Farokhzad and Langer 2009). Passive targeting is based on the enhanced permeability and retention (EPR) effect and involves the design of long-circulating nanotheragnostic agents. On the contrary, active targeting strategies implicate surface functionalization of the nanoplatforms for ligand- or receptor-mediated delivery, and/or formulation of the nanotheragnostic agent with stimuli-sensitive materials.

Passive Delivery Strategies

As is usually the case with drug nanocarriers, nanotheragnostic agents are expected to exhibit a deep interaction with the reticuloendothelial system (RES) upon administration. This could be an important advantage when Küpffer cells and related organs of the RES (e.g., spleen, liver, lungs, bone marrow) are the target: the theragnostic nanoparticle would naturally concentrate in such organs to display combined therapeutic and diagnostic activities. However, this is a severe disadvantage when other tissues and organs are the objective, because plasma clearance of the nanoparticles by macrophages will prevent accumulation at the targeted site (plasma half life < 5 min) (Arias 2009; Maeda 2009).

Passive targeting of theragnostic nanoplatforms is based on an enhanced capillary permeability typical of tumor mass, inflammatory tissues (e.g., arthritic joints), and infectious sites (Maeda 2009). As previously indicated, an extended biodistribution is only possible if theragnostic nanoparticles are characterized by a mean size < 100 nm and a spherical shape, hydrophilicity, and a surface electrical charge null or almost negligible. Extended biodistribution is ensured if hydrophilic chains (i.e., PEG, poloxamers, poloxamines, or polysaccharides) are incorporated either by physical adsorption or chemical conjugation onto the nanoparticle surface. The resulting shell of hydrophilic moieties will slow down the recognition (by opsonization) and plasma clearance of the theragnostic agent (Arias 2009; Maeda 2009).

A very interesting example of the utility of passive targeting strategies in the formulation of theragnostic nanoparticles has been recently described (Kenny et al. 2010). In this research report, PEGylated liposomes were loaded with anti-Survivin small interfering RNA (siRNA), and labeled with a fluorophore (Alexa Fluor 488) and an MRI agent [Gadolinium^{3+} 2-(4,7-bis-carboxymethyl-10-((N,N-distearylamidomethyl-$N\alpha$-amido-methyl)-1,4,7,10-tetra azacyclododec-1-yl)-acetic acid (Gd.DOTA.DSA)]. The nanoparticles very efficiently delivered functional anti-Survivin siRNA to the targeted malignant tissue in OVCAR-3 tumor-bearing mice, leading to a significant reduction in both Survivin expression and tumor growth when compared to controls. To understand the significance of these results, we have to keep in mind that the gene Survivin, a member of the inhibitor of apoptosis family, is upregulated in many cancers, but not expressed in normal tissue. The nanoparticles were shown to accumulate in xenograft tumors by MRI contrast enhancements 24 h post administration. Fluorescence microscopy corroborated these results and simultaneously demonstrated co-localization of nanoparticles and siRNA within the tumor interstitium. It was concluded that the theragnostic nanoparticles were able to induce a very significant antitumor effect and allow real time monitoring of the delivery by MRI in combination with fluorescence microscopy.

The efficient accumulation of nanotheragnostic agents into the tumor tissue taking advantage of the EPR effect has recently permitted simultaneous diagnosis, drug delivery, and therapeutic monitoring by chitosan-based nanoparticles (CNPs) containing an NIRF dye (Cy5.5) and an anticancer drug (paclitaxel, PTX) (Kim et al. 2010). Non-invasive optical fluorescence imaging revealed that the theragnostic nanoparticles were more efficiently localized in the tumor tissue (thanks to the EPR effect) compared to controls (water-soluble polymer and polymeric beads). The spherical chitosan-based theragnostic nanoplatforms (average diameter ≈ 250 nm) were characterized by great stability in serum, deformability, and rapid uptake by tumor cells. Cy5.5-labeled PTX-loaded CNPs exhibited significantly increased tumor-homing ability with low non-specific uptake by other tissues in SCC7 tumor-bearing mice. As a result, a greater antitumor effect was obtained (Fig. 4).

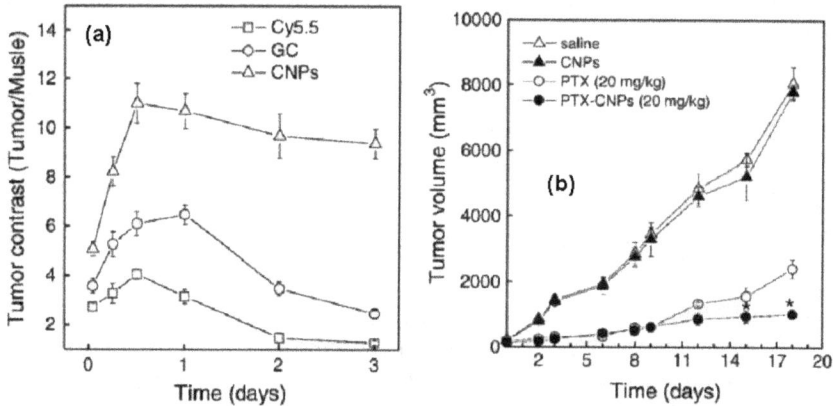

Fig. 4. (a) Tumor to background (muscle) ratio as a function of time after administration of Cy5.5, Cy5.5-labeled water-soluble glycol chitosan (GC), and Cy5.5-labeled CNPs in SCC7 tumor-bearing mice. (b) Comparative therapeutic efficacy of PTX-CNPs in SCC7 tumor-bearing C57BL/6 male mice (n = 10, tumor diameter ≈ 8 mm). *data points for the PTX-CNP-treated group that were statistically significant compared to controls by ANOVA at 95% confidence interval. All data represent mean value ± standard error. The superior tumor specificity of the nanomedicine dramatically increased the therapeutic PTX concentration into tumor tissues, thus leading to a promising anticancer effect. Adapted with permission from Kim et al. (2010). Copyright Elsevier (2010).

Active Delivery Strategies

Although long-circulating theragnostic nanoparticles generally determine high accumulation of imaging agents and drugs in targeted tissues, sometimes they could fail because of poor extravasation, rapid escape from tissue interstitium, and/or negligible accumulation in non-healthy cells. As an alternative, active targeting (or specific targeting) of theragnostic nanoplatforms is possible by selective delivery through a specific recognition mechanism (ligand- or receptor-mediated targeting) or by using stimuli-sensitive nanomaterials.

Ligand-mediated targeting is based on the conjugation of targeting molecules onto the theragnostic nanoparticle that can recognize and bind to unique ligands of targeted cells. The targeting tool is intended to improve the selective accumulation of the nanotheragnostic agent at the site of action and its direct interaction with the targeted cell or tissue. This approach is based on molecular recognition processes (ligand-receptor interactions) leading to receptor-mediated cell internalization. It is expected that such strategies will (1) modify the pharmacokinetic and tissue distribution profile of the theragnostic nanotool, enhancing the residence time of the drug molecule and imaging agent in the circulation and in the targeted place, while minimizing their adverse effects, and (2) improve intracellular penetration and distribution. These very promising possibilities have notably activated the research efforts on the subject (Fang and Zhang 2010).

For instance, polyacrylamide-based theragnostic nanoplatforms (containing SPION as contrast agents, and photofrin as photosensitizer) have been formulated for MRI enhancement and PDT of brain cancer. PEG chains and the F3 peptide, which binds to nucleolin expressed on tumor endothelium and cancer cells, were conjugated onto the nanomedicine surface and labeled with Alexa Fluor 594 for supplementary optical imaging. Very interestingly, *in vitro* results in MDA-MB-435 human breast cancer cells showed that F3-targeted nanoparticles were internalized and concentrated within tumor cell nuclei. Even more, *in vivo* studies established that PDT (based on F3-targeted photofrin-containing NPs) produced a significant improvement in treatment outcome (Reddy et al. 2006). Another example comes from the use of theragnostic nanoparticles against cardiovascular diseases. Perfluorocarbon nanoparticles have been loaded with paramagnetic Gd^{3+} and the antiproliferative agent rapamycin. This theragnostic nanoplatform has been successfully used in the detection of atherosclerotic lesions when targeted to $\alpha_v\beta_3$-integrin (an upregulated cell surface receptor in these lesions) via a peptidomimetic vitronectin antagonist. Interestingly, the theragnostic nanoparticles offered the possibility of non-invasively assessing target site accumulation and the inhibition of stenosis (McCarthy 2010).

The selectivity and specificity for disease sites can be further enhanced by using externally controlled theragnostic agents. These nanoplatforms are capable of altering their physical properties under exposure to a specific external stimulus, thus triggering the release of the signal emitter and the drug molecule exclusively into the targeted site (Arias 2009; Rai et al. 2010).

Light-triggered theragnosis combines imaging and photoactivation of therapeutic agents. Recent investigations have reported the use of theragnostic nanoparticles for combined diagnostic and photo-triggered chemotherapy of diseases. Selected materials could be sensitive to electromagnetic radiation applied at well-defined sites of the body. Consequently, these materials are irreversibly damaged by a single light dose or, more interestingly, behave as multi-switchable carriers by undergoing reversible structural changes when cycles of light/dark are applied (pulsatile drug release). Light can be further used to release endocytosed macromolecules into the cytosol, and to activate cytotoxic molecules. Finally, it has been hypothesized that light-based theragnostics could overcome resistance against antibiotics, an important problem in the treatment of infectious diseases (Rai et al. 2010).

pH-sensitive theragnostic nanoparticles can be formulated to degrade exclusively under exposure to acidic environments, being stable at the physiological pH 7.4. To that aim, nanoparticle engineering should involve the introduction into the nanoparticle structure of pH-sensitive functional groups, e.g., sulphonamide, to facilitate drug release under acid conditions (Arias 2009). pH-sensitive fluorescence probes (i.e., borondipyrromethene fluorophore and derivatives) can also be introduced to obtain pH-activatable images (Rai et al. 2010). A recent investigation was focused on the preparation of PEGylated SPION surface decorated with the fluorescent dye 5-carboxyfluorescein (5-FAM) and the monoclonal antibody (MAb) HuCC49$\Delta C_H 2$ (a humanized $C_H 2$ domain-deleted anti-TAG-72 MAb) (mean size ≈ 45 nm). PEGylation was responsible for a greater accumulation of the nanomedicine into the tumor tissue (thanks to the EPR effect: passive targeting). Meanwhile, the incorporation of the MAb led to a ligand-mediated targeting mechanism that determined the receptor-mediated internalization of the nanomedicine into cancer cells overexpressing the tumor-associated glycoprotein 72 (TAG-72, a human mucin-like glycoprotein complex) (active targeting). The antitumor drug doxorubicin was satisfactorily incorporated to the nanoplatform (loading capacity ≈ 3.5%). Cancer targeting and imaging was monitored using MRI and fluorescent microscopy in a LS174T colon cancer cell line. For instance, SPION (mean diameter ≈ 10 nm) were used as MRI contrast agent (they decreased signal intensity on a T_2-weighted image: from ≈ 120 ms to ≈ 55 ms in LS174T cells). Interestingly, the drug was released in acidic lysosomes and diffused into cytosol and nuclei, because of the protonation of the primary amine of doxorubicin, which dramatically increased drug solubility in aqueous solution. Thus, a lower half maximal inhibitory concentration 50 (IC_{50}) than drug-loaded non-specific IgG-SPION was obtained (1.44 µM *vs.* 0.44 µM). It was concluded that the theragnostic nanoparticles were highly suitable for MRI and fluorescence imaging of cancer cells, and pH-dependent intracellular drug release (Zou et al. 2010).

Theragnostic nanoparticles could be formulated to be disrupted by enzymes exclusively overexpressed into the targeted site (e.g., phospholipase C, secretory phospholipase A_2, alkaline phosphatase, elastase, transglutaminase, or sphingomyelinase). Alternatively, the formulation of theragnostic nanomedicines based on temperature-sensitive polymers [i.e., poly(N-isopropylacrylamide), and derivatives or copolymers] could help in controlling the delivery of drug molecules and imaging agents to targeted tissues (Fig. 5) (Arias 2009; Böhmer et al. 2009).

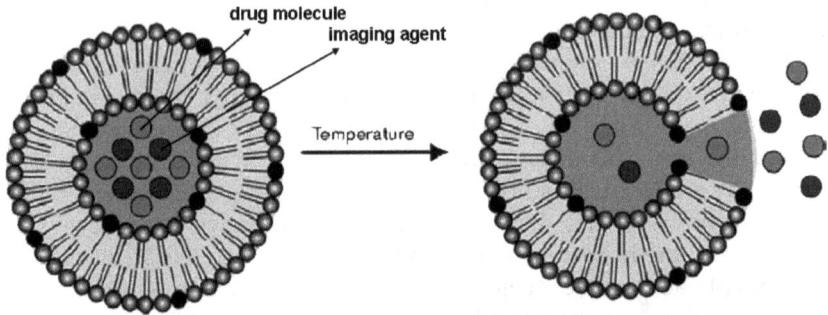

Fig. 5. Temperature-sensitive liposome containing a drug and an imaging agent that allows visualizing and quantifying drug delivery. The superior tumor specificity of the nanomedicine dramatically increased the therapeutic drug concentration into tumor tissues (and the anticancer effect). Adapted with permission from Kim et al. (2010). Copyright Elsevier (2010).

Ultrasound-mediated delivery of nanotheragnostic agents is based on the exposition of targeted organs or tissues to a given frequency of ultrasounds, which leads to (1) enhanced extravasation and cellular uptake of drug molecules, contrast agents, and/or theragnostic nanoparticles, thanks to the alteration of the cell membrane permeability, and (2) nanocarrier disruption and specific release of the drug and the imaging agent into the targeted site. This is the consequence of the *in vivo* effects of ultrasounds: cavitation, local tissue heating, and radiation force. The oscillating ultrasound pressure waves and local tissue heating will disrupt the theragnostic nanoplatforms sensitive to mechanical forces and temperature, respectively. Preclinical studies have pointed out the possibilities of combined MRI and ultrasound imaging in bringing to clinic ultrasound-triggered drug (and gene) delivery (Böhmer et al. 2009; Deckers and Moonen 2010).

Finally, theragnostic nanoparticles based on SPION could be easily guided to non-healthy tissues by an applied magnetic gradient, keeping them there until the drug molecule is entirely released (magnetic targeting). This will minimize the systemic biodistribution of the nanomedicine and, thus, the severe drug side effects on healthy cells. In this nanoplatform, SPION would further act as very efficient MRI contrast agents, as previously indicated (Arias 2010; McCarthy 2010).

CONCLUSIONS

The incorporation of pharmacologically active molecules and contrast agents into a single nanomedicine formulation can allow the prediction and improvement of the therapeutic outcome, and to visualize and better understand important aspects of drug delivery. In the near future, these multifunctional nanoplatforms will provide information on the disease location, target drug release leading to more effective therapies, and optimize drug dosing with fewer administrations, leading to increased patient compliance.

Future clinical use of nanotheragnosis rely on (1) a better knowledge of the biological disorders causing disease, (2) the design and formulation of new biocompatible materials to increase the number of safe nanotheragnostic compounds, and (3) an extensive *in vivo* evaluation of the multifunctional nanoplatforms, giving special attention to the related nanotoxicity. Finally, the complete clinical use of theragnostic nanomedicines will further involve the development of exceptionally sensitive imaging systems for precise detection of microscopic diseases.

APPLICATIONS TO OTHER AREAS OF HEALTH AND DISEASE

PDT has been reported to be significantly improved by theragnostic nanoparticles. These nanoplatforms would allow a precise localization of the interstitial lesion with diagnostic imaging (e.g., by MRI; even more, photosensitizer molecules are inherently fluorescent), which could provide guidance for light irradiation to the lesions to optimize the therapy (Lu et al. 2007; Rai et al. 2010).

Image-guided PTT by theragnostic nanoparticles is expected to enhanced cancer diagnosis and therapy because (1) selective imaging will facilitate tumor localization, (2) spatial and temporal changes in temperature and tissue morphology during PTT will be monitored, and (3) the tumor response to therapy will be immediately evaluated after therapy (von Maltzahn et al. 2009; Rai et al. 2010). PDT- and PTT-based theragnosis have been also proposed to treat infectious diseases (Rai et al. 2010).

Similar results are expected if theragnosis is applied to hyperthermia. Hyperthermia can increase the concentration of theragnostic nanoparticles in the targeted region by increasing blood flow and vessel permeability and has been reported to enhance drug toxicity in multi-drug resistant cancer cells (Rai et al. 2010).

Finally, significant research efforts should be concentrated on the development of theragnostic agents to be administered orally. The delivery of drug molecules and contrast agents through the blood-brain barrier could also be possible by means of theragnostic technology.

Key Facts

- Key facts of poor drug (or contrast agent) loading into nanoplatforms. Low loading values determine that a very high amount of the carrier material must be administered, engendering toxicity or side effects (Arias 2009).
- Key facts of burst release from nanoparticles upon administration. A rapid release determines that a significant fraction of the drug or imaging probe will be free before reaching non-healthy tissues. This leads to poor activity and severe side effects (Esmaeili et al. 2010).
- Key facts of state-of-the-art in the development of theragnostic nanoparticles. An important growth has been reported in peer-reviewed manuscripts and projects funded by the National Institutes of Health on nanotheragnosis (MacKay and Li 2010).
- Key facts of cardiovascular disease. This is the leading cause of mortality in the United States (1800 deaths per day); heart diseases are responsible for more deaths each year than cancer ($\approx 27\% \, vs \approx 23\%$) (McCarthy 2010).
- Key facts of the use of nanotechnology for biomedical purposes. Nanomedicine is making a great impact in disease imaging and diagnosis, drug delivery, and as reporters of therapeutic efficacy and of disease pathogenesis (Rai et al. 2010).
- Key facts of photodynamic therapy and photothermal therapy. The former technique involves the administration of a photosensitizer and its activation with localized light irradiation. The technique is limited by the requirement of accurate light irradiation of the target. Photothermal therapy involves the irradiation of diseased region with electromagnetic radiation to cause thermal damage.

Definitions

Active delivery (specific targeting) of imaging and therapeutic molecules: Selective delivery of imaging and therapeutic molecules to targeted sites through a specific recognition mechanism (ligand- or receptor-mediated targeting), or by their introduction into stimuli-sensitive nanoplatforms (Jabr-Milane et al. 2008; Arias 2009). In the former case, targeting molecules [i.e., MAb, peptides (integrins), transferrin, aptamers, and folic acid] are conjugated onto the nanoparticle surface, which can recognize and bind to specific and unique ligands of non-healthy cells, leading to receptor-mediated cell internalization. Stimuli-sensitive nanocarriers can alter their physical properties (e.g., swelling/deswelling, disruption/aggregation) under the influence of an external stimulus. As a result, the release of the signal emitter and the drug molecule can be triggered specifically into the desired site and/or, alternatively, this property could allow the accumulation of the imaging or therapeutic agent into the disease site. For instance, active targeting strategies include acid-triggered release, hyperthermia-induced delivery, magnetic targeting, light-triggered release, enzyme-triggered release, and ultrasound-mediated delivery.

Aptamer: Nucleic acid ligands (DNA or RNA oligonucleotides) capable of selectively binding to targeted antigens (Arias 2009). These biomolecules could be used for ligand-mediated targeting of nanotheragnostic agents.

Hyperthermia: Promising technique to increase tissue permeability and enhance tissue uptake of drugs, imaging molecules, and multifunctional nanomedicines. It involves locally heating the non-healthy region at $\approx$ 42°C, inducing an increase in the microvascular pore size and the tissue blood flow. As a result, the extravasation of nanoparticles into the site of action will be greatly enhanced. This technique is also widely investigated to trigger drug release exclusively into the targeted site (mainly from thermosensitive nanoplatforms). Hyperthermia itself has been shown to be cytotoxic to tumor cells (Arias 2009).

Passive delivery of molecules: Nanoparticle therapy and imaging can be clearly improved by the formulation of long-circulating particles. The incorporation of hydrophilic polymers (i.e., PEG) has made it possible to exploit structural abnormalities in the vasculature of particular pathologies (inflammation, cancer, and infections, to cite just a few). Long-circulating nanomedicines will undergo a specific extravasation and accumulation into the targeted tissue microenvironment thanks to the leaky vasculature (which has greater permeability to colloids than healthy tissue), and to the dysfunctional lymphatic drainage system. Even more, enhanced nanoparticle retention into the interstitial space will take place through the EPR effect (Maeda 2009; Arias 2009; Farokhzad and Langer 2009).

Theragnostic nanoparticles: Multifunctional agents with dual roles as diagnostics and therapeutics. The concept was firstly employed a decade ago to describe diagnostic tests developed to guide personalized pharmacotherapies (Fang and Zhang 2010; MacKay and Li 2010).

Transfection: The process of delivering nucleic acids (and proteins) into cells (Arias 2009).

Summary Points

- The conjugation of imaging agents or drugs to multifunctional nanoparticles improves their pharmacokinetics and tissue targeting efficiency, resulting in more accurate disease detection and better therapeutic efficacy.
- Nanotheragnostics and image-guided drug delivery are expected to enable "personalized" medicine.
- Any given nanotheragnostic tool must be made of a therapeutic molecule, an imaging agent, and a biodegradable nanoplatform. Moreover, it should be engineered according to both passive and active delivery strategies.
- Theragnostic agents are under extensive development for the treatment of cancer and cardiovascular diseases.
- The future of theragnostic nanoparticles must involve the formulation and clinical use of multifunctional nanoplatforms that integrate the capabilities of drug delivery, diagnostic imaging, and post-treatment monitoring. A deep evaluation of biocompatibility and toxicity is further needed. New molecular targets and biodegradable nanocarrier materials should further maximize nanomedicine accumulation into the targeted site, even into a particular subcellular compartment.

Abbreviations

AuNP-Hhep	:	heparin-immobilized gold nanoparticles
CNPs	:	chitosan-based nanoparticles
CT	:	computed tomography
Cy5.5	:	cyanine 5.5
DOTA	:	1,4,7,10-tetraazadodecane-N,N',N'',N'''-tetraacetic acid
EPR	:	enhanced permeability and retention
5-FAM	:	5-carboxyfluorescein
FMT	:	fluorescence-mediated tomography
GC	:	glycol chitosan
Gd^{3+}	:	gadolinium
Gd.DOTA.DSA	:	Gadolinium^{3+} 2-[4,7-bis-carboxymethyl-10-[(N,N-distearylamidomethyl-$N\alpha$-amido-methyl]-1,4,7,10-tetra azacyclododec-1-yl]-acetic acid
HSA	:	human serum albumin
IC_{50}	:	half maximal inhibitory concentration 50
MAb	:	monoclonal antibody
MRI	:	magnetic resonance imaging
NIRF	:	near-infrared fluorescence
PAT	:	photoacoustic tomography

PDT : photodynamic therapy
PTX : paclitaxel
PEG : poly(ethylene glycol)
PET : positron emission tomography
PTT : photothermal therapy
RES : reticuloendothelial system
RGD : arginine-glycine-aspartic acid
siRNA : small interfering RNA
SPECT : single photon emission computed tomography
SPION : superparamagnetic iron oxide nanoparticles
TAG-72 : tumor associated glycoprotein 72

References

Arias, J.L. 2009. Micro- and nano-particulate drug delivery systems for cancer treatment. pp. 1–85. *In*: P. Spencer, and W. Holt. (eds.). Anticancer Drugs: Design, Delivery and Pharmacology. Nova Science Publishers Inc., New York.

Arias, J.L. 2010. Drug Targeting by Magnetically Responsive Colloids. Nova Science Publishers Inc., New York.

Böhmer, M.R., A.L. Klibanov, K. Tiemann, C.S. Hall, H. Gruell, and O.C. Steinbach. 2009. Ultrasound triggered image-guided drug delivery. Eur. J. Radiol. 70: 242–253.

Couvreur, P., and C. Vauthier. 2006. Nanotechnology: Intelligent design to treat complex disease. Pharm. Res. 23: 1417–1450.

Deckers, R., and C.T.W. Moonen. 2010. Ultrasound triggered, image guided, local drug delivery. J. Control. Release DOI: 10.1016/j.jconrel.2010.07.117.

Esmaeili, F., R. Dinarvanda, M.H. Ghahremanib, S.N. Ostadc, H. Esmailyc, and F. Atyabi. 2010. Cellular cytotoxicity and *in-vivo* biodistribution of docetaxel poly(lactide-co-glycolide) nanoparticles. Anticancer Drugs 21: 43-52.

Fang, C., and M. Zhang. 2010. Nanoparticle-based theragnostics: Integrating diagnostic and therapeutic potentials in nanomedicine. J. Control. Release 146: 2–5.

Farokhzad, O.C., and R. Langer. 2009. Impact of nanotechnology on drug delivery. ACS Nano 3: 16–20.

Jabr-Milane, L.S., L.E. van Vlerken, S. Yadav, and M.M. Amiji. 2008. Multi-functional nanocarriers to overcome tumor drug resistance. Cancer Treat. Rev. 34: 592–602.

Janib, S.M., A.S. Moses and J.A. MacKay. 2010. Imaging and drug delivery using theranostic nanoparticles. Adv. Drug Deliv. Rev. 62:1052–1063.

Kenny, G.D., N. Kamaly, T.L. Kalber, L.P. Brody, M. Sahuri, E. Shamsaei, A.D. Miller, and J.D. Bell. 2010. Novel multifunctional nanoparticle mediates siRNA tumour delivery, visualisation and therapeutic tumour reduction *in vivo*. J. Control. Release DOI: 10.1016/j.jconrel.2010.09.020.

Kim, K., J.H. Kim, H. Park, Y.-S. Kim, K. Park, H. Nam, S. Lee, J.H. Park, R.-W. Park, I.-S. Kim, K. Choi, S.Y. Kim, K. Park, and I.C. Kwon. 2010. Tumor-homing multifunctional nanoparticles for cancer theragnosis: Simultaneous diagnosis, drug delivery, and therapeutic monitoring. J. Control. Release 146: 219–227.

Lammers, T., F. Kiessling, W.E. Hennink, and G. Storm. 2010. Nanotheranostics and image-guided drug delivery: Current concepts and future directions. Mol. Pharm. DOI:10.1021/mp100228v.

Lanza, G.M., X. Yu, P.M. Winter, D.R. Abendschein, K.K. Karukstis, M.J. Scott, L.K. Chinen, R.W. Fuhrhop, D.E. Scherrer, and S.A. Wickline. 2002. Targeted antiproliferative drug delivery to vascular smooth muscle cells with a magnetic resonance imaging nanoparticle contrast agent: implications for rational therapy of restenosis. Circulation 106: 2842–2847.

Lee, K., H. Lee, K.H. Bae, and T.G. Park. 2010. Heparin immobilized gold nanoparticles for targeted detection and apoptotic death of metastatic cancer cells. Biomaterials 31: 6530–6536.

Lu, Z.-R., F. Ye, and A. Vaidya. 2007. Polymer platforms for drug delivery and biomedical imaging. J. Control. Release 122: 260–277.

Maeda, H., and G.Y. Bharate and J. Daruwalla. 2009. Polymeric drugs for efficient tumor-targeted drug delivery based on EPR-effect. Eur. J. Pharm. Biopharm. 71: 409–419.

MacKay, J.A., and Z. Li. 2010. Theranostic agents that co-deliver therapeutic and imaging agents? Adv. Drug Deliv. Rev. 62: 1003–1004.

McCarthy, J.R., F.A. Jaffer, and R. Weissleder. 2006. A macrophage-targeted theranostic nanoparticle for biomedical applications. Small 2: 983–987.

McCarthy, J.R. 2010. Multifunctional agents for concurrent imaging and therapy in cardiovascular disease. Adv. Drug Deliv. Rev. 62: 1023–1030.

Raatschen, H.J., Y. Fu, D.M. Shames, M.F. Wendland, and R.C. Brasch. 2006. Magnetic resonance imaging enhancement of normal tissues and tumors using macromolecular Gd-based cascade polymer contrast agents: preclinical evaluations. Invest. Radiol. 41: 860–867.

Rai, P., S. Mallidi, X. Zheng, R. Rahmanzadeh, Y. Mir, S. Elrington, A. Khurshid, and T. Hasan. 2010. Development and applications of photo-triggered theranostic agents. Adv. Drug Deliv. Rev. 62: 1094–1124.

Reddy, G.R., M.S. Bhojani, P. McConville, J. Moody, B.A. Moffat, D.E. Hall, G. Kim, Y.E. Koo, M.J. Woolliscroft, J.V. Sugai, T.D. Johnson, M.A. Philbert, R. Kopelman, A. Rehemtulla, and B.D. Ross. 2006. Vascular targeted nanoparticles for imaging and treatment of brain tumors. Clin. Cancer Res. 12: 6677–6686.

von Maltzahn, G., J.H. Park, A. Agrawal, N.K. Bandaru, S.K. Das, M.J. Sailor, and S.N. Bhatia. 2009. Computationally guided photothermal tumor therapy using long-circulating gold nanorod antennas. Cancer Res. 69: 3892–3900.

Wagner, D.S., N.A. Delk, E.Y. Lukianova-Hleb, J.H. Hafner, M.C. Farach-Carson, and D.O. Lapotko. 2010. The *in vivo* performance of plasmonic nanobubbles as cell theranostic agents in zebrafish hosting prostate cancer xenografts. Biomaterials 31: 7567–7574.

Xie, J., K. Chen, J. Huang, S. Lee, J. Wang, J. Gao, X. Li, and X. Chen. 2010. PET/NIRF/MRI triple functional iron oxide nanoparticles. Biomaterials 31: 3016–3022.

Zong, Y., J. Guo, T. Ke, A.M. Mohs, D.L. Parker, and Z.R. Lu. 2006. Effect of size and charge on pharmacokinetics and in vivo MRI contrast enhancement of biodegradable polydisulfide Gd(III) complexes. J. Control. Release 112: 350–356.

Zou, P., Y. Yu, Y.A. Wang, Y. Zhong, A. Welton, C. Galbán, S. Wang, and D. Sun. 2010. Superparamagnetic iron oxide nanotheranostics for targeted cancer cell imaging and pH-dependent intracellular drug release. Mol. Pharm. DOI: 10.1021/mp100273t.

9

Nanotechnology in Controlling Infectious Disease

Rehab Amin

ABSTRACT

Nanomaterials are the leading requirement of the rapidly developing field of nanomedicine and bionanotechnology. Nanomicrobiology research is gaining in importance. In the treatment of infections, nanoparticles are being used either as therapeutic or as diagnostic tools. Understanding the properties of nanoparticles and their effect on the microbes is essential to their clinical application. Our interest in this chapter is to explore the great impact of nanotechnology in either early diagnosis or treatment of infectious disease. Recent innovative developments on nanomaterials with unique optical and magnetic properties and their promising applications in controlling infectious disease were reviewed. This chapter gives an overview of the field of nanomicrobiology; in this field, nanoparticles could be used for antimicrobial chemotherapy either directly as antimicrobial agents, or indirectly as targeting transport of active substances to the infectious sites (drug delivery system). Moreover, nanoparticles could be used for early detection and diagnosis of infectious diseases.

National Institute of Laser Enhanced Sciences, Cairo University, Giza, Egypt;
Email: rehabamin@niles.edu.eg

List of abbreviations after the text.

INTRODUCTION

Nanotechnology is considered one of the key technologies of the future, and scientists have high expectations from it. According to the National Nanotechnology Initiative, nanotechnology is defined as the use of structures of nanometer size for the construction of materials, devices or systems with novel or significantly improved properties, whereas, at this size, atoms and molecules work differently and provide a variety of surprising and interesting uses. Nanotechnology not only produces small structures, it can also allow for inexpensive control of the structure of matter, which is beneficial for the environment and healthcare, and for the manufacture of hundreds of commercial products. Because of their huge economic potential, nanomaterials have attracted not only scientists and researchers but also businesses.

The unique size-dependent properties of nanoparticles make these materials superior and indispensable in many applications, such as medicine, electronics, biomaterials and energy production. One of the most promising areas of nanotechnology application is medicine. The term *nanomedicine* is used to describe the hybrid field between nanoscience and medicine. This integration in the field of nanomedicine has led to the development of diagnostic devices, contrast agents, analytical tools, physical therapy applications, and drug delivery systems. Our interest in this chapter is to explore particularly the integration between nanomaterials and microbiology since nanomicrobiology is an already established scientific discipline and many research topics dealing with this field have been reported.

Infectious disease is the main cause of mortality in the world and the rapid increase of antibiotic resistance among pathogenic bacteria is a serious public health problem. Antimicrobial resistance is a factor in virtually all hospital-acquired infections and it is expected that some bacterial infections may soon become untreatable. These concerns have led to major research efforts to discover alternative strategies that could be used to allow rapid detection and curative treatment of infections, one of which is the use of nanotechnology. This chapter gives an overview about the field of nanomicrobiology to better elucidate the interaction between nanoparticles and microbiology. It also highlights the role of nanoparticles in detection, diagnosis and treatment of various sites of infection.

NANOPARTICLES AS ANTIMICROBIAL AGENTS

Nanoscale materials have received attention as novel antimicrobial agents because of their high surface area and their unique physicochemical properties. The interaction of nanoparticles with microorganisms is an expanding field of research. Within this field, an area that has been largely applied is the interaction of metal nanoparticles with microbes and their antimicrobial effects.

Silver Nanoparticles

Among noble-metal nanomaterials, silver nanoparticles (AgNPs) have received considerable attention because of their good conductivity, chemical stability, catalytic and antibacterial activity. AgNPs have been shown to be promising antimicrobial material (Yacoby and Benhar 2008). AgNPs have an extremely large relative surface area, thus increasing their contact with bacteria or fungi, improving their bactericidal and fungicidal effectiveness. It was reported that chemically prepared silver nanoparticles by the reduction of metallic ions in the presence of biocompatible capping agent; polyvinyl pyrrolidone (4%, w/v) showed one absorption band due to surface plasmon resonance (SPR). The SPR of metal particles depends on the particle size and the dielectric constant of the metal itself. This strong absorption is due to the oscillation of the surface electron around the ionic core once it is exposed to visible light. Spherical silver particles showed SPR band at 400 nm (Fig. 1). Moreover, results showed that the antibacterial effect of AgNPs was dose-dependent: results indicate that an increase in the concentration of silver nanoparticles reduced the survival of bacteria (Fig. 2) (Amin et al. 2009). The antibacterial effect of AgNPs is not only dose-dependent but also shape-dependent: it was found that different shapes of AgNPs could have different effect on bacteria. Binding of silver particles to the bacterial membrane depends on the surface area available for interaction, so that smaller nanoparticles have higher bactericidal effect than larger particles (Pale et al. 2007).

In addition, the antibacterial effect of AgNPs depends on the gram status of the microorganism. The difference between the outer wall of gram positive and gram negative bacteria is given by the degree of permeability, with an exclusion limit for substances with a molecular weight of more than about 600 Dalton for the gram negative cells. The outer wall of gram negative bacteria acts as a permeability barrier because of the presence of a lipopolysaccharide layer that is able to exclude macromolecules and hydrophilic substances, thereby being responsible for the intrinsic resistance of gram negative bacteria (Nikaido 2003). On the other hand, this lipopolysaccharide could favor the passage of only lipid-soluble materials, so lipophilicity is an important factor that controls the antimicrobial activity.

The mechanisms of the antimicrobial activity of nano-scaled silver are not yet fully elucidated, but several mechanisms are suggested. It is

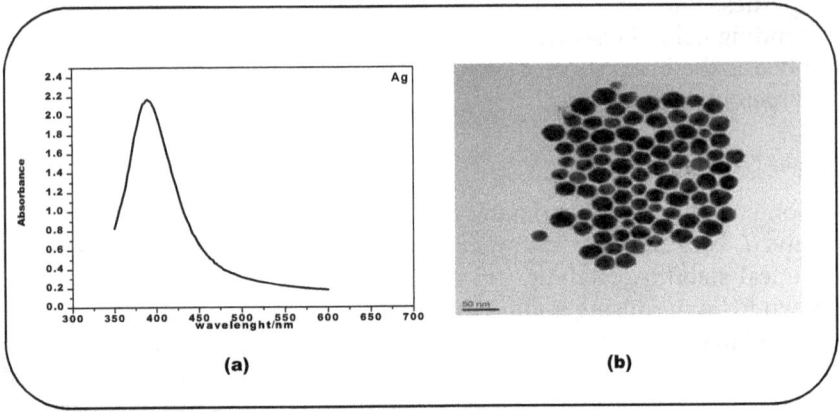

Fig. 1. Characterization of silver nanoparticles (AgNPs). (a) The UV-Vis spectra of spherical AgNPs. (b) Transmission electron microscopy (TEM) image of AgNPs.

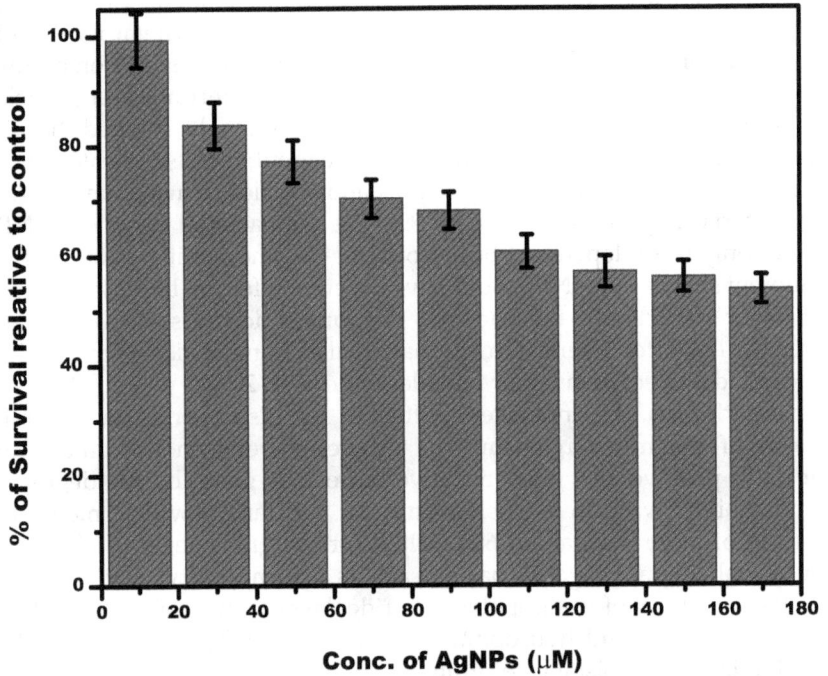

Fig. 2. Antimicrobial effect of silver nanoparticles (AgNPs). The histogram shows the effect of different concentrations AgNPs on the survival rate of *Staphylococcus capitis*. AgNPs, silver nanoparticles; OD, optical density; Cont., control.

thought that AgNPs could accumulate in the cell membrane, affecting the membrane permeability and suppressing respiration, the basal metabolism of the electron transfer system, and the transport of the substrate into the microbial cell membrane. Also, it is thought that AgNPs could release free silver ions, which may affect DNA replication, interact with membrane proteins affecting their correct functions and/or promote formation of reactive oxygen species (ROS), and this generated ROS may also affect DNA, cell membrane, and membrane proteins (Marambio-Jones and Hoeck 2010). Since AgNPs have the ability to integrate with protein, it is thought that this integration could block the bacterial enzymes required for oxygen metabolism, destabilize the cell membrane, and block cell division (Choi et al. 2008). Accordingly, it was found that silver nanoparticles could also inhibit the *in vitro* production of hepatitis B virus RNA and extracellular virions because of direct interaction between these nanoparticles and viral particles, inducing antiviral effect (Lu et al. 2008).

Metal nanoparticles that have antiseptic effect can be introduced into medical implants in order to overcome infection, which is considered a frequent complication with implants. One of the established applications of this is the use of an antiseptic layer based on AgNPs on cochlear implants to avoid post-operative complications (Heidenau et al. 2005).

Gold Nanoparticles

One of the most promising metal nanoparticles is gold nanoparticles (AuNPs). Chemically prepared spherical gold nanoparticles by citrate reduction method, capped with a biocompatible capping material polyvinyl pyrrolidone (1.0 g), which is biologically safe and exhibits no apparent toxicity, have one absorption band at around 520 nm depending on the particle size (Fig. 3). Although it was reported that AuNPs have no significant effect on cell viability since they did not accumulate or bind living cells; however, they can be bound to a wide variety of biochemically functional groups and made several potential applications for controlling infectious diseases.

Application of small-molecule coated gold nanoparticles as effective inhibitors of human immunodeficiency virus (HIV) fusion was demonstrated, and results showed that therapeutically inactive monovalent small organic molecules may be converted into highly active drugs by conjugating them to gold nanoparticles (Bowman et al. 2008).

AuNPs coated with multiple copies of an amphiphilic sulfate-ended ligand are able to bind the HIV envelope glycoprotein gp120 as measured by SPR and inhibit *in vitro* the HIV infection of T-cells at nanomolar concentrations. A 50% density of sulfated ligands on approximately 2 nm nanoparticles is enough to achieve high anti-HIV activities. This result opens up the possibility of tailoring both sulfated ligands and other anti-HIV molecules on the same gold cluster, thus contributing to the development of non–cocktail-based multifunctional anti-HIV systems (Di-Gianvincenzo et al. 2010).

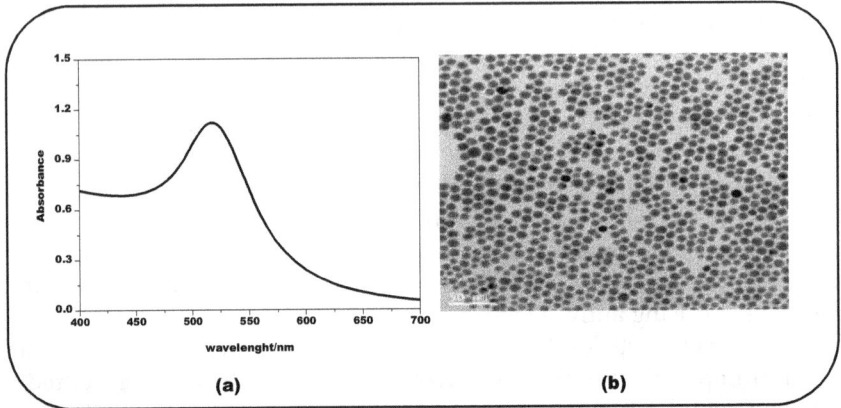

Fig. 3. Characterization of gold nanoparticles (AuNPs). (a) The UV-Vis spectra of spherical AuNPs. (b) Transmission electron microscopy (TEM) image of AuNPs.

Gold nanoparticles capped with mercaptoethanesulfonate (Au-MES) nanoparticles are used as effective inhibitors of Herpes simplex virus type 1 infection for their ability to mimic cell-surface-receptor heparan sulfate. Mechanistic studies showed that Au-MES nanoparticles interfere with viral attachment, entry, and cell-to-cell spread, thereby preventing subsequent viral infection in a multimodal manner. The ligand multiplicity achieved with carrier nanoparticles is crucial in generating polyvalent interactions with the virus at high specificity, strength, and efficiency. Such multivalent-nanoparticle-mediated inhibition is a promising approach as alternative antiviral therapy (Baram-Pinto et al. 2010).

A new approach to create physical damage to the pathogenic organisms was discovered using a combination of light energy and absorbing metallic nanoparticles. Metallic nanoparticles absorb light energy, which is quickly transferred into heat that leads to bacterial damage. Gold nanoparticles in different modifications (e.g., spheres, rods, shells) are the most promising candidates for such a photothermal effect, since they are photo-stable and nontoxic and seem to induce varying cellular damage (Pitsillides et al. 2003).

Bacteria conjugated to oval-shaped gold nanoparticles irradiated with near-infrared radiation showed a highly significant reduction in viability due to photothermal lysis (Wang et al. 2010).

Metal Oxide Nanoparticles

Nanoparticle metal oxides represent a new class of important materials that are increasingly being developed for use in research and health-related applications. Highly ionic metal oxides are interesting in the field of medicine because of their antibacterial activity. Zinc oxide (ZnO) nanoparticles have a potential application as a bacteriostatic agent in visible light and may have future applications in the development of derivative agents to control the spread and infection of a variety of bacterial strains (Jones et al. 2008).

Also, it was demonstrated that the bactericidal efficacy of ZnO nanoparticles increases with decreasing particle size. It is proposed that the surface oxygen species of ZnO nanoparticles promote the bactericidal properties of ZnO nanoparticles (Padmavathy and Vijayaraghavan 2008).

Titanium dioxide (TiO_2) is a photocatalyst once it is illuminated by light with energy higher than its band gaps. The negative electrons and oxygen combine into active oxygen; the positive electric holes and water generate hydroxyl radicals. These reactive species are able to decompose bacteria and viruses inducing cytotoxic effect; this cascade reaction is called oxidation-reduction (Yang et al. 2004) (Fig. 4). Therefore, in the presence of ultra violet (UV) light, the TiO_2 nanoparticles can be applicable to medical facilities where the potential for infection should be controlled.

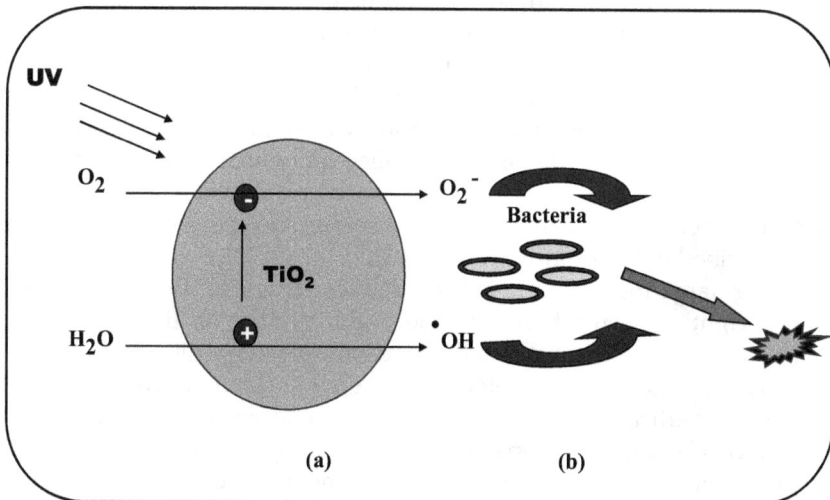

Fig. 4. Diagram showing photocatalysis activity of TiO_2NPs for microbial treatment. (a) TiO_2 nanoparticles illuminated by UV light forming electron (-) and hole (+), which interact with oxygen (O_2) and water (H_2O) respectively, forming reactive radicals [reactive oxygen (O_2^-) and hydroxyl radicals (OH)]. (b) The formed reactive species decompose microbes through oxidation-reduction reaction leading to cell damage. TiO_2, titanium oxide.

Moreover, a photocatalyst thin film that has strong antibacterial action in visible light has been developed. It was suggested that the coating technology can be applied effectively to surfaces with different degrees of roughness, where it is suitable for protecting both human health and the natural environment. Therefore, nano-sized metallic ions and metallic compounds are widely applied to a wide range of healthcare products such as dressings for burns, scalds, skin donor and recipient sites as well as in the textile industry (Tsuang et al. 2008).

NANOPARTICLES AS DRUG CARRIER

Numerous antimicrobial drugs have been prescribed to kill or inhibit the growth of microbes such as bacteria, fungi and viruses. The therapeutic efficacy of these drugs has been well established, but inefficient delivery could result in inadequate therapeutic index and local and systemic side effects including cutaneous irritation, peeling, scaling and gut flora reduction. In order to overcome these issues, alternative antimicrobial drug delivery strategies have been proposed. Although the application of nanotechnology to drug delivery appears to be relatively recent, the basic nanotechnology approaches for medical application date back several decades. A hundred years ago, the "magic bullet" concept, first theorized by the German microbiologist Paul Ehrlich in 1891, represents the first early description of the drug-targeting paradigm. This idea was taken up again at the end of the 1960s and researchers have been developing such drug delivery systems (Allen and Cullis 2004).

The first description of nanoscale size was lipid vesicles, later known as liposomes in 1965; the first controlled polymers for the sustained release of proteins and macromolecules was described in 1976 and the first long-circulating polymeric nanoparticle as site-specific drug delivery was described in 1994 (Shrivastava 2008).

Drug targeting is designed to deliver drugs to the right place, at the right concentration, for the right period of time. As drug characteristics differ substantially in chemical composition, molecular size, hydrophilicity and protein binding, the essential characteristics that identify efficacy became highly complex. Recently, encapsulation of antimicrobial drugs in nanoparticle systems has emerged as an innovative and promising alternative that enhances therapeutic effectiveness and minimizes undesirable side effects of the drugs and this is due to the unique physicochemical properties of nanoparticles such as ultrasmall and controllable size, large surface area to mass ratio, high reactivity, and functionalized structure. Extensive studies have demonstrated that nanoparticles such as liposomes, polymeric nanoparticles and solid lipid nanoparticles are able to facilitate antimicrobial delivery to microbial infection sites.

Liposomes are the most widely used antimicrobial drug delivery system. One of the distinguishing features of liposomes is its lipid bilayer structure, which mimics cell membranes and can directly fuse with microbial membranes, the drug payloads of liposomes can be released to the cell membranes or the interior of the microbial cells (Zhang et al. 2010).

Loading drugs into nanoparticles through physical encapsulation, adsorption, or chemical conjugation, the pharmacokinetics and therapeutic index of the drugs can be significantly improved in contrast to the free drug counterparts. Many advantages of nanoparticle-based drug delivery have been recognized, including improving serum solubility of the drugs, prolonging the systemic circulation lifetime, releasing drugs at a sustained and controlled manner, preferentially delivering drugs to the tissues and cells of interest, and concurrently delivering multiple therapeutic agents to the same cells for combination therapy. Moreover, drug-loaded nanoparticles can enter host cells through endocytosis and then release drug payloads to treat microbes-induced intracellular infections. Therefore, a number of nanoparticle-based drug delivery systems have been approved for clinical uses to treat a variety of diseases and many other therapeutic nanoparticle formulations are currently under various stages of clinical tests (Zhang et al. 2008).

In addition, the important technological advantages of nanoparticles used as drug carriers are high stability, high carrier capacity, feasibility of incorporation of both hydrophilic and hydrophobic substances, and feasibility of variable routes of administration including oral application and inhalation. Nanoparticles can also be designed to allow controlled sustained drug release from the matrix. These properties of nanoparticles enable improvement of drug bioavailability and reduction of the dosing frequency, and may resolve the problem of non-adherence to prescribed therapy, which is one of the major challenges in the control of tuberculosis epidemics (Gelperina et al. 2005).

Nitric oxide (NO) is a critical component of the natural host defense against pathogens such as *Staphylococcus aureus*, but its therapeutic applications have been limited by lack of effective delivery options. The efficacy of a NO-releasing nanoparticle system (NO-rNS) in methicillin-resistant *Staphylococcus aureus* (MRSA) abscesses in mice was tested; results showed that the NO-rNS enhance antimicrobial activity against MRSA *in vitro* and in abscesses. Topical or intradermal NO-rNS treatment of abscesses reduces the involved area and bacterial load while improving skin architecture (Han et al 2009). In addition, nanoparticles showed antimicrobial activity against MRSA in a murine wound model. Therefore, it was suggested that NO-rNS have the potential to serve as a novel class of topically applied antimicrobials for the treatment of cutaneous infections and wounds. (Martínez-Avila et al. 2009).

NANOPARTICLES FOR EARLY DIAGNOSIS

There is a need and urgency for sensitive, specific, accurate, easy-to-use diagnostic tests to identify trace amounts of infectious pathogens rapidly, accurately and with high sensitivity in order to prevent epidemics and loss of life. Nanomaterials are promising in diagnosing the infectious pathogens. The large surface area of nanomaterials enables attachment of a large number of target-specific molecules of interest for ultra-sensitive detection. Surface attachment of biorecognition elements (such as proteins, antibodies or DNA) to nanoparticles is usually required for obtaining appropriate surface functional groups for bioconjugation purposes. With such capability, diagnosis at the molecular and single cell level is possible. Because of this high sensitivity, nanotechnology enables detection of a few microorganisms or target molecular analytes specific to pathogens (Tallury et al. 2010). In addition, unique optical and magnetic properties of nanomaterials conjugated with ligand such as antibodies could allow rapid and real-time detection of pathogens (Fig. 5). Nanotechnology based on fluorescent, magnetic and metallic nanoparticles has been successfully used to track and detect various infectious microorganisms.

Quantum dots (QD) are fluorescent semiconductor nanocrystals that have attracted much attention as fluorescence imaging probes owing to their unique optical properties such as high quantum yield, high molar extinction coefficients, narrow emission spectra, size-dependent tunable emission, and high photostability. The novel optical properties of QD make them ideal nanomaterials for optical sensing. Luminescence of QD nanomaterials is very sensitive to their surface states; therefore, chemical or physical interactions occurring at the surface of the QD nanoparticles change the efficiency of the radiative recombination, leading to photoluminescence activation or quenching. The changes induced by the direct interaction between the analyte and the QD surface, unmodified or functionalized with a given ligand, have supported the selective detection of many targets (Clapp et al. 2005). To be applied to biological detection and imaging applications, QD have to be conjugated to molecules (e.g., peptide ligands, carbohydrate, antibodies, small molecule ligands) that can specifically recognize the biological target under study. QD nanoparticles showed unique fluorescence properties that have been explored for

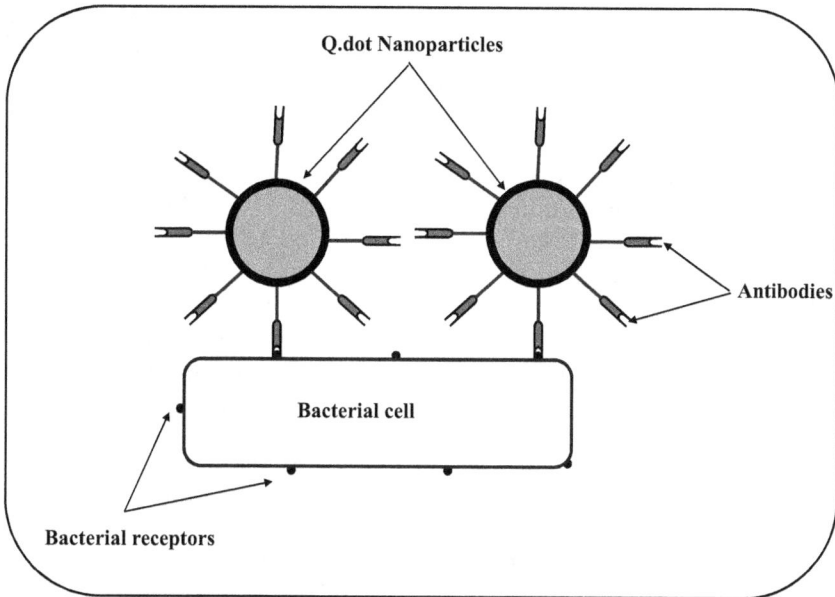

Fig. 5. Diagram showing the use of nanotechnology for diagnosis of infectious pathogens. This diagram illustrates the interaction between functionalized nanoparticles with antibodies and microbial receptors "antigen", which allow specific detection of microbes.

their application to virus detection when combined with direct antibody conjugation or streptavidin-biotin binding systems. QD nanoparticles have an advantage over many traditional fluorophores because their fluorescence properties are tuned and resistant to photobleaching. The development of these nanoparticle-based detection strategies holds the potential to be a powerful method for quick and easy detection of respiratory virus infection (Halfpenny and Wright 2010).

One often-used strategy for the development of QD nanosensors is based on energy flow such as transfer of electronic excitation energy between the components of such nano assemblies. This can occur when light energy absorbed by QD (donor) is transferred to a nearby acceptor species, such as an organic fluorophore (acceptor), in a process called Förster (Fluorescence) Resonance Energy Transfer (FRET). The rate of energy transfer depends on the distance between the donor and the acceptor, their relative orientations, and the spectral overlap. Therefore, the energy flow at the nanoscale can be altered, set up or disrupted by small perturbations such as specific interactions due to molecular binding or cleavage events.

There are numerous examples of FRET-based QD biosensors that include self-assembled nanocomplexes for detecting biological target and enzyme activity. In these FRET-based QD nanosensors, multiple copies of the FRET acceptor were often present on one QD, which may result in self-quenching and lead to low emission from the FRET acceptor.

Dye-doped FRET silica nanoparticles have been developed for imaging and sensing of infectious diseases; when the number and the ratio of amine dyes encapsulated in these silica nanoparticles were varied and they were excited using a single wavelength, unique colors were emitted. These silica nanoparticles were conjugated to antibodies specific for certain pathogens or for multiple target detection. Therefore, it was reported that luminescent silica nanoparticles demonstrate the advantages of greater sensitivity, photostability and ease of functionalization, over regular fluorescent dyes. Moreover, each silica nanoparticle can encapsulate tens of thousands of fluorescent dye molecules, providing highly amplified and reproducible signals. These advanced optical features make silica nanoparticles attractive for bioimaging and detection as well as real-time tracking and monitoring of infectious diseases (Wang et al. 2007).

Since the first study demonstrating the energy transfer from QD to organic chromophores, many studies have been developed using QD as a scaffold for FRET assays. However, FRET efficiency from an organic fluorophore (donor) to QD (acceptor) is not expected because QD are poor acceptors of energy. Upon exciting the donor, there is a coincident excitation of QD due to their broad excitation spectrum; thus, the reduced energy transfer rate prevents the detection of QD-FRET enhancement. In addition, the relatively slow QD decay rate also reduces QD ability as acceptor (Clapp et al. 2005). Bioluminescence Resonance Energy Transfer (BRET) is a principle ideally suited for luminescent QD because it eliminates the difficulties encountered in using QD as acceptor fluorophores. BRET exploits the photon generating process from a chemical reaction to transfer the excitation energy non-radiatively to a proximal fluorescent acceptor. When a light-emitting protein (donor) non-radiatively transfers energy to QD (acceptor), self-illuminating QD-conjugates are created with no requirement for external excitation light for QD to fluoresce. Rao and his team have developed groups of self-illuminating quantum dots, known as QD-BRET conjugates; they are a new class of quantum dot bioconjugates that do not need external light for excitation. Instead, light emission relies on the bioluminescence resonance energy transfer from the attached Renilla luciferase enzyme, which emits light upon the oxidation of its substrate. QD-BRET combines the advantages of the QD (such as superior brightness, photostability, tunable emission, high quantum yields, large Stokes shifts and long wavelength emission) as well as the high sensitivity of bioluminescence imaging, thus holding the promise for improved deep tissue *in vivo* imaging (Xia and Rao 2009). Schematic diagrams were plotted to represent the difference between OD, QD-FRET and QD-BRET as different types of biosensors used for detection and diagnosis of infectious pathogens (Fig. 6).

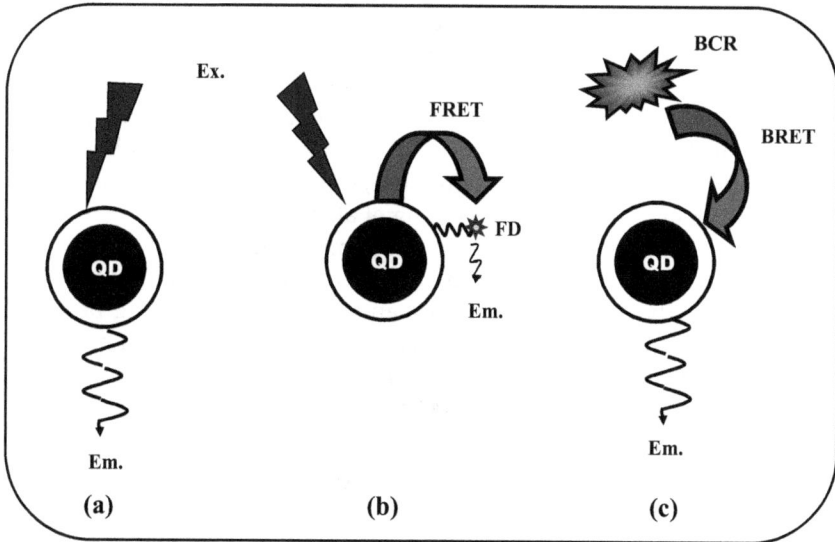

Fig. 6. Diagram of various types of QD sensors. (a) QD illuminated by light with excitation energy (Ex) and emitted fluorescence light with energy (Em). (b) QD illuminated by light with excitation energy (Ex), where this excitation energy is transferred from QD as a (donor) to the flourescence dye (acceptor) in a process called FRET. (c) QD illuminated by light emitted from a bioluminescent chemical reaction (donor) to transfer the excitation energy non-radiatively to QD (acceptor) in a process called BRET. QD, quantum dot nanoparticles; FRET, Fluorescence Resonance Energy Transfer; BRET, Bioluminescence Resonance Energy Transfer; FD, flourescence dye; BCR, bioluminescent chemical reaction.

Nanowires coated with antibodies can be used to detect viruses in a blood sample, where the binding between single virus and antibody results in a change in the nanowire's electric conductance. The method is extremely sensitive, which means that an infection can be detected at a very early stage (Larkin 2004). Gold NP-based assay could be used for improving detection sensitivity of HIV-1 antigen. Results showed that gold NP-based assay offers 100–150-fold enhancement in the detection limit over the traditional colorimetric methods. Furthermore, this assay detected HIV infection 3 d earlier than traditional methods did. These results indicate that the universal labeling technology based on NPs and its application may provide a rapid and sensitive testing platform for clinical diagnosis and laboratory research (Tang and Hewlett 2010).

CHALLENGES OF NANOTECHNOLOGY IN CONTROLLING INFECTIOUS DISEASE

There is a need to increase the specificity of nanoparticle-based antimicrobial drug delivery systems to distinguish infectious cells from healthy cells. It would be beneficial for infection treatment if antimicrobial nanoparticles could be modified with specific antimicrobial ligands.

Moreover, premature drug release from the antimicrobial loaded nanoparticles remains another major challenge, especially for treating systemic and intracellular infections. To minimize drug loss before the nanoparticles arrive at the infectious sites, infection-sensitive drug release nanoparticles can be developed (Zhang et al. 2010).

On the other hand, self-illuminating nanoparticles used for imaging required further studies to increase their selectivity. Although studies have demonstrated the superior sensitivity and deep tissue imaging potential of self-illuminating conjugates, the stability of this conjugate in biological environment needs to be improved for long-term imaging studies such as *in vivo* cell tracking (Xia and Rao 2009).

CONCLUSION

Nano-microbiology is a multidisciplinary science in which chemistry, physics, materials science, and microbiology come together. In this chapter we tried to explore the significance of nanoscience in the field of microbiology for improving human health. Nanotechnology has strong potential for treating and diagnosing infectious diseases and this chapter demonstrates various assays for treatment, diagnosis, and detection of trace amount of infectious pathogens. Nanoparticles can be applied to facilitate the administration of antimicrobial drugs, thereby overcoming some of the limitations in traditional antimicrobial therapeutics. In addition, it was expected that nanotechnology will play an important role in the future for the detection of infectious pathogens as newly developed biosensors.

Definitions

Liposomes: Nanosized, artificially made vesicle made out of the same material as a cell membrane. Liposomes can be filled with hydrophobic and hydrophilic drugs without chemical modification and used to deliver drugs for many diseases.

MRSA: Methicillin-resistant *Staphylococcus aureus*, a type of bacteria that has become resistant to many antibiotics, including methicillin, penicillin, amoxicillin, and cephalosporins.

Nanostructured biomaterials: Materials whose structural elements—clusters, crystallites or molecules—have dimensions in the 1 to 100 nm range. These nanoscale particles show unique physicochemical properties such as ultrasmall and controllable size, large surface area to mass ratio, high reactivity, and functionalizable structure.

Quantum dots (QD): A special class of materials known as semiconductors, which are crystals composed of periodic groups of II-VI, III-V, or IV-VI materials. Quantum dots are a unique class of semiconductor because they are so small, ranging from 2 to 10 nm (10–50 atoms) in diameter. At these small sizes materials behave differently, giving QD unprecedented tunability and enabling many applications.

Quantum dot fluorescence energy transfer (QD-FRET): A process that occurs when light energy absorbed by QD is transferred to a nearby acceptor species, such as an organic fluorophore; the distance between donor (QD) and acceptor (flourophore) should be less than 10 nm.

Quantum dot–based bioluminescence resonance energy transfer (QD-BRET): A process that uses the photon generating process from a chemical reaction—generated by bioluminescent reaction (donor)—to transfer the excitation energy non-radiatively to proximal fluorescent QD (acceptor), where self-illuminating QD-conjugates are created.

Summary Points

- Nanoparticles as therapeutic tool could be used directly as antimicrobial agent or indirectly as drug carrier to deliver drug into the infectious sites.
- Silver nanoparticles can be used as effective growth inhibitors in various microorganisms, making them applicable to diverse medical devices and antimicrobial control systems.
- Therapeutically inactive small organic molecules may be converted to highly active antiviral drugs after conjugation with gold nanoparticles.
- Irradiated metallic nanoparticles lead to photothermal effect that induces varying cellular damage of microbial cells.
- Metal oxide nanoparticles have a potential application as a bacteriostatic agent in the presence of UV and/or visible light.
- Nanoparticle-based drug delivery systems have emerged as a promising alternative that enhances therapeutic effectiveness and minimizes undesirable side effects of the drugs.
- Nanotechnology based on fluorescent, magnetic and metallic nanoparticles may provide rapid, sensitive and real time detection of various infectious microorganisms.

Abbreviations

AgNPs	:	silver nanoparticle
AuNPs	:	gold nanoparticle
Au-MES	:	mercaptoethanesulfonate
BRET	:	Bioluminescence Resonance Energy Transfer
FRET	:	Fluorescence Resonance Energy Transfer
HIV	:	human immunodeficiency virus
MNP	:	magnetic nanoparticles
MRS	:	magnetic relaxation sensors
MRSA	:	methicillin-resistant *Staphylococcus aureus*
NO	:	nitric oxide
NO-rNS	:	nitric oxide releasing nanoparticle system
QD	:	quantum dots
ROS	:	reactive oxygen species
SPR	:	surface plasmon resonance
TiO_2	:	titanium dioxide
UV	:	ultraviolet
ZnO	:	zinc oxide

References

Allen, T.M., and P.R. Cullis. 2004. Drug delivery systems: Entering the mainstream. Sci. 303: 1818–1822.

Amin, R.M., M.B. Mohamed, M.A. Ramadan, T. Verwanger, and B. Krammer. 2009. Rapid, sensitive micro-plate assay for screening the effect of silver and gold nanoparticles on bacteria. Nanomedicine 4: 637–643.

Baram-Pinto, D., S. Shukla, A. Gedanken, and R. Sarid. 2010. Inhibition of HSV-1 attachment, entry, and cell-to-cell spread by functionalized multivalent gold nanoparticles. Small 6: 1044–1050.

Bowman, M.C., T.E. Ballard, C.J. Ackerson, D.L. Feldheim, D.M. Margolis, and C. Melander. 2008. Inhibition of HIV fusion with multivalent gold nanoparticles. J. Am. Chem. Soc. 130: 6896–6897.

Choi, O., K.K. Deng, N.J. Kim, L.J. Ross, R.Y. Surampalli, and Z. Hu. 2008. The inhibitory effects of silver nanoparticles, silver ions, and silver chloride colloids on microbial growth. Water Res. 42: 3066–3074.

Clapp, A.R., I.L. Medintz, B.R. Fisher, G.P. Anderson, and H. Mattoussi. 2005. Can luminescent quantum dots be efficient energy acceptors with organic dye donors? J. Am. Chem. Soc. 127: 1242–1250.

Di Gianvincenzo, P., M. Marradi, O.M. Martínez-Avila, L.M. Bedoya, J. Alcamí, and S. Penadés. 2010. Gold nanoparticles capped with sulfate-ended ligands as anti-HIV agents. Bioorg. Med. Chem. Lett. 20: 2718–2721.

Gelperina, S., K. Kisich, M.D. Iseman, and L. Heifets. 2005. The potential advantages of nanoparticle drug delivery systems in chemotherapy of tuberculosis. Am. J. Respir. Crit. Care Med. 172: 1487–1490.

Halfpenny, K.C., and D.W. Wright. 2010. Nanoparticle detection of respiratory infection. Wiley Interdiscip. Rev. Nanomed. Nanobiotechnol. 2: 277–290.

Han, G., L.R. Martinez, M.R. Mihu, A.J. Friedman, J.M. Friedman, and J.D. Nosanchuk. 2009. Nitric oxide releasing nanoparticles are therapeutic for Staphylococcus aureus abscesses in a murine model of infection. PLoS ONE 4: e7804. doi:10.1371/journal.pone.0007804.

Heidenau, F., W. Mittelmeier, R. Detsch, M. Haenle, F. Stenzel, G. Ziegler, and H. Gollwitzer. 2005. A novel antibacterial titania coating: metal ion toxicity and in vitro surface colonization. J. Mater. Sci. Mater. Med. 16: 883–888.

Jones, N., B. Ray, K.T. Ranjit, and A.C. Manna. 2008. Antibacterial activity of ZnO nanoparticle suspensions on a broad spectrum of microorganisms. FEMS Microbiol. Lett. 279: 71–76.

Larkin, M. 2004. Nanowires show potential as virus detectors. Lancet Infect. Dis. 4: 656. doi:10.1016/S0140-6736(08)61345-8.

Lu, L., R.W. Sun, R. Chen, C.K. Hui, C.M. Ho, J.M. Luk, G.K. Lau, and C.M. Che. 2008. Silver nanoparticles inhibit hepatitis B virus replication. Antivir. Ther.13: 253–262.

Marambio-Jones, C., and E.V. Hoek. 2010. A review of the antibacterial effects of silver nanomaterials and potential implications for human health and the environment. J. Nanopart. Res. 12: 1531–1551.

Martínez-Avila, O., K. Hijazi, M. Marradi, C. Clavel, C. Campion, C. Kelly, and S. Penadés. 2009. Gold manno-glyconanoparticles: multivalent systems to block HIV-1 gp120 binding to the lectin DC-SIGN. Chemistry 15: 9874–9888.

Nikaido, H. 2003. Molecular basics of bacterial outer membrane permeability revisited. Microbiol. Mol. Biol. Rev. 67: 593–656.

Padmavathy, N., and R. Vijayaraghavan. 2008. Enhanced bioactivity of ZnO nanoparticles—an antimicrobial study. Sci. Technol. Adv. Mater. 9: 1–7.

Pal, S., Y.K. Tak, and J.M. Song. 2007. Does the antibacterial activity of silver nanoparticles depend on the shape of the nanoparticle? A study of the gram-negative bacterium Escherichia coli. Appl. Environ. Microbiol. 73: 1712–1720.

Pitsillides, C.M., E.K. Joe, X. Wei, R.R. Anderson, and C.P. Lin. 2003. Selective cell targeting with light-absorbing microparticles and nanoparticles. Biophys. J. 84: 4023–4032.

Shrivastava, S. 2008. Nanofabrication for drug delivery and tissue engineering. Digest J. Nanomater. Biostruct. 3(4): 257–263.

Tallury, P., A. Malhotra, L.M. Byrne, and S. Santra. 2010. Nanobioimaging and sensing of infectious diseases. Adv. Drug Deliv. Rev. 62: 424–437.

Tang, S., and I. Hewlett. 2010. Nanoparticle-based immunoassays for sensitive and early detection of HIV-1 capsid (p24) antigen. J. Infect. Dis. 201: S59–64.

Tsuang, Y.H., J.S. Sun, Y.C. Huang, C.H. Lu, W.H. Chang, and C.C. Wang. 2008. Studies of photokilling of bacteria using titanium dioxide nanoparticles. Artif. Organs 32: 167–174.

Wang, L., W.J. Zhao, M.B. O'Donoghue, and W.H. Tan. 2007. Fluorescent nanoparticles for multiplexed bacteria monitoring. Bioconjug. Chem. 18: 297–301.

Wang, S., A.K. Singh, D. Senapati, A. Neely, H. Yu, and P.C. Ray. 2010. Rapid colorimetric identification and targeted photothermal lysis of Salmonella bacteria by using bioconjugated oval-shaped gold nanoparticles. Chemisty 16: 5600–5606.

Xia, Z., and J. Rao. 2009. Biosensing and imaging based on bioluminescence resonance energy transfer. Curr. Opin. Biotech. 20: 1–8.

Yacoby, I., and I. Benhar. 2008. Antibacterial nanomedicine. Nanomedicine 3: 329–334.

Yang, H.Y., S.K. Zhu and N. Pan. 2004. Studying the mechanisms of titanium dioxide as ultraviolet-blocking additive for films and fabrics by an improved scheme. J. Appl. Polym. Sci. 92: 3201–3210.

Zhang, L., D. Pornpattananangkul, C.-M.J. Hu, and C.-M. Huang. 2010. Development of nanoparticles for antimicrobial drug delivery. Curr. Med. Chem. 17: 585–594.

Zhang, L., F.X. Gu, J.M. Chan, A.Z. Wang, R.S. Langer, and O.C. Farokhzad. 2008. Nanoparticles in medicine: therapeutic applications and developments. Clin. Pharmacol. Ther. 83: 761–769.

Virus-based Nanoparticles as Tools for Biomedicine

Stefan Franzen,[1,3,]* *Steven A. Lommel*[2,3] *and*
Bruce Oberhardt[3]

ABSTRACT

Plant viruses have great potential as delivery vectors for chemotherapy agents. First, the interior of the icosahedral capsids of several plant viruses can be loaded with anticancer agents at a density equal to that of commercially available untargeted formulations such as Doxil. Second, size and surface structure of individual viruses are extremely well conserved, enabling an extremely high degree of uniformity at the nanoscale. Third, the protein exterior of the virus capsid is readily modified with targeting agents such as folate or targeting peptides for tumor- or tissue-specific targeting. While the combination of these three features is impressive, creation of a plant virus-based formulated chemotherapy agent is still at an early stage. There are several plant viruses that have demonstrated applications such as vascular imaging, incorporation of contrast agents, drug loading and targeting to cancer cells. For example, the *Red clover necrotic mosaic virus* (RCNMV) is a robust plant virus that is transmitted through

[1]Department of Chemistry, North Carolina State University, Raleigh, NC 27695.
[2]Department of Plant Pathology, North Carolina State University, Raleigh, NC 27695.
[3]NanoVector, Inc. POB 98385, Raleigh, NC 27624-8385.
*Corresponding author

List of abbreviations after the text.

the soil. RCNMV can be reversibly loaded with the anticancer drug doxorubicin and its protein shell or capsid functionalized to target cancer cells. The internal genomic RNA cage of RCNMV also can be manipulated to design imaging and targeting agents, as demonstrated by the use of the RNA to encapsidate a variety of solid nanoparticles in the hollow interior of the virus. In any of these applications the plant virus functions as a nanoparticle, leading to the description of these modified capsids as plant virus nanoparticles (PVNs).

INTRODUCTION

In nature, viruses appear to be ubiquitous. Most viruses are harmless to humans. Relatively little is known about the viral biosphere. The kingdoms of life, often divided into six types (Animalia, Plantae, Fungi, Protoctista, Monera, and Archea) do not include viruses, which are much smaller and less complex than cells. Most viruses are unknown; known viruses range in size from about 20 to about 400 nm. Based on their size and biological function, viruses are quintessential nanoparticles. They have evolved to invade host cells and to evade cell and organism defenses, ultimately delivering their cargo to cells. Used as tools for biomedicine, viruses could theoretically deliver drugs, other small molecules, proteins, RNA, DNA, or other payloads to targeted cells. However, viruses can also cause an infection or an acute immune reaction that can be fatal. Despite the progress made in the field of adenovirus and other targeting viruses, it may prove impractical to use these mammalian infecting virus vectors for medical applications, due to the potential risks. This is a major reason to consider plant viruses as a potential alternative for biomedical applications. Plant viruses do not infect mammalian cells. On the other hand, they have desirable features that may be harnessed for drug delivery, gene therapy, diagnostic imaging and a range of other biomedical applications. In addition, targeting ability may be added to plant viruses by chemical modification, genetic mutation or adaptor binding proteins to enable certain biomedical applications.

Like all organisms, viruses possess a genome that stores genetic information. Viruses can have genomes of single- or double-stranded DNA or RNA and at least one protein surrounded by a protein shell or capsid. Some viruses also have an outer envelope of lipids and proteins. Viruses exhibit a wide range of simple to complex geometric and non-geometric morphologies. Some viruses are rod-shaped, and some icosahedral, such as the *Red clover necrotic mosaic virus* (RCNMV) shown in Fig. 1. Others have more complex shapes with a "head" and a "tail".

It is intriguing and fortuitous for nanotechnology applications that for a particular virus species the individual particle size and morphology are tightly, and perhaps absolutely, conserved. Unlike synthetic nanoparticles that are polydisperse, each virus species is monodisperse. It is, however, possible to engineer, manufacture and assemble viral capsids in different allowable shapes based upon pre-existing geometries.

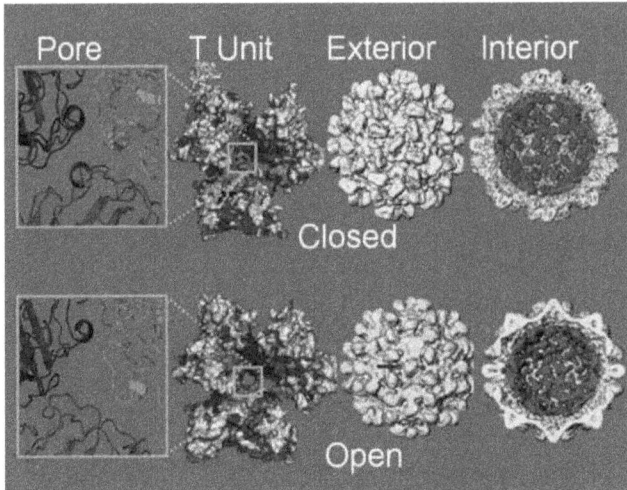

Fig. 1. Representation of two structural forms of the plant virus RCNMV determined by cryo-electron microscopy. The closed form is observed at high divalent ion concentration (Ca^{2+} and Mg^{2+}) and is likely the form of the virus as it moves through the soil from one host to another. The open form of the virus consists of 60 holes that lead from the exterior of the capsid to the hollow interior of the "cargo chamber". The open form is observed when the divalent ion concentration is low. This form is present in the cytosol of cells.

Color image of this figure appears in the color plate section at the end of the book.

Outside of a living cell, a virus is a dormant particle. However, within an appropriate host cell, a virus becomes an active entity capable of subverting the cell's metabolic machinery for its reproductive purposes. Animal viruses enter host cells by a cell membrane transport process called endocytosis. Plant viruses, in contrast, enter through wounds in the cell's outer cell wall, e.g., through environmentally induced abrasions or through punctures made by insects.

A plant virus nanoparticle (PVN) may be ideal for nanomedicine delivery for the following reasons: plant viruses do not infect mammalian cells; they have an extremely low probability of genetic recombination with animal viruses; and they have no inherent targeting capability but can be modified to target selected cells and enter via endocytosis. In addition, secondary targeting is possible to target a particular cell organelle, such as the nucleus, upon cell entry.

CAPSID STRUCTURE

RCNMV form T = 3 icosahedral virions from 180 CP subunits, with 2-, 3- and 5-fold axes of symmetry. The capsid protein (CP) has been divided into four domains. The N-terminus forms a flexible region composed of two domains: the RNA-binding (R) domain at the N-terminus (66 amino acids) and the arm (a) region (35 amino acids), which connects the R domain to the S domain. The R domain contains many basic amino acids and extends into the virion. This region is thought to allow dense packing of the virion RNA by neutralizing the charged phosphates on the RNA. The globular S domain (approximately 168 amino acids) forms the face of the virion and is composed of two sets of four-stranded antiparallel sheets. The protruding (P) domain (approximately 113 amino acids) forms antiparallel β-sheet structures in a jellyroll conformation, with one six-stranded α sheet and one four-stranded β-sheet. The P domain forms the surface projections that give the virion a rough appearance. Each projection is composed of two P domains. Finally, there is a small hinge (h) sequence (5 amino acids) that connects the S and P domains. It makes it possible for the S and P domains to adopt two different configurations relative to one another by varying the angle between the domains. Cryo-electron microscopy of RCNMV has provided insight into the arrangement of the RNA within virions (Sherman et al. 2006). The virion is now shown to be composed of four layers with the outer layer being predominantly the S and P domains. The layer just beneath this outer layer is composed mostly of RNA that adopts dodecahedral symmetry.

When the RCNMV capsid is exposed to low concentrations of calcium and magnesium, as occurs in the cellular cytoplasm, there is a conformational change in capsid proteins resulting in the opening of 60 pores, each approximately 13 Å in diameter. These pores lead from the exterior of the viral capsid to the hollow interior of the virus. The pores are thought to provide openings for the viral RNA to exit the capsid. This mechanism for the release of the RNA genome appears to differ from that of many other plant viruses that disassemble into individual small aggregates of the capsid protein (CP) subunits. RCNMV appears to remain intact, at least initially, providing the pores as exit routes for the RNA.

In RCNMV, the viral RNA is an essential structural component providing an internal scaffold for the capsid. It has been suggested that the RNA is extracted from the opened viral capsid by a process driven by plant cell ribosomes attaching to the viral RNA termini and translating the RNA out. However, for the purposes of nanotechnology applications, the RNA is not removed and at no point in the process does it spontaneously emerge from the capsid. In this regard RCNMV is different from many of the other plant virus capsids that have been used in biomedical applications.

DESCRIPTION OF RCNMV

RCNMV is a plant virus that infects a wide variety of plants, including legumes, Solanaceae and Rosaceae plants in temperate regions of the world. RCNMV may be found in cherries and has been found in municipal water supplies in temperate climates. This virus cannot replicate above 27°C, a temperature well below the homeostatic core temperature of 37°C for humans. The virus has a unique life cycle, in that it is primarily a root virus and moves from plant to plant by being released directly into the soil and transported by water. Consequently, this virus has an unusually robust capsid. The virus cannot infect any mammal including humans and is only mildly immunogenic. RCNMV has a capsid diameter of 36 to 37 nm, depending upon where it is measured. The capsid is icosahedral in shape and consists of 180 copies of a single 38 kilodalton capsid protein. Every RCNMV virion nanoparticle is identical in size and shape and has an identical surface.

OVERVIEW OF A PLANT VIRUS NANOPARTICLE

The structure of the RCNMV viral genome within the infected cell and within the assembled virions is key to the life-cycle of the virus and the assembly and stability of the virion. The RNA structure and conformational changes in the structure serve as a structural scaffold of the capsid, the origin of assembly of the virus as well as a key regulator of gene expression. In a very simple genome it is necessary to prevent molecular collisions between translation, replication, subgenomic RNA (sgRNA) synthesis, and packaging during the virus life cycle. Figure 2A depicts the two genomic RNAs of RCNMV interacting with each other to serve as a key molecular switch in the virus life cycle. The RNA-RNA interaction initiates transcription of the capsid protein subgenomic RNA (Fig. 2C). This interaction is required to form biologically active virions containing both RNA-1 and RNA-2. The RNA-2 component of the RNA-RNA interaction consists of a 34-nucleotide stem-loop structure termed the *trans*-activator (TA), which is shown in Fig. 2C. Collectively, the RNA-2 TA and its interactions with the RNA-1 *trans*-activator binding site (TABS) constitute a multifunctional temporal regulator of the RCNMV life cycle (Guenther et al. 2004).

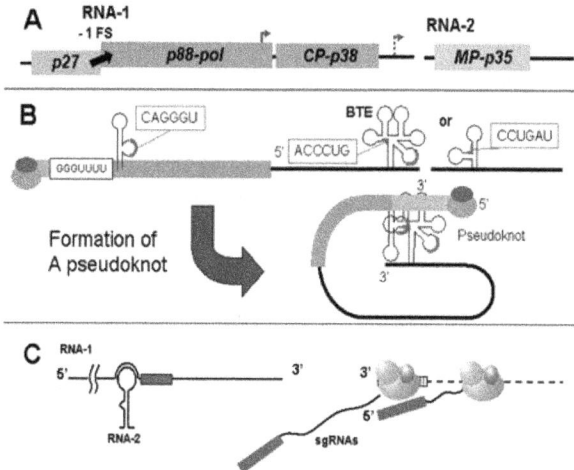

Fig. 2. RNA genome of RCNMV and long-range interactions. A. The four genes are shown on RNA-1 and RNA-2. B. The ribosomal frame shift element interacts with a distal structure to form a long-distance pseudoknot that stimulates ribosomal frameshifting. C. RNA-2 and RNA-1 interact to create the transactivator (TA) region.

Color image of this figure appears in the color plate section at the end of the book.

APPLICATIONS OF PLANT VIRUSES TO TUMOR CELL TARGETING

Over the past 5 years several plant viruses have been studied as potential drug delivery vehicles for chemotherapy. These viruses include *Cowpea chlorotic mosaic virus* (CCMV), *Cowpea mosaic virus* (CPMV), *Hibiscus chlorotic ringspot virus* (HCRSV), and *Red clover necrotic mosaic virus* (RCNMV). Each of these plant viruses has a unique set of features that imbue them with excellent nanoparticle properties. Each one has potential advantages, and yet no plant virus has yet emerged as a generally accepted standard for the field. The CCMV capsid can be manufactured in yeast, as the capsid can assemble spontaneously even in the absence of RNA (Brumfield et al. 2004). CPMV does not disassemble easily, and for this reason CPMV is being explored for imaging applications (Brunel et al. 2010). It is remarkable that although the host range for replication of CPMV is confined to plants, mammalian cells can bind and internalize CPMV in significant amounts (Singh et al. 2006). This binding appears to be mediated by a conserved 54-kDa protein found on the plasma membranes of both human and murine cell lines (Koudelka et al. 2007).

Both HCRSV (Ren et al. 2007) and RCNMV (Loo et al. 2008) are capable of loading small molecules, including cancer drugs, into the internal hollow cavity of their respective capsids. Both of these small icosahedral viruses are in the same virus family and are relatively stable. In addition, both can be labeled with targeting agents to permit tumor cell targeting. The possible targeting agents for plant virus technology are often similar to those used for diagnostic and therapeutic applications using radionuclides (de Rosales et al. 2009) and gene therapy (Wagner et al. 2004).

THE PLANT VIRUS AS NANOMETER-SCALE SCAFFOLD

The packaged RCNMV genomic RNA can be used as a structural component in nanotechnology strategies as shown in studies of the incorporation of gold, magnetic iron oxide, and quantum dot nanoparticles in the interior of the capsid using the TABS structure as a "hook" as shown in Fig. 2 (Loo et al. 2006, 2007). The function of the RNA-2 sequence attached to a nanoparticle is to bind other RNA, which could include the entire RNA-1 gene, in order to create an origin of assembly. The process shown in Fig. 3 constitutes *in vitro* assembly of the capsid protein around a nanoparticle. This strategy has been used to encapsidate spherical nanoparticles ranging from 4–15 nm in diameter. The larger particles approach the same size as the 17 nm diameter central cavity in the RCNMV capsid. Package particles closely filling the central cavity form more regularly-shaped capsids as judged by transmission electron microscopy (TEM). Similar types of methods have been developed for CCMV (Liepold et al. 2007) as well as another plant virus, *Brome mosaic virus* (BMV) (Huang et al. 2007).

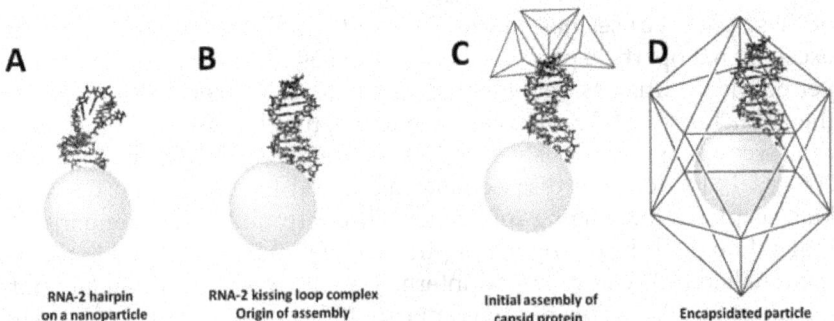

A	**B**	**C**	**D**
RNA-2 hairpin on a nanoparticle	RNA-2 kissing loop complex Origin of assembly	Initial assembly of capsid protein	Encapsidated particle

Fig. 3. Schematic illustration of a method for encapsidation of a solid (e.g., gold) nanoparticle inside a plant virus nanoparticle. A. The target nanoparticle is conjugated to RNA-2. B. The RNA-2 is permitted to react with genomic RNA-1 to form the origin of assembly. C. The origin of assembly is exposed to capsid proteins. D. The capsid proteins assemble around the particle.

Color image of this figure appears in the color plate section at the end of the book.

THE REVERSIBLE LOADING OF FLUOROPHORES
AND DRUGS IN PVNS

When RCNMV is in an environment low in calcium (Ca^{2+}) and magnesium (Mg^{2+}), pores open extending through the capsid (Fig. 3). Based on the icosahedral symmetry of the virus, there are 60 pores, whose opening permits the diffusion of charged molecules into and out of the capsids (Basnayake et al. 2006). Figure 1 shows both open and closed forms. Under conditions of high Ca^{2+} and Mg^{2+}, the RCNMV capsid is extremely stable over a wide range of solution conditions. The open form can be induced *in vitro* by the addition of divalent ion chelators such as ethylene diamine tetraacetic acid (EDTA). What is remarkable about the RCNMV capsid is that the channel opening is reversible and that the virus does not disassemble. This property permits implementation of a reversible process that involves opening, loading and then closing. In this way, modulation of the divalent ion concentration can be used to permit infusion of small molecules into the virion in a process that takes place in the laboratory rather than in a plant cell. The pores can be closed once again by restoring the concentration of Ca^{2+} and Mg^{2+} to greater than 1 μM. The infusion is the first step toward converting RCNMV into a PVN. The cytosol of plant and animal cells has sufficiently low Ca^{2+} and Mg^{2+} concentration that the infused molecules are released inside the cell, as is observed using fluorescence detection, confocal microscopy or flow cytometry. Thus, RCNMV offers unique opportunities as a nanoparticle, since it has an inherent sensor-actuator system for loading a molecular cargo and releasing it in a targeted cell.

The infusion process exploits a natural mechanism apparently employed by the virus to release its genome upon entry into a newly infected cell. Molecular infusion was tested using three dyes, Rhodamine (positive charge), luminarosine (neutral) and fluorescein (negative charge), which were infused into opened virions (Loo et al. 2008). After incubation, the PVNs were closed by addition of Ca^{2+} and Mg^{2+} combined with a titration to pH 7. Following dialysis, the fluorescence of the PVN samples was lowered to near background levels and the dyes internalized in the virion are self-quenched. This dye self-quenching phenomenon is consistent with what is observed with loading into liposomes. The PVN dye load was determined to be 90, 76 and 1 molecules, respectively, for the three dyes. While some of the infused dye is released by re-opening pores with EDTA, complete capsid disruption at pH 10 and EDTA results in significantly greater dye release. The higher loading of positively charged and neutral dye molecules is likely due to the negative charge of the RNA genome as it forms a cage-like lining inside the capsid (Fig. 1).

RCNMV may have distinct advantages over other PVNs. Since RCNMV is a soil-borne virus, the unique capsid structural dynamics that is an essential feature of its life cycle may be exploited for nanocarrier applications. Neither CCMV nor CPMV uses the pore opening mechanism observed in RCNMV (Sherman et al. 2006; Guenther et al. 2004). CCMV swells significantly at pH > 7 (Tama and Brooks 2002), whereas CPMV appears to disassemble partially to fully in the cytosol. Fortunately, the cytosol of a plant cell has essentially the same pH and divalent ion concentration as that of an animal cell, enabling exploitation of the pore opening dynamics of the PVN for mammalian delivery and release. Prior to cell entry, extracellular leakage of drug from the PVN has not been observed in simple *in vitro* experiments. Loading levels can be quite high. In the case of doxorubicin, a cancer chemotherapy drug used as a benchmark in nanoparticle studies, up to 1000 molecules have been successfully loaded in the virus interior (Loo et al. 2008). This loading density is slightly higher than the commercial liposome preparation of doxorubicin known as Doxil.

ATTACHMENT OF TARGETING PEPTIDES TO THE PLANT VIRUS NANOPARTICLE

A PVN based on RCNMV can therefore be equipped with targeting molecules and can carry an internal payload that is hidden from the external milieu until the PVN attaches to the targeted cell, is internalized, and senses the cytoplasm via the loss of Ca^{2+} from its capsid proteins. For this sequence to be completed, the PVN, which may enter the cell via an endosome, must escape from the endosome and make contact with the cytoplasm, a low divalent ion medium. Endosomal escape may be mediated by certain peptide sequences that could be attached to the PVN. Many infectious viruses have evolved peptide sequences that alter their conformation in the low pH (~5.9) of the endosome and thereby can interact with the endosomal membrane and disrupt it. This mechanism of cell entry has been imitated in a number of targeting strategies for gene therapy or targeted nanoparticle delivery. Thus, the targeting of the cell may consist of a peptide that binds to a cell-surface receptor, combined with a peptide that can disrupt the endosomal membrane. Further targeting in the form of a nuclear localization signal is also possible. In the following section, we briefly consider the determinants of targeting and the chemical attachment of targeting peptides to the surface of RCNMV, as an example of plant viral targeting technology.

There are two major types of mammalian cell targeting ligands: (1) peptides with sequences derived from natural proteins that target signaling, growth or structural membrane receptor proteins and (2) metabolites or metabolite carriers that target membrane transport receptor proteins. Peptide targeting of hormonal, cytokine, endocrine and neural receptors has been widely used in the field of therapeutic and diagnostic radionuclides (de Rosales et al. 2009). The coupling chemistry and surface charge effects of peptide labeling in nanoparticle drug delivery strategies have proven difficult to control. Despite the large number of potential targets that may lead to specific cell targeting, the majority of nanoparticle-based cell targeting has thus far relied on folate, which is a small molecule. Colloidal stability is a major issue in the nanotechnology field that deserves greater attention in the design of formulations. The attachment of large numbers of charged or hydrophobic moieties can destabilize a formulation, subsequently precipitating it from its initial suspended (colloidal) state.

Bioconjugate chemistry uses various linker molecules to attach folate or peptides to amino acids via covalent bonding, which usually is restricted to lysine and cysteine (Hermanson 2008). Lysine and cysteine each present amine and thiolate functional groups. These groups can react with each respective end of a heterobifunctional linker. In the plant virus field, the most common chemical couplings are based on Sulfo-SMCC and ethynyl-azide reactivity, so-called "click" chemistry (Kolb et al. 2001). The SMCC group shown in Fig. 6A presents two interacting ends that can bind a peptide to a PVN. At one end of the molecule the sulfide of a cysteine can react with the succinimide moiety by a Michael addition. On the other end of the molecule the N-hydroxy succinimide (NHS) ester can react with the lysine amine group. These reactions have reasonably high yield, but they are pH-dependent. There can be efficiency losses due to hydrolysis that compete with successful conjugate. The optimum pH for the NHS ester reaction is around 7.5, a pH at which only a minority of the primary amines of lysine are deprotonated (pKa = 9 for lysine). However, at high pH the hydroxide concentration from solvent water is sufficiently high that hydrolysis competes with conjugation. Therefore, the chemistry is time-sensitive as well, and the reagents must be prepared fresh and used quickly. The solubility of SMCC is quite low. Therefore, a sulfate group is attached to the NHS ester to prepare Sulfo-SMCC (not shown in Fig. 6). Even with this sulfate group, dimethyl sulfoxide (DMSO) is recommended for use as a co-solvent. DMSO can affect protein folding and stability and should be used sparingly. Depending upon the complexity of surface groups there is a possibility of side reactions. Moreover, attachment by SMCC coupling is only useful when a peptide has a single cysteine or lysine. Otherwise, multiple reactions are possible. To circumvent these problems, Sharpless and co-workers developed the reaction shown in Fig. 6B, known as "click chemistry" (Kolb et al. 2001). Click chemistry shown in Fig. 6B has been adapted for use in conjugation of targeting groups to CPMV (Gupta et al. 2005). This reaction must be catalyzed by Cu(I), which can present some complications in subsequent purification of the plant virus conjugate. Regardless of the precise chemistry used, one must contend with the difficulty of achieving complete saturation of binding sites (incomplete reaction) and the effect of the coupling reaction itself on PVN colloidal stability.

The attachment of peptides using heterobifunctional linker chemistry based on Sulfo-SMCC has proven effective both *in vitro* and *in vivo*. However, recent mass spectrometry studies suggest that there are a number of undesirable side reactions that occur in the crosslinking step on a viral capsid (data not shown). One must use the accessible amino acids. As shown in Fig. 5A, the most accessible surface lysines on RCNMV are K219 and K208.

These are on the protruding P-domain shown in cyan in Fig. 5A. Most frequently, cysteine-terminated peptides have been used and thus, these are the sites of binding. However, lysine-terminated peptides can be used when lysine is not a crucial amino acid in the targeting sequence. In such cases, C154 and C267 can be used for attachment. While these methods have been shown to work, one of the advantages of a plant virus capsid is that the multiple CPs are arranged in a regular geometric array that can be bound by antibodies, single chain antibodies and phage or yeast display proteins. Such methods provide a robust method for attachment of targeting molecules using avidity, rather than covalent attachment, as the mechanism for targeting. Alternatively, the RCNMV CP protein can be genetically engineered to either include a peptide targeting signal or remove, add or change the location of reactive lysines or cysteines. The concept introduction of targeting sequences by mutation has been demonstrated in CCMV (Gillitzer et al. 2002), CPMV (Chatterji et al. 2004) and *Tomato bushy stunt virus* (TBSV) (Joelson et al. 1997).

CELL TARGETING

When the PVN enters the cytoplasm of a targeted cell, divalent ions are leached out of the capsid and pores form. This process may take 1–4 h, allowing time for the PVN to transit the cytoplasm and home in on intracellular targets before unloading its cargo. This delayed release is another natural desirable feature of the RCNMV capsid that provides the opportunity to target the release of the cargo to specific intracellular structures, such as the nucleus, or mitochondria. For this purpose, secondary targeting molecules may be employed and the PVN surface is decorated with cell targeting and intracellular targeting signals.

Cancer cells often over-express certain receptors that are found on the surfaces of normal cells. One challenge with trying to target molecules that are over-expressed on cancer cells is that normal cells that often exhibit a low expression of the same molecules become targets as well. This potential targeting of normal cells thus limits the selectivity of targeted cancer therapeutics, such as in the case of affinity binding, e.g., by antibody molecules. A key advantage of using peptide array targeting is that each peptide binds only weakly to its target molecule. Collectively, the peptide bonds can result in "avidity binding". Avidity binding is achieved when many peptides bind simultaneously. With avidity binding one can theoretically optimize the density of a peptide array to bind to cancer cells with over-expressed receptors but not to normal cells with low expression of the same targeted molecules.

The avidity binding constant (K_A) for a peptide array on the surface of a PVN is proportional to K^N, where K is the equilibrium constant for each individual peptide bond to a target molecule on the cell surface and N is the number of such bonds. The bond energy of the bond between one peptide and a target molecule is very weak. K_A increases exponentially with the number of peptide-cell surface molecule bonds. When N is high, enough peptide bonds form to generate an overall binding strength that exceeds the disruptive thermal energy forces responsible for Brownian motion ($k_B T$, where k_B is Boltzmann's constant and T the absolute temperature). Sufficient peptide bonding will allow the PVN to stably bind to the cell, as with high receptor expression that tends to occur in aggressive cancers. For the same receptors with low expression on healthy cells, N will be low, and the PVN will probably not be capable of stably binding to the cell.

Therefore, the PVN can theoretically be designed to differentiate between cancer cells that over-express certain receptors and normal cells that do not. The PVN can do this, since every PVN is identical; being composed of 180 copies of the identical 38kD CP, and with the same surface and an essentially identical peptide array. There are 360 surface-exposed lysines and 180 cysteines on the P-domain available for peptide coupling. However, there are numerous lysines and cysteines in the S- and R-domains, which can also be modified by bioconjugate chemistry (Hermanson 2008). The resulting nanoparticle with displayed peptides will tend therefore to be more uniform than chemically synthesized nanoparticles composed of polymers, liposomes or solid silica or gold nanoparticles, where there is a polydispersity in the particle diameter and variation in surface charge and hydrophobicity. While the potential advantage of plant viruses is significant, there can still be variations in the bioconjugate chemistry that must be controlled. For this reason antibody, phage or yeast display methods are perhaps essential for discovering binding proteins. Because of the regular surface of the plant virus such methods have significant potential application on plant viruses. The technological advancement of binding peptides on solid nanoparticles has been pursued for a number of years, but low colloidal stability and lack of reproducible and controlled synthesis conditions have hampered applications in the nanomedicine field thus far.

RCNMV AS A CHEMOTHERAPEUTIC DELIVERY AGENT

Cancer is one of the most common targets in the nanotechnology field. The goal of targeted delivery that kills the cancer but not the healthy surrounding tissue drives a great deal of innovation in the field. The effects of avidity and multifunctional targeting both have great potential for achieving the kind of specificity needed to target tumors. The phenotype of cancer cells is often significantly altered relative to normal tissue. Growth factor receptors, hormonal receptors, and certain cytokine receptors are upregulated in cancer. Their over-expression on the cell surface provides an opportunity for targeting by macromolecules and nanoparticles, provided that the targeting agent can be made bioavailable and the circulation time is sufficiently long to enable meaningful targeting of a sufficient number of nanoparticles to exhibit a biological effect. Small molecule pumps in the class of the ABC transporters are often also upregulated in cancer. These efflux pumps remove small molecule drugs and chemotherapeutics, and lead to multi-drug resistance. Multi-drug resistance, along with metastasis, is a significant problem in cancer, for which there is not yet a simple solution. One goal of targeted nanoparticle therapy is to carry the payload to the cell in such a manner as to evade efflux pumps and to deliver the chemotherapeutic payload to the intracellular target, for example, the cell nucleus. Finally, cancer cells often over-express metabolic receptors because of their rapid growth. These receptors also provide targets for nanoparticles. The most common of these is the folate receptor, which has become one of the major targeting molecules in the nanoparticle field (Ren et al. 2007; Destito et al. 2007). We have used targeting peptides to attempt to find greater specificity for specific refractory cancers.

Figure 7 shows the potential of PVNs as targeting agents. The data shown comprise a dose-response curve for the cytotoxic delivery of doxorubicin to HT-1299 lung cancer cells. The PVNs were prepared by infusion of ~900 doxorubicin molecules following the protocol outlined in Fig. 4 above. The targeting peptide used is the ADH304 peptide, which originally designed as a surrogate for the drug ADH-1, which targets N-Cadherin (Kelland 2007). N-cadherin is expressed in mesenchymal cell types. One abnormal trait of cancer cells is that they undergo the epithelial-mesenchymal transition and express N-Cadherin in a form that can bind targeting agents. Normal tissues express E-Cadherin, with the exception of nerve cells, in which N-Cadherin is not exposed. Thus, N-Cadherin has been identified as a target for chemotherapy.

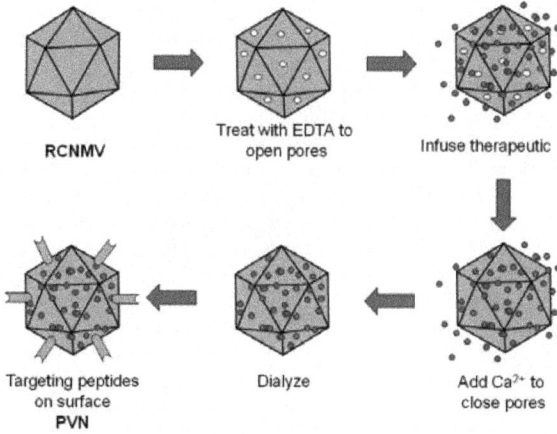

Fig. 4. Schematic representation of the approach for loading a cargo inside the plant virus RCNMV. The process is initiated by addition of excess EDTA to sequester the divalent ions, Ca^{2+} and Mg^{2+}. The cargo is then loaded under these conditions. Subsequently, Ca^{2+} is added back and the excess drug is dialyzed away. Then targeting peptides are added to the surface of the virus to complete the PVN formulation.

Color image of this figure appears in the color plate section at the end of the book.

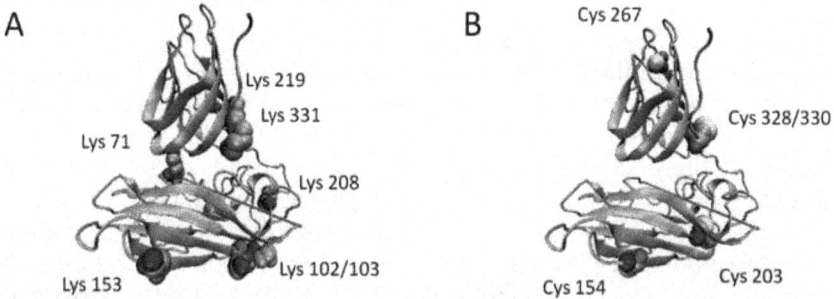

Fig. 5. Capsid structure with locations of lysines and cysteines specified by number. The coordinates were determined using a combination of electron diffraction and homology modeling as described in the text. A. The lysines in the structure are shown. B. The cysteines in the structure are shown.

Color image of this figure appears in the color plate section at the end of the book.

Figure 7 shows a dose-response curve obtained at three incubation times of 24, 28 and 72 h. Doxorubicin clearly has an increased effect as the cells are exposed for a longer period of time. The EC_{50} decreases from 3

A   B

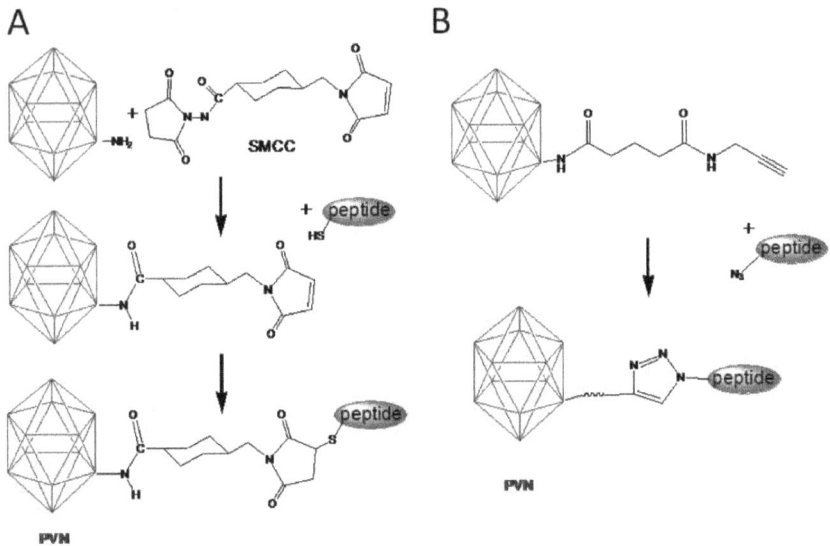

Fig. 6. Two of the most important bioconjugation methods used on plant viruses. A. Bioconjugation using SMCC is shown. The reaction shown assumes that a surface lysine on the PVN is reactive in the first step. Alternatively, a cysteine (-SH) on the PVN could react first, followed by reaction with a lysine (-NH2) on the peptide. B. Click chemistry is shown. A surface lysine is functionalized to present an ethynyl group (C_2H). The peptide is synthesized to present an azide group. Cu(I) must be used as a catalyst (not shown in figure). The reverse presentation is possible, azide (-N_3) on the PVN and the ethynyl group on the peptide.

μM to 0.4 μM over the course of the study. Over this same time the EC_{50} of the PVN formulation decreases from 7.6 μM to 4.6 μM (blue curve). In this particular instance, the cytotoxicity of the PVN formulation is less than that observed for free doxorubicin. However, this is not the case for all cell lines. The degree of uptake *in vitro* depends on the number of receptors (in this case N-Cadherin) on the particular cell type. These data are orphan data that were not used for *in vivo* studies. To move forward to a murine model, it is desirable to pass the breakeven point, where the PVN formulation is at least as cytotoxic as the free drug. Limited experience suggests that cytotoxicity *in vivo* is greater for the PVN than indicated in the *in vitro* studies. *In vitro* studies are inherently limited to relatively short exposure times. In the case of targeted delivery of a sequestered drug, the pharmacokinetics may be delayed since it may take time for the drug to be released from the PVN once it has reached the target.

The experiments shown in Fig. 7 were conducted on HeLa (human cervical cancer) cells purchased from American Type Culture Collection (Rockville, MD). HeLa and HepG2 cells were cultured in Minimum Essential Medium Eagle media containing 10% fetal bovine serum

(Bio-Whittaker, Inc., Walkersville, MD) and 1% penicillin-streptomycin at 37°C in an atmosphere containing 5% CO_2. To plate the cells into a well plate, the culture media was removed, and cells were washed with phosphate buffer twice before harvesting using 5% trypsin solution (Gibco Laboratories, Gaithersburg, MD). Then cells were seeded ~10,000 cells/well in a 6-well plate and incubated at 37°C for 24 h before experiments were performed. The cells were exposed to a 100 μL solution containing the PVN formulation suspended in phosphate buffer saline. Control cells were given only phosphate buffer saline solution.

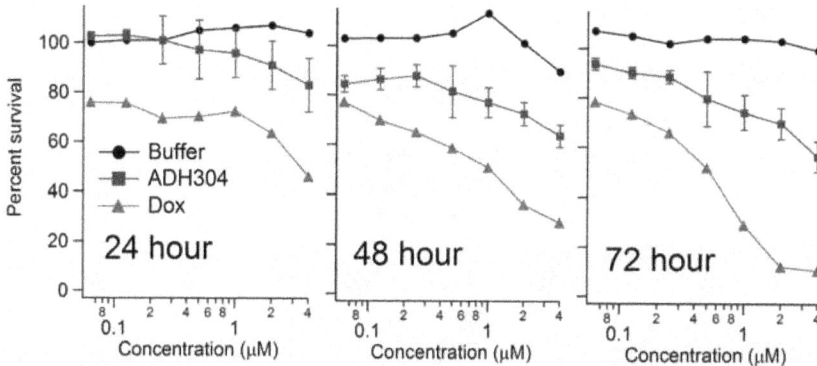

Fig. 7. Dose response curves for the delivery of doxorubicin to HT-1299 cells. HT-1299 is a non-small cell lung carcinoma cell line. The percentage survival of the cells is plotted vs. concentration of the doxorubicin. The buffer control is an aliquot of buffer added to the media at the same volume used for the doxorubicin and $^{DOX}PVN^{ADH304}$ formulation.

THE PVN AS A SEQUESTRATION AGENT

Figure 7 shows that the PVN can target cells with a cytotoxic effect. As with *in vitro* cell lines the effect of a PVN *in vivo* depends upon whether a targeting agent is attached to the surface. While some plant viruses target specific mammalian cell types, e.g., CPMV targets endothelial cells, plant viruses do not have cell targeting sequences by design, since plants do not have cell surface receptors of the type found on most mammalian cells (Koudelka et al. 2007). We have observed that when no targeting agent is present on the nanoparticle surface, there is no noticeable effect *in vivo*. For example, in a recent animal study the mice were given more than 3 times the maximum tolerated dose (MTD) of doxorubicin (Steed et al. 2009). In this study of a murine model without targeting, the doxorubicin was apparently sequestered inside the plant virus sufficiently well that it did not affect the mice at the relatively high dose given. This aspect of PVN technology is quite promising for targeted delivery and agrees with results obtained on HCRSV.

APPLICATION TO AREAS OF HEALTH AND DISEASE

The use of plant viruses to target tumors is a nascent field that has a great deal of potential. Plant viruses are robust protein shells that contain small genomes. In some cases the genome can be removed and the plant virus spontaneously assembles to provide a loosely bound shell. RCNMV is a plant virus that requires its RNA genome for structure. The consequence is that RCNMV is very robust. It can be loaded with chemotherapeutic drugs, such as doxorubicin, in high yield. It can be functionalized on the exterior using targeting peptides.

Key Facts

- Plant viruses are robust protein shells that are found widely in food and in our environment. Yet, they are incapable of infecting humans. Their relatively low immunogenicity and toxicity has made them an important in vaccine development. Natural modes of transmission do not involve targeting.
- Many plant viruses have an icosahedral (nearly spherical) structure with hollow interiors. Plant viruses have dynamic transitions that developed naturally for release of DNA or RNA in cells. Release of cargo in cells is triggered by intracellular conditions, including acidity, and concentration of ions such as calcium and magnesium, but appears to differ for each type of plant virus. Swelling and pore-opening can be used to load plant viruses with therapeutic agents in the laboratory.
- Plant viruses have a regular exterior structure. They can be modified with targeting agents, such as peptides, by bioconjugate chemistry, but also by genetic modification or via binding proteins derived from antibodies, yeast or phage display.
- Plant viruses have a robust structure stable to extremes of pH, temperature and high ionic strength. Red clover necrotic mosaic virus is a soil-borne plant virus that is exceptionally stable because of its adaptation to the soil environment. It has an exceptionally high calcium binding affinity, which increases its stability in various media, and decreases the possibility of drug leakage in a formulation.
- Targeted therapy can provide the specificity of antibodies without necessarily involving a single strong interaction. Avidity of binding is widely used by viruses to attach to cells. Avidity refers to binding by multiple interactions, which may individually be weak but are strong in combination.

Definitions

Avidity: Combined binding strength due to multiple interactions of the same interacting species pair.

Capsid: The protein shell that surrounds the viral genome.

Encapsidation: Triggered assembly of the capsid around a cargo.

Heterobifunctional linker: A chemical agent that can crosslink two different chemical groups, often a targeting agent to a macromolecule, nanoparticle or, in the present application, a plant virus.

Infusion: Triggered opening of surface pores to permit diffusion or a drug or fluorophore into the plant virus interior.

Multi-drug resistance: The mutation and natural selection of cancer cells that express various types of efflux pumps that remove chemotherapy drugs from the cytosol and render them ineffective.

Plant virus nanoparticle: A chemically altered plant virus that contains either interior or external modifications for cell targeting applications.

Summary Points

- RCNMV is a plant virus readily loaded with chemotherapeutic cargo, fluorescent agents or other drugs. The plant virus can also be modified using radioactive tracers, external fluorophores and targeting peptides to create a plant virus nanoparticle (PVN). This approach has the advantage of regular geometry, dynamic transitions for uptake and release of cargo that are "built-in", and relatively high colloidal stability. The fact that humans are exposed to plant viruses in the food supply without acute immune responses or infection suggests that the PVN is a relatively benign protein-nucleic acid shell. Clearly, it is biodegradable. Preliminary indications suggest that the PVN can be taken up by living organisms with relatively little impact.

Abbreviations

CCMV	:	cowpea chlorotic mosaic virus
CP	:	capsid protein
CPMV	:	cowpea mosaic virus
DMSO	:	dimethyl sulfoxide
EDTA	:	ethylene diamine tetraacetic acid
HRCSV	:	hibiscus chlorotic ringspot virus
PVN	:	plant virus nanoparticle
RCNMV	:	red clover necrotic mosaic virus
TA	:	trans-activator

SMCC   :   succinimidyl-4-(N-maleimidomethyl)cyclohexane-1-carboxylate

TABS   :   trans-activator binding site

RNA   :   ribonucleic acid

RME   :   receptor mediated endocytosis

References

Basnayake, V.R., T.L. Sit, and S.A. Lommel. 2006. Virology 345: 532.

Brumfield, S., D. Willits, L. Tang, J.E. Johnson, T. Douglas and M.J. Young. 2004. Gen. Virol. 85: 1049.

Brunel, F.M., J.D. Lewis, G. Destito, N.F. Steinmetz, M. Manchester, H. Stuhlmann, and P.E. Dawson. 2010. Nano Lett., 10: 1093.

Chatterji, A., W.F. Ochoa, M. Paine, B.R. Ratna, J.E. Johnson, and T.W. Lin. 2004. Chem. Biol. 11: 855.

de Rosales, R.T.M., E. Arstad, and P.J. Blower. 2009. Targeted Oncol., 4: 183.

Destito, G., R. Yeh, C.S. Rae, M.G. Finn, and M. Manchester. 2007. Chem. & Biol. 14: 1152.

Gillitzer, E., D. Willits, M. Young, and T. Douglas. 2002. Chem. Comm. 2390.

Guenther, R.H., T.L. Sit, H.S. Gracz, M.A. Dolan, H.L. Townsend, G.H. Liu, W.H. Newman, P.F. Agris, and S.A. Lommel. 2004. Nucleic Acids Res. 32: 2819.

Gupta, S.S., J. Kuzelka, P. Singh, W.G. Lewis, M. Manchester, and M.G. Finn. 2005. Bioconj. Chem. 16: 1572.

Hermanson, G.T. 2008. Bioconjugate Techniques. Elsevier, London.

Huang, X.L., L.M. Bronstein, J. Retrum, C. Dufort, I. Tsvetkova, S. Aniagyei, B. Stein, G. Stucky, B. McKenna, N. Remmes, D. Baxter, C.C. Kao, and B. Dragnea. 2007. Nano Lett. 7: 2407.

Joelson, T., L. Akerblom, P. Oxelfelt, B. Strandberg, K. Tomenius, and T.J.J. Morris. 1997. Gen. Virol. 78: 1213.

Kelland, L. 2007. Drugs of the Future. 32: 925.

Kolb, H.C., M.G. Finn, and K.B. Sharpless. 2001. Angw. Chem. Intl. Ed. 40: 2004.

Koudelka, K.J., C.S. Rae, M.J. Gonzalez, and M.J. Manchester. 2007. Virology 81: 1632.

Liepold, L., S. Anderson, D. Willits, L. Oltrogge, J.A. Frank, T. Douglas, and M. Young. 2007. Magn. Resonance Med. 58: 871.

Loo, L., R.H. Guenther, V.R. Basnayake, S.A. Lommel, and S.J. Franzen. 2006. Am. Chem. Soc. 128: 4502.

Loo, L., R.H. Guenther, S.A. Lommel, and S. Franzen. 2008. Chem. Comm. 88.

Loo, L., R.H. Guenther, S.A. Lommel, and S.J. Franzen. 2007. Am. Chem. Soc. 129: 11111.

Ren, Y., S.M. Wong, and L.Y. Lim. 2007. Bioconj. Chem. 18: 836.

Sherman, M.B., R.H. Guenther, F. Tama, T.L. Sit, C.L. Brooks, A.M. Mikhailov, E.V. Orlova, T.S. Baker, and S.A. Lommel. 2006. J. Virol. 80: 10395.

Singh, P., M.J. Gonzalez, and M. Manchester. 2006. Drug Dev. Res. 67: 23.

Steed, P.M., B. Oberhardt, C. Luft, M. Hu, S.A. Lommel, S. Franzen, D. Guenther, D. Lockney, S.D. Harrison, K. Meshaw, and E. Rainbolt. 2009. American Association for Cancer Research 100th Annual Meeting: Poster 3790.

Tama, F., and C.L. Brooks. 2002. J. Mol. Biol. 318: 733.

Wagner, E., R. Kircheis, and G.F. Walker. 2004. Biomed. Pharmacother. 58: 152.

11

Safety of Nanoparticles in Medicine

Maria Dusinska,[1,]* Lise Marie Fjellsbø,[1] Zuzana Magdolenova,[1] Solveig Ravnum,[1] Alessandra Rinna[1] and Elise Rundén-Pran[2]

ABSTRACT

Among the beneficial applications of nanotechnology, nanomedicine offers perhaps the greatest potential for improving human conditions and quality of life. Nanomaterials (NMs) have unique properties and applications when it comes to drug delivery and imaging, and they have the potential to improve diagnostics and therapy of many human disorders, including neurodegenerative disorders, by their ability to cross the blood-brain-barrier (BBB). The nanosize and the large surface area of nanoparticles (NPs) reflects increased reactivity, and even an inert bulk compound, such as gold, may elicit a response in humans when administered as a NM.

However, concern has been raised that the properties that make NMs unique and so useful could also be coupled to unintentional effects on human health. Thus, it is important to

[1]Norwegian Institute for Air Research (NILU), Health Effects Group, Centre for Ecology and Economics, Instituttvn. 18, N-2027 Kjeller, Norway.

[2] Norwegian Institute for Air Research (NILU), Health Effects Group, Department for Environmental Chemistry, Instituttvn. 18, N-2027 Kjeller, Norway.

*Corresponding author: Email: maria.dusinska@nilu.no; maria.dusinska@szu.sk

List of abbreviations after the text.

obtain information about the potential toxicity of NMs to discover and prevent serious unwanted human effects. The goal must be to realize the great opportunities and benefits of NMs while at the same time minimizing the risk related to their applications. There is a great knowledge gap between nanotechnology and potential toxicity of NMs. There is only scant knowledge about cellular uptake, transport across biological barriers, distribution within the body and possible mechanisms of toxicity. Despite this lack of knowledge, NMs are widely used in research, industry, and medicine. To develop efficient and safe NMs for use in nanomedicine, care must be taken to demonstrate a balance between therapeutic efficacy and safety of the NMs, with the benefits higher compared to risks.

INTRODUCTION

Nanotechnology, by development of engineered nanomaterials, is a rapidly growing industry with great potential and applications in many areas, including nanomedicine. The use of NMs has revolutionized therapeutics and diagnostics. NPs are generally defined as particles with dimensions in the range of 1–100 nm (SCENIHR 2007), and owing to their nanosize they have physico-chemical properties different from those of the bulk material (Table 1).

With increased use of NMs in medicine, major improvements in aspect of human health are expected. However, concern has been raised that the properties that make NMs so useful could also lead to unintended effects on human health. One example is gold, which in bulk form is non-toxic, but in nanosize shows toxic properties (Chen et al. 2008). It is therefore important to find a balance between opportunities and benefits of NMs while at the same time minimizing the risk related to their applications.

Table 1. Physico-chemical properties of NMs.

Characteristics	Explanation	Techniques
Size and size distribution	Size in dry state and when added to media must be measured. Several diameters can be measured, e.g., visual, mobility, hydrodynamical.	TEM, DLS, SEM

Table 1. contd....

Table 1. contd....

Characteristics	Explanation	Techniques
Agglomeration/ Aggregation/ Dispersibility	An agglomerate is a group of particles or aggregates held together by relatively weak forces. The resulting external surface area is similar to the sum of each of the surfaces. An aggregate is a group of strongly bound particles. The resulting external surface area is smaller than the sum of each of the surfaces	SEM, TEM, DLS, FFF
Solubility, persistence and stability	Measured in medium to be sure of what is applied to the test system.	DLS, GF-AAS, ICP-MS, TEM
Chemical composition including impurities	When determining the toxicity, the impurities might have as large impact as the intended material to be tested. This must always be addressed.	EA, ICP-MS, EDX
Shape and morphology	NMs are created in a number of shapes for different use, e.g., spherical, rods and fibre-like.	TEM
Redox potential		Chemical titration, electro-chemical measurements
Surface area and porosity	Increasing surface area (with decreased size) is related with increasing reactivity.	BET
Surface charge and pH, ζ-Potential	Surface charge determines NP behaviour in solution (if they are dispersed in medium or form agglomerates). It also influences ability to traverse biological barriers.	Isoelectric focusing, laser-doppler anemometry
Crystal structure	A crystal structure is composed of a pattern, where a set of atoms are arranged in a particular way. Even if the composition is the same, the material's properties may change due to different crystallinity. Some nanoparticles can have different crystalline structure (e.g., for TiO2 anatase or rutile) or can be amorphous.	XRD, TEM

Scanning electron microscopy (SEM), transmission electron microscopy (TEM), X-ray diffraction (XRD), nuclear magnetic resonance (NMR), infra-red (IR) spectroscopy, elemental analysis, dynamic light scattering (DLS), zeta-potential (ζ-pot) measurements, inductively coupled plasma-mass spectrometry (ICP-MS), graphite furnace atomic absorption spectrometry (GF-AAS), isotherm N2 absorption/desorption measurements (BET), field flow fractionation (FFF), energy dispersive x-ray spectroscopy (EDX).

APPLICATIONS OF NANOMATERIALS TO AREAS OF HEALTH AND DISEASE

Nanomedicine is a field with continuous progress, introducing novel applications in many health care areas (Table 2). The underlying motivation is improvement of quality of life with economic and social benefits. Some of the most promising areas are the following (Jain 2008; Surendiran et al. 2009):

- Nanodiagnostics (molecular diagnostics, imaging using NP-based contrast materials, nanobiosensors)
- Nanopharmaceuticals (targeted drug delivery, nanotechnology-based drugs, implanted nanopumps, nanocoated stents)
- Reconstructive surgery (tissue engineering, implantation of rejection-resistant artificial tissues and organs)
- Nanorobotics (vascular surgery, detection and destruction of cancer)
- Nanosurgery (nanolasers, nanosensors implanted in catheters)
- Regenerative medicine (tissue repair)
- Ultrafast DNA sequencing

Nanocarriers have great potential in the field of drug delivery, medical diagnostics, therapeutics and molecular targeting. The specific targeting of nanocarriers to cells or organs and the controlled release of drug improves efficacy at the target and reduces toxicity in non-target organs (Duncan 2005). NPs composed of biodegradable and biocompatible materials, such as natural or synthetic lipids and polymers (e.g., PLGA), polyethylenimine, poly-L-Lysine or dendrimers, are used as nanocarriers for drug or gene delivery (Mansour 2009). Dendrimers are attractive polymer carriers because of their well-defined nanoscale scaffold. Recent advances with use of NPs, such as the organogold complexes, showed anti-tumour, anti-microbial, anti-malarial, anti-HIV and anti-rheumatoid arthritis effects in humans.

EXPOSURE TO NANOMATERIALS

Possible routes of exposure to NMs are by inhalation, skin contact or orally. However, in nanomedicine the main routes of exposure are by intravenous injection or surgical implantation of novel nanodrugs, tissue replacements and imaging agents (Fig. 1).

Groups at risk of high exposure are patients and medical staff preparing and applying NMs. Also, occupational exposure could occur during manufacture of these materials. Secondary exposure of NMs could occur through waste water after disposal/recycling and release into the environment. Proper exposure assessment is a challenge; techniques have yet to be developed for performing risk assessment for NMs.

Table 2. Examples of nanomaterials in nanomedicine and their applications.

Nanomaterial	Medical application	Description	Product	Disorder
Liposomes: spherical NPs of lipid bilayer with aqueous interior	Drug delivery	Loaded with drug in aqueous or lipid compartment. Rapid degradation can be prevented by coating	Amphotericin Doxorubicin	Fungal infection Ovarian tumour
Nanopores: wafers with high density of pores allowing entry of O_2, glucose and, e.g., insulin	Protection of transplanted tissue from the host immune system		β-pancreas cells	Diabetes
Fullerenes: Carbon allotrope (bucky balls)	Radioactive tracer in imaging	Enclosing of different atoms inside the ball, e.g., metals	Photosensitizers	Cancer
Nanotubes: Sheet of graphite rolled into a cylinder and capped at one or both ends by a bulckyball	Drug carriers	Single-walled carbon nanotubes (SWCNT) or multi-walled (MWCNT)	Indium 111 radionucleotide labelled CNTs Amphotericin	Cancer Antifungal agent
Quantum dots: Nanocrystals getting fluorescent when stimulated with light	Imaging	Tagged with biomolecules	QD conjugated with polyethylene glycol (PEG) and prostate antibody	Prostate cancer diagnosis
Nanoshell: NPs with a core of silica coated with thin metallic shell	Drug delivery Diagnostics	Can be targeted to desired tissue by immunological methods	Gold nanoshell	Chemotherapy Diabetes
Nanobubbles	Drug delivery Imaging	Form microbubbles when heated to physiological temperature and release drug upon ultrasound exposure	Doxorubicin MRX 815	Breast cancer Intravascular clot
Paramagnetic NPs	Diagnostics		Iron oxide NPs	Contrast agents in MRI
Dendrimers: Nanomolecules with regular branching structure	Drug delivery Gene therapy Imaging	Cavities within the dendrimer used for drug transport	PAMAM	Chemotherapy in cancer MRI

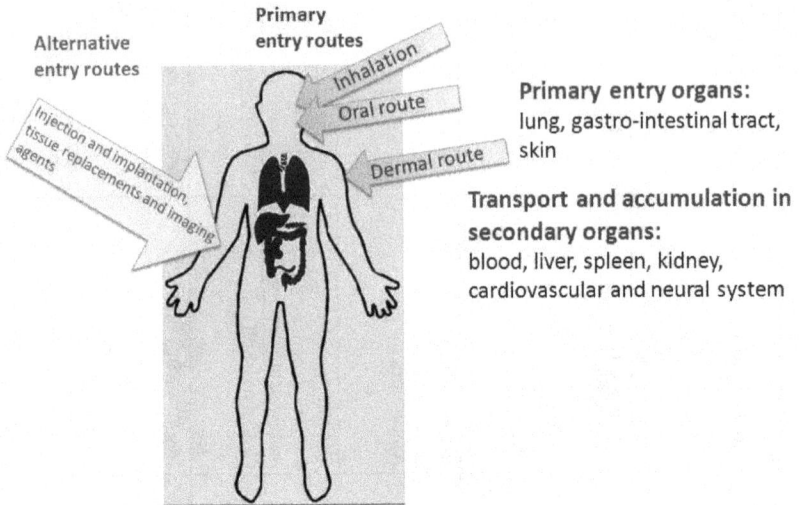

Fig. 1. Human exposure to nanomaterials. Exposure routes, transport and accumulation in organs are shown. Unpublished.

EFFICACY AND POTENTIAL HAZARD OF NANOMATERIALS IN NANOMEDICINE

There is a pressing need to understand how engineered NMs interact with organs, tissues and cells, and what is their bioavailability and biopersistence. How are NMs transported into the body, transferred across biological barriers, and translocated to different tissues? The intrinsic chemical properties of the NMs will strongly influence the kinetics in the body and subsequently target organs and toxicity. The small size and the particle shape enable uptake into blood and lymph circulation and distribution to tissues in the body that normally are protected by barriers, such as the brain by penetration of the BBB. Thus, nanocarriers can be developed and targeted to the brain to administer drugs for neurodegenerative disorders.

Although NMs are designed with high target specificity, and consequently reduced toxicity in non-target cells and tissues, it is important to explore the potential toxic properties before applying a NM to humans. A NM has to be tested at three levels, as is done for all therapeutical drugs. First, tests are performed *in vitro* in cell or tissue cultures. If these tests are promising, further testing is performed *in vivo* in animal models. Though animal testing has major ethical issues, it is essential for studying fate, toxicokinetics and target organs. However, major efforts are made to reduce the number of animals used according to the 3R strategy of replacement, refinement and reduction. The last step in the testing process is the clinical trial, where the drug is tested first on healthy humans, and then on patients.

Despite the limited knowledge on toxicity, NMs are widely used in research, industry, and medicine. An example is nanosilver-based dressings and surgical sutures that control wound infection, although nanosilver is associated with ROS-mediated apoptosis (Hsin et al. 2008).

Most toxicological testing with NMs so far has been performed in *in vitro* systems, and some in *in vivo* models. Very few long-term studies have been performed. Despite these concerns, 130 nanotech-based drugs and delivery systems and 125 devices or diagnostic tests have entered pre-clinical, clinical, or commercial development since 2005 (http:// nanobiotechnews.com).

PHYSICO-CHEMICAL PROPERTIES OF NANOMATERIALS AND THEIR CHARACTERIZATION

A number of characteristics must be determined and taken into account when testing the toxic potential of NMs. The most important properties that may be linked to adverse health effects are listed in Table 1. Physico-chemical properties of NMs might differ between batches or providers, and it is recommended always to include characterization of these materials, particularly size and composition when tested, not least so that results from different studies can be compared. Properties of NMs change in liquid suspension, and precipitation, aggregation or agglomeration could occur (Stone et al. 2010; Bouwmeester et al. 2010). Thus, it is very important to establish standard protocols for physio-chemical characterization of NMs in the test system.

UPTAKE AND MECHANISMS OF TOXICITY

Uptake and Organ Specificity

NPs can penetrate cell membranes and epithelial or endothelial barriers in the body, including the BBB. Cellular uptake mechanisms of NPs include diffusion, phagocytosis, pinocytosis and receptor-mediated endocytosis (Fig. 2). Uptake is highly dependent upon physico-chemical properties and surface characteristics, such as the extent of partition across biological membranes, which is directly related to the polarity of a molecule; nonpolar or lipophilic molecules more easily cross membranes than polar compounds (Faraji and Wipf 2009).

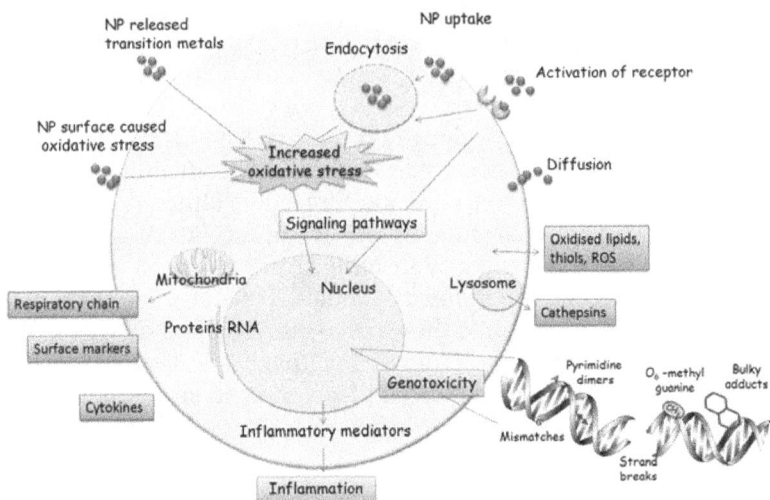

Fig. 2. Schematic illustration of cellular uptake and possible mechanisms of action of nanomaterials. NP, nanoparticle; ROS, reactive oxygen species. Unpublished.

Color image of this figure appears in the color plate section at the end of the book.

NPs can be transported through the blood and accumulate in secondary target tissues and organs such as liver, spleen, kidney, the cardiovascular system and the central nervous system (CNS), where they may cause adverse effects (Oberdorster et al. 2005a).

Transport through Blood-Brain-Barrier and Neurotoxicity

The BBB is constituted by endothelial cells connected by tight junctions, and it regulates transport of solutes and other substances into and out of the brain. It also maintains the homeostasis of the brain microenvironment, which is important for proper neuronal activity and function of the CNS. Conventional drugs are unable to pass the BBB and reach the CNS (Bhaskar et al. 2010). Advances in the field of drug delivery with NMs to the CNS are very important for treatment and diagnostics in neurology. Thus, nanomedicine offers an opportunity for treatment of neurodegenerative disorder (e.g., stroke, epilepsy, Parkinson's disease and Alzheimer's disease), as well as tumours. For some CNS disorders (e.g., HIV, dementia and meningitis), treatment depends upon getting a higher concentration of the drug to the target tissue by increased penetration of the BBB (Bhaskar et al. 2010).

Many conditions, such as size, biocompatibility, specificity, stability in blood and controlled drug release, need to be satisfied for successful delivery to the target. The nanocarrier must also avoid the reticuloendothelial system and activation of the immune system (Olivier 2005). For this purpose NPs can be conjugated with surface biomolecules. An example is PEGylated NPs. Some of the most prominent candidates for biomolecules are siRNAs for genetic therapy and fluorescent dyes for markers. Coated PBCA NPs have been shown to cross the BBB and deliver drugs such as the anti-tumour antibiotic doxorubicin and the NMDA receptor antagonist MRZ 2/576 (De Juan et al. 2006). PEG PHDCA NPs have been used as vectors for drug delivery in an experimental model of prion disease (Calvo et al. 2001). Polymeric NPs have also been used for drug delivery in Alzheimer's disease (Wilson et al. 2010). Stem cells for treatment of stroke and spinal cord injury can be labelled with superparamagnetic iron-oxide NPs and this offers promising perspectives for future treatment (Kubinova and Sykova 2010).

Oxidative Stress and Inflammation

The generation of ROS is the key mechanism by which NPs exert their pro-inflammatory and pro-atherogenic effects on the respiratory and cardiovascular tracts, and this can account also for NP toxicity (Nel et al. 2005). During NP exposure an excess of ROS production can occur and the antioxidant defences available can be overwhelmed. Glutathione is depleted and oxidized glutathione (GSSG) accumulates, resulting in a drop in the GSH/GSSG ratio. Oxidative stress may initiate an inflammatory response, modified cellular redox signaling as well as perturbed mitochondrial function and cell death. Inflammation is induced by upregulation of redox-sensitive transcription factors such as NFkB, AP1, Nrf2 and MAPKs such as ERK, JNK and p38 (Eom and Choi 2010; Hsin et al. 2008).

The potential for inflammatory and pro-oxidant activity is largely dependent on NM surface chemistry and *in vivo* surface modifications, and it has been shown that polymeric NPs composed of PLGA and a novel PLGA derivative provoke a reduced inflammatory response in Balb-C mice (Dailey et al. 2006). As another example, complexes of SWCNTs with single-stranded DNA can be used for very sensitive and specific detection of single-nucleotide polymorphisms (Rajendra et al. 2005) but at the same time SWCNTs were reported to induce inflammation and oxidative stress (Shvedova et al. 2005).

Genotoxicity/Carcinogenicity

Genotoxicity may be produced by direct interaction of NPs with the genetic material, indirectly by ROS production, or by toxic ions released from soluble NPs. Secondary genotoxicity is a result of oxidative DNA attack by ROS via activated phagocytes (neutrophils, macrophages) during NP-elicited inflammation (Stone et al. 2009).

NPs that cross cellular membranes may reach the nucleus through diffusion across the nuclear membrane or transportation through the nuclear pore complexes, and interact directly with DNA. In dividing cells, access is also provided by dissolution of the nuclear envelope during mitosis. For instance, functionalized SWCN or silver NPs have been reported to enter the cell nucleus, and C60 NPs can bind to and deform nucleotides (Singh et al. 2009; Liang et al. 2008).

ROS can induce oxidation of bases in DNA (e.g., 8-oxoG) and strand breaks. These lesions can give rise to mutations and thus be potentially carcinogenic. Transition metal ions, such as Fe^{2+}, Ag^+, Cu^+, Mn^{2+}, Cr^{5+} and Ni^{2+} released from soluble NPs may also contribute to DNA damage by ROS production via the Fenton reaction. Coating of NPs with redox-cycling organic chemicals (such as quinones), and metal impurities in carbon nanotubes can amplify chemical changes in the NM environment. A reaction of H_2O_2 with silver NPs is proposed to cause formation of Ag^+ *in vivo* (AshaRani et al. 2009).

The number of studies on mechanisms of genotoxicity of NPs is increasing, but there are many conflicting results in the literature (Table 3). Different NP characteristics (size, shape, surface properties, composition, solubility, aggregation/agglomeration), NP uptake, and presence of mutagens and transition metals affiliated with the NPs have to be taken into consideration (Table 1). Various other factors such as exposure scenarios (e.g., dose, exposure time, cell type) could influence upon the results of the genotoxicity testing.

In vivo Toxicity

In vivo studies have demonstrated toxic effects of NPs in nanocarrier systems such as drug delivery or gene delivery. Administration of AuNPs in mice by oral and intra-peritoneal routes induced decreases in body weight, spleen index, red blood cells, and hematocrit (Zhang et al. 2010). Exposure of AuNPs to rabbits demonstrated an increased concentration of white blood cells and a distribution of the NPs to the spleen, liver, and implanted liver tumours, but not to other tissues (Glazer et al. 2010). Studies in rats and mice showed that AgNPs administered by inhalation, ingestion or intra-peritoneal injection were detected in blood and caused toxicity in many organs including the brain (Ahamed et al. 2010). A comparative study in *Daphnia magna* demonstrated that AgNPs were much more toxic than AuNPs (Li et al. 2010). Rats exposed to Fe_2O_3 or ZnO NPs showed severe damage in liver and lung tissue and significantly decreased levels of serum enzymes (Wang et al. 2010). The possible toxicity of cadmium-

Table 3. Genotoxicity testing of different nanomaterials used in nanomedicine.

Nanoparticle	Genotoxicity testing method	Result	Cells / organism	References
Single-wall carbon nanotubes (SWCNT)				
SWCNT	Comet assay	+	Primary mouse embryo fibroblast cells	Yang et al. 2009
SWCNT	Comet assay	−	FE1-Muta™Mouse lung epithelial cell line	Jacobsen et al. 2008
	Modified comet assay (FPG)	+		
	CII mutation analysis	+/−		
SWCNT	Alkaline comet assay	+	Chinese hamster lung fibroblast (V79) cell line	Kisin et al. 2007
	Micronucleus assay	−	Chinese hamster lung fibroblast (V79) cell line	
	Bacterial reverse mutation assay	−	*Salmonella typhimurium* strains YG1024/YG1029	
SWCNT	Comet assay	+	Broncho-alveolar lavage (BAL) fluid obtained from apolipoprotein E knockout mouse	Jacobsen et al. 2009
SWCNT	Comet assay	+	BEAS 2B cells, MeT 5A cells	Lindberg et al. 2008
SWCNT	Oxidatively damaged DNA (8-oxoG and dG measured by HPLC with electrochemical and UV detection)	+ in liver and lung − in colon mucosa	Female Fisher 344 rats from Taconic	Folkmann et al. 2009
	mRNA expression of *HO1, MUTYH, NEIL1, NUDT1,* and *OGG1* *OGG1* repair activity	−		
Multi-wall carbon nanotubes (MWCNT)				
MWCNT	Comet assay	+	Human lung epithelial cell line A549	Karlsson et al. 2008
	Modified comet assay (FPG)	−		

Table 3. contd....

Table 3. contd....

Nanoparticle	Genotoxicity testing method	Result	Cells / organism	References
MWCNT	Micronucleus assay _ex vivo_	+	Type II pneumocytes (AT-II)	Muller et al. 2008
	Cytokinesis block micronucleus assay	+	Rat lung epithelial (RLE) cell line	
	MN test in combination with fluorescent _in situ_ hybridization technique	+	Human epithelial cells (MCF-7)	
MWCNT	Comet assay	+	BEAS 2B cells, MeT 5A cells	Lindberg et al. 2008
Baytubes® macrosized	Chromosomal aberration test	-	Chinese hamster lung fibroblasts V79	Wirnitzer et al. 2009
MWCNT agglomerates	Bacterial reverse mutation assay	-	_S. typhimurium_ (strains TA 1535, TA 100, TA 1537, TA 98 and TA 102)	
MWCNT	Western blot analysis (OGG1, Rad 51 and XRCC4)	+	Mouse embryonic stem cells J11 and Aprt+/-3C4	Zhu et al. 2007
	Mutation frequency analysis	+		
C60 fullerenes				
C60 fullerenes	Comet assay	-	FE1-Muta™Mouse lung epithelial cell line	Jacobsen et al. 2008
	Modified comet assay (FPG)	+		
	CII mutation analysis	+/-		
mixture of C60 and C70 fullerite	Bacterial reverse mutation assay	-	_S. typhimurium_ TA100, A1535, TA98,TA1537, _E. coli_ WP2uvrA/ pKM101	Mori et al. 2006
	Chromosomal aberration test	-	Chinese hamster lung cells (CHL/IU)	
C60 fullerenes suspensions: aqu/nC60 EtOH/nC60	Alkaline comet assay	+	Human lymphocytes	Dhawan et al. 2006
		+		
C60 fullerenes	_gpt_ delta assay	+	_gpt_ delta transgenic MEF cells	Xu et al. 2009

Material	Assay/Test	-/+	Cell/organism	Reference
Fullerenes C60	Comet assay	-/+	Broncho-alveolar lavage (BAL) fluid obtained from apolipoprotein E knockout mouse	Jacobsen et al. 2009
C60 fullerenes	Oxidatively damaged DNA (8-oxoG and dG measured by HPLC with electrochemical and UV detection)	+ in liver + in lung - in colon	Female Fisher 344 rats from Taconic	Folkmann et al. 2009
	mRNA expression of *HO1, MUTYH, NEIL1, NUDT1,* and *OGG1*	+ *OGG1* in liver		
	OGG1 repair activity			
C60 fullerenes	Bacterial reverse mutation test	-	*S. typhimurium* TA98, TA100,TA1535, TA1537, *E. coli* WP2*uvrA*/pKM101	Shinohara et al. 2009
	In vitro chromosomal aberration test	-	Chinese hamster lung cells	
	In vivo micronucleus test	-	Male and female mice (bone marrow cells)	
Water-soluble polymer-wrapped fullerenes	Bacterial reverse mutation assay	-	*S. typhimurium* TA98, TA100, TA1535, TA1537, *E. coli* WP2*uvrA*/pKM	Aoshima et al. 2010
	Chromosomal aberration assay	-	Chinese hamster lung cells (CHL/IU)	
Iron NPs				
Fe_3O_4 (magnetite)	Comet assay	-	Human lung epithelial cell line A549	Karlsson et al. 2008
	Modified comet assay (FPG)	+		
$CuZnFe_2O_4$	Comet assay	+	Human lung epithelial cell line A549	Karlsson et al. 2008
	Modified comet assay (FPG)	+		
Fe_2O_3 particles	Comet assay	-	Human lung epithelial cell line A549	Karlsson et al. 2008
	Modified comet assay (FPG)	-		
DMSA-coated maghemite (nano-γFe_2O_3)	Alkaline comet assay	-	Human dermal fibroblasts	Auffan et al.2006
Magnetite NPs	Micronucleus assay	+	Polychromatophilic erythrocytes from female Swiss mice	Freitas et al. 2002

Table 3. contd....

Table 3. contd....

Nanoparticle	Genotoxicity testing method	Result	Cells / organism	References
Iron-platinum (FePt) NPs capped with tetramethylammonium hydroxide	Bacterial reverse mutation assay	Only in TA109 -S9 mix was weakly +	*S. typhimurium* TA98, TA100, A1535, A1537, *E. coli* WP2uvrA/pKM101	Maenosono et al.2007
Magnetite nanoparticles surface coated with polyaspartic acid	Micronucleus assay	+	Polychromatophilic erythrocytes from Swiss mice	Sadeghiani et al. 2005
Silica-coated magnetic NPs containing rhodamine B isothiocyanate	Bacterial reverse mutation assay	-	*S. typhimurium* TA97, TA98, TA100, TA102	Kim et al. 2006
	Chromosomal aberration test	-	Chinese hamster lung fibroblast cells	
Fe_2O_3-NP	Comet assay	+	Human diploid fibroblasts (IMR-90) and SV-40 virus-transformed human bronchial epithelial cells (BEAS-2B)	Bhattacharya et al. 2009
	Oxidative DNA-damage (8-OHdG): 8-OHdG detection by ELISA technique	-	IMR-90	
FePt NPs capped with 2-aminoethanethiol	Bacterial reverse mutation assay	-	*S. typhimurium* (TA98, TA100, TA1535, TA1537), and *E. coli* WP2uvrA⁻	Maenosono et al. 2009
	Chromosomal aberration test	False positive?	Chinese hamster lung fibroblasts (CHL/IU)	
Silver NPs				
Ag NP	*In vivo* rat bone marrow micronucleus test, according to (OECD 476)	-	Sprague-Dawley male and female rats	Kim et al. 2007
Coated (polysaccharide surface functionalized) Ag NPs / Uncoated (non-functionalized Ag NPs	Imunoblot Expression of p53 protein; Phospho-p53 (ser15); Rad51; Phospho-H2AX-Ser-139	Coated more severe damage than uncoated	Mouse embryonic stem (mES) cells Mouse embryonic fibroblasts (MEF) cells	Ahamed et al. 2008

NanoAg AgNO$_3$	Light scattering, absorption spectra and transmission electron microscopy	Weak	Calf thymus DNA (ctDNA)	Chi et al. 2009
NanoAg–CPB AgNO$_3$	Light scattering, absorption spectra and transmission electron microscopy	+	Calf thymus DNA (ctDNA)	Chi et al. 2009
Ag NP - starch capped	Comet assay Cytokinesis block micronucleus assay	+ +	Human lung fibroblast cells (IMR-90) and human glioblastoma cells (U251)	AshaRani et al. 2009
Ag NPs coated with 0.2% PVP (poly vinyl pyrrolidone)	DNA adducts (^{32}P postlabelling)	+	A549 human lung carcinoma epithelial-like cell line	Foldbjerg et al. 2010
Gold NPs				
Nanosized gold particles	Comet assay	-/+	Broncho-alveolar lavage (BAL) fluid obtained from apolipoprotein E knockout mouse	Jacobsen et al. 2009
Quantum dots (QDs)				
CdSe/ZnS QDs (CdSe QDs capped with ZnS shell)	Gel electrophoresis (plasmid nicking assay)	+	Supercoiled double strands of DNA	Green et al. 2005
QDs negatively and positively charged CdTe QDs	Comet assay	+	Broncho-alveolar lavage (BAL) fluid obtained from apolipoprotein E knockout mouse	Jacobsen et al. 2009
Silica NPs				
Crystalline UF-SiO$_2$	Cytokinesis block micronucleus assay Comet assay HPRT gene mutation assay	+ - +	WIL2-NS cells	Wang et al. 2007
SiO$_2$ NPs	Comet assay	+	Primary mouse embryo fibroblast cells	Yang et al. 2009

Table 3. contd....

Table 3. contd....

Nanoparticle	Genotoxicity testing method	Result	Cells / organism	References
TMR- and RuBpy-doped luminescent silica NP	Alkaline comet assay	-	Human lung epithelial cells (A549)	Jin et al. 2007
	Pulsed field gel electrophoresis	-		
	Western blot analysis of DNA adducts	-		
	Agarose gel electrophoresis	-		
	DNA repair enzyme activity assay	-		
Glantreo (30, 80, 400 nm)	Comet assay	-	3T3-L1 fibroblasts	Barnes et al. 2008
Nanofibres (NFs) from natural materials - the cellulose NFs				
Bacterial cellulose NFs	Bacterial reverse mutation assay	-	*S. tryphimurium* TA97a, TA98, TA100, A102	Moreira et al. 2009
	Comet assay	-	CHO cells	

derived QDs has been reviewed (Ghaderi et al. 2010), and emergence of dissolved Cd^{2+} in solution in an exposure study with *Daphnia magna* indicates that the toxicity is likely due to Cd poisoning (Pace et al. 2010).

Promising carriers, such as dendrimers, cause cytotoxicity in human cell lines, haematological toxicity in rats and increased liver enzyme activity in mice (Jain 2010). Also, use of cationic liposomes was limited because of pulmonary toxicity. Careful selection of lipids resulted in nanoliposomes that when tested in murine models proved to be safe and more efficient (Ozpolat et al. 2009).

CNTs are being explored for drug delivery and diagnostic applications, but *in vitro* and *in vivo* studies showed that CNTs may induce oxidative stress or prominent pulmonary inflammation (Kayat et al. 2010). Animal exposure studies in mice and rats indicated that the main pathologic effect is pulmonary fibrosis, due to the physico-chemical properties of CNT. Similarities in mechanisms of toxicity have been suggested with the pathogenic properties of asbestos fibres (Shvedova et al. 2009).

The dependence of NP toxicity on dimension, size, shape and surface chemistry, especially coating, was recently shown in studies where PEG coating significantly reduced the pulmonary toxicity of gold/silica NPs in rats (Adiseshaiah et al. 2009; Goodman et al. 2004). Suitable biochemical functionalization of the CNTs, as well as choice of the appropriate surfactant for dispersion of the CNTs, make the CNTs less cytotoxic (Kayat et al. 2010).

TOXICITY TESTING, VALIDATED TESTS AND REFERENCE MATERIAL

One of the main challenges encountered in toxicity testing of NPs is characterization of NMs, since validated methods and appropriate SOPs are lacking. Additionally, well-characterized and defined NMs to be used as certified reference standards need to be developed as quality controls.

Reference Standards

Both positive and negative controls are needed when toxicity of conventional chemicals is assessed; this applies even more for NM testing. There are several initiatives currently focusing on a selection of NMs and characterization properties to be recommended as reference standards. The OECD list comprises fullerenes, SWCNTs and MWCNTs, carbon black, polystyrene, dendrimers, nanoclays, and NPs of silver, iron, titanium dioxide, cerium oxide, zinc oxide, and synthetic amorphous silica (OECD 2008). Reference standards for nanomedicine products relevant to their application are needed. It is suggested that NMs already available on the market and used in nanomedicine should be chosen as benchmark/ negative control or reference standards. Silver NPs were suggested as potential positive control for nanotechnology generally, as they appear to be associated with relatively high toxicity in a number of studies (Stone et al. 2010).

Validated *In Vitro* and *In Vivo* Models

Toxicity tests should address key physico-chemical properties of the NPs (Table 1), biopersistence, uptake, transport through biological barriers and possible mechanisms of toxicity, such as ROS generation, cell activation, inflammation, immunotoxicity, genotoxicity and other generic endpoints such as reproductive and developmental toxicity. They also should provide target cell–specific endpoints addressing specific organs or tissues to identify the safe dose-range.

Both efficacy and safety tests should be conducted with relevant *in vitro* and *in vivo* models, while at the same time minimizing the number of animals used, following 3R principles. The safe dose-range needs to be established for all endpoints relevant to human health. However, for ADME toxico- and pharmaco-kinetic studies only *in vivo* models can be used.

The biological effects associated with different NPs should be studied using relevant model systems at cellular or tissue level derived from the target organs or *in vivo* directly in specific organs and tissues (Liang et al. 2008). Human model systems derived from human cells are recommended for *in vitro* assays, because they can probably better predict toxicity of NPs in humans (Oberdörster et al. 2005b).

Though *in vitro, ex vivo* and *in vivo* animal experiments are performed to assess efficacy/safety of NMs to increase our understanding of different NMs, risk assessment always needs to address the uncertainty underlying intra- and interspecies differences when interpolating from animal models to human safety, as well as interindividual susceptibility.

REGULATORY PERSPECTIVE WITHIN NANOMEDICINE

At the moment, no specific regulations exist for NMs. In the European Union, substances in nanoform are regulated as chemical by the main chemicals legislation, REACH, which addresses all types of substances, independently of their size, shape or physical state.

Legislation relating to nanopharmaceuticals and nanomedical products and devices is based on established legal principles pertaining to pharmaceuticals, medical products, devices and other biomedical issues.

European legislation for nanomedical products and devices follows the existing regulatory directives of EMEA. However, none of these regulations or directives was written with nanomedical applications in mind, and they will need to be revised to take into account specific risks and challenges posed by NMs in medicine (EMEA 2006).

In the US, NMs are regulated following FDA legislation. The FDA is organized into several centres that specialize in regulating particular types of products. The CDER, CBER and CDRH regulate drugs, devices, and biologics respectively, and are also primarily responsible for regulating nanomedical products.

Within nanomedicine, the unique physico-chemical properties coupled with the nanosize of the materials makes it impossible to predict hazardous potential from data on other compounds, and thus a case-by-case approach is recommended. However, though existing regulatory frameworks in principle cover all important aspects of production and products, different behaviour of NPs from the corresponding bulk material highlights the need for specific nano-policies or adaptation of existing frames.

Key Facts

- Application of nanotechnology in the nanopharma industry and nanomedicine has great potential as it can considerably improve screening, diagnosis and treatment of diseases.
- The unique and beneficial properties of NMs, such as size, increased surface reactivity, chemical composition, bioavailability and adsorption, are also a cause of concern because of possible negative impacts on human health.
- NMs can enter cells, pass through biological barriers and induce cytotoxicity, oxidative stress, inflammation, immunotoxicity, neurotoxicity, genotoxicity and other unintentional effects.
- Biodistribution and mechanisms underlying toxicity are still largely unknown, and dependent upon size as well as chemical composition of the NP.
- There is lack of validated analytical and biological methods as well as certified reference standards for risk assessment for establishing safe doses of NMs.
- There is no special legislation for regulation of NMs. Regulations of nanopharmaceuticals and nanomedical products and devices are based on established EMEA and FDA and national legislations for medical products.

Definitions

Nanogenotoxicology: A subdiscipline of nanotoxicology dealing with genotoxicity of nanomaterial.

Nanomaterial: Collective term for any discrete piece of material with one or more external dimensions in the nanoscale (≤ 100 nm) (SCENIHR 2007).

Nanomedicine: Medical application of nanotechnology using nanoscale material and nanoscale tools for screening, diagnosis, prevention and treatment of diseases and for the understanding of mechanisms and pathophysiological processes of disease.

Nanoparticles: Material that has all three dimensions in the nanoscale (of the order of 100 nm or less) (ISO technical committee 229 (SCENIHR 2007)).

Nanotoxicology: A new discipline of bioscience, a branch of toxicology studying toxicity of nanomaterial.

Reference material: According to ISO, a material "used for calibration, method validation, the establishment of metrological traceability, method development, and other quality control purposes". A certified reference material is a material "accompanied by a certificate, one or more of whose property values are certified by a procedure which establishes traceability to an accurate realization of the unit in which the property values are expressed, and for which each certified value is accompanied by an uncertainty at a stated level of confidence" (ISO/IEC Guide 99:2007).

Summary Points

- Nanomedicine brings humans into direct contact with NPs. The unique properties of NMs are being explored for its beneficial use in nanomedicine, but the same unique physical and chemical properties may be associated with potentially deleterious effects on human health. Despite the lack of knowledge, NMs are widely used in research, industry, and medicine.
- It is essential for ethical, social and regulatory acceptance and public confidence as well as for the nanomedicine industries that appropriate risk assessments be undertaken in relation to health and safety.
- To develop efficient and safe NMs there must be balance between therapeutic efficacy and safety with high ratio of benefits to risks.
- Validated analytical methods (for characterization of NMs) and biological methods (for efficacy/safety assessment), and certified NPs/NMs reference standards (for quality assurance) together with appropriate *in vitro* and *in vivo* models have to be developed.
- Though existing regulatory frameworks in principle cover all important aspects of production and products, NPs often display

different chemical, physical, and biological characteristics from those of the bulk form of the same substance, highlighting the need for specific nano-policies.

Abbreviations

ADME	:	absorption, distribution, metabolism, and excretion
AP1	:	activating protein1
BBB	:	blood-brain barrier
BET	:	Brunauer–Emmett–Teller
C60	:	fullerenes
CBER	:	Center for Biologics Evaluation and Research
CDER	:	Center for Drug Evaluation and Research
CDRH	:	Center for Devices and Radiological Health
CNS	:	central nervous system
CNT	:	carbon nanotubes
CHMP	:	Committee for Medicinal Products for Human Use
DLS	:	dynamic light scattering
EMEA	:	European Medicines Agency
ERK	:	extracellular-signal-regulated kinases
FDA	:	Food and Drug Administration
GSH	:	glutathione
GSSG	:	oxidized glutathione
ISO	:	International Organisation for Standardization
JNK	:	c-Jun N-terminal kinases
MAPK	:	mitogen-activated protein kinases
MRI	:	magnetic resonance imaging
MWCNTs	:	multi-walled carbon nanotubes
NFκB	:	nuclear factor kappa B
NMs	:	nanomaterials
NPs	:	nanoparticles
Nrf2	:	NF-E2 related factor 2
OECD	:	Organisation for Economic Co-operation and Development
PBCA	:	poly(butylcyanoacrylate)
PEG	:	polyethylene glycol
PHDCA	:	polyhexadecyl cyanoacrylate
PLGA	:	poly(lactic-co-glycolic acid)
QDs	:	quantum dots
ROS	:	reactive oxygen species
REACH	:	Registration, Evaluation, Authorisation and Restriction of Chemicals

3R   :   replacement, refinement and reduction
SOPs   :   standard operating procedures
SWCNTs   :   single-walled carbon nanotubes

References

Adiseshaiah, P.P., J.B. Hall, and S.E. McNeil. 2009. Nanomaterial standards for efficacy and toxicity assessment. Nanotechnology 2: 99–112.

Ahamed, M., M.S. Alsalhi, and M.K. Siddiqui. 2010. Silver nanoparticle applications and human health. Clin. Chim. Acta. 411(23–24): 1841–1848.

AshaRani, P.V., G. Low Kah Mun, M.P. Hande, and S. Valiyaveettil. 2009. Cytotoxicity and genotoxicity of silver nanoparticles in human cells. ACS Nano. 3: 279–290.

Bhaskar, S., F. Tian, T. Stoeger, W. Kreyling, J.M. de la Fuente, V. Grazú, P. Borm, G. Estrada, V. Ntziachristos, and D. Razansky. 2010. Multifunctional Nanocarriers for diagnostics, drug delivery and targeted treatment across blood-brain barrier: perspectives on tracking and neuroimaging. Part Fibre Toxicol. 7: 3.

Bouwmeester, H., I. Lynch, H.J.P. Marvin, K.A. Dawson, M. Berges, D. Braguer, H.J. Byrne, A. Casy, G. Chambers, M.J.D. Clift, G. Elia, T. Fernandes, L.M. Fjellesbo, P. Hatto, L. Juillerat, C. Klein, W. Kreyling, C. Nickel, M. Riediker, and V. Stone. 2011. Minimal analytical characterisation of engineered nanomaterials needed for hazard assessment in biological matrices. Nanotoxicology: 5(1): 1–11.

Calvo, P., B. Gouritin, I. Brigger, C. Lasmezas, J. Deslys, A. Williams, J.P. Andreux, D. Dormont, and P. Couvreur. 2001. PEGylated polycyanoacrylate nanoparticles as vector for drug delivery in prion diseases. J. Neurosci. Methods 111: 151–155.

Chen, P.C., S.C. Mwakwari, and A.K. Oyelere. 2008. Gold nanoparticles: From nanomedicine to nanosensing. Nanotechnol. Sci. Appl. 1: 45–66.

Dailey, L.A., N. Jekel, L. Fink, T. Gessler, T. Schmehl, M. Wittmar, T. Kissel, and W. Seeger. 2006. Investigation of proinflammatory potential of biodegradable nanoparticle drug delivery systems in the lung. Toxicol. Appl. Pharmacol. 215: 100–108.

De Juan, B.S., H. Von Briesen, S.E. Gelperina, and J. Kreuter. 2006. Cytotoxicity of doxorubicin bound to poly(butyl cyanoacrylate) nanoparticles in rat glioma cell lines using different assays. J. Drug Target. 14: 614–622.

Duncan, R. 2005. Targeting and intracellular delivery of drugs. pp 163–204. *In:* R.A. Meyers (ed.). 2005. Encyclopedia of Molecular Cell Biology and Molecular Medicine. Wiley-VCH Verlag, GmbH & Co. KGaA, Weinheim, Germany.

EMEA. 2006. Reflection Paper on Nanotechnology-Based Medicinal Products for Human Use, EMEA/CHMP/79769/2006. www.emea.europa.eu/pdfs/human/genetherapy/7976906en.pdf

Eom, H.J., and J. Choi. 2010. p38 MAPK activation, DNA damage, cell cycle arrest and apoptosis as mechanisms of toxicity of silver nanoparticles in Jurkat T cells. Envron. Sci. Technol. 44(21): 8337–8342.

Faraji, A.H., and P. Wipf. 2009. Nanoparticles in cellular drug delivery. Bioorg. Med. Chem. 17: 2950–2962.

Ghaderi, S., B. Ramesh, and A.M. Seifalian. 2010. Fluorescence nanoparticles "quantum dots" as drug delivery system and their toxicity: a review. J. Drug Target. Oct 22. doi:10.3109/1061186X.2010.526227

Glazer, E.S., C. Zhu, A.N. Hamir, A. Borne, C.S. Thompson, and S.A. Curley. 2010. Biodistribution and acute toxicity of naked gold nanoparticles in a rabbit hepatic tumor model. Nanotoxicology (In press).

Goodman, C.M., C.D. McCusker, T. Yilmaz, and V.M. Rotello. 2004. Toxicity of gold nanoparticles functionalized with cationic and anionic side chains. Bioconjug. Chem. 15: 897–900.

Hsin, Y.H., C.F. Chen, S. Huang, T.S. Shih, P.S. Lai, and P.J. Chueh. 2008. The apoptotic effect of nanosilver is mediated by a ROS- and JNK-dependent mechanism involving the mitochondrial pathway in NIH3T3 cells. Toxicol. Lett. 179: 130–139.

Jain, K.K. 2008. Nanomedicine: application of nanobiotechnology in medical practice. Med. Princ. Pract. 17: 89–101.

Kayat, J., V. Gajbhiye, R.K. Tekade, and N.K. Jain. 2011. Pulmonary toxicity of carbon nanotubes: a systematic report (Review article). Nanomedicine: 7(1): 40–9.

Kubinova, S., and E. Sykova. 2010. Nanotechnology for treatment of stroke and spinal cord injury. Nanomedicine (Lond) 5: 99–108.

Li, T., B. Albee, M. Alemayehu, R. Diaz, L. Ingham, S. Kamal, M. Rodriguez, and S.W. Bishnoi. 2010. Comparative toxicity study of Ag, Au, and Ag-Au bimetallic nanoparticles on Daphnia magna. Anal. Bioanal. Chem. 398(2): 689–700.

Liang, X.J., C. Chen, Y. Zhao, L. Jia, and P.C Wang. 2008. Biopharmaceutics and therapeutic potential of engineered nanomaterials. Curr. Drug. Metab. 9: 697–709.

Mansour, H.M., Y-S. Rhee, and X. Wu. 2009. Nanomedicine in pulmonary delivery. Int. J. Nanomed. 4: 299–319.

Nel, A. 2005. Atmosphere air pollution-related illness: Effects of particles. Science 308: 804–806

Oberdörster, G., E. Oberdörster, and J. Oberdörster. 2005a. Nanotoxicology: an emerging discipline evolving from studies of ultrafine particles. Environ. Health Persp. 113: 823–839.

Oberdörster, G., A. Maynard, K. Donaldson, V. Castranova, J. Fitzpatrick, K. Ausman, J. Carter, B. Karn, W. Kreyling, D. Lai, S. Olin, N. Monteiro-Riviere, D. Warheit, H. Yang, and ILSI Research Foundation/Risk Science Institute Nanomaterial Toxicity Screening Working Group. 2005b. Principles for characterizing the potential human health effects from exposure to nanomaterials: elements of a screening strategy. Part Fibre Toxicol. 2: 8.

OECD. 2008. OECD Work on the Safety of Manufactured Nanomaterials. Environment, Health and Safety Division. Environment Directorate. OECD. www.oecd.org/dataoecd/54/27/41567645.ppt

Olivier, J.C. 2005. Drug transport to brain with targeted nanoparticles. NeuroRx 2: 108–119.

Ozpolat, B., A.K. Sood, and G. Lopez-Berestein. 2009. Nanomedicine based approaches for the delivery of siRNA in cancer. J. Intern. Med. 267: 44–53.

Pace, H.E., E.K. Lesher, and J.F. Ranville. 2010. Influence of stability on the acute toxicity of CdSe/ZnS nanocrystals to Daphnia magna. Environ. Toxicol. Chem. 29(6): 1338–1344.

Rajendra, J., and A. Rodger. 2005. The binding of single-stranded DNA and PNA to single-walled carbon nanotubes probed by flow linear dichroism. Chemistry 11: 4841–4847.

Scientific Committee on Emerging and Newly-Identified Health Risks (SCENIHR). 2007. The existing and proposed definitions relating to products of nanotechnologies. Technical Guidance Documents for new and existing substances for assessing the risks of nanomaterials. European Commission, Brussels, Belgium.

Shvedova, A.A., E.R. Kisin, R. Mercer, A.R. Murray, V.J. Johnson, A.I. Potapovich, Y.Y. Tyurina, O. Gorelik, S. Arepalli, D. Schwegler-Berry, A.F. Hubbs, J. Antonini, D.E. Evans, B.K. Ku, D. Ramsey, A. Maynard, V.E. Kagan, V. Castranova, and P. Baron. 2005. Unusual inflammatory and fibrogenic pulmonary responses to single-walled carbon naotubes in mice. Am. J. Physiol. Lung Cell Mol. Physiol. 289: 698–708.

Shvedova, A.A., E.R Kisin, D. Porter, P. Schulte, V.E. Kagan, and B. Fadeel. 2009. Mechanisms of pulmonary toxicity and medical applications of carbon nanotubes: two faces of Janus. Pharmacology & Therapeutics 121: 192–204.

Singh, N., B. Manshian, G.J. Jenkins, S.M. Griffiths, P.M. Williams, T.G. Maffeis, C.J. Wright, and S.H. Doak 2009. NanoGenotoxicology: the DNA damaging potential of engineered nanomaterials. Biomaterials 30: 3891–3914.

Stone, V., H. Johnston, and R.P. Schins. 2009. Development of *in vitro* systems for nanotoxicology: methodological considerations. Cri. Rev. Toxicol. 39: 613–626.

Stone, V., B. Nowack, A. Baun, N. van den Brink, F. von der Kammer, M. Dusinska, R. Handy, S. Hankin, M. Hassellöv, E. Joner, and T.F. Fernandes. 2010. Nanomaterials for environmental studies: Classification, reference material issues, and strategies for physico-chemical characterization. Sci. Total Environ. 408(7): 1745–1754.

Surendiran, A., S. Sandhiya, S.C. Pradhan, and C. Adithan. 2009. Novel applications of nanotechnology in medicine. Indian J. Med. Res. 130: 689–701.

Wang, L., L. Wang, W. Ding, and F. Zhang. 2010. Acute toxicity of ferric oxide and zinc oxide nanoparticles in rats. J. Nanosci. Nanotechnol. 10(12): 8617–8624.

Wilson, B., M.K. Samanta, K. Santhi, K.P. Kumar, M. Ramasamy, and B. Suresh. 2010. Chitosan nanoparticles as a new delivery system for the anti-Alzheimer drug tacrine. Nanomedicine 6: 144–152.

Zhang, X.D., H.Y. Wu, D. Wu, Y.Y. Wang, J.H. Chang, Z.B. Zhai, A.M. Meng, P.X. Liu, L.A. Zhang, and F.Y. Fan. 2010. Toxicologic effects of gold nanoparticles *in vivo* by different administration routes. Int. J. Nanomed. 5: 771–781.

Section 2: Applications to Health

12

Pulmonary Drug Delivery with Nanoparticles

Moritz Beck-Broichsitter,[1] Thomas Schmehl,[1]
Werner Seeger[1] and Tobias Gessler[1,]*

ABSTRACT

The human respiratory system comprises the conducting airways responsible for air transport and the respiratory zone where gas exchange takes place. By nature, the respiratory system is directly accessible from the outside. Therefore, drug administration via inhalation offers the possibility to treat respiratory diseases locally. Aerosol therapy possesses several advantages over other routes of drug application, like rapid onset of drug action, high local drug concentration, minimized systemic exposure and reduced side effects. Moreover, because of its large alveolar surface and the thin epithelial barrier with extensive vascularization, the lung is an ideal avenue for non-invasive systemic drug therapy.

Disadvantages of conventional inhalation therapy are the short duration of drug action and the insufficient selectivity of drug effects at the organ and cellular level. At this point, nanotechnology opens new perspectives in the design of drug delivery vehicles for pulmonary application. Nano-sized drug carriers with various

[1]Medical Clinic II, Department of Internal Medicine, Justus-Liebig-Universität Giessen, Klinikstrasse 36, D-35392 Giessen, Germany.

*Corresponding author; Email: tobias.gessler@innere.med.uni-giessen.de

List of abbreviations after the text.

novel physical, chemical and biological properties facilitate targeting of an organ, tissues, cells or subcellular compartments as well as modification of the duration and intensity of pharmacological effects. Among such carriers, biocompatible polymeric nanoparticles have become an attractive concept owing to their advantageous drug delivery properties, including ability to be aerosolized and stability during nebulization. Systematic manipulation of nanoparticle composition and characteristics, including functionality of polymers, size of particles, surface charge, chemistry and area, way of drug loading, enables the design of "intelligent" carriers specific for targeting to different tissues or cells. Diverse manufacturing techniques are known for the preparation of drug-loaded polymeric nanoparticles, with the particular choice depending on the physicochemical properties of the polymeric nanoparticle matrix material, the active compound to be encapsulated, and the therapeutic goals to be reached. Different preclinical models are used to investigate the mechanisms of nanoparticle action at the target site; *ex vivo* and *in vivo* models allow the measurement of lung-specific pharmacokinetics and pharmacodynamics. Pulmonary toxicity of nanoparticles is related to total surface area and surface characteristics, and further investigations of acute and chronic toxicological effects of inhaled therapeutic nanoparticles are necessary.

In summary, polymeric nanoparticles are promising tools for lung application representing a solid basis for future advancement in nanomedicine strategies for pulmonary drug delivery. In order to further optimize the design of polymeric nanoparticles, profound knowledge of chemistry and pharmaceutical technology, lung morphology, physiology and pathology, as well as aerosol technology and aerosol physics is an essential prerequisite.

INTRODUCTION

Inhalation therapy allows direct application of a drug to the respiratory system. The local deposition of the administered drug enables a site-specific treatment of respiratory diseases and offers several advantages over other routes of administration, like rapid onset of drug action, high local drug concentration and lower systemic exposure with reduced side effects. As an example, the delivery of β_2-agonists and corticosteroids to the airways by inhalation has improved the treatment of well-known respiratory diseases like bronchial asthma and chronic obstructive pulmonary disease (Courrier et al. 2002). Furthermore, prostacyclin analogues have recently been established for inhalation therapy of life-threatening pulmonary hypertension providing superior pulmonary selectivity compared to oral or intravenous administration of these vasodilators (McLaughlin et al. 2010). In addition, inhalation represents a non-invasive alternative for systemic delivery of pharmaceuticals. The large alveolar surface area and the thin epithelial air-blood barrier allow rapid adsorption of drugs deposited in the alveolar region.

The short duration of drug action at the target site is considered a significant disadvantage of aerosol therapy. Moreover, conventional inhalation therapy does not permit targeted drug delivery to tissues or cells, and a modification of drug distribution at the organ and cellular level is only poorly achievable. Avoiding short-term and non-selective effects of conventional inhalation therapy, the development of pulmonary controlled release formulations would improve the convenience and compliance of patients (Gessler et al. 2008).

In this context, nanomedicine has attracted much attention for controlled and targeted drug delivery. Nanotechnology offers novel prospects in the design of more sophisticated drug delivery vehicles that facilitate targeting of an organ, tissues, cells or subcellular compartments. Moreover, nano-sized drug carriers have been shown to modulate the duration and intensity of pharmacological effects at the target site (Farokhzad and Langer 2009; Beck-Broichsitter et al. 2011). Among the large number of drug delivery systems conceived for pulmonary application, nanoparticles are of special interest as drug carriers for the treatment of respiratory diseases, as they enable a prolonged drug release and a cell-specific targeted drug delivery (Kurmi et al. 2010). A successful drug delivery system needs to demonstrate sufficient association of the therapeutic agent with the carrier particles, controlled and targeted drug release properties as well as biocompatibility and low toxicity of the employed excipients. Nanoparticles composed of polymers with particular physicochemical and biological properties, such as degradability and compatibility, have been predominantly used. In addition, nanoparticles intended for pulmonary delivery need to meet further standards, such as protection of the drug against degradation, ability to be transferred into an aerosol and aerosolization stability. Nanoparticles made from biodegradable polymers fulfil the stringent requirements placed on these delivery systems (Lebhardt et al. 2010; Beck-Broichsitter et al. 2011).

This chapter gives an overview of the application of polymeric nanoparticles for pulmonary drug delivery. Anatomy and physiology of the lung as well as methods for nanoparticle preparation and characterization are discussed. Current results from *ex vivo* and *in vivo* studies are included to underline the unique potential of nanoparticles as drug delivery vehicles for the treatment of respiratory diseases.

STRUCTURE AND FUNCTION OF THE LUNG

In order to develop successful drug delivery systems for pulmonary applications, it is essential to understand and consider the nature of the lung in its healthy as well as various diseased states (Yu and Chien 1997). The respiratory tract is divided into the conducting airways and the respiratory zone. The conducting airways act as air transport system and include the nasal cavity, pharynx, larynx, trachea, bronchi and bronchioles. In the respiratory zone, i.e., respiratory bronchioles and alveoli, gas exchange takes place (Fig. 1). The conducting airways exhibit 16 bifurcations, followed by another six bifurcations of the respiratory bronchioles representing the passage to the respiratory zone where the alveolar ducts with alveolar sacs finally branch off (Weibel 1963). The surface area of the respiratory zone has been shown to be enormous. Furthermore, the thickness of the air-blood barrier in the alveolar region

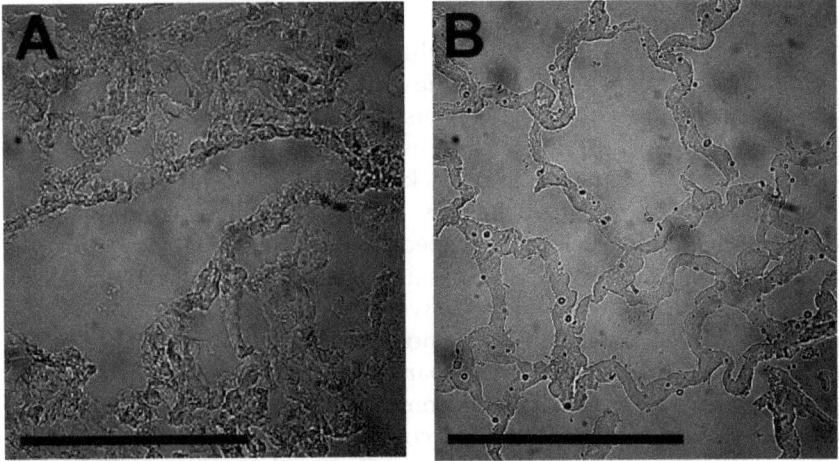

Fig. 1. Light photomicrographs of native lung slices showing the morphology present in respiratory region of the lung (respiratory bronchioles (A) and alveoli (B)). Frozen parts of the lung tissue were cut into thin slices using a microtome and imaged using a light microscope. Scale bars: 100 µm. (Unpublished material of the authors.)

(≤ 1 µm) is much smaller than that in the tracheo-bronchial airways (≥ 10 µm) facilitating rapid gas exchange with the blood (Patton and Byron 2007). The airways play an essential role for warming, humidifying and cleaning of the inhaled air. Inhaled particles may deposit on the mucus layer that coats the walls of the conducting airways. The mucus is secreted by goblet and submucosal gland cells and forms an adhesive, viscoelastic layer consisting of glycoproteins and water as the major components. Ciliated cells are the most abundant cells in the bronchial epithelia at all levels of the conducting region. Their main function is the transport of mucus out of the lung by ciliary motion (mucociliary clearance, bronchiotracheal escalator) (Marriot 1990).

The alveolar epithelium is coated by a complex surfactant lining that reduces the surface tension to prevent collapse of the alveoli during breathing. Pulmonary surfactant contains approximately 90% lipids and 10% proteins. The lipids in the surface lining material consist mainly of phospholipids (80–90%). About half of the protein mass of the alveolar lining layer is composed of the surfactant-associated proteins SP-A, SP-B, SP-C and SP-D. While the high molecular weight hydrophilic proteins SP-A and SP-D play a fundamental role in host defense, the low molecular weight hydrophobic proteins SP-B and SP-C are responsible for the decrease of surface tension in the alveolar region to values of ~0 mN/m during compression/expansion cycles (Creuwels et al. 1997). Such extremely low surface tensions can only be achieved by a complex interaction between phospholipids and SP-B and SP-C (Fig. 2).

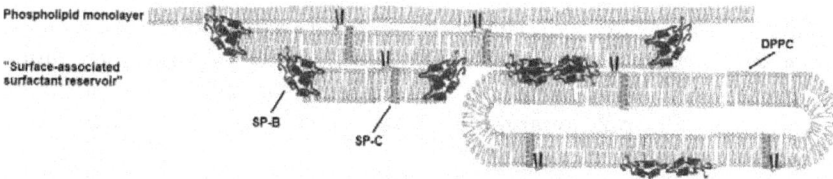

Fig. 2. Model of a multilayer interfacial surfactant film. During expiration (surfactant film compression) a monolayer highly enriched with phospholipids (dipalmitoylphosphatidylcholine (DPPC)) is formed by a "squeeze out" process of excess surfactant material into the bulk phase (monolayer purification). The removal of surfactant components generates a "surface-associated surfactant reservoir", which is maintained in close contact with the interface through an interaction between phospholipids surfactant-associated proteins (SP-B and SP-C). SP-B initiates protrusion of different membrane layers while SP-C sustains close association of the different surfactant layers, thereby allowing a rapid re-entry of surfactant upon inspiration (surfactant film expansion). (Modified from Perez-Gil 2008 with permission of the copyright holder.)

Pulmonary surfactant is the secretory product of cuboidal type II pneumocytes, which regulate alveolar fluid balance, coagulation and fibrinolysis, as well as host defence. These cells can proliferate and differentiate into type I cells (Fehrenbach 2001). Type I pneumocytes are thin cells with a large extension, covering over ~95% of the alveolar epithelial surface, thus forming the primary diffusion barrier between air and blood. While small hydrophobic solutes are known to pass this barrier by diffusion, it is poorly permeable for large hydrophilic substances (peptides and proteins). Macromolecules are absorbed by active transport mechanisms (Patton and Byron 2007). In addition to epithelial cells, the alveoli contain freely roaming macrophages that remove solid particles from the alveolar region (macrophage clearance) (Brain 1992).

APPLICATIONS TO OTHER AREAS OF HEALTH AND DISEASE

Progress in biotechnology, medicine and pharmacy has provided a variety of new macromolecular therapeutics. Despite these tremendous advances, the systemic application of peptides and proteins is limited owing to their known instability and low permeability when administered through the widely preferred oral route. As a result, most protein and peptide formulations on the market are for administration via injection. The parenteral route of application, however, generally does not meet with convenience and compliance of patients, in particular because the indication for the use of these agents is usually treatment of a chronic disease requiring frequent painful injections. Furthermore, a sometimes necessary continuous venous access is associated with possible side effects and complications such as bleeding, mispuncture or infections. Inhalation therapy offers a promising non-invasive alternative for systemic drug therapy. The human lung provides a large alveolar area of ~150 m^2 that is enveloped by a capillary network less than 1 μm beneath the epithelial surface, from which many agents can be readily absorbed to the bloodstream (Patton and Byron 2007). However, the systemic application of macromolecules via inhalation has suffered set-backs, as for example demonstrated for pulmonary-administered insulin (Exubera®) which was withdrawn from the market in 2008.

PREPARATION OF POLYMERIC NANOPARTICLES

Due to considerable interest in nanoparticles as drug delivery systems over the past decades, different strategies have been proposed for nanoparticle production. Nanoparticles used for drug delivery applications are solid or semisolid colloidal particles ranging in sizes from approximately 10 to 1000 nm, composed of macromolecular materials. The particular choice of the nanoparticle preparation technique essentially depends on the physicochemical properties of the polymer and on the drug to be encapsulated in the nanoparticles, as well as on the therapeutic goal to be reached. Conventionally, the techniques used to prepare polymeric nanoparticles are classified into two groups. The first technique involves nanoparticle preparation through various polymerization reactions of monomers, whereas the second is based on precipitation of natural, semisynthetic or synthetic polymers. Emulsification solvent evaporation, emulsification solvent diffusion, salting out, and nanoprecipitation are typical examples of the second group (Vauthier and Bouchemal 2009). Besides nanoparticle preparation in aqueous media, additional methods have been conceived to prepare submicron polymeric particles by spray drying techniques (Fig. 3). Therefore, polymer solutions are atomized into small droplets that are dried using a warmed gas. The breakdown can either be performed by perforated membranes that are actuated by piezoelements ("vibrating mesh nozzles") or by electrohydrodynamic atomization ("electro-spraying") (Peltonen et al. 2010).

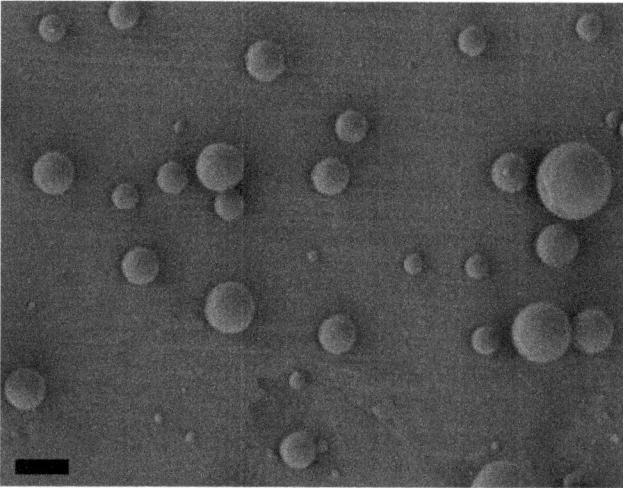

Fig. 3. Scanning electron photomicrograph of spray-dried polymeric particles. Monolithic submicron poly(lactide-*co*-glycolide) particles were prepared from a polymer solution in methylene chloride. Scale bar: 1 µm. (Unpublished material of the authors.)

The employed manufacturing technique determines the gained nanoparticle structure (Fig. 4). Nanocapsules have a core containing the drug surrounded by a polymeric shell (core-shell structure). These reservoir systems can be prepared by interfacial polymerization, interfacial polymer deposition or a layer-by-layer approach. Nanospheres represent matrix-systems that are fabricated either by emulsification or direct precipitation of a polymer solution. For targeting purposes the surface of nanoparticles may be coated by hydrophilic polymers (passive targeting) or decorated with antibodies (active targeting) (Vauthier and Bouchemal 2009).

Criteria such as biocompatibility and degradability determine the selection of an appropriate polymeric nanoparticle matrix material. Aliphatic polyesters like poly(lactide-*co*-glycolide) (PLGA) are the most extensively used materials for biomedical applications owing to their known biocompatibility and biodegradability (Anderson and Shive 1997). Under physiological conditions PLGA is hydrolytically degraded to lactic and glycolic acid that are eliminated through the Krebs cycle. The degradation rate is affected by the polymer composition, the molecular weight, the particle size and the environmental conditions. However,

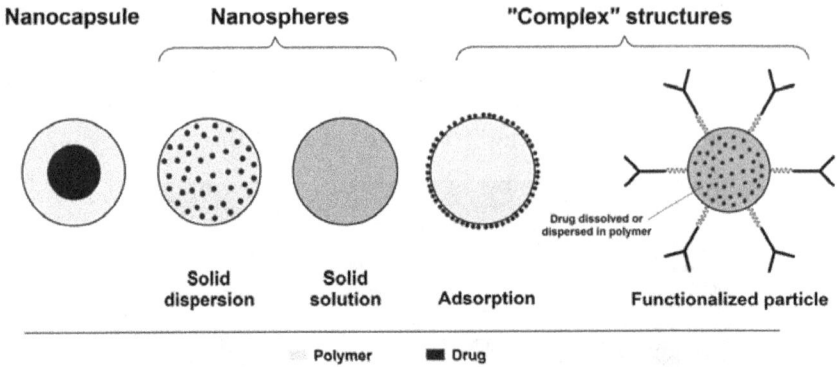

Fig. 4. Different types of drug-loaded nanoparticles. The large number of techniques proposed for nanoparticle preparation allows an extensive modulation of nanoparticle characteristics. (Unpublished material of the authors.)

a common disadvantage of linear polyesters that limits their use as nanoparticle matrix materials is the slow degradation rate (weeks to months), leading to an unwanted accumulation in the body when repeated administrations are needed. Therefore, branched polyesters with fast degradation rates have been synthesized for drug delivery applications (Dailey et al. 2005).

Moreover, for an effective nanoparticulate drug delivery system, sufficient drug loading and controlled drug release over a predetermined period of time must be ensured. Nanoparticles prepared from hydrophobic polymers (PLGA) often incur the drawback of poor drug incorporation due to the low affinity of the drug compounds to the polymers. Improving the design of nanoparticulate carriers can be accomplished by the synthesis of functional polymers that promote interactions with drugs, e.g., electrostatic interactions or host-guest complexes (Dailey et al. 2005).

The active compound can be associated with the colloidal carrier in various physical states, depending on the employed manufacturing technique. Nanoparticle production in the presence of the drug yields solid solutions or solid dispersions of the drug in the polymer matrix. Alternatively, the association of drug and polymer is achieved by subsequent sorption of the drug to unloaded nanoparticles (Soppimath et al. 2001). The characteristics of drug release from drug delivery systems are influenced by the type of employed encapsulation technique and the physicochemical properties (interaction) of drug and polymer. If the drug is dissolved or dispersed in the polymer matrix, its release characteristics may mainly depend on the biodegradation rate of the polymer (Fig. 5).

The release of drugs from biodegradable polymers is controlled by diffusion of the drug through the matrix and by biodegradation of the polymer. Thus, the release rate of drugs from biodegradable nanoparticles can be varied according to the physicochemical properties of the employed polymer (degradation rate). The release rate of drugs from nanoparticles is also strongly influenced by the biological environment as nanoparticles are known to interact with biological components like proteins and cells. As a consequence, the *in vitro* drug release characteristics may not predict the release situation *in vivo* (Soppimath et al. 2001).

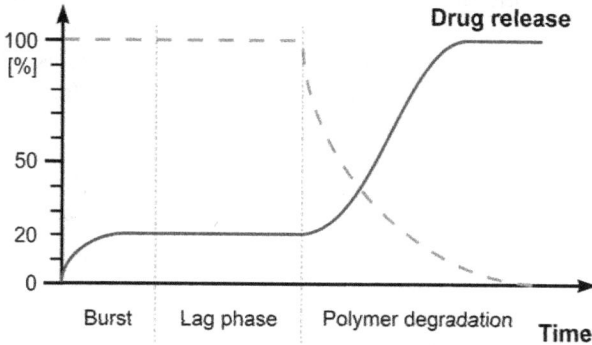

Fig. 5. Schematic illustration of the drug release profile from biodegradable nanoparticles. The figure illustrates the relationship between polymer degradation and drug release kinetics. (Unpublished material of the authors.)

PULMONARY DRUG DELIVERY USING POLYMERIC NANOPARTICLES

Pulmonary drug delivery offers the chance for a targeted treatment of lung diseases, with avoidance of high-dose systemic exposures. This basic concept of local therapy has been used for a long time in the therapy of airway diseases. Disadvantages of aerosol therapy, e.g., short-term and non-selective effects, necessitate the development of more sophisticated drug delivery systems (Yu and Chien 1997). Meanwhile, diverse carrier systems have been considered for the controlled delivery of therapeutic compounds to the lung. While most of the products being developed are composed of microparticles, interest in polymeric nanoparticles has grown owing to their advantageous pulmonary drug delivery properties (Kurmi et al. 2010).

FORMULATION ASPECTS AND AEROSOL CHARACTERISTICS

The advantageous drug delivery properties of polymeric nanoparticles encouraged researchers to develop suitable application forms for inhalation purpose. The appropriate technology to generate nanosized aerosols out of polymeric nanoparticle formulations, however, is yet insufficiently developed. Therefore, most approaches with inhaled therapeutic nanoparticles employ conventional medical aerosol devices to generate aerosols suitable for pulmonary deposition.

Following inspiration, a particle can be transported towards the walls of the airways, where it will be deposited upon contact with the airway surfaces. The pulmonary deposition of inhaled particles is governed by the following three physical mechanisms: inertial impaction, gravitational sedimentation and Brownian diffusion. Therefore, it is evident that deposition not only depends on particle characteristics (e.g., size, density, shape) but also on the breathing pattern determining gas volume and flow rate as well as residence time of a particle in the respiratory tract. The aerodynamic diameter (d_a) of an aerosol particle is defined in terms of its aerodynamic behavior with consideration of shape, size and density. In the range below d_a of approximately 150 nm, deposition predominantly occurs because of diffusional transport. For particles with d_a larger than 150 nm, sedimentation becomes an increasingly effective deposition mechanism. Impaction of particles increases with aerodynamic diameter and gas flow rate, governing deposition of particles with d_a >10 µm in the upper respiratory tract. To target smaller airways and the alveolar region, d_a between 1 µm and 5 µm is preferable. Moreover, the deposition pattern within the respiratory tract, i.e., regional distribution and concentration of inhaled particles, is also influenced by lung morphology (Jaafar-Maalej et al. 2009).

Drug-loaded nanoparticles for inhalation can be generated by nebulization of nanosuspensions or aerosolization of nanoparticle-containing microparticles (composite microparticles). Nebulizers can produce aerosol droplet sizes from nanosuspensions suitable for deposition in the deeper lung. Vibrating mesh nebulizers are preferred for the delivery of biodegradable nanoparticles because they prevent nanoparticle aggregation during the aerosolization process (Lebhardt et al. 2010; Beck-Broichsitter et al. 2011). For dry powder application, polymeric nanoparticles need to be encapsulated into composite microparticles with defined aerodynamic properties to obtain peripheral lung deposition (Fig. 6). The delivery of nanoparticles as part of microparticles has been intensively investigated for several reasons. Common obstacles that limit the use of biodegradable nanoparticles are their chemical (hydrolytic degradation and drug leakage) and physical (aggregation) instability in aqueous suspension. The most commonly used method to stabilize biodegradable nanoparticles is spray drying in the presence of stabilizers like sugars or polymers. Moreover, this technique offers the advantage that nanoparticles are transformed into respirable microparticles in a one-step process. Composite microparticles release the unaffected nanoparticles when they get into contact with water (Lebhardt et al. 2010; Beck-Broichsitter et al. 2011).

Fig. 6. Scanning electron photomicrograph of spray- dried polymeric nanoparticle containing microparticles (composite microparticles). Composite microparticles were prepared from an aqueous nanosuspension containing 100 nm poly(styrene) nanoparticles. Scale bar: 1 µm. (Unpublished material of the authors.)

Several examples for the delivery of nanoparticles as part of composite microparticles can be found in the literature. Tomoda et al. (2009) encapsulated drug-loaded biodegradable nanoparticles into inhalable trehalose microparticles using a spray drying technique. A study performed by Ohashi et al. (2009) investigated the encapsulation of drug-loaded nanoparticles into mannitol microparticles using a similar method. Porous nanoparticle-aggregate particles (PNAP) prepared by spray drying were introduced by Tsapis et al. (2002) as dry powder delivery vehicles for pulmonary application.

PARTICLE DISTRIBUTION AFTER PULMONARY APPLICATION

Concerning the tissue distribution of polymeric nanoparticles after pulmonary application, size and form of the particles play an important role (Gill et al. 2007). In general, microparticles are rapidly cleared from the lung. Depending on the deposition pattern, microparticles are eliminated by the bronchiotracheal escalator or engulfed by macrophages. In contrast, nanoparticles with particle sizes below 250 nm circumvent mucociliary and macrophage clearance and show long residence times until degradation or translocation by epithelial cells. Several studies indicate that the majority of polymeric nanoparticles remain in the lung after pulmonary delivery and only small amounts are translocated to the systemic circulation (Nemmar et al. 2005; Liu et al. 2009). The shape of polymeric nanoparticles may also be a factor that determines lung distribution. Polymeric fibers enable a regionally targeted deposition in the respiratory tract due to their geometry, since they have the additional deposition mechanism of interception, contrary to spherical particles (Jaafar-Maalej et al. 2009). Fibers could be preferentially deposited in different regions of the respiratory tracts, e.g., alveolar or tracheo-bronchial region, depending on their aspect ratio. With geometrically suitable dimensions of fibers, engulfment by alveolar macrophages would be ruled out and prolonged particle retention in the lung can be achieved (Kurmi et al. 2010).

PULMONARY TOXICITY OF POLYMERIC CARRIERS

The same properties that make nanoparticulate drug delivery systems so attractive may on the other hand be problematic when regarding the safety and toxicology of these systems after pulmonary application. While the chemical composition and concentration of inhaled particles obviously play an essential role in the induced toxicity, it has recently been shown that the total surface area of particles is also important (Dailey et al. 2006; Gill et al. 2007). Moreover, positively charged nanoparticles showed a higher tendency to provoke pulmonary responses than negatively charged nanoparticles (Hamoir et al. 2003). These observations raise further questions as to whether material surface properties or surface area play the more important role in inducing inflammatory responses in the lung. PLGA is one of the most common biodegradable polymers used for drug delivery purposes. However, due to its extremely slow rate of biodegradation, PLGA is considered unsuitable for pulmonary drug delivery, especially when frequent dosing is required (Dailey et al. 2005, 2006; Lebhardt et al. 2010; Beck-Broichsitter et al. 2011). The adjustable properties of branched polyesters make them highly suitable for pulmonary formulations, especially with regard to biodegradation rates and pulmonary toxicity (Dailey et al. 2005, 2006).

Despite many studies focusing on the toxicological responses of pulmonary administered polymeric nanoparticles, their influence on the biophysical properties of pulmonary surfactant is largely unknown (Gill et al. 2007). Having the importance of the lung surfactant layer in mind, factors affecting surfactant function might cause severe side effects. Not surprisingly, a number of inhaled substances were shown to cause considerable disturbances of pulmonary surfactant function (Zuo et al. 2008). The mechanisms involved in surfactant inhibition by particulate matter are not well understood. Nanoparticles were shown to alter the structure and function of pulmonary surfactant *in vitro* and provoke surfactant dysfunction by interaction with surfactant constituents. The total surface area of nanoparticles appears as important as surface charge (ζ-potential) and surface hydrophobicity for the observed surfactant inhibition. The ability of polymeric nanoparticles to disturb pulmonary surfactant function is thought to be due to an adsorption process of the surfactant-associated proteins (SP-B and SP-C) to the surface of nanoparticles. Adsorbed surfactant proteins are unable to organize the surfactant structure at the air-water interface and thus are unable to reduce surface tension in a similar way (extent and time course) as native surfactant material (Beck-Broichsitter et al. 2010b).

EVALUATION OF INHALED NANOPARTICLE PERFORMANCE AT THE TARGET SITE

A basic concern in the field of nanomedicine is the development of successful nanoparticulate controlled release formulations with the aim to improve the characteristics of the therapeutic agent at the target site. Biological environments are known to strongly influence the release properties of nano-sized drug delivery vehicles. As a consequence, conventional *in vitro* drug release studies may have very little in common with the delivery and release situation *in vivo* and the development of more sophisticated controlled drug release carriers to the lung is precluded (Beck-Broichsitter et al. 2011).

Different preclinical models are used to investigate the fate of nanoparticles after pulmonary administration. These range from *in vitro* cell culture models used as primarily screening tools to *in vivo* methods that provide fundamental pharmacokinetic and pharmacodynamic information about the applied drug formulation at the site of action (Sakagami 2006; Beck-Broichsitter et al. 2011).

Ex vivo isolated, perfused and ventilated lung models (IPL) allow the measurement of lung-specific pharmacokinetic effects of inhaled therapeutics. The basic techniques of IPL for pharmacokinetic measurements include lung isolation, perfusion and ventilation, the delivery of the formulation to the air-space and an adequate sample analysis. These preparations maintain structural integrity of the lung tissue and allow careful control of the experimental regimen of the isolated lung. Drugs can be administered directly to the respiratory tract in a quantitative and reproducible manner by instillation or inhalation, and simple sampling and analysis of perfusate provide the absorptive profile (Sakagami 2006; Beck-Broichsitter et al. 2011).

The IPL technique was employed to study the pulmonary absorption and distribution characteristics of a hydrophilic model drug after aerosolization as solution or entrapped into polymeric nanoparticles (Beck-Broichsitter et al. 2009). Nebulization of the nanosuspension was performed using a vibrating mesh nebulizer. The nebulization process generated adequate droplet sizes for deposition in the respiratory region of the IPL, and no signs of nanoparticle aggregation were observed during the aerosolization process. Although the *in vitro* drug release from nanoparticles was fast, a pulmonary retention of the encapsulated model drug was observed as reflected by a reduced concentration in the perfusate after aerosol deposition of equal amounts of the model compound.

The performance of novel salbutamol-loaded biodegradable nanoparticles intended for pulmonary application has been characterized in an IPL (Beck-Broichsitter et al. 2010a). Salbutamol and salbutamol-loaded nanoparticles were administered to the IPL by nebulization. The drug absorption from the air space into the perfusate during the IPL experiments was characterized by two distinct phases for both formulations (Fig. 7): an initial continuous increase of salbutamol concentration was followed by a slower second absorption phase. The drug absorption profile during the initial phase from salbutamol-loaded nanoparticles hardly differed from those of salbutamol solution. The following second absorption phase, however, disclosed differences between the two formulations. Significant slower drug absorption from salbutamol-loaded nanoparticles compared to salbutamol solution was observed and added up to retention of salbutamol in the lung tissue. The absorption profile and the significantly decreased total recovery for salbutamol from nanoparticles (49%) compared to the total recovery for salbutamol from solution (75%) have been attributed to internalization of salbutamol-loaded nanoparticles into lung cells (Beck-Broichsitter et al. 2010a).

In addition, a number of *in vivo* studies underline the potential of nanoparticles as controlled drug delivery vehicles to the lung (Table 1) (Lebhardt et al. 2010). Rytting et al. (2010) prepared salbutamol-loaded nanoparticles from a new type of branched polyester. The *in vitro* release of salbutamol from nanoparticles was completed within 2.5 h. Drug transport studies across cell monolayers revealed a delayed transport of salbutamol for drug-loaded nanoparticles. Finally, a prolonged bronchoprotection was observed *in vivo* for salbutamol-loaded nanoparticles. Ohashi et al. (2009) encapsulated rifampicin-loaded nanoparticles into composite microparticles (3.2 μm) using a spray-drying technique. Nanoparticles released from composite microparticles were retained in rat lungs for prolonged periods allowing recognition of nanoparticles by alveolar macrophages. Sung et

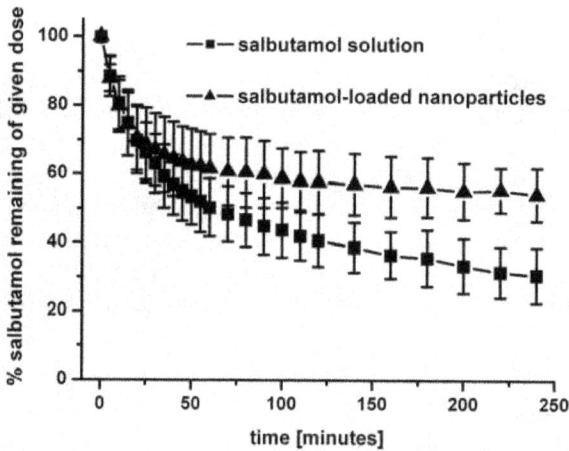

Fig. 7. Pulmonary retention profile for salbutamol solution and salbutamol-loaded nanoparticles in the isolated, perfused and ventilated rabbit lung. The time course clearly illustrates a significant retention of salbutamol in the lung tissue for salbutamol-loaded nanoparticles compared to salbutamol solution. (Data from Beck-Broichsitter et al. 2010a, unpublished figure.)

Table 1. Animal studies employing nanoparticles as drug delivery systems to the lung.

Nanoparticle material	Drug	Formulation	Application method	Biological responses	Ref.
Branched polyester	Salbutamol	Nanosuspension	Nebulization	Prolonged effect of 2 h *in vivo*	Rytting et al. 2010
Poly(lactide-*co*-glycolide)	Rifampicin	Composite microparticles	Powder insufflation	Sustained particle retention in lungs, increased particle uptake by alveolar macrophages	Ohashi et al. 2009
Poly(lactide-*co*-glycolide)	Rifampicin	Composite microparticles (PNAP)	Powder insufflation	Retention of rifampicin in lungs, systemic levels up to 8 h	Sung et al. 2009

Drug-loaded nanoparticles intended for pulmonary application preferentially consist of biodegradable polymers like polyesters. Pulmonary delivery of nanoparticles can be performed by instillation or nebulization when nanoparticles are applied from nanosuspensions, or by dry powder insufflation or inhalation when nanoparticles are incorporated into composite microparticles. Parameters for monitoring the nanoparticle effects are drug or particle retention in the lung, sustained systemic drug levels or improved pharmacodynamics.

al. (2009) prepared PNAP composed of rifampicin-loaded nanoparticles for pulmonary drug delivery. Composite microparticles had suitable aerodynamic properties for efficient deposition in the respiratory tract. Although the *in vitro* drug release exhibited an initial burst of rifampicin, pharmacokinetic studies performed *in vivo* demonstrated an increased retention of rifampicin in the lung tissue. Delivery of rifampicin-loaded PNAP by inhalation achieved systemic drug levels detected for up to 8 h.

Overall, these results demonstrate that polymeric nanoparticles are promising tools for lung application and present a solid basis for future advancement in nanomedicine strategies for pulmonary drug delivery.

Summary Points

- The human respiratory system is classified into two parts; the conducting airways responsible for air transport and the respiratory zone where gas exchange takes place. Pulmonary surfactant, containing mainly phospholipids and surfactant-associated proteins, decreases the surface tension in the alveolar region of the lung.
- Two distinct clearance mechanisms of the respiratory system (bronchiotracheal escalator in the ciliated airways and alveolar macrophages in the non-ciliated respiratory region) are responsible for the removal of deposited particulate matter.
- The respiratory system is by nature directly accessible. Therefore, drug administration via inhalation offers the possibility to treat respiratory diseases locally.
- Because of its large alveolar surface and the thin, highly permeable epithelial barrier with extensive vascularization, the lung is also an ideal avenue for systemic non-invasive drug therapy.
- Aerosol therapy possesses several advantages over other routes of drug administration, like rapid onset of drug action, high local drug concentration, lower systemic exposure and reduced side effects.
- To overcome the drawbacks of conventional inhalation therapy (short duration of drug action and non-selectivity at the organ and cellular level), nanoparticles have been proposed for the controlled and targeted delivery of therapeutic and diagnostic compounds to the desired site of action.
- Biodegradable polymeric nanoparticles combine biocompatibility with necessary physicochemical properties for aerosol delivery (e.g., sufficient drug encapsulation, stability and preserved drug association during nebulization) and desired pharmacological effects.
- Diverse manufacturing techniques are known for the preparation of drug-loaded polymeric nanoparticles. PLGA are the most

extensively used polymers for nanoparticle production owing to their biocompatibility and biodegradability.

- The use of branched and functional polymers may improve drug loading efficiency and degradability as well as sustained and controlled release properties of nanoparticles designed for pulmonary application. Diffusion and biodegradation govern the process of drug release from polymeric nanoparticles.

- Deposition of inhaled particles in the respiratory tract is mainly governed by inertial impaction, gravitational sedimentation and Brownian diffusion. Therefore, particle characteristics (e.g., size, density, shape) and a subject's breathing pattern play an important role. To target the respiration region, aerosol particles should have an aerodynamic diameter between 1 and 5 μm.

- Drug-loaded nanoparticles are presently applied to the lung by nebulization of nanosuspensions and by aerosolization of composite microparticles. Composite microparticles are prepared to stabilize nanoparticles for extended periods of time using standard techniques like spray drying.

- Microparticles are rapidly cleared from the lung; nanoparticles circumvent clearance mechanisms of the lung and show prolonged residence times in the lung.

- Pulmonary toxicity of nanoparticles is related to total surface area and surface characteristics. Nanoparticles interact with the pulmonary surfactant system by adsorption of surfactant compounds. Further investigations of acute and chronic toxicological effects of inhaled therapeutic nanoparticles are necessary.

- Different preclinical models are used to investigate the mechanisms of nanoparticle action at the target site. *Ex vivo* (IPL) and *in vivo* models allow the measurement of lung-specific pharmacokinetic and pharmacodynamic effects of inhaled therapeutics.

- Numerous studies document the unique properties and the potential of polymeric nanoparticles for controlled and targeted drug delivery to the lung.

Key Facts about Inhalation Therapy

- The respiratory tract is directly accessible for therapy with airborne drugs.

- Inhalation therapy is a common method to treat respiratory diseases like bronchial asthma, chronic obstructive pulmonary disease or pulmonary hypertension.

- In addition, biopharmaceuticals can successfully be delivered to the systemic circulation via inhalation, as demonstrated for insulin.

- Devices for aerosolization of drugs comprise pressurized metered dose inhalers (pMDI, drug solution or suspension in liquid propellant, breath- or patient-actuated release of a predefined dose through a nozzle), dry powder inhalers (DPI, micronized drug particles in a powder reservoir, dispersion of the particles by the gas flow of the inhaling patient) and nebulizers (jet, ultrasonic and vibrating mesh nebulizers, aqueous drug solution or suspension aerosolized either by compressed air through a nozzle, mechanical oscillations from a piezoelectric crystal or by a vibrating plate with micron-sized holes).
- Extent and site of drug deposition is mainly determined by particle-related and physiological factors.
- Relevant particle characteristics are size, shape, density, charge and hygroscopicity. Physiological parameters include a subject's breathing pattern and the morphology of the respiratory tract.
- Drug action at the target site is dependent on the specific drug and applied formulation. In addition, clearance mechanisms (mucociliary and macrophage clearance) influence the pharmacokinetics and pharmacodynamics of inhaled and deposited pharmaceuticals.
- Advantages of inhalation therapy are rapid onset of drug action, high local drug concentration and lower systemic exposure with reduced side effects.

Definitions

Aerodynamic diameter: The geometric diameter of a sphere of unit density (1 g/cm^3) that has the same settling velocity as the particle itself.

Aerosol: Solid, liquid or composite particles dispersed in a gas. Aerosols are two-phase systems comprising both the particles and the gas in which the particles are dispersed.

Controlled release: Regulation of the spatial and temporal release of therapeutic agents from a carrier system.

Drug targeting: The directed delivery of a pharmaceutical to a confined area within the body.

Polymeric nanoparticles: Solid or semisolid colloidal particles ranging from approximately 10 to 1000 nm in diameter, composed of macromolecular materials.

Abbreviations

d_a	:	aerodynamic diameter
DPPC	:	dipalmitoylphosphatidylcholine
IPL	:	isolated, perfused and ventilated lung

PLGA   :   poly(lactide-*co*-glycolide)
PNAP   :   porous nanoparticle-aggregate particles
SP   :   surfactant-associated protein

References

Anderson, J.M., and M.S. Shive. 1997. Biodegradation and biocompatibility of PLA and PLGA microspheres. Adv. Drug Deliv. Rev. 28: 5–24.

Beck-Broichsitter, M., J. Gauss, C.B. Packhäuser, K. Lahnstein, T. Schmehl, W. Seeger, T. Kissel, and T. Gessler. 2009. Pulmonary drug delivery with aerosolizable nanoparticles in an *ex vivo* lung model. Int. J. Pharm. 367: 169–178.

Beck-Broichsitter, M., J. Gauss, T. Gessler, W. Seeger, T. Kissel, and T. Schmehl. 2010a. Pulmonary targeting with biodegradable salbutamol-loaded nanoparticles. J. Aerosol Med. 23: 47–57.

Beck-Broichsitter, M., C. Ruppert, T. Schmehl, A. Günther, T. Betz, U. Bakowsky, W. Seeger, T. Kissel, and T. Gessler. 2010b. Biophysical investigation of pulmonary surfactant surface properties upon contact with polymeric nanoparticles *in vitro*. Nanomedicine. Epub ahead of print, doi:10.1016/j.nano.2010.10.007.

Beck-Broichsitter, M., T. Schmehl, W. Seeger, and T. Gessler. 2011. Evaluating the controlled release properties of inhaled nanoparticles using isolated, perfused and ventilated lung models. J. Nanomaterials (in press).

Brain, J.D. 1992. Mechanisms, measurement, and significance of lung macrophage function. Environ. Health Perspect. 97: 5–10.

Courrier, H.M., N. Butz, and T.F. Vandamme. 2002. Pulmonary drug delivery systems: recent developments and prospects. Crit. Rev. Ther. Drug Carrier Syst. 19: 425–498.

Creuwels, L.A.J.M., L.M.G. Van Golde, and H.P. Haagsman. 1997. The pulmonary surfactant system: biochemical and clinical aspects. Lung 175: 1–39.

Dailey, L.A., M. Wittmar, and T. Kissel. 2005. The role of branched polyesters and their modifications in the development of modern drug delivery vehicles. J. Control. Release 101: 137–149.

Dailey, L.A., N. Jekel, L. Fink, T. Gessler, T. Schmehl, M. Wittmar, T. Kissel, and W. Seeger. 2006. Investigation of the proinflammatory potential of biodegradable nanoparticle drug delivery systems in the lung. Toxicol. Appl. Pharmacol. 215: 100–108.

Farokhzad, O.C., and R. Langer. 2009. Impact of nanotechnology on drug delivery. ACS Nano 3: 16–20.

Fehrenbach, H. 2001. Alveolar epithelial type II cell: defender of the alveolus revisited. Respir. Res. 2: 33–46.

Gessler, T., W. Seeger, and T. Schmehl. 2008. Inhaled prostanoids in the therapy of pulmonary hypertension. J. Aerosol Med. 21: 1–12.

Gill, S., R. Löbenberg, T. Ku, S. Azarmi, W. Roa, and E.J. Prenner. 2007. Nanoparticles: characteristics, mechanism of action, and toxicity in pulmonary drug delivery—a review. J. Biomed. Nanotechnol. 3: 107–119.

Hamoir, J., A. Nemmar, D. Halloy, D. Wirth, G. Vincke, A. Vanderplasschen, B. Nemery, and P. Gustin. 2003. Effect of polystyrene particles on lung microvascular permeability in isolated perfused rabbit lungs: role of size and surface properties. Toxicol. Appl. Pharmacol. 190: 278–285.

Jaafar-Maalej, C., V. Andrieu, A. Elaissari, and H. Fessi. 2009. Assessment methods of inhaled aerosols: technical aspects and applications. Expert Opin. Drug Deliv. 6: 941–959.

Kurmi, B.D., J. Kayat, V. Gajbhiye, R.K. Tekade, and N.K. Jain. 2010. Micro- and nanocarrier-mediated lung targeting. Expert Opin. Drug Deliv. 7: 781–794.

Lebhardt, T., S. Rösler, M. Beck-Broichsitter, and T. Kissel. 2010. Polymeric nanocarriers for drug delivery to the lung. J. Drug Deliv. Sci. Tech. 20: 171–180.

Liu, Y., A. Ibricevic, J.A. Cohen, J.L. Cohen, S.P. Gunsten, J.M.J. Frechet, M.J. Walter, M.J. Welch, and S.L. Brody. 2009. Impact of hydrogel nanoparticle size and functionalization on *in vivo* behavior for lung imaging and therapeutics. Mol. Pharmaceutics 6: 1891–1902.

Marriott, C. 1990. Mucus and mucociliary clearance in the respiratory tract. Adv. Drug Deliv. Rev. 5: 19–35.

McLaughlin, V.V., R.L. Benza, L.J. Rubin, R.N. Channick, R. Voswinckel, V.F. Tapson, I.M. Robbins, H. Olschewski, M. Rubenfire, and W. Seeger. 2010. Addition of inhaled treprostinil to oral therapy for pulmonary arterial hypertension: a randomized controlled clinical trial. J. Am. Coll. Cardiol. 55: 1915–1922.

Nemmar, A., J. Hamoir, B. Nemery, and P. Gustin. 2005. Evaluation of particle translocation across the alveolo-capillary barrier in isolated perfused rabbit lung model. Toxicology 208: 105–113.

Ohashi, K., T. Kabasawa, T. Ozeki, and H. Okada. 2009. One-step preparation of rifampicin/poly (lactic-co-glycolic acid) nanoparticle containing mannitol microspheres using a four-fluid nozzle spray drier for inhalation therapy of tuberculosis. J. Control. Release 135: 19–24.

Patton, J.S., and P.R. Byron. 2007. Inhaling medicines: delivering drugs to the body through the lungs. Nature Rev. Drug Discov. 6: 67–74.

Peltonen, L., H. Valo, R. Kolakovic, T. Laaksonen, and J. Hirvonen. 2010. Electrospraying, spray drying and related techniques for production and formulation of drug nanoparticles. Expert Opin. Drug Deliv. 7: 705–719.

Perez-Gil, J. 2008. Structure of pulmonary surfactant membranes and films: The role of proteins and lipid-protein interactions. Biochim. Biophys. Acta. 1778: 1676–1695.

Rytting, E., M. Bur, R. Cartier, T. Bouyssou, X. Wang, M. Krüger, C.M. Lehr, and T. Kissel. 2010. *In vitro* and *in vivo* performance of biocompatible negatively-charged salbutamol-loaded nanoparticles. J. Control. Release 141: 101–107.

Sakagami, M. 2006. *In vivo, in vitro* and *ex vivo* models to assess pulmonary absorption and disposition of inhaled therapeutics for systemic delivery. Adv. Drug Deliv. Rev. 58: 1030–1060.

Soppimath, K.S., T.M. Aminabhavi, A.R. Kulkarni, and W.E. Rudzinski. 2001. Biodegradable polymeric nanoparticles as drug delivery devices. J. Control. Release 70: 1–20.

Sung, J.C., D.J. Padilla, L. Garcia-Contreras, J.L. VerBerkmoes, D. Durbin, C.A. Peloquin, K.J. Elbert, A.J. Hickey, and D.A. Edwards. 2009. Formulation and pharmacokinetics of self-assembled rifampicin nanoparticle systems for pulmonary delivery. Pharm. Res. 26: 1847–1855.

Tomoda, K., T. Ohkoshi, K. Hirota, G.S. Sonavane, T. Nakajima, H. Terada, M. Komuro, K. Kitazato, and K. Makino. 2009. Preparation and properties of inhalable nanocomposite particles for treatment of lung cancer. Colloids Surf. B. 71: 177–182.

Tsapis, N., D. Bennett, B. Jackson, D.A. Weitz, and D.A. Edwards. 2002. Trojan particles: large porous carriers of nanoparticles for drug delivery. Proc. Natl. Acad. Sci. USA. 99: 12001–12005.

Vauthier, C., and K. Bouchemal. 2009. Methods for the preparation and manufacture of polymeric nanoparticles. Pharm. Res. 26: 1025–1058.

Weibel, E.R. 1963. Morphometry of the Human Lung. Springer, Berlin.

Yu, J., and Y.W. Chien. 1997. Pulmonary drug delivery: physiologic and mechanistic aspects. Crit. Rev. Ther. Drug Carrier Syst. 14: 395–453.

Zuo, Y.Y., R.A.W. Veldhuizen, A.W. Neumann, N.O. Petersen, and F. Possmayer. 2008. Current perspectives in pulmonary surfactant—Inhibition, enhancement and evaluation. Biochim. Biophys. Acta. 1778: 1947–1977.

Carbon Nanotubes and Infectious Diseases

Yitzhak Rosen,[1], Brandon Mattix,[2] Apparao Rao[3]*
and Frank Alexis[2]

ABSTRACT

Carbon nanotubes (CNTs) have unique physical and chemical properties. Many of the therapeutic entities that are carried with CNTs cannot enter cells by themselves. CNTs along with these delivered entities have shown the ability to cross the plasma membrane into the cytoplasm relatively easily and allow the delivered entities to perform their desired intracellular function. Delivering multiple and diverse payloads of molecules is an important characteristic that may be critical for dealing with the challenges of infectious diseases. These challenges include drug resistance, toxicity of delivered products, and inability to deliver therapeutic agents into hard-to-reach and targeted areas. This chapter reviews the unique advantages of CNTs, their relevance to drug delivery, their applications to infectious diseases, and toxicity.

[1]Superior NanoBioSystems LLC, 1725 T Street, suite 31, Washington DC, 20009, USA.;
Email: yitzhakrosen@yahoo.com

[2]Department of Bioengineering. Rhodes Research Center, #203. Clemson University, Clemson, South Carolina, 29634.

[3]Department of Physics. 202C Kinard Labs, Clemson University. Clemson, SC 29634-097.

*Corresponding author

List of abbreviations after the text.

INTRODUCTION

Carbon nanotubes (CNTs) possess many unique characteristics making them potentially ideal for a variety of applications. CNTs can be functionalized, which allows them to act as drug delivery systems that can be customized to carry multiple payloads that may allow an effective therapy. This ability may be applicable for challenges seen in infectious diseases and cancer (Rosen et al. 2009).

Infectious diseases have a huge morbidity and mortality toll. There are many critical challenges in treating infectious diseases that need to be addressed. Some of the challenges refer to the unique characteristics of the infectious agent, while others refer to the patient. It is important to integrate both as effective agents for certain infectious agents that may be too toxic for the patient (Fauci et al. 2008). An effective drug delivery system represents an interface that may integrate both kinds of challenges.

A critical challenge in developing therapeutics for infectious diseases is drug resistance such as that due to the constant modification mechanisms such as gene modifications and genetic entity transfers from different sources. This is further exacerbated as drug resistance can occur very quickly, while the introduction of novel therapeutic agents to reduce or prevent resistance may take many years. For infections that have rapid, life-threatening manifestations, the delay in introducing new medications may be disastrous (Rosen et al. 2009; Fauci et al. 2008; Arias et al. 2008; Cohen et al. 1992).

Another challenge is creating vaccines to induce a desired and broad array of antibodies especially at appropriate levels, as not all vaccines can achieve these objectives. A classic example is the HIV vaccine, which has yet to reach these objectives (Mandell et al. 2004; Fauci et al. 2008).

Drug delivery represents another important challenge particularly in fungal infections as many of the current drugs are highly toxic. Furthermore, delivering medications into desired areas such as areas of the brain via the blood-brain barrier, sequestered abscesses, cysts, or microabscesses represents an important challenge (Rosen et al. 2009; Fauci et al. 2008).

Another important challenge is the delivery of host defense peptides, which are involved in the direct killing of bacteria as well as having roles in antiviral and immunomodulatory functions, particularly in patients who have scant supply of them (Rosen et al. 2009; Agerberth et al. 2006).

CNTs can represent a model that may be used to address these challenges as they have the ability for targeted delivery of combinational and synergistic payloads; however, it must be noted that CNTs have toxicity issues and a pharmacological profile that require further characterization and assessment before they are implemented ubiquitously in the clinical milieu (Rosen et al. 2009; Kostarelos 2008; Lacerda et al. 2008a). While it is beyond the scope of this chapter to discuss all of these topics in great detail, this chapter provides a discussion of the unique characteristics of CNTS, the solutions that they can provide as drug delivery systems in infectious disease, and some of the recent work on toxicity issues.

APPLICATIONS TO OTHER AREAS OF HEALTH AND DISEASE

While this chapter focuses on CNTs and their application in infectious diseases, many of the challenges seen in infectious diseases, such as multiple drug resistance, can be seen in other pathological states, such as cancer (Fauci et al. 2008). In addition, many of the therapeutic agents used in cancer are toxic when administered systemically. Moreover, some of these agents are not able to accumulate into tumors. Therefore, an effective drug delivery system may be critical in addressing these challenges. (Bianco et al. 2008; Liu et al. 2007).

Delivering genetic entities such as small interference RNAs (siRNAs) is an emerging and potentially promising field that is being applied in a variety of fields such as cancer and infectious disease. Delivering this entity requires an effective, targeted drug delivery system. CNTs have been shown to effectively deliver siRNAs (Al-Jamal et al. 2010; Podesta et al. 2009; Krajcik et al. 2008). Moreover, CNTs can be surface functionalized with a targeting molecule such as antibody, so that targeted delivery can occur (Bianco et al. 2008).

PHYSICAL PROPERTIES OF CARBON NANOTUBES

Carbon is a remarkable element that co-exists in several stable forms such as graphite, diamond, fullerenes, nanotubes (CNTs), graphene, and nanohorns (Fig. 1). The topology and dimension of these carbon forms, also known as allotropes of carbon, result in unique physical and chemical properties that continue to draw significant interest from the scientific community. Until 1985, carbon was thought to co-exist only as sp^2 bonded graphite, or sp^3 bonded diamond. The carbon atoms are arranged in covalently bonded 2D (graphite) and 3D (diamond) networks in these allotropes. In 1985, a remarkable discovery of a third allotrope of carbon was reported by a group of researchers led by Richard Smalley and Robert Curl (Rice University) and Harry Kroto (University of Sussex). Based on mass spectroscopic studies of their laser-vaporized soot, they reported a third allotrope of carbon that comprised 60 covalently bonded carbon atoms, and resembled the shape of a soccer-ball with 12 pentagonal and

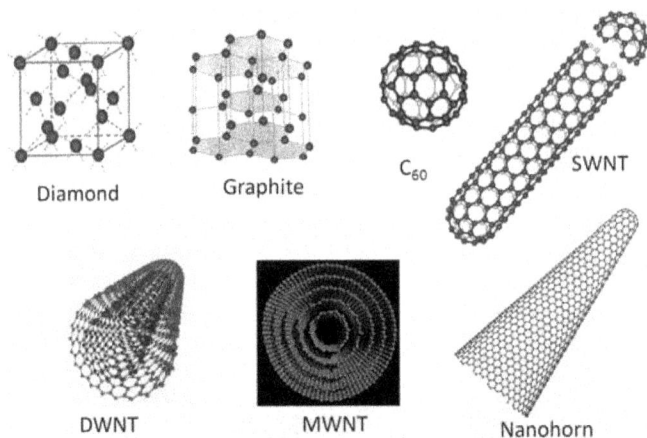

Fig. 1. Schematic representation of allotropes of carbon. These include a variety of entities that manifest as different physical structures as indicated in this figure.

Color image of this figure appears in the color plate section at the end of the book.

20 hexagonal faces (see Fig. 1) (Kroto et al. 1985). Interestingly, the C_{60} molecule exhibits sp^x hybridized properties, where $2 < x < 3$, and forms a covalently bonded face-centered-cubic solid. Subsequently, other related molecules such as C_{70}, C_{76} and C_{84} (collectively called fullerenes) were also discovered and viewed as elongated versions of the C_{60} molecule. In 1991, Iijima provided experimental evidence for the existence of extremely elongated tubular fullerene $C_{1,000,000}$, more popularly referred to as a carbon nanotube (cf. Fig. 1) (Iijima 1991). An intriguing feature of a carbon nanotube is that it is only a single atomic layer in thickness and can be viewed as a sheet of graphite (a hexagonal lattice of carbon) rolled into a seamless cylinder with tube diameter of the order of 1 nm. Depending on the synthesis conditions, carbon nanotubes form either single-walled (SWNT), double-walled (DWNT) or multi-walled nanotubes (MWNT) (cf. Fig. 1).

A member of the fullerene family, CNTs are tiny cylinders of carbon atoms that have 100 times the strength of steel (at 1/6th its weight) and have been proven to conduct electricity better than copper. Owing to its nanometer-size diameter and the presence of strong covalent bonds, carbon nanotubes acquire unique physical and chemical properties. When compared to steel on a similar length scale, nanotubes are six times as light and 100 times as strong as steel. Not only are SWNTs extremely strong, they also exist in two forms: metallic or semiconducting SWNTs. Therefore, considerable research is underway to explore their fundamental properties for targeted applications. For example, metallic SWNTs conduct electricity better than copper and are regarded as promising nanomaterials in composite materials, nano-electronics, and energy-related materials. The semiconducting SWNTs possess an energy band gap of ~ 1 eV and exhibit excellent photoluminescence in the infrared region. This spectral region falls within the tissue transparency window and therefore SWNTs serve as ideal markers for fluorescence imaging in biological systems. Furthermore, the tubular morphology of nanotubes and the ability to functionalize the walls of nanotubes have led researchers to explore nanotubes as plausible drug-delivery vehicles.

SYNTHESIS OF CARBON NANOTUBES

There are three widely used methods for synthesizing carbon nanotubes. The first method is the electric arc discharge, in which a catalyst-impregnated graphite electrode is vaporized by an electric arc (~100 A) in an inert atmosphere. Typically, the condensed carbonaceous deposits on the inner surface of the water-cooled arc chamber are rich in SWNT content (Journet et al. 1997). The second method, commonly referred to as the pulsed laser vaporization method, uses a pulsed laser beam (Nd:YAG, 1064 nm) that is tightly focused onto a catalyst-impregnated rotating graphite target (Fig. 2). The ablated material from the target generates soot that is collected on a water-cooled cold finger (Thess et al. 1996). In the electric arc discharge or the pulsed laser vaporization method, the as-prepared soot contains 60–70% SWNTs and the remainder is largely composed of amphorous carbon and nanoparticles. The third synthesis method is chemical vapor deposition (CVD) and is the most promising method for large-scale synthesis of carbon nanotubes. In the CVD method, nanotubes are prepared from a thermal decomposition (700–1200°C) of

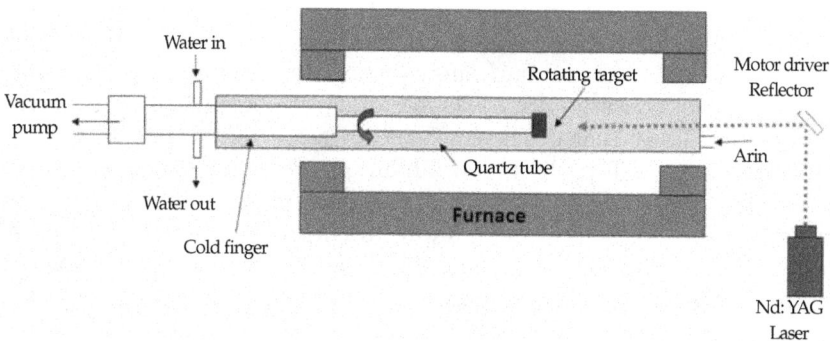

Fig. 2. Schematic of the pulsed laser ablation method for the synthesis of single-walled carbon nanotubes. A rotating graphite target (impregnated with transition metal catalyst) is ablated with the Nd-YAG laser. The ablated soot is carried by the inert gas from the center of the furnace (operating temperature ~1200°C) to the water-cooled finger, where it condenses as SWNTs.

a hydrocarbon, e.g., methane, in the presence of nanosized catalyst particles (typically transition metal particles such as Fe, Co, Ni) in an inert atmosphere. The CVD process is usually the preferred method of choice for preparing MWNTs, although reports have shown that SWNTs can also be grown (Andrews et al. 1999; Bachilo et al. 2003; Hata et al. 2004).

INTERACTIONS OF CARBON NANOTUBES IN THE BIOLOGICAL ENVIRONMENT AND ENVIRONMENTAL/OCCUPATIONAL HAZARDS

The interaction of CNTs with a biological environment is dependent on many factors, including CNT dispersion, type, surface properties, concentration, and purity (Helland et al. 2007). Environmental and occupational interactions also need to be assessed. Environmentally, CNTs are non-biodegradable and insoluble in water. This can lead to accumulation in the food chain, eventually leading to CNT aggregation at toxic levels. Furthermore, the large surface area of CNTs allows for easy binding of pollutants and other toxic chemicals to the CNTs. This can cause the undesirable molecules to be carried further up the food chain, leading to potentially toxic effects. Care must be taken when handling CNTs occupationally because studies have shown that CNTs can become airborne and ingested (Helland et al. 2007). Mechanical stimulation was shown to cause the CNTs to become airborne at a concentration less than 53 $\mu g/m^3$. Another occupational aspect tested was CNT adherence to gloves. After handling samples, workers were found to have 0.2–0.6 mg of CNTs per hand (Helland et al. 2007). Though this does pose a risk, CNTs are typically handled with great care. Production is usually carried out in small volumes and samples are handled carefully to reduce product loss. Therefore, no adverse or severe occupational hazards have been discovered as long as standard safety procedures are followed.

A primary concern with CNTs in biological uses is their cytotoxicity. It has been found that CNTs can demonstrate cytotoxic effects such as oxidative stress, cell proliferation inhibition, and cellular apoptosis (Liu et al. 2009). These negative biological effects originate from the impurities introduced during fabrication. Impurities include amorphous carbon, copper, iron, nickel, and molybdenum. However, these can be removed by modification techniques. The amount of impurities present depends strongly on the production method used. Of the methods, CVD is the best for use of CNTs in a biological environment. It has less carbonaceous impurities than the others, while also creating more homogenous size and length distributions (Foldvari et al. 2008). However, catalytic metal residues must be removed after CVD fabrication to improve its biocompatibility.

In comparison, SWCNTs have been proven to be more cytotoxic than MWCNTs. Their smaller size allows more rapid clearance, decreased inflammatory response, and granuloma formation (Kostarelos et al. 2009). Furthermore, the length of CNTs also contributes to their cytotoxicity. Longer MWNTs have been shown to be more cytotoxic and carcinogenic.

CNTs are naturally hydrophobic and insoluble in most organic or aqueous environments. Therefore, they readily aggregate upon exposure to an aqueous environment. This is a negative characteristic of pristine CNTs since a primary characteristic is their nano size. Once aggregated, CNTs form larger clumps or rope-like structures that can lead to physical blockings of the renal system, blood vessels, and airways (Helland et al. 2007).

Interactions with biological systems vary depending on the type of CNT used, but a few general characteristics exist. CNTs are able to clear the renal system fairly rapidly (Liu et al. 2009). This can be beneficial because rapid clearance can mean rapid delivery. However, it can lead to difficulty in maintaining the injected drug in blood circulation for prolonged periods. Since raw CNTs do not function desirably in a biological environment, steps need to be taken to improve their biocompatibility. This is achieved by creating functionalized CNTs with covalent and non-covalently bound molecules grafted to the CNT surface. Chemicals and molecules used to date include polymers, amines, phenyl-carboxyls, hydroxyls, and glucosamines. The attachment of these groups alters the physical properties of the CNTs, making them soluble and more dispersed. Functionalization allows the biological interactions of CNTs to be more controlled and also reduces their cytotoxicity (Liu et al. 2009). Parameters such as accumulation site, excretion method, blood circulation half-life, and degradation time can be predicted by the functional group. Studies have been performed to analyze the effects of certain molecules on these parameters. Relevant results and information from selected studies are highlighted in Table 1. CNT backbone can be modified to make it biologically inert and resistant to non-specific protein binding. This decreases its uptake and removal from the reticuloendothelial system. In contrast to functionalized ones, commercial pristine, unmodified CNTs have been shown to have a circulation half-life of 1 h (Cherukuri et al. 2006). This short circulation time is attributed to the rapid adherence of blood serum proteins to the CNT surface, thereby marking it for removal. Common sites of CNT accumulation include the kidneys, liver, and stomach.

With regard to interaction with cells, CNTs have been shown to cross the plasma membrane into the cytoplasm. Cell types shown to internalize CNTs include fibroblasts, epithelial, cancer cells, and phagocytes. The primary methods for internalization include phagocytosis, passive diffusion, and mediated endocytosis, which can be used to deliver drugs intracellularly (Bianco et al. 2008).

Table 1. Biological information for varying CNT functional groups.

Functional group	Site of accumulation	Primary excretion method	Blood circulation	Additional relevant information
Hydroxyl (Wang et al. 2004)	Stomach, kidneys, bone	N/A	N/A	N/A
Amine (Sing et al. 2006)	Kidneys, muscle, skin, lungs	Urine	Half life 3 h	SWNTs, diameter 30–38 nm
Glucosamine— General (Guo et al. 2007)	Stomach	Feces and urine	Half life 5.5 h	MWNTs, diameter 20–40 nm
Glucosamine— N-acetylglucosamine (Hong et al. 2010)	Lungs	N/A	< 1% in circulation after 3 min	SWNTs, NaI encapsulated inside with CNT ends sealed, showed no tissue fibrosis or necrosis after 30 d
PEG—linear (Liu et al. 2007)	Liver, spleen	Biliary (feces)	Half life 2 h	SWNTs, lengths 50–150 nm, shorter PEG chains showed larger accumulation, performed over 3 mon
PEG—branched (Liu et al. 2008)	Liver, spleen	Biliary (feces)	Half life 15 h	SWNTs, lengths 50–150 nm, performed over 3 mon

Experimental results from studies performed analyzing the biological interactions of various functionalized CNTs. The primary biological interaction characteristics presented include primary accumulation site, excretion method, blood circulation properties, and CNT properties.

DRUG DELIVERY AND CARBON NANOTUBES

Many different types of biological and synthetic molecules can be bound to a CNT backbone. These molecules are bound to the CNT via covalent, non-covalent, and hydrogen bonds, as well as through van der Waals forces. They have also been shown to adsorb both ionized and non-ionized drugs, and can be targeted for specific cells or tissues via peptide attachment. This targeting allows for more accurate and efficient drug delivery (Ilbasmis-Tamer et al. 2010).

Drugs can be loaded onto CNTs in many ways. Simplified visual representations of these attachment methods can be noted in Fig. 3. For encapsulation, the drug is inside the CNT and protected by the backbone. The drug can also be adsorbed to the CNT surface. Additionally, a drug can be conjugated to the CNT backbone via a linking molecule. Finally, the drug can be adsorbed to the CNT backbone and then encapsulated by another type of molecule.

The capacity for cell internalization makes CNT a very desirable drug-carrying molecule. A recent study demonstrated that grafting methotrexate (MTX), a cancer drug, onto MWNTs improved the cellular interactions of

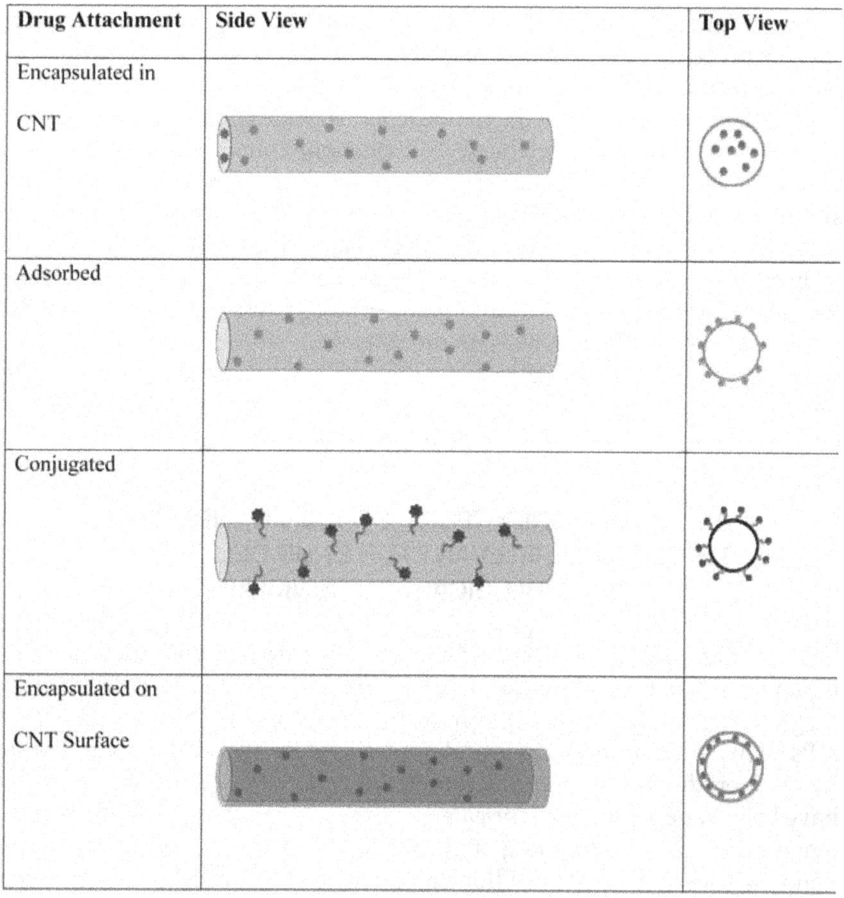

Drug Attachment	Side View	Top View
Encapsulated in CNT		
Adsorbed		
Conjugated		
Encapsulated on CNT Surface		

Fig. 3. CNTs as drug-carrying molecules. A very general visual representation of the ways in which CNTs can be used as drug delivery vesicles. The primary methods include encapsulation, adsorption, conjugation, and encapsulation on the CNT surface. Side and top views are presented for each method.

MTX. The MWNTs were functionalized with amino groups first, followed by MTX attachment to the amino groups. Alone, MTX is not readily internalized into the cell. However, when grafted onto CNTs, the MTX-CNT molecules readily crossed into the cytoplasm after 1 h of treatment (Pastorin et al. 2006). Furthermore, the results showed that MTX-CNT concentration in the cytoplasm was proportional to the dosage given.

CNTs have also been used as targeted transport vesicles for delivery of cancer drugs. Cisplatin is a platinum-based cancer drug that has been discontinued from treatment at times because of drug resistance and reduced cellular uptake. Platinum anticancer drugs are difficult to use because they can become inactivated before reaching their target. Lippard's group aimed to create a SWNT nanocarrier to transport cisplatin specifically to targeted cancer cells (Dhar et al. 2008). Many types of cancer cells over-express the α-folate membrane receptor. To target these cells, the study attached folic acid to the nanocarrier to bind with α-folate receptors. The carrier vessel contained the SWNT backbone conjugated with cisplatin and folic acid was then attached to the cisplatin. *In vitro* analysis showed significant accumulation of platinum inside the cellular nucleus of cancer cells. This accumulation showed that cisplatin was successfully released from the SWNT once inside the cytoplasm. Furthermore, once released in the cytoplasm, cisplatin diffused into the nucleus, where it interacted with DNA, causing cytotoxic effects on the cells.

Only a few studies that have used CNTs for drug delivery are highlighted here. Many other studies have shown or are in the process of analyzing the use of CNTs as targeted drug delivery vesicles.

PHYSICAL ATTRIBUTES OF CARBON NANOTUBES POTENTIALLY RELEVANT FOR DRUG DELIVERY IN THE BIOLOGICAL ENVIRONMENT

The diversity of NPs used include lipid, polymer, metallic, and ceramic-based particles. Overall, most of these NPs are spherical or rod-like with lengths and diameters of a few hundred nanometers. Only two types of NPs have demonstrated stable and potentially useful rod-like shapes with nanometer diameters and micrometer lengths. Polymeric fillomicelles have been experimentally fabricated with diameters from 22 to 60 nm and lengths from 2 to 8 μm (Gerg et al. 2007). This study compared spherical nanoparticles to the rod-like fillomicelles. Results showed that fillomicelles remained in blood circulation for 1 wk, 10 times as long as their spherical counterparts (Gerg et al. 2007). However, cellular uptake was higher for the spherical nanoparticles. Studies have shown that microparticles larger than 5 µm tend to get trapped in the capillary beds of the liver, leading to accumulation and poor circulation. Microparticles between 1 and 5 µm tend to localize in the liver, where they are eventually phagocytized. However, nanoparticles less than 1 µm and larger than 200 nm have been shown to easily filter through the spleen (Caldorera-Moore et al. 2010).

Many other types of drug delivery molecules exist, but what is it about CNTs that makes them desirable? In contrast to fillomicelles, CNTs have a very small diameter that could increase its internalization into cells. Diameters for SWNTs can range from 1 to 50 nm and 2.5 to 100 nm for MWNTs (Foldvari et al. 2008). Lengths can range from 100 nm to 9 µm for both SWNTs and MWNTs. Additionally, their rod-like shape has been shown to increase blood circulation times allowing for prolonged drug delivery. Physical dimensions of CNTs vary between SWNTs and MWNTs.

Compared to other drug delivery nanocarriers, it is clear that CNTs possess many unique characteristics, making them ideal for certain uses. A general comparison with other nanocarriers used for drug delivery can be found in Table 2.

Table 2. Comparison of nanocarrier molecules.

Nanocarrier	Pros	Cons
CNTs (Liu et al. 2009)	• Large surface area • Rod-like shape • Easy uptake into cells • Variety of surface modifications possible • Deliver various types of drugs	• Cytotoxicity dependent on variety of factors • Require extensive purification • Require optimized coating • Mostly surface-coated drugs
Liposomes (Doshi et al. 2009)	• FDA-approved for clinical use • Self-assemble into lipid bilayer • Controlled size • Lipid composition control • Deliver hydrophobic and hydrophilic drugs	• Poorly sustained drug release
Lipid nanoparticles (Doshi et al. 2009)	• FDA-approved for clinical use • Already in use for cosmetics and dermatology • Large surface area • Prolonged drug release profile • Rapid uptake by cells • Variety of production and administration methods	• Low stability • Deliver mostly hydrophobic drugs

Table 2. contd....

Table 2. contd....

Nanocarrier	Pros	Cons
Polysaccharide nanoparticles (Doshi et al. 2009)	• High stability • Natural polymers • Safe, nontoxic • Hydrophilic • Biodegradable • Abundantly available • Contain large number of reactive groups	• Scaling up • Purity • Poorly sustained drug release
Micelles (Trivedi et al. 2010)	• Low toxicity • Deliver hydrophilic and hydrophobic drugs • Spherical and rod-like shape • Can be modified to target specific cells • Synthetic polymers • In clinical trials	• Low stability • Poorly sustained drug release • Storage
Polymeric nanoparticles (Trivedi et al. 2010)	• Synthetic polymers • Long blood circulating time • In clinical trials • Biodegradable • Sustained release • Controlled size	• Deliver mostly hydrophobic drugs • Poor storage stability

CNT AS NON-VIRAL VECTORS

In light of limitations with viral vectors, the introduction of non-viral delivery of genetic products may an important step to address infectious disease challenges. This may be applicable to the design of future vaccines or simply as therapeutic vectors to target infectious agents such as viruses, bacteria, mycobacteria, helminths, and protozoa by attacking their genetic components (Rosen et al. 2009; Fauci 2008; Mandell et al. 2004).

Non-viral vectors are an attractive alternative to viral vectors, which have limitations including immunogenic reactions, inflammatory reactions that may cause transient transgene expression, and potential oncogenic effects that may occur at a later time. Genetic entities such as siRNAs and DNAs are among the potential genetic entities that may be delivered with non-viral vectors. While non-viral vectors have potential use for these challenges, non-viral vectors do require a minimum therapeutic level of gene expression to be effective. Non-viral vectors have the advantages of assembly in cell-free systems from well-defined components. This may provide important advantages over viral vectors for both safety and manufacturing (Rosen et al. 2009; Kam et al. 2006; Walther et al. 2000).

TARGETING BACTERIAL AND MYCOBACTERIUM INFECTIONS

As noted, drug resistance represents an important challenge when targeting bacterial infections because of the multiple mechanisms involved, which may occur very rapidly. For example, β-lactamases, produced by certain bacteria, disrupt the β-lactamase ring of penicillin. Alteration of the antibiotic binding site can be seen in the ubiquitous and dangerous methicillin-resistant *Staphylococcus aureus* strains. Another drug resistance mechanism is the implementation of alternative metabolic pathways by bacteria (Rosen et al. 2009; Banerjee et al. 2008; Arias et al. 2008; Fauci et al. 2008; Pitout 2008; Mandell et al. 2004; Cohen 1992).

Combinational therapy, that is, concomitant therapy consisting of different therapeutic agents, is a typical clinical method that can be used to target such resistance. Combinational therapy is particularly important in reducing drug-resistant mycobacterium infections, such as tuberculosis. It should be noted that while combinational therapy is considered an important strategy to limit drug resistance, it too can be susceptible to the development of drug resistance (Fauci et al. 2008; Mandell et al. 2004). Using this logic, a drug delivery system such as CNTs can be functionalized to deliver multiple payloads and/or specific genetic entities that can interfere with drug resistance activity, genetic components and/or related products (Rosen et al. 2009; Krajcik et al. 2008; Kostarelos 2008).

CNTs themselves, as stand-alone entities, have been shown to exhibit antibacterial effects without any functionalization. Moreover, their dimensional characteristics play a significant role and therefore further elaboration is needed to assess their effects on both bacteria and host cells (Kang et al. 2007, 2008a, b; Elimelech 2007). Elimelech et al. suggested that antibacterial effects of CNTs are most likely due to cell membrane damage by direct contact with CNTs (Elimelech 2007).

VIRAL INFECTIONS AND CARBON NANOTUBES DELIVERY OF SIRNAS

Eradicating many of the viral infections represents a major challenge in therapeutics. Some viruses, such as HIV, carry effective methods to evade the host immune system. In addition, their structural properties are far different than typical eukaryotic cells, although they may implement entities from the host to further their presence (Fauci et al. 2008; Mandell et al. 2004).

One area of research that is rapidly developing is delivering siRNA to target undesired genomic expression. It has been suggested as a potentially promising field in targeting a variety of viruses and even cancer

(Al-Jamal et al. 2010; Ladeira et al. 2010; Podesta et al. 2009; Krajcik et al. 2008; Kam et al. 2005). The SWCNT nanocomplex showed the ability to carry extracellular signal-regulated kinase (ERK) siRNA and was able to suppress expression of the ERK target proteins in primary cardiomyocytes by about 75% (Krajcik et al. 2008).

CAN PRIONS BE TARGETED?

Prions, which are thought to consist mainly or entirely of misfolded prion proteins (PrPs) and cause fatal neurodegenerative diseases, are particularly challenging to target (Fauci et al. 2008). However, Toupet et al. demonstrated that chronic injections of dominant negative lentiviral vectors into the brain may be a promising approach for a curative treatment of these devastating diseases (Toupet et al. 2008). Since a viral vector was used, then a non-viral vector, such as a targeted nanocomplex delivery system, should be tested as well (Rosen et al. 2008).

VACCINE EFFICACY ENHANCEMENT

Enhancing the efficacy of a vaccine may be achieved by enhancing neutralizing antibody response (Pantarotto et al. 2003). Developing an effective delivery system that can carry multiple payloads may help achieve this objective. Targets that evade the immune system, which include helminths, protozoa such as malaria, HIV and other agents, may be finally susceptible if multiple payloads can be carried, especially if they include gene silencers (Al-Jamal et al. 2010; Ladeira et al. 2010; Podesta et al. 2009; Krajcik et al. 2008; Fauci et al. 2008; Kam et al. 2005). The delivery system must retain a natural confirmation, or a near natural confirmation, of the delivered entity, which in this case can be the antigen that would elicit a desired immune response. Furthermore, the delivered entity must also retain its activity. An immunogenic response to the delivery system is undesirable (Pantarotto et al. 2003).

Pantarotto et al. developed a delivery system that consisted of virus-specific neutralizing antibody responses to a vaccine delivery system using covalent linkage to a neutralizing B cell epitope from the Foot and Mouth Disease Virus to mono- and bis-derivatized CNTs as a vaccine complex. Results of this study indicated high levels of virus-neutralizing antibodies elicited by the mono-derivatized CNTs vaccine complex while addressing the objectives indicated above (Pantarotto et al. 2003).

CONCLUSION

CNTs have unique physical and structural properties that give them targeted, multiple payload drug delivery capabilities. CNTs may be used as a model in drug delivery research and development, particularly in addressing key challenges of infectious diseases (Rosen et al. 2009; Kostarelos 2008; Lanone et al. 2006; Kam et al. 2005). However, there are concerns about toxicity issues as noted in this chapter and elsewhere that might be addressed with surface functionalization. Its resemblance to asbestos requires very careful handling (Rosen et al. 2009; Lacerda et al. 2008a, b; Tagaki et al. 2008; Bottini 2006).

Summary Points

- CNTs have unique physical and structural properties that give them targeted, multiple payload drug delivery capabilities (Ladeira et al. 2010; Podesta et al. 2009; Rosen et al. 2009; Kostarelos et al. 2008; Lanone et al. 2006).
- Infectious disease, just like cancer, represents a therapeutic challenge due to drug resistance, toxicity of current drugs, and inability of therapeutic agents to reach a variety of targets (Fauci et al. 2008; Mandell et al. 2004).
- CNTs have toxicity issues, including pulmonary, renal, and immunotoxicity. These, along with their pharmacological profile, require further elucidation before they can be widely used in clinical medicine (Rosen et al. 2009; Lacerda et al. 2008a, b; Tagaki et al. 2008; Bottini 2006).

Key Facts

- CNTs can cross the plasma membrane into the cytoplasm to deliver their payloads (Kostarelos et al. 2009; Bianco et al. 2008).
- CNTs can deliver multiple payloads to overcome key challenges of infectious disease. Multiple payloads can include having an antibody for targeted delivery along with several agents to allow for a synergistic, combinational therapeutic effect (Rosen et al. 2009; Kostarelos et al. 2009).
- CNTs have toxicity issues that require a more in-depth analysis before their implementation in clinical medicine. However, recent data shows that surface functionalization can reduce CNT toxicity (Rosen et al. 2009; Lacerda et al. 2008a, b; Bottini et al. 2006).

Definitions

Carbon nanotubes: Tiny cylinders of carbon atoms that have 100 times the strength of steel (at 1/6th its weight). They are members of the fullerene family. A carbon nanotube is only a single atomic layer in thickness and can be viewed as a sheet of graphite (a hexagonal lattice of carbon) rolled into a seamless cylinder with tube diameter of the order of 1 nm (Journet et al. 1997; Iijima 1991; Kroto et al. 1985).

Combinational therapy: A strategy that involves the concomitant administration of different therapeutic agents, which can have a synergistic effect. It is a typical clinical method that can combat drug resistance and/or limit its development (Fauci et al. 2008; Arias et al. 2008; Mandell 2004).

Drug resistance: A phenomenon that occurs when a targeted entity such as an infectious agent is no longer affected by drugs or other related entities that are used to eradicate it, prevent its growth or limit its effects. Drug resistance can often occur in infectious disease and cancer (Fauci et al. 2008).

Infectious disease: A disease caused by pathological entities, including bacteria, mycobacteria, viruses, helminths, protozoa, and prions. The manifestations of infectious diseases can be dependent on both patient and infectious agent (Fauci et al. 2008).

Small interference RNA: A short strand of double-stranded RNA used to interfere with the expression of targeted genes. It may be applicable in cancer and infectious disease. CNTs have been shown to deliver it into a number of targeted cells (Al-Jamal et al. 2010; Podesta et al. 2009; Krajcik et al. 2008).

Abbreviations

CNTs	:	carbon nanotubes
CVD	:	chemical vapor deposition
MWCNTs /MWNTs	:	multi-walled carbon nanotubes
NP	:	nanoparticles
RNA	:	ribonucleic acid
siRNA	:	small interference RNA
SWCNTs /SWNTs	:	single-walled carbon nanotubes
f-CNTs	:	functionalized carbon nanotubes
EAD	:	electric-arc discharge
MTX	:	methotrexate

References

Agerberth, B., and G.H. Gudmundsson. 2006. Host antimicrobial defence peptides in human disease. Curr. Top Microbiol. Immunol. 306: 67–90.

Al-Jamal, K.T., F.M. Toma, A. Yilmazer, H. Ali-Boucetta, A. Nunes, M.A. Herrero, B. Tian, A. Eddaoui, W.T. Al-Jamal, A. Bianco, M. Prato, and K. Kostarelo. 2010. Enhanced cellular internalization and gene silencing with a series of cationic dendron-multiwalled carbon nanotube:siRNA complexes. FASEB J. 24(11): 4354–4365.

Andrews, R., D. Jacques, A.M. Rao, F. Derbyshire, D. Qian, X. Fan, E.C. Dickey, and J. Chen. 1999. Continuous production of aligned carbon nanotubes: a step closer to commercial realization. Chem. Phys. Lett. 303(5–6): 467–474.

Arias, C.A., and B.E. Murray. 2008. Emergence and management of drug-resistant enterococcal infections. Expert Rev. Anti-Infect. Ther. 6(5): 637–655.

Bachilo, S.M., L. Balzano, J.E. Herrera, F. Pompeo, D.E. Resasco, and R.B. Weisman. 2003. Narrow (n,m)-distribution of single-walled carbon nanotubes grown using a solid supported catalyst, J. Am. Chem. Soc. 125: 11186–11187.

Banerjee, R., G.F. Schecter, J. Flood, and T.C. Porco. 2008. Extensively drug-resistant tuberculosis: new strains, new challenges. Expert Rev. Anti–Infect. Ther. 6(5): 713–724.

Bianco, A., R. Sainz, S. Li, H. Dumortier, L. Lacerda, K. Kostarelos, S. Giordani, and M. Prato. 2008. Biomedical applications of functionalised carbon nanotubes. In: F. Cataldo and T. da Ros (eds.). Medicinal Chemistry and Pharmacological Potential of Fullerenes and Carbon Nanotubes. Springer, Netherlands. pp. 23–50.

Bottini, M., S. Bruckner, K. Nika et al. 2006. Multi-walled carbon nanotubes induce T lymphocyte apoptosis. Toxicol. Lett. 160(2): 121–126.

Cherukuri, P., C.J. Gannon, T.K. Leeuw, H.K. Schmidt, R.E. Smalley, S.A. Curley, and R.B. Weisman. 2006. Mammalian pharmacokinetics of carbon nanotubes using intrinsic near-infrared fluorescence. Proc. Natl. Acad. Sci. USA. 103: 18882–18886.

Cohen, M.L. 1992. Epidemiology of drug resistance: implications for a post-antimicrobial era. Science 257(5073): 1050–1055.

Dhar, S., Z. Liu, J. Thomale, H. Dai, and S.J. Lippard. 2008. Targeted single-wall carbon nanotube-mediated Pt(IV) prodrug delivery using folate as a homing device. J Am. Chem. Soc. 130: 11467–11476.

Doshi, N., and S. Mitragotri. 2009. Designer biomaterials for nanomedicine. Adv. Funct. Mater. 19: 3843–3854.

Elimelech, M. 2007. Carbon nanotubes kill bacteria. Mater. Technol. 3: 79–179.

Foldvari, M., and M. Bagonluri. 2008. Carbon nanotubes as functional excipients for nanomedicines: I. Pharmaceutical properties. Nanomed. Nanotechnol. Biol. Med. 4: 173–182.

Geng, Y., P. Dalhaimer, S. Cai, R. Tsai, M. Tewari, T. Minko, and D.E. Discher. 2007. Shape effects of filaments versus spherical particles in flow and drug delivery. Nature Nanotechnol. 2: 249–255.

Guo, J., X. Zhang, Q. Li, and W. Li. 2007. Biodistribution of functionalized multiwall carbon nanotubes in mice. Nucl. Med. Biol. 34: 579–583.

Harrison's principles of internal medicine, 16th (ed.). 2008. In: A.S. Fauci, E. Braunwald, S. Hauser, D. Longo, J.L. Jameson and J. Loscalzo (eds.). McGraw-Hill Professional, 17th edition.

Hata, K., D.N. Futaba, K. Mizuno, T. Namai, M. Yumura and S. Iijima. 2004. Water-assisted highly efficient synthesis of impurity-free single-walled carbon nanotubes. Science 306: 1362–1364.

Helland, A., P. Wick, A. Koehler, K. Schmid and C. Som. 2007. Reviewing the environmental and human health knowledge base of carbon nanotubes. Environ. Health Persp. 115: 1125–1131.

Hong, S.Y., G. Tobias, K.T. Al-Jamal, B. Ballesteros, H. Ali-Boucetta, S. Lozano-Perez, P.D. Nellist, R.B. Sim, C. Finucane, S.J. Mather, M.L.H. Green, K. Kostarelos, and B.G. Davis. 2010. Filled and glycosylated carbon nanotubes for in vivo radioemitter localization and imaging. Nature Mater. 9: 485–490.

Iijima, S. 1991. Helical microtubules of graphitic carbon. Nature 354: 56–58.

Ilbasmis-Tamer, S., S. Yilmaz, E. Banoglu, and I.T. Degim. 2010. Carbon nanotubes to deliver drug molecules. J. Biomed. Nanotechnol. 6: 20–27.

Journet, C., W.K. Maser, P. Bernier, A. Loiseau, M. Lamy de la Chapelle, S. Lefrant, P. Deniard, R. Lee, and J.E. Fischer. 1997. Large-scale production of single-walled carbon nanotubes by the electric-arc technique. Nature 388: 756–758.

Kam, N.W., Z. Liu, and H. Dai. 2005. Functionalization of carbon nanotubes via cleavable disulfide bonds for efficient intracellular delivery of siRNA and potent gene silencing. J. Am. Chem. Soc. 127(36): 12492–12493.

Kang, S., M.S. Mauter, and M. Elimelech. 2008a. Physicochemical determinants of multiwalled carbon nanotube bacterial cytotoxicity. Environ. Sci. Technol. 42(19): 7528–7534.

Kang, S., M. Herzberg, D.F. Rodrigues, and M. Elimelech. 2008b. Antibacterial effects of carbon nanotubes: size does matter. Langmuir 24(13): 6409–6413.

Kang, S., M. Pinault, L.D. Pfefferle, and M. Elimelech. 2007. Single-walled carbon nanotubes exhibit strong antimicrobial activity. Langmuir 23(17): 8670–8673.

Kostarelos, K., A. Bianco, and M. Prato. 2009. Promises, facts and challenges for carbon nanotubes in imaging and therapeutics. Nature Nanotechnol. 4: 627–633.

Kostarelos, K. 2008. The good, the bad and ugly nanomaterials in biology—learning from the carbon nanotube experience. 2nd International Conference on Nanotoxicology, Zurich, Switzerland. September 7–10, p. 6, section 1.1.

Krajcik, R., A. Jung, A. Hirsch et al. 2008. Functionalization of carbon nanotubes enables non-covalent binding and intracellular delivery of small interfering RNA for efficient knock-down of genes. Biochem. Biophys. Res. Commun. 369(2): 595–602.

Kroto, H.W., J.R. Heath, S.C. O'Brien, R.F. Curl, and R.E. Smalley. 1985. C60: Buckminsterfullerene, Nature 318: 162.

Lacerda, L., H. Ali-Boucetta, M.A. Herrero et al. 2008a. Tissue histology and physiology following intravenous administration of different types of functionalized multiwalled carbon nanotubes. Nanomedicine 3(2): 149–161.

Lacerda, L., M.A. Herrero, K. Venner et al. 2008b. Carbon nanotube shape and individualization critical for renal excretion. Small 4(8): 1130–1132.

Ladeira, M.S., V.A. Andrade, E.R. Gomes, C.J. Aguiar, E.R. Moraes, J.S. Soares, E.E. Silva, R.G. Lacerda, L.O. Ladeira, A. Jorio, P. Lima, M.F. Leite, R.R. Resende, and S. Guatimosim. 2010. Highly efficient siRNA delivery system into human and murine cells using single-wall carbon nanotubes. Nanotechnology 21(38): 385101.

Lanone, S., and J. Boczkowski. 2006. Biomedical applications and potential health risks of nanomaterials: molecular mechanisms. Curr. Mol. Med. 6(6): 651–663.

Liu, Z., C. Davis, W. Cai, L. He, X. Chen, and H. Dai. 2008. Circulation and long-term fate of functionalized, biocompatible single-walled carbon nanotubes in mice probed by Raman spectroscopy. Proc. Natl. Acad Sci. USA 105: 1410–1415.

Liu, Z., S. Tabakman, K. Welsher, and H. Dai. 2009. Carbon nanotubes in biology and medicine: *in vitro* and *in vivo* detection, imaging and drug delivery. Nano Res. 2: 85–120.

Liu, Z., W. Cai, L. He, N. Nakayama, K. Chen, X. Sun, X. Chen, and H. Dai. 2007. *In vivo* biodistribution and highly efficient tumour targeting of carbon nanotubes in mice. Nature Nanotechnol. 2: 47–52.

Mandell, G.L., J.E. Bennett and R. Dolin. 2004. Principles and Practice of Infectious Diseases, 6th ed. Churchill Livingstone.

Pantarotto, D., C.D. Partidos, J. Hoebeke et al. 2003. Immunization with peptide-functionalized carbon nanotubes enhances virus-specific neutralizing antibody responses. Chem. Biol. 10(10): 961–966.

Pastorin, G., W. Wu, S. Wieckowski, J.P. Briand, K. Kostarelos, M. Prato, and A. Bianco. 2006. Double functionalization of carbon nanotubes for multimodal drug delivery. Chem. Comm. (Cambridge, England) 11: 1182–1184.

Pitout, J.D. 2008. Multiresistant Enterobacteriaceae: new threat of an old problem. Expert Rev. Anti–Infect. Ther. 6(5): 657–669.

Podesta, J.E., K.T. Al-Jamal, M.A. Herrero, B. Tian, H. Ali-Boucetta, V. Hegde, A. Bianco, M. Prato, and K. Kostarelos. 2009. Antitumor activity and prolonged survival by carbon-nanotube-mediated therapeutic siRNA silencing in a human lung xenograft model. Small 5(10): 1176–1185.

Rosen, Y., and N.M. Elman. 2009. Carbon nanotubes in drug delivery: focus on infectious diseases. Expert Opin. Drug Deliv. 6(5): 517–530.

Schipper, M.L., N. Nakayama-Ratchford, C.R. Davis, N.W. Kam, P. Chu, Z. Liu, X. Sun, H. Dai, and S.S. Gambhir. 2008. A pilot toxicology study of single-walled carbon nanotubes in a small sample of mice. Nature Nanotechnol. 3: 216–221.

Singh, R., D. Pantarotto, L. Lacerda, G. Pastorin, C. Klumpp, M. Prato, A. Bianco, and K. Kostarelos. 2006. Tissue biodistribution and blood clearance rates of intravenously administered carbon nanotube radiotracers. Proc. Natl. Acad. Sci. USA. 103: 3357–3362.

Takagi, A., A. Hirose, T. Nishimura et al. 2008. Induction of mesothelioma in p53+/– mouse by intraperitoneal application of multi-wall carbon nanotube. J. Toxicol. Sci. 33(1): 105–116.

Thess, A., R. Lee, P. Nikolaev, H. Dai, P. Petit, J. Robert, C. Xu, Y.H. Lee, S.G. Kim, A.G. Rinzler, D.T. Colbert, G.E. Scuseria, D. Tománek, J.E. Fischer, and R.E. Smalley. 1996. Crystalline ropes of metallic nanotubes. Science. 273: 483.

Toupet, K., V. Compan, C. Crozet et al. 2008. Effective gene therapy in a mouse model of prion diseases. PLoS ONE 3(7): e2773.

Trivedi, R., and U.B. Kompella. 2010. Nanomicellar formulations for sustained drug delivery: strategies and underlying principles. Nanomedicine (London). 5: 485–505.

Walther, W., and U. Stein. 2000. Viral vectors for gene transfer: a review of their use in the treatment of human diseases. Drugs 60(2): 249–271.

Wang, H., J. Wang, X. Deng, H. Sun, Z. Shi, Z. Gu, Y. Liu, and Y. Zhaoc. 2004. Biodistribution of carbon single-wall carbon nanotubes in mice. J. Nanosci. Nanotechnol. 4: 1019–1024.

Radiation Protection with Nanoparticles

Cheryl H. Baker

ABSTRACT

At the onset of radiation exposure, free radicals are formed through ionizing reactions that are then capable of destroying normal tissues. While cells release a level of protective molecules, such as glutathione and metallothionine, they are not capable of blocking all damage, thus resulting in the death of normal tissues. In an effort to combat the harmful effects of radiation, various free radical scavengers have been tested for their ability to protect normal cells and tissues. The most effective free radical scavenger to date is amifostine (Ethyol), whose active free thiol metabolite WR-1065 has been shown to prevent both radiation-induced cell death and mutagenesis while facilitating the repair of normal cells. Although amifostine is the only clinically relevant compound, this drug has a short half-life in serum and has serious side effects, rendering it difficult and costly to administer. Studies have shown the effects of engineered cerium oxide nanoparticles for protection against radiation-induced damage in a variety of tissue types. The role of nanoparticles as radioprotectants is a cutting-edge development in decades of scientific interest regarding the protection of normal cells and tissues from radiation. The

Department of Burnett School of Biomedical Sciences, University of Central Florida, 6900 Lake Nona Boulevard, Orlando, Florida 32827; Email: cherylhbaker@gmail.com

List of abbreviations after the text.

chemistry of engineered cerium oxide nanoparticles supports a potential role as a biological free radical scavenger or antioxidant. The work presented in this chapter addresses the effectiveness of cerium oxide nanoparticles in radioprotection of a variety of cells and in animal models during radiation exposure. It is hoped that it will encourage the development of innovative and new approaches to radiation protection using nanotechnology.

INTRODUCTION

At the onset of radiation exposure, free radicals are formed through ionizing reactions, such as the photoelectric, Compton and Auger effects. These free radicals react with DNA and RNA, causing molecular alterations, improper segregation of chromosomes during mitosis, and radiation-induced mitotic death (mitotic catastrophe) (Cohen-Jonathan et al. 1999; Nair et al. 2001). Furthermore, radiation-induced cellular oxidative damage is initiated by the generation of reactive oxygen species (ROS), which are known to change the oxidative status of cells, resulting in changes in mitochondrial function and activation/inactivation of various proteins involved in the apoptosis (cell death) process (Pradhan et al. 1973). When healthy (normal) cells are exposed to radiation, they ameliorate the damaging effect of free radicals by the release of innate protective molecules such as superoxide dismutase (SOD), glutathione, and metallothionine, which increase and intensify DNA repair mechanisms (Pradhan et al. 1973). Nonetheless, while these protective and repair mechanisms for cells are efficient, they are not capable of blocking all of the damage, which ultimately leads to normal tissue death.

In an effort to combat the harmful effects of radiation, various free radical scavengers have been tested for their ability to protect normal cells and tissues. Free radical scavengers such as Amifostine, Vitamin E, ascorbate, carotenes, melatonin and lipoic acid derivatives are the subject of many reviews (Beckman et al. 1998). However, many of these free radical scavengers were found to have limited success due to short half-lives (hours or even minutes), lack of penetration to the site of radical production, and daily dosing requirements. This report discusses a novel approach for the protection of normal cells against radiation-induced cell damage by using cerium oxide (CeO_2) nanoparticles.

CeO_2 nanoparticles have been tested for their ability to serve as free radical scavengers (Chen et al. 2006; Tarnuzzer et al. 2005; Rzigalinski et al. 2003) to render protection against chemical, biological and radiological insults that promote the production of free radicals. The chemistry of engineered CeO_2 nanoparticles supports a potential role as a biological free radical scavenger or antioxidant. It was suggested that the unique structure of CeO_2 nanoparticles, with respect to valence and oxygen defects, promotes cell longevity and decreases toxic insults by virtue of the antioxidant properties that occur when the nanoparticles enter the cells (Patil et al. 2007), prevent the accumulation of ROS and thereby prevent the activation of the apoptotic response and death of the cells (Chen et al. 2006).

In this report, CeO_2 nanoparticles are shown to confer protection against radiation-induced cell damage *in vitro* and *in vivo*, suggesting that CeO_2 nanoparticles are an effective radioprotectant for normal tissues.

RADIOTHERAPY SIDE EFFECTS

No cancer treatment is without side effects. Following radiotherapy, many patients experience side effects such as mild neutropenia, swelling or pain, and telangiectasia (a sunburn-type appearance of the skin); however, these early side effects usually disappear within several weeks. Early side effects occur in rapidly proliferating tissues, are generally not dose-limiting factors, and have minimal long-term impact on the quality of life of the patient. Of far greater concern is the emergence of late-reacting tissue damage in organs such as the lungs, skin and spinal cord; radiation damage to such tissues manifests itself weeks to months after the completion of therapy. These severe normal tissue reactions cause extensive discomfort to the affected individuals and limit the radiation dose that can be delivered to the entire patient population.

CeO_2 NANOPARTICLES AS RADIOPROTECTANTS

Nanotechnology is a multidisciplinary field that involves the design and engineering of objects <100 nanometers (nm) in size. Nanoparticles are a new generation of free radical scavengers. The role of nanoparticles as radioprotectants is a cutting-edge development addressing decades of scientific interest regarding the protection of normal cells and tissues from radiation. The chemistry of engineered CeO_2 nanoparticles supports a potential role as a biological free radical scavenger or antioxidant. Current studies highlighted in this chapter suggest that nanoparticles may be a therapeutic regenerative material that will scavenge ROS that are responsible for radiation-induced cell damage.

CeO$_2$ Nanoparticles in Biological Applications

While there are some concerns about the toxicity of nanoparticles, there are very few reports regarding the biologically detrimental effects of CeO$_2$ nanoparticles. In an article published in *Toxicology*, Park et al. conclude that CeO$_2$ nanoparticles (15–45 nm; 5–40 µg/ml) induced oxidative stress and cell death in cultured human lung epithelial cells (Park et al. 2008). It is important to note that these particles are significantly larger than the nanoparticles used in the experiments discussed because the size of a nanoparticle affects the free radical scavenging ability of the particle by modifying the ratio of cerium (III) to cerium (IV). Furthermore, Park et al. exposed the cells to CeO$_2$ nanoparticle doses ~1000 times the effective radioprotective dose (Tarnuzzer et al. 2005).

Need for Better Radioprotective Compound

Free radical scavengers such as Amifostine, Vitamin E, ascorbate, carotenes, melatonin and lipoic acid derivatives possess few active sites per molecule. A subsequently investigated antioxidant, C60, may be able to scavenge comparatively more radicals than the currently available antioxidants (Gharbi et al. 2005). But, because of the limited number of free radical scavenging sites, repeated dosing is required to replace molecular species that were used in free radical reduction. However, CeO$_2$ nanoparticles offer many active sites for free radical scavenging because of their large surface to volume ratio and, more importantly, because of their mixed valence states for unique redox chemistry. SOD-mimetic activity of CeO$_2$ has been reported (Heckert et al. 2008). Additionally, the free radical scavenging property of CeO$_2$ nanoparticles is regenerative (Tarnuzzer et al. 2005), which is not the case for other antioxidants. It is believed that, because of the chemical nature of CeO$_2$ nanoparticles, there is an auto-regenerative reaction cycle (Ce^{3+} → Ce^{4+} → Ce^{3+}) continuing on the surface of ceria nanoparticles and this is thought to be the current mechanism by which it provides the material with an unprecedented free radical scavenging ability (Fig. 1A, B).

CeO$_2$ Nanoparticles Exhibit *In Vitro* Free Radical Scavenging Ability

The chemistry of engineered CeO$_2$ nanoparticles supports their potential role as free radical scavengers, antioxidants, in biological systems (Gharbi et al. 2005). It was suggested that the unique surface chemistry of CeO$_2$ nanoparticles, with respect to valence and oxygen defects, decreases oxidative insults by virtue of its antioxidant properties and promotes cell

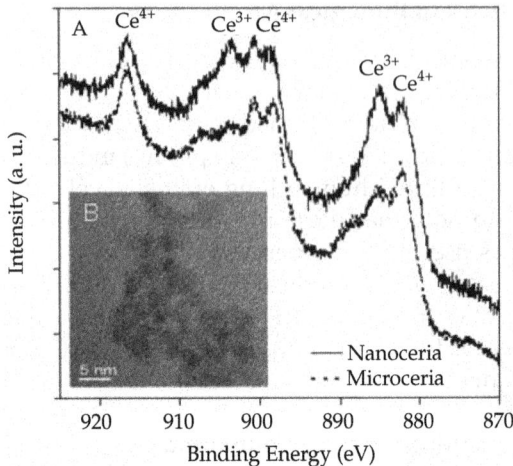

Fig. 1. Characterization of CeO₂ nanoparticles. (A) X-ray photoelectron spectroscopy (XPS) spectra indicates high concentration of Ce^{3+} in CeO_2 compared to microceria particles. Peaks at 882.1 and 886 eV correspond to Ce^{+4} and Ce^{+3} peaks. Peaks at 918 eV correspond to satellite peaks indicating the presence of Ce+4 peak. B. High resolution transmission electron microscopy (HRTEM) image of the synthesized particles indicating the particle size of 3–5 nm with fluorite lattice structure. With permission from C.H. Baker. 2009. Protection from radiation-induced pneumonitis using cerium oxide nanoparticles. Nanomedicine 5: 225–231.

longevity. Thus far, studies have shown that CeO_2 nanoparticles enter mammalian cells (Patil et al. 2007), decrease the accumulation of ROS, and prevent the activation of the ROS-induced apoptosis (Chen et al. 2006). Since cells produce ROS after being exposed to radiation (Korsvik et al. 2007), the antioxidant capability of CeO_2 nanoparticles has been suggested as the key mechanism by which CeO_2 nanoparticles confers radioprotection (Tarnuzzer et al. 2005). Furthermore, a study concluded that CeO_2 nanoparticles exhibited SOD-mimetic activity (Korsvik et al. 2007). Results supporting the antioxidant properties of CeO_2 nanoparticles are mounting, and many studies suggest that these nanoparticles act as free radical scavengers (Tarnuzzer et al. 2005; Rzigalinski et al. 2003; Chen et al. 2008) and may render protection against chemical insults that promote the production of free radicals (Perez et al. 2008). Thus, it has been proposed that CeO_2 nanoparticles may confer radioprotection by scavenging the free radical produced during radiotherapy (Tarnuzzer et al. 2005).

CeO₂ Nanoparticles Protect Mice from Total Body Irradiation

Balb/C mice were randomized into two groups (n=10). Group 1 was injected with saline (control group). Group 2 received a total CeO_2 nanoparticle dose of 0.005 mg/kg. On day 5, all animals received 12.5

Gy of X-ray radiation. No animals died in the CeO_2 nanoparticles group during the first 60 d post irradiation. In sharp contrast, 20% of the control animals died (Fig. 2A). During the experiment we observed that many of the control animals exhibited skin desquamation, while the CeO_2 nanoparticle-treated animals had little skin damage (Fig. 2B). These results suggest that CeO_2 nanoparticles are able to protect mice from a single dose of radiation, and support their role as radioprotectant (Colon et al. 2009).

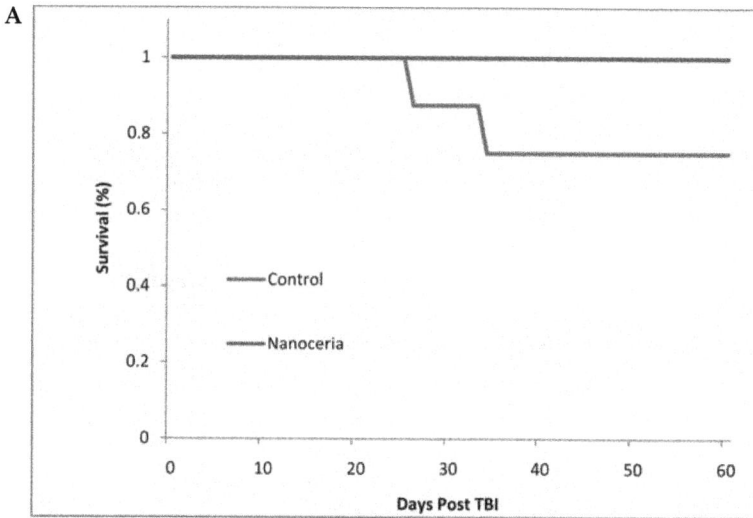

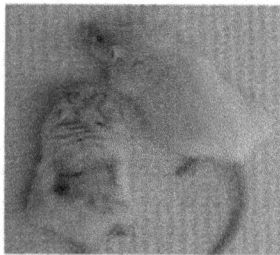

Control CeO_2

Fig. 2. CeO_2 nanoparticles protect mice from total body irradiation. A. Survival curve. B. Mice treated with CeO_2 nanoparticles had significantly less skin desquamation than untreated mice (control) 26 d after total body irradiation (12.5 Gy). Unpublished data from C.H. Baker.

CeO_2 Nanoparticles is Well Tolerated in Athymic Mice

To investigate the acute toxicity of CeO_2 nanoparticles, athymic nude mice were randomized into five groups. Each group received a total nanoparticle dose in the range of 0 (saline), 0.135 mg/kg. 1.35 mg/kg, 13.5 mg/kg, or

135 mg/kg. The mice were observed over a 3 wk period. No mice died or experienced notable side effects during the treatment. At the end of the treatment, the mice were sacrificed. During necropsy no abnormal pathologies were observed. This indicates that CeO_2 nanoparticles are well tolerated in mice up to 3 million times the effective dose. Therefore, it was suggested that CeO_2 nanoparticles cause limited toxicity and side effects in mice (Colon et al. 2009).

APPLICATIONS TO AREAS OF HEALTH AND DISEASE

When biological systems are under high energy exposure, ROS are produced at high levels and cellular components can be damaged. These ROS can be used by biological systems as a defense mechanism against microorganisms and can act as signal transduction and transcription agents in development, stress responses, and programmed cell death. Oxidative stress arises from the strong cellular oxidizing potential of excess ROS, or free radicals. In addition, elevated levels of oxidative damage are related to increased risks for cataracts, cardiovascular disease, and cancer.

Therefore, the potential benefit of radioprotection using CeO_2 nanoparticles is of great significance on multiple levels—the most important is its potential impact on human life. This research is relevant to the health and quality of life of humans worldwide who are exposed to radiation environments such as those listed below:

- Patients receiving radiation treatments for cancer
- Astronauts in NASA exposed to particle radiation
- Military and civilians potentially exposed to radiation in battle, terrorism or occupational exposure

Verification of the effectiveness of nanoparticles as radioprotectors opens the field for future studies that would examine in depth the mechanism, tissue distribution and safety of CeO_2 nanoparticles prior to their use in Phase I clinical trials. In the end, these studies may lead to faster recovery and improved quality of life for patients suffering from radiation damage.

PROTECTION OF RADIATION-INDUCED PNEUMONITIS USING CeO₂ NANOPARTICLES

Radiotherapy as a Treatment for Lung Cancer

Radiotherapy is an effective treatment option for lung cancer. However, lung tissue is particularly sensitive to radiation. Thus, the efficacy of radiotherapy is limited by the low tolerance of lung tissue to radiation exposure, and medical professionals seek to optimize the ratio of tumor debulking to lung toxicity. Unfortunately, 30% of patients who receive radiation during their treatment for lung cancer experience clinically significant lung injury (Robnett 2000), and there is no effective therapeutic available for the prevention of acute or chronic radiation-induced pneumonopathy (Lee et al. 2008). The availability of a radioprotective therapeutic that selectively protects normal lung tissue from radiation-induced damage would significantly improve the ability of medical professionals to treat patients with lung cancer.

CeO_2 Nanoparticles Exhibit Selective Radioprotection of Lung Fibroblasts *In Vitro*

Normal lung fibroblasts (CCL-135), pretreated with CeO_2 nanoparticles (10 nM) were exposed to 20 Gy. A Cell Titer-Glo Luminescent Cell Viability Assay (which signals the presence of metabolically active cells) was performed 48 h after irradiation, and the irradiated normal lung fibroblasts that received CeO_2 nanoparticles pretreatment had increased viability when compared to irradiated normal cells that did not receive CeO_2 nanoparticles treatment (Fig. 3A). When the same experiment was performed on a non-small cell lung cancer cell line (A549), there was no protection (Fig. 3B) (Colon et al. 2009).

In a similar study, normal lung fibroblast (CCL 135) and lung cancer cells (A549) were pretreated with 10 nM CeO_2 nanoparticles for 24 h. Cells were then irradiated with 20 Gy and incubated for 48 h and assayed for Caspase3/7 activity, which is a protein that is activated during apoptosis. In the presence of CeO_2 nanoparticles, normal cells did not undergo radiation-induced apoptosis (Fig. 4A). In sharp contrast, CeO_2 nanoparticles did not protect the A549 cells from radiation-induced apoptosis (Fig. 4B) (Colon et al. 2009).

Radiation-induced damage and oxidative stress are closely tied. Irradiated cells produce damaging ROS. Previous studies show that CeO_2 nanoparticles exhibit SOD-mimetic activity. To investigate whether CeO_2 nanoparticles can decrease intracellular ROS post irradiation, normal lung fibroblasts were treated with CeO_2 nanoparticles (10 nM) for 24 h and then irradiated (20 Gy). Intracellular ROS was imaged using the Image-iT Live Green Reactive Oxygen Species Detection Kit. Control cells were irradiated in the absence of CeO_2 nanoparticles (Fig. 5A). Results show that CeO_2 nanoparticles decreased the radiation-induced accumulation of ROS (Fig. 5B). These *in vitro* results show that CeO_2 nanoparticles selectively conferred protection against radiation-induced cell death in normal cells (and not cancer cells) (Colon et al. 2009).

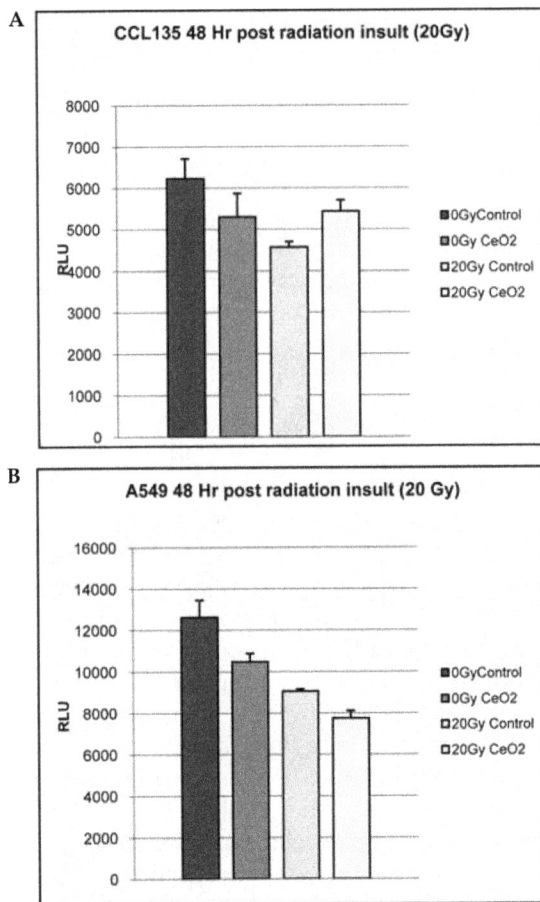

Fig. 3. CeO$_2$ nanoparticles exhibit selective protection of lung fibroblasts. Protection of (A) normal lung cells (CCL 135) by CeO$_2$ nanoparticles. (B) No protection observed in lung cancer cells (A549). With permission from C.H. Baker 2009. Protection from radiation-induced pneumonitis using cerium oxide nanoparticles. Nanomedicine 5: 225–231.

CeO$_2$ Nanoparticle Treatment Decreases Radiation-Induced Pneumonitis in Murine Model

Radiation pneumonitis and subsequent pulmonary fibrosis can significantly decrease the quality of life of humans exposed to radiation. In an attempt to administer nanoparticles to live animals and to evaluate the radiation protection activity of CeO$_2$ nanoparticles, the survival of non–tumor-bearing athymic nude mice was measured. Non–tumor-bearing athymic nude mice were exposed to fractionated doses of 30 Gy radiation (weekly administration of 5 Gy) in the presence or absence

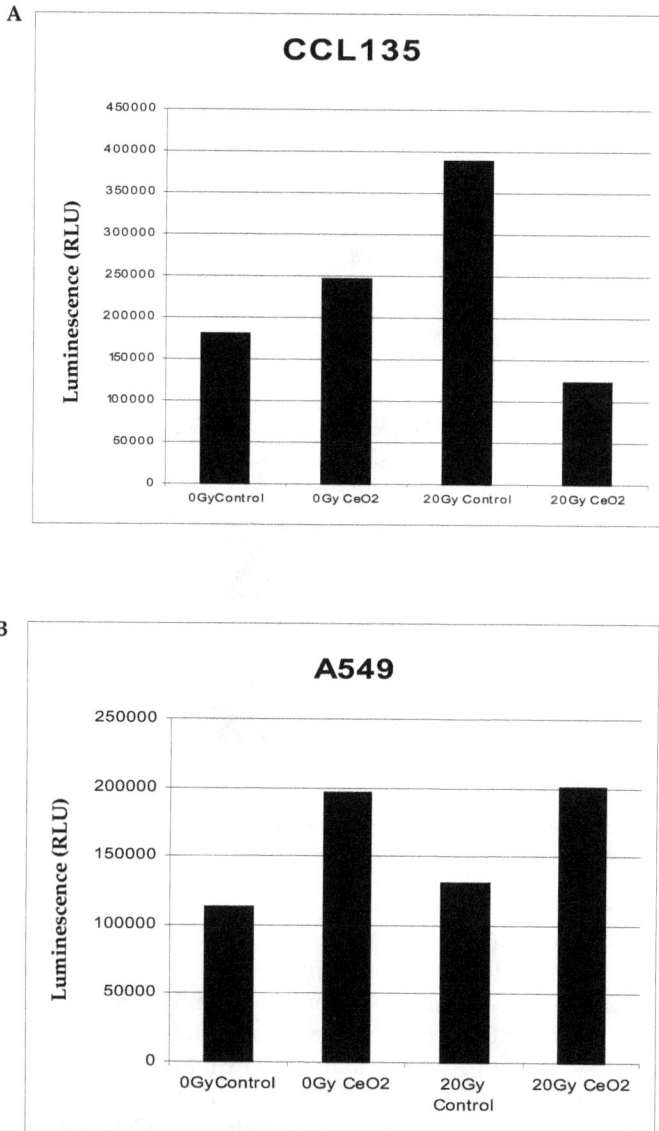

Fig. 4. Protection of radiation-induced apoptosis by CeO_2 nanoparticles in normal lung cells. Radiation-induced apoptosis of (A) normal lung cells (CCL 135) and (B) lung cancer cells (A549). Cells were exposed to 20 Gy radiation in the absence or presence of 10 nM CeO_2 nanoparticles and Caspase 3/7 activity was measured by the Caspase-Glo 3/7 assay. Luminescence is proportional to the amount of caspase activity present. With permission from C.H. Baker. 2009. Protection from radiation-induced pneumonitis using cerium oxide nanoparticles. Nanomedicine 5: 225–231.

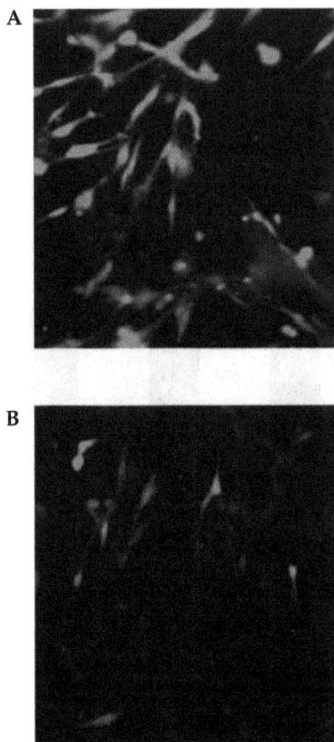

Fig. 5. ROS expression in irradiated normal lung fibroblasts. 4 hours post radiation, the levels of ROS were detected in (A) irradiated normal lung fibroblasts and (B) irradiated normal lung fibroblasts pretreated with CeO_2. Unpublished data from C.H. Baker.
Color image of this figure appears in the color plate section at the end of the book.

of twice weekly i.v. injections of CeO_2 nanoparticles or i.p. injections of Amifostine 30 min prior to radiation. Results show (Fig. 6) that CeO_2 nanoparticles are well tolerated by athymic nude mice and protect mice from radiation-associated death. All control mice lived until termination date of 231 d. In mice treated with CeO_2 nanoparticles alone, 20% were sacrificed on day 150 for histology analysis. The remaining 80% were alive until the termination date of 231 d. After treatment with radiation alone, Amifostine alone, and a combination of radiation and CeO_2 nanoparticles, or radiation and Amifostine, the median survival time was 132, 119, 225, and 81 d, respectively (control versus radiation, $P < 0.019$; control versus CeO_2, $P < 0.66$; control versus Amifostine, $P < 0.0370$; radiation versus radiation and CeO_2, $P < 0.0041$; radiation versus radiation and Amifostine, $P < 0.0432$). In contrast, Amifostine was highly toxic, as shown by the significant difference in median survival time (as compared to control mice). In summary, these results suggest that CeO_2 nanoparticles are well tolerated by mice and have a significant advantage over the clinically used Amifostine (Colon et al. 2009).

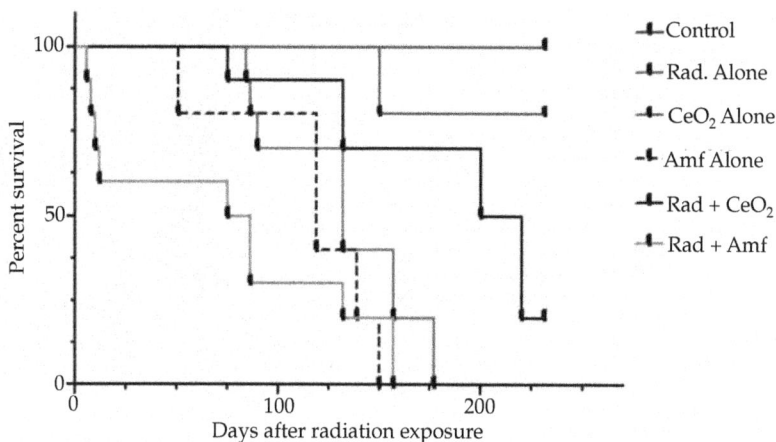

Fig. 6. Tolerability of CeO_2 nanoparticles in mice. CeO_2 were well tolerated by mice and the median survival of radiated mice was significantly increased in mice pretreated with 15 nM (0.00001 mg/kg) CeO_2 (50% alive on day 225) as compared to mice treated with radiation alone (50% alive on day 132) or pretreated with 150 mg/kg Amifostine before radiation (50% alive on day 81). Note that 20% of mice treated with CeO_2 alone were terminated on day 150 for histology analysis. With permission from C.H. Baker 2009. Protection from radiation-induced pneumonitis using cerium oxide nanoparticles. Nanomedicine 5: 225–231.

Color image of this figure appears in the color plate section at the end of the book.

To determine the degree of radiation-induced pneumonitis, the lungs were harvested and processed for histology and hematoxylin and eosin (H&E) staining. The lungs from mice in the control group (radiation alone) showed visible pneumonitis, with extensive macrophage invasion, whereas the lungs from irradiated mice receiving CeO_2 nanoparticles showed no visible pneumonitis and appeared normal (Fig. 7). In addition, the amount of fibrosis and collagen deposition (indicative of chronic lung conditions) was measured in the lungs of control mice (no radiation/normal lungs), or in lungs of those mice treated with radiation alone, radiation plus CeO_2, or radiation plus Amifostine, using Masson's Trichrome stain. The histology analyses show that fibrosis and collagen deposition were common in the irradiated lungs of those mice given radiation alone and of those mice given a pretreatment of Amifostine (Fig. 6). Furthermore, analysis indicated that collagen deposits were relatively recent, due to the faint blue stain, as compared to dark blue staining of older, more cross-linked collagen seen in human chronic lung diseases. In sharp contrast, no significant Trichrome staining was observed in normal lungs (control) or in those irradiated lungs of mice treated with CeO_2 (Colon et al. 2009).

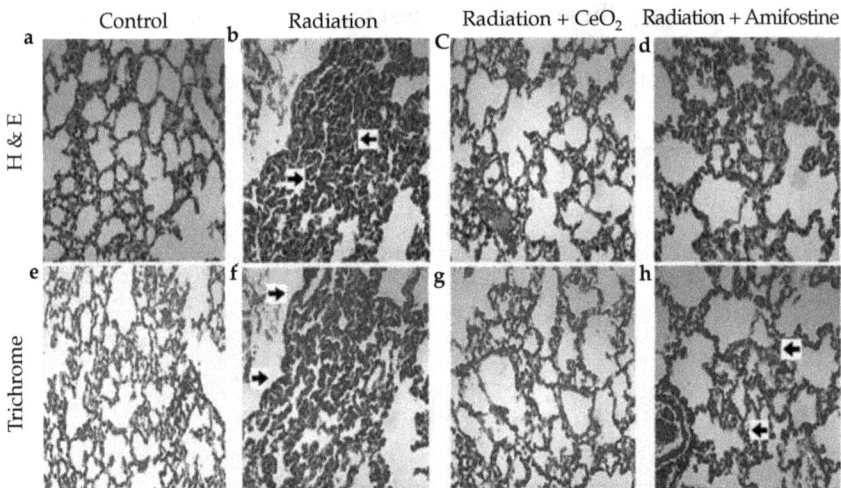

Fig. 7. CeO_2 nanoparticles protect lungs from radiation-induced pneumonitis. Hematoxylin and Eosin (H&E) stains to assess lung damage in normal lungs (a), lungs from mice treated with radiation alone (b), lungs from mice treated with radiation plus CeO_2 (c) and lungs from mice treated with radiation plus Amifostine (d). The H&E stains show significant lung damage in mice treated with radiation (b). Radiation-induced cell damage is protected in lungs of mice treated with radiation in combination with CeO_2 (c) and these lungs appear normal in control (a). The amount of fibrosis and collagen deposition (indicative of chronic lung conditions) was measured by using Masson's Trichrome stain. Results show that fibrosis and collagen deposition (indicated by arrows) were common in the lungs of those mice given radiation alone (f) and in lungs of those mice given a pretreatment of Amifostine (h). The amount of fibrosis and collagen deposition in lungs of mice treated with radiation in combination with CeO_2 (g) was minimal and these lungs appeared normal (e).With permission from C.H. Baker 2009. Protection from radiation-induced pneumonitis using cerium oxide nanoparticles. Nanomedicine 5: 225–231.

Color image of this figure appears in the color plate section at the end of the book.

CeO_2 Nanoparticle Treatment Reduces Overexpression of TGF-β, a Marker for Fibrosis

Athymic mice were randomized into two groups. Group 1 received 0.005 mg/kg of CeO_2 nanoparticles prior to irradiation, while group 2 received saline. The mice were irradiated in the ventral thorax with 30 Gy X-rays (fractionated into 5 doses over 2 wk). The mice were sacrificed 120 d after irradiation and the lungs extracted for immunohistochemistry. Slides of lung tissue were stained using a primary antibody (monoclonal mouse anti-mouseTGF-β1 and secondary antibody (goat anti-mouse HRP), and the slides were counterstained with hematoxylin. The stained slides were imaged with light microscope using oil immersion at 1000x (Fig. 8A, B). The images demonstrate a significant level of TGF-β expression in lungs of the untreated animals. Since high levels of TGF-β expression are linked to lung fibrosis and pneumonopathy (Lee et al. 2008), the decrease in TGF-β expression in the animals that received CeO_2 nanoparticles treatment (as compared to control) indicates that CeO_2 nanoparticles protected the mice from radiation-induced pneumonopathy (Colon et al. 2009).

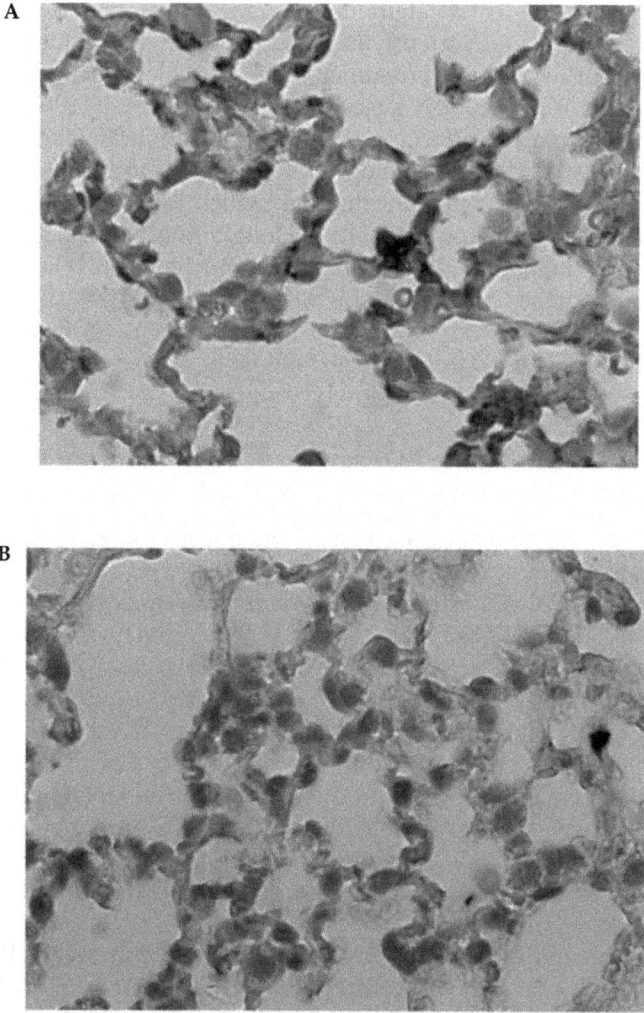

Fig. 8. CeO_2 nanoparticles reduce TGF-β expression post radiation. 120 d after XRT(30 Gy) fractionated over 5 doses and 2 wk, mice that received nanoceria treatment had significantly less TGF-β deposition. A. Lung tissue from untreated animal. B. Lung tissue from treated animal (0.005 mg/kg). Unpublished data from C.H. Baker.

Color image of this figure appears in the color plate section at the end of the book.

CeO₂ NANOPARTICLES PROTECT GASTROINTESTINAL EPITHELIUM FROM RADIATION-INDUCED DAMAGE BY REDUCTION OF ROS AND UP-REGULATION OF SOD-2

In the context of colorectal carcinomas, damage on surrounding healthy cells that have been inadvertently exposed to ionizing radiation has been exacerbated during radiation treatment since the colon is untethered and mobile, making it particularly susceptible to physical perturbation, such as bladder filling or breathing, which may cause unintended radiation exposure to nearby tissue. Ionizing radiation insult to the tissue causes DNA damage and free radical formation, which leads to stress-induced programmed cell death-apoptosis. In the long term, this damage leads to bowel obstruction, fistula, perforation, or hemorrhage, and these injuries often require further treatment, in particular, more invasive surgery (Meissner 1999). This study is the first to show that CeO_2 nanoparticles confer radioprotection on colon intestinal cells by exerting free radical scavenger properties and SOD-mimetic properties.

CeO_2 Nanoparticles Reduce ROS levels and Protect Normal Human Colon Cells from Radiation-Induced Cell Death *In Vitro*

In order to investigate the effects of CeO_2 nanoparticles on ROS production, normal human colon cells (CRL 1541) were exposed to increasing concentrations of CeO_2 nanoparticles 24 h prior to a single exposure of 20 Gy radiation. ROS production was measured using the Image-iT LIVE™ green ROS detection kit. Results show that when radiation was administered as single therapy, the qualitative production of ROS was significantly increased. However, when CeO_2 nanoparticles were administered 24 h prior to radiation, the presence of CeO_2 nanoparticles significantly decreased the ROS production, in a dose-dependent manner (Fig. 9A). There was no observable difference in ROS production between the control (non-irradiated cells) and the non-irradiated cells treated in combination with increasing concentrations of CeO_2 nanoparticles (Fig. 9A) (Colon et al. 2010).

In another set of experiments, normal human colon cells (CRL 1541) were exposed to increasing concentrations of CeO_2 nanoparticles added 24 h prior to a single exposure of 20 Gy. Ninety-six hours later, cell viability was measured. Results show that when radiation was administered as single therapy, the number of viable cells in culture was significantly decreased as compared to control (15%). However, when 1, 10 or 100 nM of CeO_2 nanoparticles were administered 24 h prior to radiation, the CeO_2 nanoparticles significantly protected the cells from radiation-induced cell death (3% for 1 nM, 1% for 10 and 100 nM) (Fig. 9B) (Colon et al. 2010).

A. Production of ROS post radiation treatment on normal colon CRL 1541 cells

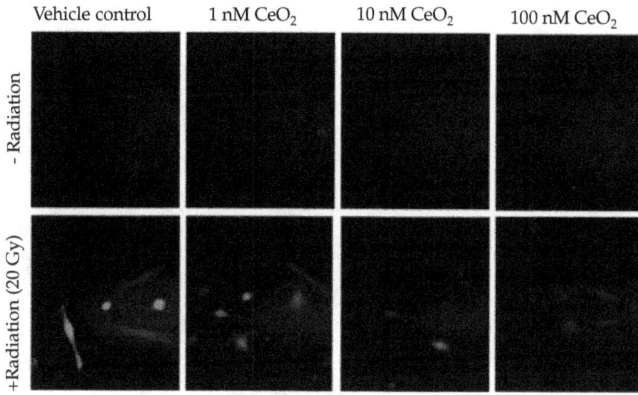

B. ATP assay 96 hrs post radiation treatment on
normal colon CRL 1541 cells

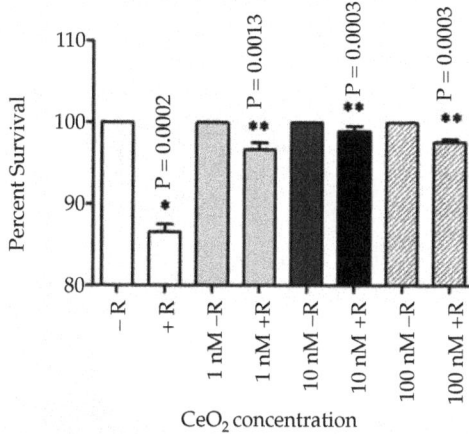

CeO$_2$ concentration

*Control (0 Gy) vs. Radiation (20 Gy)

** Radiation (20 Gy) vs. Radiation + CeO$_2$

Fig. 9. CeO$_2$ nanoparticles protect normal colon cells against radiation-induced cell damage. A. ROS production of normal human colon cells (CRL 1541) immediately following 20 Gy radiation exposure with pretreatment of 1, 10, or 100 nM CeO$_2$ nanoparticles was significantly reduced as compared to cells exposed to radiation alone. B. CRL 1541 cells were exposed to 20 Gy radiation in the absence or presence of 1, 10, or 100 nM CeO$_2$ and 96 h after exposure cell viability was measured by Cell Titer-Glo Luminescent Cell Viability Assay (cell number correlates with luminescent output (RLU). With permission from C.H. Baker 2010. Cerium oxide nanoparticles protect gastrointestinal epithelium from radiation-induced damage by reduction of reactive oxygen species and upregulation of superoxide dismutase 2. Nanomedicine 5: 698–705.

Color image of this figure appears in the color plate section at the end of the book.

CeO₂ Nanoparticles Induce SOD-2 Expression in Normal Human Colon Cells *In Vitro*

The effect of CeO_2 nanoparticles (added 24 h before radiation) on SOD-2 protein expression on CRL1541 cells growing in normal growth media was measured. Western blot analysis show increased levels of SOD-2 in normal colon cells in the presence of CeO_2 nanoparticles and in a dose-dependent fashion, the band intensity of SOD-2 in 100 nM CeO_2 nanoparticles-treated cells was roughly 2-fold that of non-treated control cells. The cells exhibited increased SOD-2 expression with the addition of increasing concentrations of CeO_2 nanoparticles (Fig. 10) suggesting that CeO_2 nanoparticles increased normal colon cell SOD-2 expression when added 24 h before radiation, conferring cytoprotection from the radiation

Protein expression of SOD-2 24 hrs post CeO_2 treatment on normal colon CRL 1541 cells

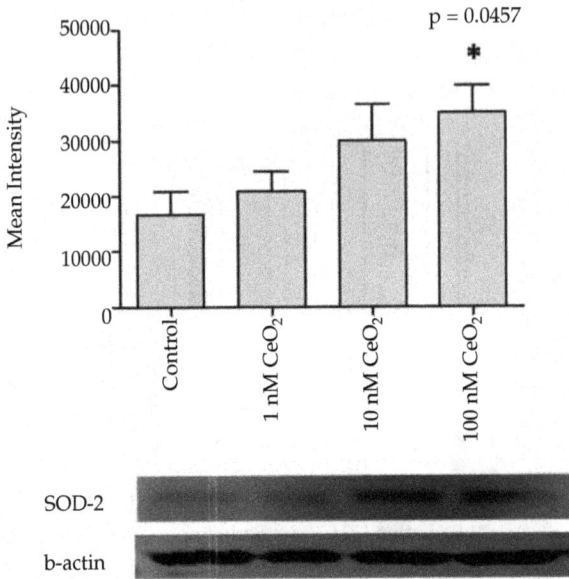

*Control (untreated) vs. CeO_2 (100 nM)

Fig. 10. CeO_2 nanoparticles induce protein SOD-2 expression. The effect of CeO_2 nanoparticles on SOD-2 protein expression on CRL1541 cells growing in normal growth media. The cells exhibited a dose-dependent increase in protein expression of SOD-2 with the addition of increasing concentrations of CeO_2 nanoparticles. The protein band intensity of SOD-2 in cells incubated with 100 nM CeO_2 nanoparticles was roughly 2-fold that of cells incubated in media alone. With permission from C.H. Baker 2010. Cerium oxide nanoparticles protect gastrointestinal epithelium from radiation-induced damage by reduction of reactive oxygen species and upregulation of superoxide dismutase 2. Nanomedicine 5: 698–705.

insult. This phenomenon is corroborated by a corresponding increase in cell survival rates when normal colon cells are treated with increasing doses of CeO_2 nanoparticles (Colon et al. 2010).

CeO_2 Nanoparticles Reduce Apoptotic Cell Death in Gastrointestinal Mice Cells *In Vivo*

In an attempt to investigate the ability of CeO_2 nanoparticles to protect the gastrointestinal epithelium of mice against radiation-induced damage, mice were randomized and colon tissues were harvested and processed 4 h post radiation. The colonic crypt cells from mice treated with CeO_2 nanoparticles in combination with radiation exhibited a significant decrease in apoptotic colon cryptic cells (as measured by TUNEL) and Caspase-3 expression as compared to the colonic crypt cells from radiated (no CeO_2) mice (Fig. 11). The number of TUNEL and Caspase-3 positive cells in each colonic crypt decreased by 50% in mice treated with a combination

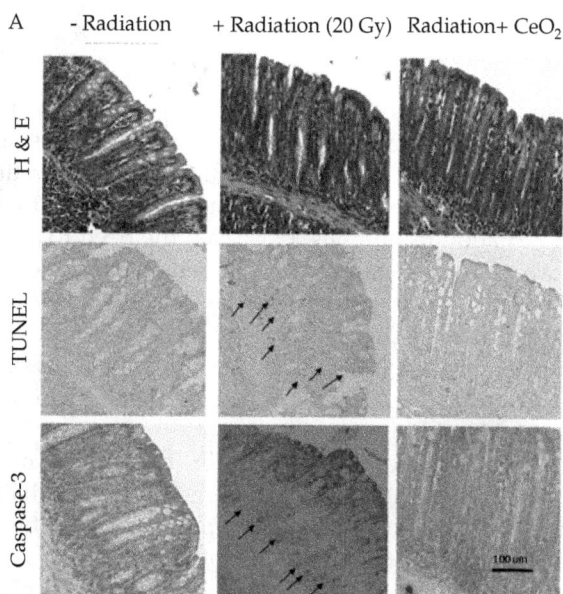

Fig. 11. CeO_2 nanoparticles protect normal human colon tissue from radiation-induced cell death. Hematoxlin and eosin (H&E) stains of murine colons 4 h post a single dose of 20 Gy radiation. Radiation was administered to the bowel of non–tumor-bearing athymic nude mice pretreated with four i.p. treatments of CeO_2 nanoparticles. Results show a significant decrease in apoptotic colon cryptic cells (as measured by TUNEL) and Caspase-3 expression as compared to the colonic crypt cells from mice treated with radiation alone. With permission from C.H. Baker 2010. Cerium oxide nanoparticles protect gastrointestinal epithelium from radiation-induced damage by reduction of reactive oxygen species and upregulation of superoxide dismutase 2. Nanomedicine 5: 698–705.

Color image of this figure appears in the color plate section at the end of the book.

of CeO_2 nanoparticles and radiation, as compared to mice treated with radiation alone. It is interesting to note the decrease in Caspase-3 in mice treated with CeO_2 nanoparticles as compared to control (normal) mice, which could be explained by the fact that CeO_2 may reduce the normal intrinsic cell death pathway and/or normal metabolic ROS, as reviewed by Rzigalinksi et al. (2003).

To demonstrate the ability of the CeO_2 nanoparticles to induce the overexpression of SOD-2, colons from mice were sectioned 24 h after a single injection of CeO_2 nanoparticles and 10 random crypts per mouse from five different mice per group were stained for SOD-2 expression (Fig. 12A). The colonic crypt cells from mice treated with CeO_2 nanoparticles exhibited a 40% increase in SOD-2 expression as compared to untreated (normal) mice (Fig. 12B). Immunohistochemical analysis of normal colon from mice treated with CeO_2 nanoparticles show an increase in SOD-2 expression (Colon et al. 2010).

DISCUSSION

The field of radiation oncology has worked diligently over the last decade to improve radiation delivery techniques in order to spare sensitive structures from the effects of ionizing radiation. These techniques have resulted in improved functional outcomes compared to prior, more rudimentary, radiation techniques. However, the need to attain adequate tumor coverage and the exquisite radiosensitivity of certain normal structures are intrinsic limitations to the magnitude of function and quality of life that can be preserved with these techniques. Hence, even with the implementation of these techniques many patients still experience significant acute and late toxicity after radiation treatment that adversely impacts their quality of life.

To further mitigate radiation-induced toxicity we must continue to develop strategies to protect normal tissues from radiation-induced damage. One such strategy is the development of radiation protectors. Several compounds have been described, but amifostine remains the only agent currently in clinical use (Citrin et al. 2010). Major limitations to the clinical use of amifostine are its short half-life, daily dosing requirements, toxicity based on route of administration, and cost (Beckman et al. 1998; Tarnuzzer et al. 2005; Colon et al. 2009; Citrin et al. 2010). Hence, there remains a substantial clinical need for a radioprotective agent that can be delivered with relative ease, is long lasting and well tolerated, and can protect a spectrum of sensitive normal tissues in which damage can cause a significant reduction in quality of life.

A.

B.

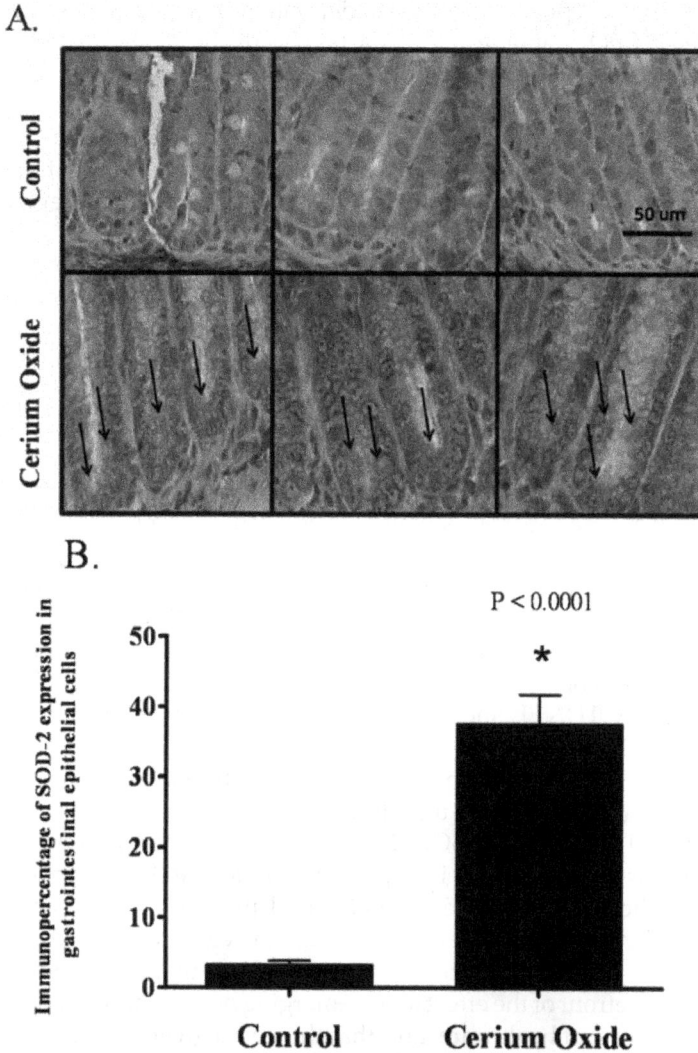

Fig. 12. CeO_2 nanoparticles induce SOD-2 expression in normal colon. A. Representative sections of SOD-2 expression (brown staining) in colonic crypts in mice treated with CeO_2 nanoparticles or in normal (control) mice. Colons were collected 24 h post a single injection of CeO_2 nanoparticles. B. The immunopercentage of SOD-2 expression increased by 40% in mice treated with CeO_2 nanoparticles as compared to control mice. Each data point represents the mean +/- SEM from analyzing 10 random crypts per mouse from five different mice, which has been expressed as percentage of crypt cells staining positive for SOD-2. With permission from C.H. Baker 2010. Cerium oxide nanoparticles protect gastrointestinal epithelium from radiation-induced damage by reduction of reactive oxygen species and upregulation of superoxide dismutase 2. Nanomedicine 5: 698–705.

Color image of this figure appears in the color plate section at the end of the book.

The above report lends a great deal of credence to the argument for the use of CeO_2 nanoparticles in a therapeutic setting as a free radical scavenger, especially in the context of therapeutic ionizing radiation. As mentioned above, CeO_2 nanoparticles, because of their large surface energy derived from a high surface to volume ratio and unique valence state oscillations, contain many oxygen vacancies that allow them to be much more efficient than endogenous antioxidants, and to be regenerative in their enzymatic activity, which we hypothesize to be due to the valence reversing from +3 to +4 valence states. Additionally, mice administered with CeO_2 nanoparticles experience no serious side effects, demonstrating the low toxicity of CeO_2 nanoparticles (Colon et al. 2009).

Elevated ROS levels have long been implicated in numerous diseases such as kidney fibrosis (Kim et al. 2009), chronic inflammation and organ dysfunction, especially when induced by ionizing radiation (Zhao et al. 2007). It is now widely accepted that ROS can interfere in intracellular processes that cause the above-mentioned injuries. Thus, the therapeutic value of CeO_2 nanoparticles may be due to their free radical scavenging properties. Furthermore, CeO_2 nanoparticles, as scavenging enzymes, are many times more efficient than SOD, which may be due to the large surface to volume ratio, as well as the ratio of Ce^{3+}/Ce^{4+} (Rzigalinski et al. 2003). The *in vivo* experiments also reinforce the conclusion that CeO_2 nanoparticles confer significant protection from ionizing radiation as evidenced by TUNEL and Caspase-3 stains, indicators of cell apoptosis (Marshman et al. 2001).

In the end, while CeO_2 nanoparticles may affect intracellular oxidative pathways, we show clearly that they are not detrimental and suspect that the elevated expression of SOD-2 contributes to an increased protection of normal cells against ROS. It is important to note that the therapeutic value of free radical scavengers extends beyond protecting against radiation-induced damage to DNA; it also results in reduction in inflammation, fibrosis and organ dysfunction. Thus, we believe that CeO_2 nanoparticles are at the forefront of the effort to use emerging nanotechnology to improve quality of life and healthcare, and that they hold great potential for future clinical trials.

Summary Points

- The potential benefit of radioprotection through use of nanoparticles could be significant. Moreover, this technology may have broad applications across the spectrum of human oncology treated with ionizing radiation. It is also important to hypothesize that this research may be relevant to the health of all individuals exposed to radiation as a result of treatment, occupation, or accident. Potential populations of humans that could derive a benefit are (1) health care workers exposed to scatter radiation from fluoroscopy or repeated imaging, (2) astronauts, (3) patients undergoing interventional procedures, and (4) military personnel and civilians exposed to radiation in battle or as a result of terrorism.

- Animal studies have demonstrated that CeO_2 nanoparticles are well tolerated in live animals.
- Lung tissues harvested after whole-lung irradiation demonstrated no histological evidence of pneumonitis and fibrosis in athymic mice treated with 15 nM CeO_2 compared to "no-nanoparticle" controls.
- Extension of this work demonstrated that CeO_2 nanoparticles protect gastrointestinal epithelium against radiation-induced damage.
- This work also suggests that CeO_2 nanoparticles confer radioprotection by acting as free-radical scavengers and by increasing the production of SOD-2.

Key Facts of Radiation Damage

- Ionizing radiation damage to normal tissue poses severe side effects that undermine its therapeutic value, and the mechanisms of this damage are being studied to resolve this issue.
- It has long been accepted that ionizing radiation results in the formation of free radicals that remain in biological systems for milliseconds and cause oxidative damage to DNA, proteins, and lipids, which combined include many of the biological effects of radiation (Altman et al. 1970; Hall 2000).
- While it had been assumed that the initial oxidative damage caused by the free radicals was responsible for the alterations in intracellular processes following irradiation, there is increasing evidence that these free radicals also play a role in intracellular metabolic oxidation/reduction reactions and might contribute to the activation of protective or damaging mechanisms that explain the damaging consequences of ionizing radiation (Spitz et al. 2004).
- Uncontrolled radical production is tied to multiple pathological diseases. The radiation damage to the macromolecules is linearly associated with the amount of radiation dose.
- The United Nation Scientific Committee's report based on the information on atomic bomb survivors' accidental radiation exposure cases as well as radiation therapy studies suggest that the LD50/60 (50% death within 60 d) is 2.5 Gy or more (bone marrow dose) when little medical assistance is available, and about 5 Gy when extensive medical care is provided.

- On a macrocosmic scale, this spectrum of ionizing radiation-induced cellular perturbation causes those symptoms commonly associated with therapeutic radiation, which include nausea, vomiting, fatigue, and hair loss.

Definitions

Apoptosis: Death of cells.

Athymic nude mice: A type of laboratory mouse that is hairless, lacks a normal thymus gland and has a defective immune system because of a genetic mutation. Athymic nude mice are often used in cancer research because they do not reject tumor cells, from mice or other species.

Balb/C mice: An albino, laboratory-bred strain of the House Mouse from which a number of common substrains are derived. Now over 200 generations from their origin in New York in 1920, BALB/c mice are distributed globally, and among the most widely used inbred strains used in animal experimentation.

Histology: Study of the anatomy of cells, tissue, plants and animals.

Free Radicals: An atom or group of atoms that has at least one unpaired electron and is therefore unstable and highly reactive. In tissues, free radicals can damage cells and accelerate the progression of cancer, cardiovascular disease, and age-related diseases.

Gastrointestinal epithelium: The lining of the stomach and intestines (includes the colon): the gastrointestinal tract.

Ionizing radiation: High-energy radiation capable of producing ionization in substances through which it passes.

Nanoparticles: Ultrafine unit with dimensions measured in nanometers (nm; billionths of a meter). Nanoparticles possess unique physical properties such as very large surface areas and can be classified as hard or soft. They exist naturally in the environment and are produced as a result of human activities. Manufactured nanoparticles can have various compositions and thus may have practical applications in a variety of areas, ranging from environmental remediation to engineering and medicine.

Pneumonitis: Inflammation of lung tissue.

Proliferate: To grow or multiply by rapidly producing new tissue, parts, cells, or offspring.

Reactive oxygen species: Chemically reactive molecules containing oxygen. During times of environmental stress (e.g., UV or heat exposure), ROS levels can increase dramatically. This may result in significant damage to cell structures, called oxidative stress. ROS are also generated by exogenous sources such as ionizing radiation.

Viable cells: Cells that are living.

Abbreviations

ROS	:	reactive oxygen species
SOD	:	superoxide dismutase
CeO$_2$	:	cerium oxide
ATP	:	adenosine triphosphate
H&E	:	hematoxylin and eosin
TGF-β	:	transforming growth factor-beta

References

Altman, K.I., G.B. Gerber, and S. Okada. 1970. Radiation Biochemistry. Academic Press, New York.

Beckman, K.B., and B.N. Ames. 1998. The free radical theory of aging matures. Physiol. Rev. 78: 547–581.

Chen, J., S. Patil, S. Seal, and J. McGinnis. 2008. Nanoceria particles prevent ROI-induced blindness. Adv. Exp. Med. Biol. 613: 53–59.

Chen, J., S. Patil, S. Seal, and J.F. McGinnis. 2006. Rare earth nanoparticles prevent retinal degeneration induced by intracellular peroxides. Nature Nanotechnol. 1: 142–150.

Citrin, D., and A.P. Cotrim, F. Hyodo, B.J. Baum, M.C. Krishna and J.B. Mitchell. 2010. Radioprotectors and mitigators of radiation-induced normal tissue injury. Oncologist 15: 360–371.

Cohen-Jonathan, E., E.J. Bernhard, and W.G. McKenna. 1999. How does radiation kill cells? Curr. Opin. Chem. Biol. 3: 77–83.

Colon, J., L. Herrera, J. Smith, S. Patil, C. Komanski, P. Kupelian, S. Seal, D.W. Jenkins, and C.H. Baker. 2009. Protection from radiation-induced pneumonitis using cerium oxide nanoparticles. Nanomedicine 5: 225–231.

Colon, J., N. Hsieh, A. Ferguson, P. Kupelian, S. Seal, D.W. Jenkins, and C.H. Baker. 2010. Cerium oxide nanoparticles protect gastrointestinal epithelium from radiation-induced damage by reduction of reactive oxygen species and upregulation of superoxide dismutase 2. Nanomedicine 5: 698–705.

Gharbi, N., M. Pressac, M. Hadchouel, H. Szwarc, S.R. Wilson, and F. Moussa. 2005. [60] fullerene is a powerful antioxidant *in vivo* with no acute or subacute toxicity. Nano Lett. 5: 2578–2585.

Hall, E.J. 2000. Radiobiology for the Radiologist. Lippincott Williams and Wilkins, Philadelphia.

Heckert, E.G., A.S. Karakoti, S. Seal, and W.T. Self. 2008. The role of cerium redox state in the SOD mimetic activity of nanoceria. Biomaterials 29: 2705–2709.

Kim, J., Y.M. Seok, K.J. Jung, and K.M. Park. 2009. Reactive oxygen species/oxidative stress contributes to progression of kidney fibrosis following transient ischemic injury in mice. Amer. J. Renal Physiol. 297: 461–470.

Korsvik, C., S. Patil, S. Seal, and W.T. Self. 2007. Superoxide dismutase mimetic properties exhibited by vacancy engineered ceria nanoparticles. Chem. Commun. 10: 1056–1058.

Lee, J.C., R. Krochak, A. Blouin, S. Kanterakis, S. Chatterjee, E. Arguiri, A. Vachani, C.C. Solomides, K.A. Chengel, and M. Christofidou-Solomidou. 2008. Dietary flaxseed prevents radiation-induced oxidative lung damage, inflammation and fibrosis in a mouse model of thoracic radiation injury. Cancer Biol. Ther. 1: 27–33.

Marshman, E., P.D. Ottewell, C.S. Potten, and A.J.M. Watson. 2001. Caspase activation during spontaneous and radiation-induced apoptosis in the murine intestine. J. Pathol. 195: 285–292.

Meissner, K. 1999. Late radiogenic small bowel damage: guidelines for the general surgeon. Dig. Surg. 16: 169–174.

Nair, C.K.K., D.K. Parida and T. Nomura. 2001. Radioprotectors in radiotherapy. J. Rad. Res. 42: 21–37.

Park, E.J., J. Choi, Y.K. Park, and K. Park. 2008. Oxidative stress induced by cerium oxide nanoparticles in cultured BEAS-2B cells. Toxicology 245: 90–100.

Patil, S., A. Sandberg, E. Heckert, W.T. Self, and S. Seal. 2007. Protein absorption and cellular uptake of cerium oxide nanoparticles as a function of zeta potential. Biomaterials 28: 4600–4607.

Perez, M.J., A. Asati, S. Nath, and C. Kaittanis. 2008. Synthesis of biocompatible dextran-coated nanoceria with pH-dependent antioxidant properties. Small 5: 552–556.

Pradhan, D.S., C.K.K. Nair, and A. Sreenivasan. 1973. Radiation injury repair and sensitization of microorganisms. Proc. Indian Nat. Sci. Acad. B. 39: 516–530.

Robnett, T.J. 2000. Factors predicting severe radiation pneumonitis in patients receiving definitive chemoradiation for lung cancer. Int. J. Rad. Oncol. Biol. Phys. 1: 89–94.

Rzigalinski, B.A., D. Bailey, L. Chow, S.C. Kuiry, S. Patil, S. Merchant, and S. Seal. 2003. Cerium oxide nanoparticles increase the lifespan of cultured brain cells and protect against free radical and mechanical trauma. Faseb J. 17: A606.

Spitz, D.R., E.I. Azzam, J.J. Li, and D. Gius. 2004. Metabolic oxidation/reduction reactions and cellular responses to ionizing radiation: A unifying concept in stress response biology. Cancer Metastasis Rev. 23: 311–322.

Tarnuzzer, R.W., J. Colon, S. Patil, and S. Seal. 2005. Vacancy engineered ceria nanostructures for protection from radiation-induced cellular damage. Nano Lett. 5: 2573–2577.

Zhao, W., D.I. Diz, and M.E. Robbins. 2007. Oxidative damage pathways in relation to normal tissue injury. Br. J. Radiol. 80: S23–S31.

Nanoparticle Therapy in Arthritis

José L. Arias

ABSTRACT

Conventional pharmacotherapy against arthritis is seriously limited by the high incidence and severity of adverse side effects (drug toxicity). This is the consequence of the little affinity of anti-arthritic molecules for damaged joints, which also leads to low drug concentrations at the site of action (drug inefficiency). Several approaches have tried to beat this challenge, e.g., chemical modifications in the drug molecule and formulation of drug delivery systems. The latter strategy can result in enhanced drug localization into the targeted region (controlled biodistribution), thus enhancing the anti-arthritic effect, while reducing systemic toxicity. Such colloidal carriers are of benefit for oral, parenteral, and transdermal drug delivery to arthritis. These nano-formulations offer remarkable possibilities when they are made of biodegradable polymers and lipid-based colloids. The biological fate (and efficacy) of these nanomedicines can be significantly improved by passive and/or active targeting strategies. In this chapter, we analyse the real possibilities and future perspectives of nanoparticle therapy in arthritis. It is expected that the application of pharmaceutical nanotechnology to arthritis treatment will revolutionize conventional pharmacotherapy against the disease.

Department of Pharmacy and Pharmaceutical Technology, Faculty of Pharmacy, University of Granada, Campus Universitario de Cartuja s/n, 18071 Granada, Spain; Email: jlarias@ugr.es

List of abbreviations after the text.

INTRODUCTION

The term *arthritis* comprises a number of diseases characterized by evolving damage (degeneration) of the joints. It is characterized by a chronic inflammation of the joint synovium and severe pain. The pathophysiological cause of the disease remains not completely clear, but inflammation is associated to tissues close to the joints attacked by the immune system. The major key factors involved in the origin and development of the degenerative joint disease are trauma, infection, and age. Arthritis in the form of osteoarthritis is the most common joint disorder, and the most frequent cause of pain, loss of function and disability in adults. In the United States of America, it is the second cause of work disability in men over 50 years old (Arden and Nevitt 2006).

Pharmacotherapy fails in preventing the damage process itself and can only help patients to feel better. Marketed medicines are merely capable of preventing joint destruction and deformity, reducing joint inflammation and pain, and maximizing joint function. Arthritis patients are treated with three general classes of drugs: non-steroidal anti-inflammatory drugs (NSAIDs), disease-modifying antirheumatic drugs (DMARD), and corticosteroids. Although these molecules are effective, their use is associated to a high incidence of serious side effects (i.e., liver and kidney, bleeding, ulcers, skin bruising, weight gain, cataract, abdominal pain, bone thinning, bone marrow depression, diabetes, and pain and possibility of infections at the injection site), particularly under extended treatment schedules. Severe cardiovascular effects (myocardial infarction, stroke, and hypertension) and renal effects (decreased renal blood flow and glomerular filtration rate) have been also described. These adverse drug reactions are the result of an extensive drug biodistribution into healthy tissues (lack of selectivity for the targeted site of action), and of the need of high drug doses to obtain an efficient therapeutic effect (drug dose leads to very low drug concentrations at the site of action) (Brigger et al. 2002; Arias 2009; Ulbrich and Lamprecht 2010). As a consequence, such medicines should not be used for prolonged periods without proper physician supervision.

The association of drug molecules to nanoparticulate drug delivery systems in arthritis treatment has been proposed as a very interesting approach for the improvement of drug efficacy, and for the minimization of the associated toxicity (Arias et al. 2009, 2010; Ulbrich and Lamprecht 2010). In this way, there is a clear connection between the requirements to fulfil in the formulation of such nanomedicines and the pharmacotherapy outcomes that are obtained (Table 1) (Brigger et al. 2002; Arias 2009; Decuzzi et al. 2009). Very interestingly, researchers have hypothesized the use of nanoparticles as potential dual-modality contrast agents (e.g., containing embedded luminophores, and surface-immobilized gadolinium chelates) for imaging of inflamed synovium in inflammatory arthritis (Kim et al. 2009). However, it should be kept in mind that nanoparticulate drug carriers must be widely characterized (e.g., toxicity, chemistry, physics, pharmacokinetics, biodistribution) before human use. Hence, each colloid must be always considered as a new formulation with unique characteristics.

Table 1. Engineering requisites in the formulation of drug-loaded nanoparticles, and outcomes to be obtained in arthritis treatment.

Engineering requisites	Pharmacotherapy outcomes
Adequate geometry (spherical shape, and size under 200 nm).	Long-circulating properties: controllable drug biodistribution and complete drug accumulation in the targeted site. Extended exposure of the damaged joints to the drug molecules. Improved pharmacokinetic profile of the pharmacotherapy agent (uniform oral absorption, long biological half-life, etc.). Clear reduction in drug toxicity.
Hydrophilicity (or negligible hydrophobicity).	
(Almost) null surface electrical charge.	
Lack of drug leakage problems *in vitro* and *in vivo*; controllable drug release exclusively into the arthritis site.	
Maximum biocompatibility and biodegradability, with negligible toxicity of breakdown products.	Minimal antigenicity and toxicity.
Physical and chemical stability.	
Delivery of adequate drug amounts, without overloading the body with unfamiliar material.	

This chapter is devoted to the analysis of the possibilities and applications of nanoparticulate drug delivery systems in the formulation of nanomedicines against arthritis. Particular attention is given to recent advances in nanoparticle engineering to control the biological fate of the active agents: passive targeting and active targeting strategies.

POLYMER-BASED NANOPARTICLES

Polymers are one of the most promising materials to be used in the formulation of nanoparticulate drug delivery systems. They can be easily engineered to assure the maximum drug-loading capacity along with an ideal sustained drug release. Maybe the most promising biodegradable polymers for nanoparticle therapy in arthritis are chitosan, poly(ε-caprolactone), poly(D,L-lactide-*co*-glycolide) (PLGA), and poly(alkylcyanoacrylate).

Chitosan

Chitosan is a semicrystalline and pseudo-natural cationic polymer that can be synthesized by deacetylation of chitin. It is characterized by a very good hydrosolubility, biocompatibility (and biodegradability) and, very importantly for oral drug delivery applications, by a very high mucoadhesivity. Chitosan nanoparticles can be easily prepared by coacervation. Briefly, the addition of sodium sulphate (under mechanical stirring) to a solution of chitosan in acetic acid determines a decreased solubility of the polymer, rapidly leading to its precipitation into nanoparticles.

Several investigations have described the formulation of chitosan-based nanomedicines with very good drug-loading and controlled drug release properties. Very interestingly, it has been also suggested as a very suitable material for gene delivery (transfection). For these reasons, it is used in the development of nanoparticulate drug delivery systems against arthritis (Arias 2009). For instance, a recent investigation proved that it is possible to knock down the tumor necrosis factor-alpha (TNF-alpha) expression in macrophages (and, hence, downregulate systemic and local inflammation) by intraperitoneal administration of anti–TNF-alpha Dicer-substrate small interfering RNA (DsiRNA)–loaded chitosan nanoparticles. The nanomedicine was able to downregulate TNF-alpha–induced inflammatory responses and, consequently, to arrest joint swelling in collagen-induced arthritic mice. Histology studies of joints showed minimal cartilage destruction and inflammatory cell infiltration in treated mice. It was hypothesized that the onset of arthritis could be delayed using a prophylactic nanomedicine dosing regime (Howard et al. 2009). Similar results have been obtained when chitosan nanoparticles loaded with the interleukin-1 receptor antagonist (IL-1Ra) were used for direct gene therapy against osteoarthritis. IL-1Ra–loaded chitosan nanoparticles were injected directly into the knee joint cavities of osteoarthritic rabbits, and a clear expression of IL-1Ra was detected in the knee joint synovial fluid (in contrast with control groups). As a consequence, a significant reduction in the severity of osteoarthritis cartilage breakdown (histologic lesions) was possible (Zhang et al. 2006).

Poly(ε-caprolactone)

Poly(ε-caprolactone) is a highly hydrophobic crystalline polymer widely used in controlled drug delivery, thanks to its biodegradability, very slow degradation rate, high drug permeability, and non-toxicity. Drug-entrapped poly(ε-caprolactone) nanoparticles are generally synthesized by the interfacial polymer disposition, or the dialysis methods. For example, when nanoparticles are prepared by interfacial polymer disposition, the polymer is dissolved in an organic solvent. This organic phase is poured by drops under mechanical stirring into an aqueous solution containing a surfactant agent (i.e., pluronic® F-68) to immediately obtain a nanoparticle suspension. Finally, the organic solvent is removed under reduced pressure (Arias 2009).

Following this procedure, poly(ε-caprolactone) nanoparticles loaded with the NSAID diclofenac sodium can be obtained (average size $\approx$ 200 nm; drug entrapment efficiency $\approx$ 45%; drug loading $\approx$ 20%) with a very appropriate drug release profile: a biphasic process with an early rapid drug release ($\approx$ 40% within 2 h), the remaining drug being slowly liberated during the next 94 h (Arias et al. 2010). It is expected that such formulation can improve the gastrointestinal tolerance to diclofenac sodium (Schaffazick et al. 2003).

Poly(D,L-lactide-*co*-glycolide)

Poly(D,L-lactide-*co*-glycolide) or PLGA may be the most important biodegradable polymer in the formulation of nanoparticles for drug delivery purposes. PLGA is a copolymer of PLA and poly(glycolide) that has been approved by the Food and Drug Administration for human use. The most widely used PLGA composition (50:50) is characterized by the fastest biodegradation rate of the polymeric family ($\approx$ 50 d). Drug-loaded nanoparticles are generally prepared by the emulsion or double-emulsion technique (followed by solvent evaporation or spray drying), using polyvinyl alcohol (PVA) as a stabilizer. Briefly, an organic solution of the polymer is emulsified in an aqueous PVA phase. Solvent evaporation is the final step of the formulation procedure (Arias 2009). PLGA-based nanoplatforms loaded with anti-arthritic drugs (e.g., ibuprofen) are a very promising approach against arthritis (Neumann et al. 2010).

As an example, betamethasone sodium phosphate–loaded PLGA nanoparticles (average size $\approx$ 100–200 nm) has been formulated by an oil in water emulsion solvent diffusion method. The nano-steroid was administered intravenously after the initial sign of arthritis to rats with adjuvant arthritis, and to mice with anti-type II collagen antibody–induced arthritis. Importantly, a 30% decrease in paw inflammation was achieved in 1 d and maintained for 1 wk after a single injection of the nanomedicine. Even more, reduced soft tissue swelling was described 7 d after the treatment. Similar results were also obtained in mice with anti-type II collagen antibody–induced arthritis. The anti-arthritic effect was hypothesized to be the consequence of a selective accumulation of the nanomedicine in the inflamed joint and the prolonged *in situ* drug release (Higaki et al. 2005). In another recent investigation, dexamethasone (DEX)-loaded PLGA nanoparticles very efficiently suppressed macrophage cell growth by $\approx$ 40% compared to controls (Fig. 1) (Lo et al. 2010).

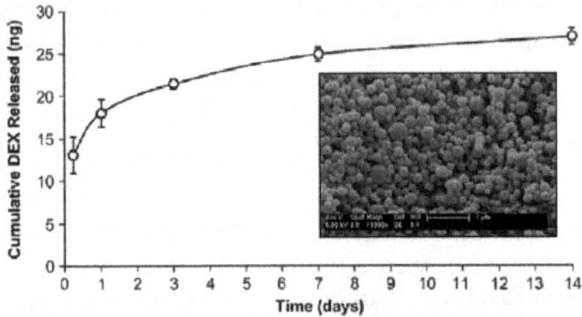

Fig. 1. DEX release from PLGA nanoparticles as a function of the incubation time in phosphate buffered saline (PBS, pH = 7.4 ± 0.1). Inset: scanning electron microscope (SEM) picture of the nanomedicine. Bar length: 1 μm. DEX-loaded PLGA nanoparticles (average size ≈ 200 nm) were satisfactorily formulated with prolonged/sustained drug release characteristics. Reprinted with permission from Lo et al. (2010). Copyright Elsevier (2010).

Poly(alkylcyanoacrylate)

Poly(alkylcyanoacrylate)-based nanomedicines are biodegradable and biocompatible nanoplatforms characterized by a low toxicity in multiple dosing. They are principally prepared by the emulsion/ polymerization method. In short, the methodology involves the addition of an alkylcyanoacrylate monomer to an acidic aqueous solution containing a surfactant agent (i.e., dextran-70 or pluronic® F-68). Under these conditions, the polymer precipitates spontaneously into nanoparticles (Arias 2009).

Arthritis treatment can take advantage of the drug delivery capabilities of these polymers. For instance, poly(ethylcyanoacrylate) nanoparticles have been formulated for oral delivery of the NSAID ketorolac (drug loading ≈ 85%). In this research report, two groups of healthy rats were given orally either a ketorolac tromethamine solution (dose: 1.5 mg/kg) or a ketorolac-loaded nanoparticle aqueous dispersion (dose: 1.6 mg/kg) (Fig. 2) (Radwan et al. 2010).

LIPID-BASED NANOPARTICLES

Classically, the use of lipid-based drug delivery systems has competed with biodegradable polymers. However, it should be considered as an alternative/complementary possibility in the development of safe and effective nanomedicines against disease. The development of lipid-based nanoparticles against arthritis has been mainly focussed on the formulation of liposomes, niosomes, and solid lipid nanoparticles.

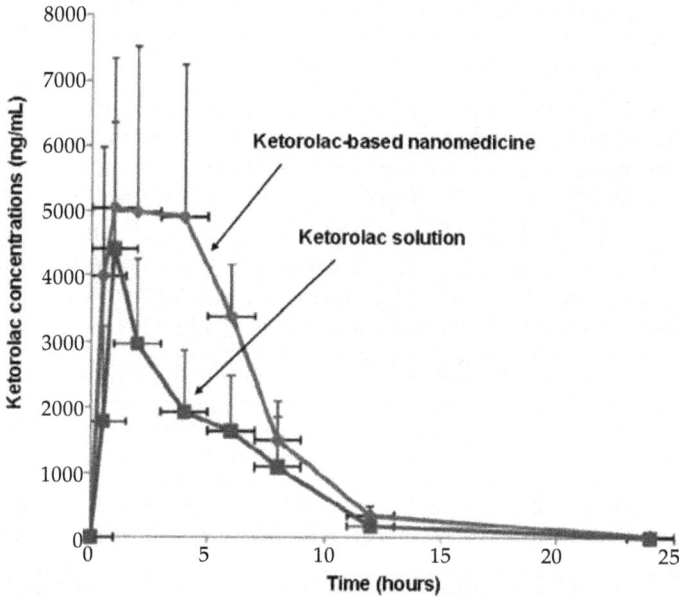

Fig. 2. Average ketorolac plasma concentration (ng/mL) *vs.* time (h) curves in rats after oral administration of 1.5 mg/kg of drug solution, and 1.6 mg/kg of the drug-loaded poly(ethylcyanoacrylate) suspension (n = 6). The area under curve of plasma concentration-time of the drug was significantly higher than that corresponding to the ketorolac solution. Reprinted with permission from Radwan et al. (2010). Copyright Elsevier (2010).

Liposomes

Liposomes are biodegradable (and biocompatible) vesicular systems made of at least one phospholipid bilayer. Interestingly, they can incorporate highly lipophilic drugs into their lipid bilayer(s) and/or encapsulate water-soluble active agents in their internal aqueous compartments (whether into the core or between lamellae). The formulation procedure starts by dissolving lipids in an organic solvent. Then, a dry lipid film is obtained by solvent evaporation, which is next dispersed into an aqueous phase. The preparation methods differ according to the procedure followed to disperse the lipids (sonication, extrusion, thin lipid film hydration, etc.) (Arias 2009).

Liposome-based formulations can reduce drug side effects while enhancing drug efficacy, by reducing the availability of active agents in systemic circulation and increasing drug accumulation and retention time at the inflammation site (Ulbrich and Lamprecht 2010). For instance, a better anti-inflammatory effect after a single dose administration has been described when the NSAID diclofenac sodium is incorporated to a liposome formulation, compared to the equivalent free drug solution. Histopathology studies showed that joints treated with the lipid-based nanomedicine had significantly ($p < 0.05$) lower scores than contra lateral control joints for inflammatory changes in the synovium. The results were further confirmed by biodistribution studies (Türker et al. 2008).

As an improvement to these formulations against arthritis, the surface functionalization of liposomes with poly(ethylene glycol) (PEG) chains has been proposed. These sterically stabilized nanometer-sized liposomes (mean diameter ≈ 100 nm) efficiently exhibit a prolonged *in vivo* circulation (see *Passive Drug Targeting to Arthritis* for further details). For example, such liposomes have been developed for the *in vivo* delivery of the glucocorticoid methylprednisolone hemisuccinate (Schroeder et al. 2008).

Niosomes

Niosomes are closed bilayer structures obtained by self-assembly of non-ionic amphiphiles in aqueous media. The larger stability, lower cost, and relative storage stability hold up the use of these vesicles as an alternative to phospholipid ones. Little research has been done to date on the development of niosome-based drug nanocarriers for arthritis therapy. Interestingly, high drug loading values and sustained drug release capacity have been described when using this nanoplatform (Arias 2009).

For instance, a niosomal gel formulation loaded with celecoxib has been prepared for enhanced skin drug accumulation, prolonged drug release, and improved site specificity of this NSAID. The niosomal gel provided 6.5-fold the drug deposition of carbopol gel (195.2 ± 8.7 and 30.0 ± 1.5 µg, respectively). Moreover, the muscle to plasma concentration ratio for niosomal gel formulation was 6 (2.16 ± 0.12 µg/g *vs.* 0.34 ± 0.01 µg/mL), while for carbopol gel the ratio was 1 (0.36 ± 0.01 µg/g *vs.* 0.43 ± 0.02 µg/mL). As a result, significant reduction of rat paw edema was obtained compared to that after application of the conventional gel (Kaur et al. 2007).

Solid Lipid Nanoparticles

Solid lipid nanoparticles (SLN) can be defined as aqueous dispersions of solid lipid matrices stabilized by surfactants, or as dry lipid powders obtained by spray drying or lyophilization. SLN are particularly characterized by their capacity of remaining in solid form even at physiological temperature, physical stability, low toxicity, sustained drug release properties, protection of labile drugs from degradation, and cheap and easy scale production. Their formulation involves the use of biodegradable (and biocompatible) lipids, e.g., hard fats (cetylpalmitate and glycerol behenate), triglycerides (trilaurin, tripalmitin, and tristearin), or lipid acids (palmitic acid and stearic acid). They must be stabilized in an aqueous phase by using compatible emulsifiers (e.g., poloxamer 188, lecithin, polysorbate 80). Independently of the formulation methodology (mostly solvent emulsification-evaporation or diffusion, high-pressure homogenization, or microemulsification), the lipid must be melted and dispersed into nanometer-sized lipid droplets to ensure the formation of nanoparticles after incorporation into an aqueous medium (Arias 2009).

In order to enhance the loading of water-soluble drugs into SLN, three charge neutralization mechanisms have been suggested: the complexation of ionic polymers and drugs, the addition of organic counterions to form ionic pairs with charged drug molecules, or the preparation of "lipid drug conjugate" nanoparticles. When highly water-soluble drugs are not surface-charged, one interesting approach is the preparation of lipophilic drug derivatives. The pharmacokinetic and pharmacodynamic profiles of drugs can be greatly enhanced when such molecules are loaded to SLN (Arias 2009). SLN has been formulated for prolonged drug release of anti-arthritic drugs, i.e., celecoxib (Thakkar et al. 2007), ibuprofen (Paolicelli et al. 2008), and DEX (Serpe et al. 2010).

PASSIVE DRUG TARGETING TO ARTHRITIS

It is a fact that any given nanoparticulate drug delivery system displays an intense interaction with the reticuloendothelial system (RES) upon administration, including Kupffer cells and related organs (spleen, liver, lungs, and bone marrow). Interestingly, anti-arthritic treatment by passive targeting of drug-loaded nanoparticles can be then achieved thanks to the immune response. Drug nanocarriers have a natural tendency to accumulate in the organs that comprise the mononuclear phagocytic system (MPS), and this can be advantageously used as a strategy for rheumatoid arthritis since the targets are the same cells (macrophages) that are responsible for nanoparticle clearance from blood (plasma half life $\approx$ 3 min). Additionally, we should keep in mind that the spleen is the major immune organ in the production of macrophages involved in the inflammatory response. Hence, spleen targeting mediated by macrophages using the passive targeting tendency of DMARD-loaded nanoparticles is expected to significantly enhance their therapeutic effect, along with an important minimization of the associated severe side effects (Ye et al. 2008; Ulbrich and Lamprecht 2010).

For example, SLN loaded with the DMARD actarit were prepared by the solvent diffusion-evaporation method (mean size $\approx$ 250 nm; average drug entrapment efficiency and loading $\approx$ 50% and 8%, respectively). It was determined that actarit-loaded SLN exhibited a longer mean retention time (MRT $\approx$ 14 h) than the corresponding actarit PEG solution (MRT $\approx$ 1 h) after intravenous injection to New Zealand rabbits. The area under curve of plasma concentration-time of actarit-loaded SLNs was 1.88 times that of the actarit PEG solution. Thanks to the use of SLN, the overall targeting efficiency of actarit was increased from $\approx$ 6% to $\approx$ 16% in spleen, and the renal distribution of the DMARD was significantly reduced in comparison to the free drug solution (Ye et al. 2008).

There is an alternative passive drug targeting strategy to specifically target the diseased joints. It is based on the locally enhanced capillary permeability of arthritic joints (Paleolog and Fava 1998). Long-circulating engineering allows anti-arthritic drug-loaded nanoparticles to exploit these structural abnormalities in the vasculature of diseased joints. Apart from the intrinsic properties of nanoparticles (geometry, surface thermodynamics, and surface electrical charge), the introduction by physical adsorption or chemical conjugation of hydrophilic polymers (i.e., PEG, poloxamers, poloxamines, or polysaccharides) onto the nanoparticle surface will provide a shell of hydrophilic and neutral chains able to repel plasma proteins (Fig. 3) (Brigger et al. 2002; Arias 2009).

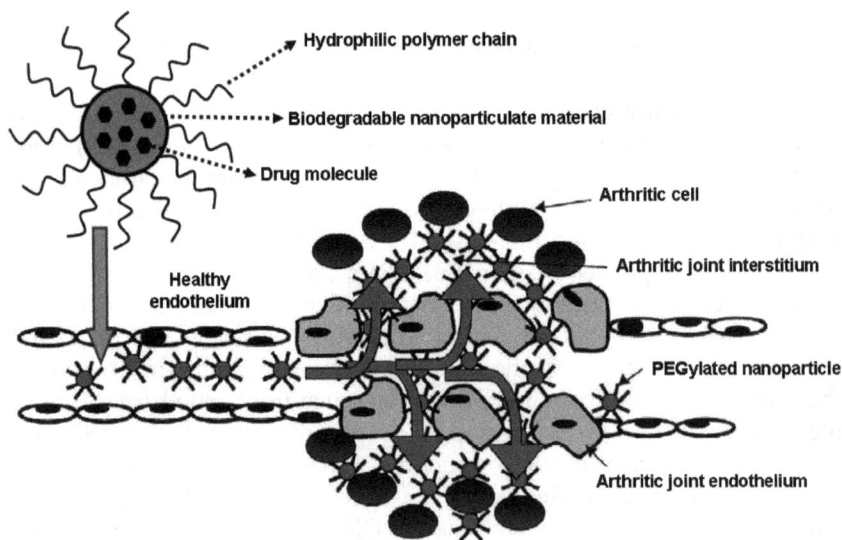

Fig. 3. Extravasation of long-circulating nanoparticles into the arthritic joint interstitium by passive diffusion or convection through the hyperpermeable endothelium. Surface functionalization with hydrophilic polymers will retard nanoparticle opsonization (and plasma clearance). This specific extravasation and accumulation of nanoparticles into the arthritic joint, known as the enhance permeability and retention effect, is based on the "leaky" vasculature of these tissues. Adapted with permission from Brigger et al. (2002). Copyright Elsevier (2002).

An interesting example on this second alternative for passive targeting to arthritis joints has been described (Ishihara et al. 2009). In this work, betamethasone disodium 21-phosphate–loaded PLGA/PLA nanoparticles and PEGylated betamethasone disodium 21-phosphate–loaded PLGA/PLA nanoparticles were evaluated in experimental arthritis models. PEGylated nanoparticles (mean size ≈ 115 nm) exhibited the highest anti-inflammatory activity. In rats with adjuvant arthritis, a 35% decrease in paw inflammation was maintained for 9 d with a single injection of long-circulating drug–loaded nanoparticles (drug dose: 40 µg). However, a weaker response was obtained when administering even superior doses of non-PEGylated drug–loaded nanoparticles, and free betamethasone solution. In mice with anti-type II collagen antibody–induced arthritis, a single injection of the PEGylated formulation (drug dose: 3 µg) resulted in complete remission of the inflammatory response after 1 wk. It was concluded that the strongest therapeutic benefit obtained with the PEGylated betamethasone disodium 21-phosphate–loaded PLGA/PLA nanoparticles was the consequence of a prolonged plasma half-life and targeting to the inflamed joint, in addition to an *in situ* sustained drug release.

ACTIVE DRUG TARGETING TO ARTHRITIS

New drug delivery strategies have tried to further enhance the accumulation of such nanomedicines in the arthritis joint, while keeping to a very minimum their systemic distribution (and severe side effects). Active drug targeting can be based on a specific recognition mechanism (ligand- or receptor-mediated targeting) and/or on the formulation of stimuli-sensitive drug nanoplatforms (Fig. 4). However, some difficulties should be overcome before the final introduction of such advanced nanomedicines in clinic (1) strongly binding ligands could make difficult the disposition of drug nanocarriers into the arthritis joint; (2) immunonanoparticles may show enhanced clearance from the body; and (3) nanoparticle internalization generally takes place by endocytosis, so the anti-arthritic drug should be able to escape from endosomes/lysosomes before degradation (Arias 2009).

DRUG DELIVERY THROUGH SPECIFIC RECOGNITION MECHANISM (LIGAND- OR RECEPTOR-MEDIATED TARGETING)

This active drug targeting strategy to arthritic joints naturally occurs by both intra-articular and intravenous administration of drug-loaded

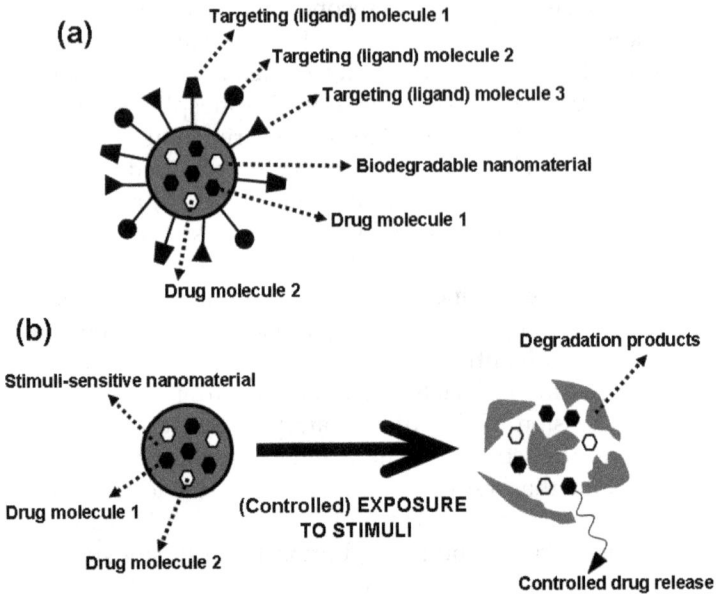

Fig. 4. Nanomedicine surface decorated with targeting moieties for ligand-mediated cell adhesion/internalization (a), and/or (b) engineered with nanomaterials sensitive to an external stimulus. Surface functionalization of the nanoparticles is typically done with folate, integrin, monoclonal antibody, and/or peptide molecules. Common external stimuli involved in the disruption of the nanoparticle (and drug release) are light, ultrasound, enzymes, pH, temperature, and/or magnetic gradient.

nanoparticles that are surface-functionalized with ligand moieties able to bind to unique targeted molecules on arthritic cells and/or inflammation vascular endothelium. The potential use of several biomolecules with that aim has been described, e.g., cell-adhesion molecules (P-selectin, E-selectin, and L-selectin), RGD (arginine-glycine-aspartic acid) peptides and integrins, folate moieties, and monoclonal antibodies (Arias 2009; Ulbrich and Lamprecht 2010).

Ligand-targeted nanotherapy has been investigated to deliver antiangiogenic agents (e.g., fumagillin) for the treatment of inflammatory arthritis. In a recent study, arthritis was established using the K/BxN mouse model of inflammatory arthritis. Then, mice received three consecutive daily doses of fumagillin-loaded nanoparticles surface-functionalized with $\alpha v \beta_3$ ligand targeting integrins. Fumagillin-loaded nanoparticles surface-decorated with $\alpha v \beta_3$ were injected to mice. As a consequence, the animals exhibited a significantly lower disease activity score (mean value ≈ 1; $p < 0.001$) and change in ankle thickness (mean increase ≈ 0.2 mm; $p < 0.001$) 7 d after arthritis induction. Controls received no treatment or blank nanoparticles surface-functionalized with $\alpha v \beta_3$. Both control groups exhibited a mean arthritic score of ≈ 9, and a mean change in ankle thickness of ≈ 1 mm. Synovial tissues from mice treated with targeted fumagillin-loaded nanoparticles presented significant decrease in inflammation and angiogenesis, and preserved proteoglycan integrity (Zhou et al. 2009).

The folate receptor can be found in synovial mononuclear cells and CD14$^+$ cells from patients with rheumatoid arthritis. The interaction of folate moieties with these folate receptors is reported to lead to nanoparticle endocytosis. Maybe the most interesting advantage of this active drug targeting strategy is the constant availability of folate receptors onto cell membrane thanks to continuous recycling (Fernandes et al. 2008; Arias 2009). Folate-mediated gene delivery has been investigated for the improvement of the protective effects of IL-1Ra (a natural blocker of the inflammatory cytokine interleukin-1) against bone damage and inflammation in a rat adjuvant-induced arthritis model. Compared to rats injected with IL-1Ra gene-loaded chitosan nanoparticles (control group), gene-loaded nanoparticles surface-decorated with folate moieties significantly reverted alterations in bone turnover in arthritic animals by modulating the osteocalcin level, as well as the activities of alkaline phosphatase and tartrate-resistant acid phosphatase. The protective effects were evident from the decrease in the expression levels of interleukine-1beta and prostaglandin E$_2$. Furthermore, gene-loaded chitosan nanoparticles surface-functionalized with folate molecules were less cytotoxic and enhanced IL-1Ra protein synthesis, compared to control formulations (Fernandes et al. 2008).

DRUG DELIVERY BY STIMULI-SENSITIVE NANOPARTICLES

Stimuli-sensitive nanocarriers are made of nanoparticulate materials that are able to alter their physical properties (e.g., disruption/aggregation, swelling/deswelling) under exposure to a specific external stimulus. Active drug targeting strategies can take advantage of this characteristic to trigger drug release exclusively into the targeted site (acid-, thermosensitive-, light-, ultrasound-, or enzyme-triggered drug release) or, otherwise, to totally accumulate the drug at the targeted site of action before allowing its release (e.g., magnetically responsive nanoparticulate materials) (Arias 2009).

Light-triggered Drug Release

Liposomes are widely used to formulate photosensitive drug delivery systems for light-triggered drug release. Different mechanisms to disrupt liposomes by light exposure have been investigated. For instance, plasmalogen photooxidation is based on enhanced membrane permeability upon photooxidative cleavage of plasmenylcholine. Alternatively, liposome photofragmentation and destabilization can be the consequence of exciting at 345 nm photocleavable derivatives of dioleoylphosphatidylethanolamine (Arias 2009). Recently, anti-arthritic magnetic nanoassemblies (mean size ≈ 15 nm) have been formulated by attaching magnetite nanoparticles to the glucocorticoid DEX through a photosensitive linker. DEX release by photo-triggered response exclusively occurred by near-infrared irradiation (Banerjee and Chen 2009).

Enzyme-triggered Drug Release

It is possible to formulate nanocarriers susceptible to enzymes specifically overexpressed at the site of action of the vehiculized drug. Under exposition to the enzyme, the drug-loaded nanoparticles are disrupted and drug release takes place. Long-circulating drug-loaded liposomes are commonly formulated to be disrupted by enzymes, e.g., secretory phospholipase A_2, elastase, sphingomyelinase, phospholipase C, alkaline phosphatase, or transglutaminase (Arias 2009).

Polyamidoamine dendrimers have been recently formulated for ibuprofen active targeting by enzyme-triggered drug release. It was determined that ibuprofen release in diluted plasma only takes place by esterase activity, when a linear methoxiPEG-ibuprofen conjugate was used to load the drug to the nanocarrier (Kurtoglu et al. 2010).

Acid-triggered Drug Release

Typically, pH-sensitive drug nanocarriers are engineered to be stable at physiological pH until they get under exposure to acidic environments, where they do degrade and drug release occurs. pH-sensitive functional groups, e.g., sulphonamide, are used for enhancing drug release under these conditions. For instance, poly(vinylpyrrolidone-co-dimethylmaleic anhydride) is used to formulate pH-sensitive polymer-based nanoparticulate drug delivery systems (Arias 2009).

In a recent investigation, ibuprofen and naproxen were very efficiently loaded to dextran via *in situ* activation of its carboxylic groups with N,N'-carbonyldiimidazole. The resulting hydrophobic derivatives self-assemble into nanoparticles with high drug loading values. Interestingly, it was observed that a defined tuning of the degree of substitution value may allow the adjustment of the pH-dependent hydrolysis rate and, as a consequence, drug release (Horning et al. 2009). Acid-triggered drug release has been described in poly(β-benzyl-L-aspartate)-*block*-poly(vinylpyrrolidone) diblock copolymeric nanoparticles loaded with prednisone acetate (mean size ≈ 70 nm; drug entrapment efficiency ≈ 35%; and drug loading ≈ 6%). It was determined that prednisone acetate release at pH 7.4 was much slower than at pH 2.1. In detail, at pH 7.4, no more than 40% of the initial prednisone acetate loaded amount was released after 50 h. In contrast, at pH 2.1, up to 68% of the initial drug loaded amount was released after 50 h (Wang et al. 2009).

Thermosensitive-triggered Drug Release

The physical behavior of temperature-sensitive polymers can be exploited to control drug delivery. These polymeric materials [e.g., poly(*N*-isopropylacrylamide), and derivatives or copolymers] are characterized by a significant difference in the hydration degree below and above their lower critical solution temperature (LCST) (Arias 2009).

For instance, amphiphilic block copolymers of hydrophilic poly(*N*-isopropylacrylamide) and hydrophobic poly(10-undecenoic acid) are characterized by a LCST of 30.8°C. When the temperature is under LCST, the block copolymer forms micelles in an aqueous medium (average size ≈ 160 nm). Interestingly, prednisone acetate release from the micelles is characterized by both a dramatic thermoresponsive switching behavior and a unique pH-responsive behavior (Fig. 5) (Wei et al. 2006).

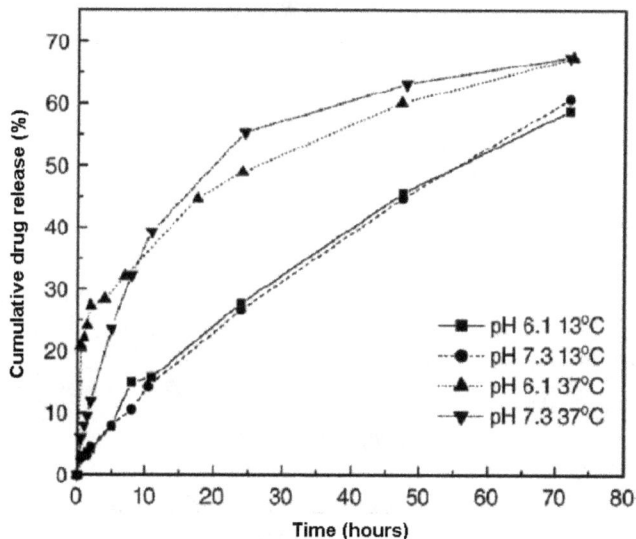

Fig. 5. Prednisone acetate release from poly(*N*-isopropylacrylamide)/poly(10-undecenoic acid) copolymeric micelles as a function of the incubation time in Michaelis buffer solutions below and above the LCST. The pH-responsive behavior relies on the thermoresponsive behavior: the micelles exclusively became pH-sensitive at a temperature above LCST. Reprinted with permission from Wei et al. (2006). Copyright Elsevier (2006).

Magnetic Drug Targeting

For drug delivery purposes, magnetic colloids are usually made of a magnetic core (iron oxide: magnetite, or maghemite) inside a biodegradable polymer-based or lipid-based matrix/shell. The former is responsible for the capacity of the magnetic nanocomposite to respond to magnetic gradients, while the latter will be in charge of delivering the drug or contrast agent. Thus, an applied magnetic gradient will drive the magnetic nanocomposite to the desired site of action, keeping it there until the anti-arthritic drug is entirely released in deep contact with the targeted cells. This will reduce the severe drug side effects on healthy tissues by minimizing its systemic biodistribution. Because the magnetic gradient decreases with the distance to the target, one of the most important limitations of magnetic targeting is associated to the strength of the external magnetic field to be applied to control the residence time of the magnetic drug-loaded nanoparticles in the targeted region. To solve the problem, internal magnets have been located in the vicinity of the targeted site by minimally invasive surgery (Arias 2009, 2010).

It has been recently hypothesized that diclofenac sodium could be efficiently delivered to the arthritic site with the help of a magnetic colloid. This is also expected to lead to a significant reduction in the amount of drug needed to bring out an adequate therapeutic response. A NSAID-based magnetic nanomedicine was formulated by an emulsion solvent evaporation process. The magnetic nanocomposites (mean size $\approx$ 450 nm) consisted of a magnetic core (iron), and a biocompatible polymeric shell (ethylcellulose, where diclofenac sodium was successfully loaded). Compared to diclofenac sodium adsorption (drug entrapment efficiency $\approx$ 12%, and drug loading $\approx$ 2%), drug entrapment in the polymeric matrix permitted higher diclofenac sodium loading (drug entrapment efficiency $\approx$ 65% and drug loading $\approx$ 7%) and a slower (prolonged) drug release profile (Fig. 6) (Arias et al. 2009).

CONCLUSIONS

Conventional pharmacotherapy against arthritis can be significantly improved by taking advantage of nanoparticulate drug delivery systems. The wide variety of biocompatible materials that can be used with great functionalization/targetability capabilities and drug vehiculization properties have permitted the formulation of advanced nanomedicines with highly promising preclinical results. Although such nanomedicines can in principle passively target arthritic joints and/or spleen, a better control of the biological fate of the drug-loaded nanoplatforms is needed to assure the optimal anti-arthritic effect. To that aim, anti-arthritic drug

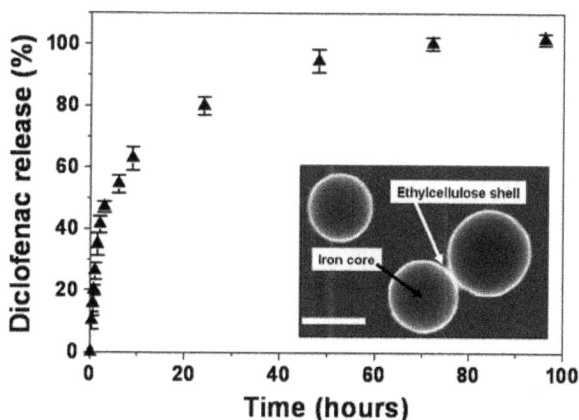

Fig. 6. Release of diclofenac sodium from iron/ethylcellulose (core/shell) nanoparticles as a function of the incubation time in PBS (pH = 7.4 ± 0.1). Inset: dark-field high resolution transmission electron microphotograph of iron/ethylcellulose nanocomposites. Bar length: 400 nm. There is clearly a biphasic process with an early rapid diclofenac sodium release within 1 h (up to ≈ 26%), the remaining NSAID being slowly liberated during the next 48 h. Adapted with permission from Arias et al. (2009). Copyright Elsevier (2009).

nanocarriers must be engineered to be capable of (1) developing specific recognition mechanisms with the targeted cells, or the vascular endothelium of the arthritic joint, and/or (2) responding to defined external stimuli exclusively located at the site of action. Future perspectives of nanoparticle therapy in arthritis rely on a better understanding of the pathology of the disease, as well as on greater advances in the formulation, characterization, and nanoengineering of such nanomedicines.

APPLICATIONS TO OTHER AREAS OF HEALTH AND DISEASE

Drug delivery by biodegradable nanoparticulate materials has become one of the most fascinating research areas, because of the potential implications in ensuring the maximum benefits in pharmacotherapy: an enhanced drug accumulation at the disease site will lead to the best pharmacological effect with negligible toxicity.

As a consequence, investigations on the engineering of drug nanocarriers have been expanded to almost the entire spectrum of pharmacotherapy molecules. For instance, several reports have highlighted the potential benefits of nanoparticulate drug delivery systems in oral chemotherapy, drug delivery to brain tumors (bypassing the blood-brain barrier), and overcoming the problem of multi-drug resistances in cancer cells.

Recent preclinical investigations have reported latent benefits in the diagnosis of severe diseases. The incorporation of contrast agents for magnetic resonance imaging (e.g., gadolinium, superparamagnetic iron oxides) or luminophores into engineered biodegradable nanoparticles will offer diagnostic benefits in the form of improved signal detection.

More interestingly, a new era of multifunctional nanoparticles has entered the disease arena. The so-called nanotheragnostic agents offer combined therapeutic and diagnostic activities. They are expected to permit early detection of the disease, identification of disease biomarkers and signals for the choice of therapy, and efficient (multi) drug delivery to the site of action.

Key Facts

- Key facts of pharmacotherapy failure in arthritis treatment: (1) drug hydrophobicity; (2) unfavorable pharmacokinetics; and (3) poor drug selectivity for targeted tissues (Brigger et al. 2002; Arias 2009; Ulbrich and Lamprecht 2010).
- Key facts of nanoparticle geometry and biological fate. Particle size determines the physical behavior and biological fate of drug nanocarriers. The larger the particles are, the shorter the plasma half life will be. Non-spherical shapes are expected to induce strongest interactions with the RES (Decuzzi et al. 2009; Arias 2010).
- Key facts of nanoparticle surface charge and biological fate. Lower cellular uptake by endocytosis is described for negatively charged nanoparticles. Shorter plasma half-lives are described for nanoparticles with very high surface charges. Surface charge values lower than a critical value will determine particle aggregation and precipitation (Arias 2010).
- Key facts of nanoparticle surface thermodynamics and biological fate. Protein adsorption is markedly enhanced if the hydrophobicity (and/ or size, and/or surface charge) of nanoparticles is high. Adsorbed proteins (specifically, opsonins) determine the capacity of RES cells to phagocyte the nanoparticles, removing them from the bloodstream toward the MPS organs. A better physical stability has been described for hydrophilic nanoparticles (Arias 2010).
- Key facts of the indication of drug vehiculization into nanoparticles. Drug entrapment efficiency (%) is calculated by the following formula: (encapsulated drug (mg)/total drug in the colloidal suspension (mg)) × 100. Drug loading (%) is determined by the following formula: (encapsulated drug (mg)/carrier (mg)) × 100 (Arias et al. 2010).

Definitions

Immunonanoparticles: Nanoparticles surface functionalized with monoclonal antibodies for targeted drug delivery.

Mononuclear phagocytic system: A part of the immune system consisting of phagocytic cells (monocytes and macrophages) embedded in a reticular connective tissue (e.g., lymph nodes, spleen, liver, bone marrow). This term is synonymous with RES and lymphoreticular system.

Nanocapsule: Vesicular systems with the drug trapped in an oily or aqueous cavity generally surrounded by a polymeric or lipidic shell.

Nanoparticle: Solid material generally made of biodegradable polymers with a mean size less than 1 μm.

Nanosphere (alternatively named nanoparticle): Matrix nanosystems in which the drug is found mainly dispersed all along a continuous particle core.

Summary Points

- *In vitro* and *in vivo* investigations have demonstrated the potential of nanoparticles in the efficient enhancement of drug accumulation in the targeted arthritic joint, and/or spleen. This will enhance the anti-arthritic drug effect, along with a considerable reduction of systemic drug toxicity.
- Oral and transdermal administration of anti-arthritic molecules can be improved by using nanoparticles.
- Biodegradable polymers and lipid-based vesicles are the most promising materials in the formulation of drug nanocarriers.
- Nanomedicines against arthritis should be formulated by taking advantage of both passive and active drug targeting strategies.
- The future of nanoparticle therapy in arthritis relies on an extensive characterization of the pathology, and on greater advances in their formulation, characterization, and nanoengineering.

Abbreviations

DMARD	:	disease-modifying antirheumatic drug
DEX	:	dexamethasone
DsiRNA	:	Dicer-substrate small interfering RNA
IL-1Ra	:	interleukin-1 receptor antagonist
LCST	:	lower critical solution temperature
MPS	:	mononuclear phagocytic system
MRT	:	mean retention time
NSAID	:	non-steroidal anti-inflammatory drug

PBS   :   phosphate buffered saline
PEG   :   poly(ethylene glycol)
PLA   :   poly(D,L-lactide)
PLGA   :   poly(D,L-lactide-*co*-glycolide)
POE   :   poly(ortho-ester)s
PVA   :   polyvinyl alcohol
RES   :   reticuloendothelial system
SEM   :   scanning electron microscope
SLN   :   solid lipid nanoparticles
TNF-alpha :   tumor necrosis factor-alpha

References

Arden, N., and M.C. Nevitt. 2006. Osteoarthritis: Epidemiology. Best. Pract. Res. Clin. Rheumatol. 20: 3–25.

Arias, J.L. 2009 . Micro- and nano-particulate drug delivery systems for cancer treatment. pp. 1–85. *In*: P. Spencer and W. Holt. (eds.). Anticancer Drugs: Design, Delivery and Pharmacology. Nova Science Publishers Inc., New York.

Arias, J.L. 2010. Drug Targeting by Magnetically Responsive Colloids. Nova Science Publishers Inc., New York.

Arias, J.L., M. López-Viota, J. López-Viota, and A.V. Delgado. 2009. Development of iron/ ethylcellulose (core/shell) nanoparticles loaded with diclofenac sodium for arthritis treatment. Int. J. Pharm. 382: 270–276.

Arias, J.L., M. López-Viota, E. Sáez-Fernández, and M.A. Ruiz. 2010. Formulation and physicochemical characterization of poly(ε-caprolactone) nanoparticles loaded with ftorafur and diclofenac sodium. Colloids Surf. B Biointerfaces 75: 204–208.

Banerjee, S.S., and D.H. Chen. 2009. A multifunctional magnetic nanocarrier bearing fluorescent dye for targeted drug delivery by enhanced two-photon triggered release. Nanotechnology 20: 185103.

Brigger, I., C. Dubernet and P. Couvreur. 2002. Nanoparticles in cancer therapy and diagnosis. Adv. Drug Deliv. Rev. 54: 631–651.

Decuzzi, P., R. Pasqualini, W. Arap, and M. Ferrari. 2009. Intravascular delivery of particulate systems: Does geometry really matter. Pharm. Res. 26: 235–243.

Fernandes, J.C., H. Wang, C. Jreyssaty, M. Benderdour, P. Lavigne, X. Qiu, F.M. Winnik, X. Zhang, K. Dai, and Q. Shi. 2008. Bone-protective effects of nonviral gene therapy with folate-chitosan DNA nanoparticle containing interleukin-1 receptor antagonist gene in rats with adjuvant-induced arthritis. Mol. Ther. 16: 1243–1251.

Higaki, M., T. Ishihara, N. Izumo, M. Takatsu, and Y. Mizushima. 2005. Treatment of experimental arthritis with poly(D, L-lactic/glycolic acid) nanoparticles encapsulating betamethasone sodium phosphate. Ann. Rheum. Dis. 64: 1132–1136.

Horning, S., H. Bunjes, and T. Heinze. 2009. Preparation and characterization of nanoparticles based on dextran-drug conjugates. J. Colloid Interface Sci. 338: 56–62.

Howard, K.A., S.R. Paluda, M.A. Behlke, F. Besenbacher, B. Deleuran, and J. Kjems. 2009. Chitosan/siRNA nanoparticle-mediated TNF-alpha knockdown in peritoneal macrophages for anti-inflammatory treatment in a murine arthritis model. Mol. Ther. 17: 162–168.

Ishihara, T., T. Kubota, T. Choi, and M. Higaki. 2009. Treatment of experimental arthritis with stealth-type polymeric nanoparticles encapsulating betamethasone phosphate. JPET 329: 412–417.

Kaur, K., S. Jain, B. Sapra, and A.K. Tiwary. 2007. Niosomal gel for site-specific sustained delivery of anti-arthritic drug: in vitro-in vivo evaluation. Curr. Drug Deliv. 4: 276–282.

Kim, J.S., H. An, W.J. Rieter, D. Esserman, K.M. Taylor-Pashow, R.B. Sartor, W. Lin, W. Lin, and T.K. Tarrant. 2009. Multimodal optical and Gd-based nanoparticles for imaging in inflammatory arthritis. Exp. Rheumatol. 27: 580–586.

Kurtoglu, Y.E., M.K. Mishra, S. Kannan, and R.M. Kannan. 2010. Drug release characteristics of PAMAM dendrimer-drug conjugates with different linkers. Int. J. Pharm. 384: 189–194.

Lo, C.T., P.R. Van Tassel, and W.M. Saltzman. 2010. Poly(lactide-co-glycolide) nanoparticle assembly for highly efficient delivery of potent therapeutic agents from medical devices. Biomaterials 31: 3631–3642.

Neumann, D., C. Merkwirth, and A. Lamprecht. 2010. Nanoparticle design characterized by in silico preparation parameter prediction using ensemble models. J. Pharm. Sci. 99: 1982–1996.

Paleolog, E.M., and R.A. Fava. 1998. Angiogenesis in rheumathoid arthritis: Implications for future therapeutic strategies. Springer Semin. Immunopathol. 20: 73–94.

Paolicelli, P., F. Cerreto, S. Cesa, M. Feeney, F. Corrente, C. Marianecci, and M.A. Casadei. 2008. Influence of the formulation components on the properties of the system SLN-dextran hydrogel for the modified release of drugs. J. Microencapsul. 12: 1–10.

Radwan, M.A., B.T. AlQuadeib, N.M. Aloudah, and H.Y. Aboul Enein. 2010. Pharmacokinetics of ketorolac loaded to polyethylcyanoacrylate nanoparticles using UPLC MS/MS for its determination in rats. Int. J. Pharm. 397: 173–178.

Schaffazick, S.R., A.R. Pohlmann, T. Dalla-Costa, and S.S. Guterres. 2003. Freeze-drying polymeric colloidal suspensions: nanocapsules, nanospheres and nanodispersion. A comparative study. Eur. J. Pharm. Biopharm. 56: 501–505.

Schroeder, A., A. Sigal, K. Turjeman, and Y. Barenholz. 2008. Using PEGylated nano-liposomes to target tissue invaded by a foreign body. J. Drug Target. 16: 591–595.

Serpe, L., R. Canaparo, M. Daperno, R. Sostegni, G. Matinasso, E. Muntoni, L. Ippolito, N. Vivenza, A. Pera, M. Eandi, M.R. Gasco, and G.P. Zara. 2010. Solid lipid nanoparticles as anti-inflammatory drug delivery system in a human inflammatory bowel disease whole-blood model. Eur. J. Pharm. Sci. 39: 428–436.

Thakkar, H., R. Kumar Sharma, and R.S. Murthy. 2007. Enhanced retention of celecoxib-loaded solid lipid nanoparticles after intra-articular administration. Drugs R. D. 8: 275–285.

Türker, S., S. Erdoğan, Y.A. Ozer, H. Bilgili, and S. Deveci. 2008. Enhanced efficacy of diclofenac sodium–loaded lipogelosome formulation in intra-articular treatment of rheumatoid arthritis. J. Drug Target. 16: 51-57.

Ulbrich, W., and A. Lamprecht. 2010. Targeted drug-delivery approaches by nanoparticulate carriers in the therapy of inflammatory diseases. J. R. Soc. Interface 7: S55–S66.

Wang, L., R. Zeng, C. Li, and R. Qiao. 2009. Self-assembled polypeptide-block-poly(vinylpyrrolidone) as prospective drug-delivery systems. Colloids Surf. B Biointerfaces 74: 284–292.

Wei, H., X.Z. Zhang, H. Cheng, W.Q. Chen, S.X. Cheng, and R.X. Zhuo. 2006. Self-assembled thermo- and pH responsive micelles of poly(10-undecenoic acid-*b*-*N*-isopropylacrylamide) for drug delivery. J. Control. Release 116: 266–274.

Ye, J., Q. Wang, X. Zhou and N. Zhang. 2008. Injectable actarit-loaded solid lipid nanoparticles as passive targeting therapeutic agents for rheumatoid arthritis. Int. J. Pharm. 352: 273–279.

Zhang, X., C. Yu, X. Shi, C. Zhang, T. Tang, and K. Dai. 2006. Direct chitosan-mediated gene delivery to the rabbit knee joints in vitro and *in vivo*. Biochem. Biophys. Res. Commun. 341: 202–208.

Zhou, H.F., H.W. Chan, S.A. Wickline, G.M. Lanza, and C.T. Pham. 2009. Alphavbeta3-targeted nanotherapy suppresses inflammatory arthritis in mice. FASEB J. 23: 2978–2985.

Phospholipid-based Nanomicelles in Cancer Nanomedicine

Hayat Onyuksel[1] and Amrita Banerjee[2]

ABSTRACT

Lipid-based drug delivery systems have received much attention in cancer therapy because of their acceptable biosafety profile. Phospholipid micelles that form an essential part of lipid-based formulations significantly improve the physico-chemical characteristics of poorly water-soluble small molecule drugs as well as large molecule polypeptide drugs. Moreover, the ability to target these micelles specifically to tumor tissues makes phospholipid micelles a smart choice of carrier for various antineoplastic agents that exhibit delivery challenges. These micelles increase therapeutic efficacy, improve biodistribution and reduce adverse effects of drugs. They have been shown effective in overcoming drug resistance in multidrug-resistant cancer cells. Phospholipid nanomicelles can also be used to deliver imaging agents that suffer similar limitations to that of poorly water-soluble drugs. Besides, these micelles are relatively stable to dilution and can be easily prepared and lyophilized for long-term storage, thereby paving the path towards easy transition of the

[1]University of Illinois at Chicago, 833 South Wood Street, (M/C 865), Room 358, College of Pharmacy, Chicago IL 60612-7231, USA.

[2]University of Illinois at Chicago, 833 South Wood Street, (M/C 865), Room 360, College of Pharmacy, Chicago IL 60612-7231, USA.

List of abbreviations after the text.

formulation to clinics. As a result, phospholipid micelles have been widely evaluated for diagnosis and treatment of cancer. This chapter deals with the development and preclinical application of phospholipid-based micelles in the diagnosis and treatment of cancer to cure or improve life expectancy and quality of life in patients suffering from cancer.

INTRODUCTION

Cancer is one of the biggest causes of deaths among people in the United States. About 7.6 million people die of cancer every year worldwide (Farmer et al. 2010). In a report published by the National Institute of Health, the total economic burden of cancer in 2010 is estimated to be $263.8 billion and it is now the costliest disease to treat. These alarming statistics are primarily due to lack of effective treatment for cancer and inability to detect the disease at its early stage. The common treatment approaches for cancer, such as surgery, radiation therapy and chemotherapy, have their own shortcomings. For instance, tumor resection through surgery often leads to incomplete cure either because of incomplete removal of malignant cells from the vicinity, or because of metastasis of the tumor to other tissues caused by damage to surrounding tissues and blood vessels during surgery, which facilitates tumor cell translocation to other sites. Radiation therapy, on the other hand, in most cases is capable of only shrinking tumor size temporarily, after which the tumor continues to grow sometimes more aggressively than before. Radiotherapy besides killing cancer cells can also damage normal tissues and organs around the tumor region and affect their normal functioning. Another common treatment approach for cancer is chemotherapy, treatment using chemical agents. These chemical agents are cytotoxic drugs that are administered to patients to kill rapidly proliferating cells such as cancer cells. However, because of non-specificity towards cancer cells, chemotherapy also kills other rapidly proliferating normal cells of the body such as blood cells and hair follicles. Adverse effects such as anemia, low white blood cell and platelet count, nausea, vomiting, hair loss, and fatigue are common in patients receiving chemotherapy. To exacerbate the issue, some currently marketed products of anticancer drugs such as Taxol (marketed formulation of paclitaxel) use co-solvents such as cremophor EL (to solubilize the drug), which is known to cause severe toxicities.

Apart from these issues, when small molecule anticancer drugs are injected intravenously, they often bind with serum albumin and have high volumes of distribution (drug mostly resides in fats, bones and less vascularized tissues). This in turn leads to two major concerns: (1) the need to administer large doses for therapy and (2) toxicity due to undesirable effect on non-cancerous normal tissues. Additionally, the narrow therapeutic indices of chemotherapeutic drugs require careful monitoring of drug concentration in plasma. Drug dosage higher than the therapeutic index can result in extensive systemic toxicity, while dosage below the therapeutic index can lead to development of multidrug resistance (MDR) in the cancers. When such tumors re-emerge they are not only resistant to the original treatment drug, but to the entire gamut of structurally related drugs, which eventually renders treatment through chemotherapy ineffective.

With regard to formulation development of anticancer drugs, a matter of concern is the high percentage of existing or newly discovered drugs that are poorly soluble in aqueous media. This raises delivery issues and hampers successful development and commercialization of new pharmaceutical products. The conventional solution to delivery problems is chemical modification of drug molecules to make them more water-soluble. However, this method can be both expensive and time-consuming, and most importantly can compromise bioactivity. *In vivo* instability issues of chemotherapeutic drugs are also an impediment to the successful development or clinical application of many anticancer drugs.

The current cancer therapy, therefore, needs to be revamped. There is a need for better drug delivery and treatment approach for cancer.

NANOMEDICINES FOR CANCER THERAPY

Nanotechnology can bring about radical changes in the diagnosis and treatment of cancer. Using this technology, cancer can be detected and treated through the use of nano-diagnostic agents and multifunctional therapeutics. This emerging field uses the concept of passive and active targeting to deliver chemotherapeutic or diagnostic agents specifically to the tumor site. By definition, nanotechnological devices are 1–100 nm in size; that is imperative for passive targeting of the carriers through leaky vasculature characteristics of pathological conditions such as solid tumors. Nanocarriers offer numerous advantages over traditional delivery systems for cancer therapy such as targeted delivery or accumulation at the tumor tissue, enhanced cellular uptake through improved intracellular delivery, protection of drug from degradation, controlled pharmacokinetics and biodistribution, and prevention of the drug from interacting with other biological environments (Peer et al. 2007). To achieve smooth translation from bench to clinics, it is important that the carrier (1) be composed of biodegradable, biocompatible materials, (2) be either soluble or colloidal aggregate in aqueous conditions, (3) demonstrate minimal toxicity of its own, (4) have long-circulating half-life, (5) demonstrate good shelf-life stability and (6) be easily prepared in a reproducible manner, sterilized and scaled up.

Some widely investigated nanotechnological delivery systems developed for the treatment of cancer are: polymer-based nanostructures such as dendrimers, polymeric micelles and polymer-drug conjugates; inorganic nanoparticles such as nanoshells and gold nanoparticles; lipid-based nanocarriers such as phospholipid micelles and liposomes; and immunoconjugates such as radio-immunoconjugates and immunotoxins (Peer et al. 2007).

In recent years, scores of nanomedicines have been approved by the Food and Drug Administration (FDA) and commercialized for cancer therapy. Examples of a few liposomal formulations marketed for cancer therapy are DaunoXome® (formulation of Daunorubicin for treatment of Kaposi's sarcoma), Doxil® (polyethyleneglycolated liposomal formulation of Doxorubicin for treatment of recurrent breast cancer, ovarian cancer and Kaposi's sarcoma), Myocet® (formulation of Doxorubicin for same indications as Doxil) and Marqibo® (formulation of vincristine sulfate for treatment of aggressive non-Hodgkin's lymphoma). Abraxane® is an example of albumin-bound paclitaxel nanoparticulate formulation for treatment of metastatic breast cancer after failure of combination chemotherapy. Examples of polymer-protein conjugate formulations marketed for cancer therapy are Oncaspar® (polyethyleneglycol [PEG] conjugated to L-asparaginase for treatment of newly diagnosed acute lymphoblastic leukemia) and Neulasta™ (PEG-granulocyte colony stimulating factor for prevention of chemotherapy-associated neutropenia). Examples of immunoconjugates currently marketed are Bexxar® (Anti-CD20 conjugated to iodine-131 for the treatment of relapsed/refractory CD20 positive, follicular non-Hodgkin's lymphoma), and Ontak® (interleukin-2 fused diphtheria toxin for the treatment of recurrent cutaneous T-cell lymphoma) (Peer et al. 2007).

There has been an unprecedented rise in research on new nanomedicines. This is encouraging as it demonstrates the promise that nanotechnology-based interventions hold in providing better treatment approaches for cancer and other diseases.

PHOSPHOLIPID-BASED MICELLES FOR CANCER NANOMEDICINES

Among the different nanocarriers discussed above, lipid-based drug delivery systems are considered relatively safe compared to others. This is because lipid nanocarriers are composed of phospholipids, cholesterol or triglycerides that are natural components of our body. Therefore, they are less likely to trigger immune responses than polymeric nanocarriers. In aqueous media, relatively insoluble phospholipids such as phosphatidylcholine

(PC) self-assemble to form bilayers, commonly known as lipid bilayers. However, more soluble amphiphilic phospholipids self-assemble to form micellar structures. By definition, a micelle is a self-aggregated complex of amphiphilic molecules. In aqueous media, above a certain concentration known as the critical micellar concentration (CMC), the molecules aggregate such that the polar parts of the molecules face water, while the non-polar parts cluster together and are oriented away from water. This is an energetically favorable arrangement and therefore micelles are formed spontaneously in a reproducible manner with uniform size and aggregation number (at a given condition). Micellar systems exist in a dynamic equilibrium where there is a rapid exchange between monomers in solution and micelles. If the CMC of a micellar species is low, the micelles are formed more readily and are more stable to dilution. In the body, lungs have a good proportion of phospholipid surfactants. The pulmonary surfactants maintain alveoli morphology and function, and decrease surface tension at the air-water interface in the alveoli (Bummer 2000). The gastrointestinal system of our body also uses phospholipid bile salt mixed micelles for solubilization and subsequent absorption of fat from the intestine.

A fundamental property of a micelle is the ability to solubilize a wide variety of organic solutes with distinct hydrophobicities. The primary assembly of a micelle in aqueous media provides an inner hydrophobic core in which lipophilic drug molecules are solubilized through hydrophobic and other physical interactions. The hydrophilic outer layer prevents interaction of the solubilized drugs with aqueous bulk phase and thus provides stability to drugs encapsulated in the core. Drugs solubilized in micelles can be targeted to their sites of action through passive and active targeting. Different types of phospholipids are used as micellar drug carriers, the most common of which are those made of PEGylated phospholipids. PEG increases the aqueous solubility of the phospholipid molecule and imparts stealth characteristics to micelles by preventing recognition by opsonins *in vivo*. This 'steric barrier' makes the micelles long-circulating. Generally, PEGylated phospholipids are quite stable to *in vivo* dilution because of a CMC that is lower by orders of magnitude (~10^{-6} M) than more conventional detergents such as sodium dodecyl sulfate and cetyltrimethylammonium bromide (Erogbogbo et al. 2008).

We have used PEGylated phospholipids such as distearoyl phosphatidylethanolamine-polyethyleneglycol$_{2000}$ (DSPE-PEG$_{2000}$) as delivery systems for anticancer drugs. These phospholipids, when dispersed in aqueous media above their CMC (0.5–1 μM), spontaneously self-associate to form sterically stabilized micelles (SSM) with an average aggregation number of 90 and diameter of ~15 nm (Fig. 1) (Ashok et al. 2004). Incorporation of 10% PC results in the formation of sterically stabilized mixed micelles (SSMM) (Fig. 2). These micelles have been used to solubilize poorly water-soluble drugs without the use of toxic solvents or detergents. The solubilized drugs are chemically unmodified and, hence, bioactivity is not compromised. Additionally, since the CMC of the micelles is in the micromolar range, high *in vivo* stability is achieved. DSPE-PEG$_{2000}$ micelles have been used as delivery systems for many anticancer drugs such as paclitaxel (Krishnadas et al. 2003), diazepam (Ashok et al. 2004), camptothecin (Sezgin et al. 2006; Koo et al. 2010), porphyrine (Torchilin 2007), 17-allylamino-17-demethoxy-geldanamycin (17-AAG) (Onyuksel et al. 2009a), and indisulam (Cesur et al. 2009).

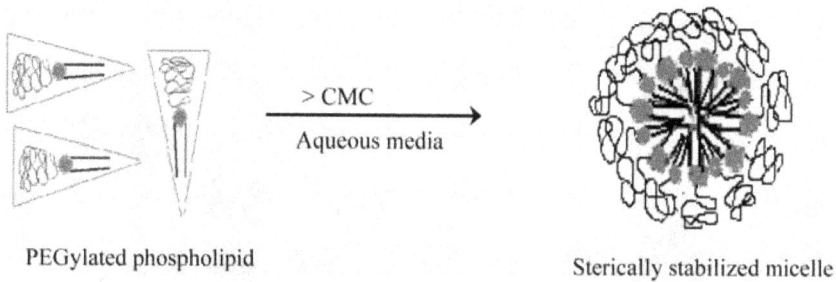

PEGylated phospholipid   Sterically stabilized micelle

Fig. 1. Diagrammatic representation of formation of sterically stabilized micelles (SSM). Distearoyl-phosphatidylethanolamine polyethyleneglycol$_{2000}$ (DSPE-PEG$_{2000}$) molecules self-assemble in aqueous media to form SSM at lipid concentration above their critical micellar concentration (CMC). (Unpublished figure.)

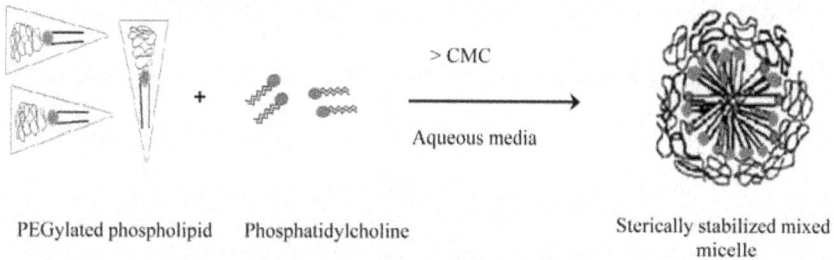

PEGylated phospholipid   Phosphatidylcholine   Sterically stabilized mixed micelle

Fig. 2. Diagrammatic representation of formation of sterically stabilized mixed micelles (SSMM). Distearoyl-phosphatidylethanolamine polyethyleneglycol$_{2000}$ (DSPE-PEG$_{2000}$) and phosphatidylcholine in aqueous media form SSMM at lipid concentration above critical micellar concentration (CMC), prepared using film-rehydration method. (Unpublished figure.)

These drug-loaded micelles are easy to prepare and can be produced in large batches. They can be lyophilized without any cryo/lyoprotectant (Fig. 3) (Lim et al. 2008). This is a signification step towards easy transitioning of the formulation to the clinics.

Other examples of lipid micelles are PEG-PE micelles (PEG conjugated to phosphatidylethanolamine) that have particle sizes between 7 and 35 nm and CMC between 6×10^{-6} and 1×10^{-5} M (Torchilin 2007). Amphiphilic polyvinyl pyrrolidone-lipid conjugates with various polymer lengths having CMC ranging from 2×10^{-6} to 2×10^{-4} M have also been developed (Torchilin 2007). Lysophosphatidylcholine (LPC) and lysophosphatidylethanolamine (LPE) micelles have been studied and characterized (Yamanaka et al. 1997). Another lipid-based micellar system is the bile salt/phospholipid mixed micelles (Onyuksel et al. 1994). These mixed micelles (also called proliposomes) spontaneously undergo micelle to vesicle transition upon dilution to form drug-loaded liposomes.

Fig. 3. Lyophilized cakes of sterically stabilized micelles (SSM). Freeze-dried cakes of SSM at different lipid concentrations (a) 5 mM, (b) 10 mM, (c) 15 mM and (d) 20 mM. (Reprinted from Lim et al. 2008 with permission from Elsevier.)

PASSIVE AND ACTIVE TARGETING WITH PHOSPHOLIPID MICELLES

Targeting anticancer drug-loaded micelles passively and actively to the tumor tissue increases drug efficacy while decreasing its adverse effects.

Passive Targeting

Phospholipid micelles due to their nano size can be passively targeted to tumor tissues by using the enhanced permeation and retention (EPR) effect. Tumors are characterized by the presence of leaky vasculature formed as a result of an acute need for providing oxygen and nutrients to rapidly proliferating tumor cells. This leads to growth and formation of blood vessels with aberrant basement membranes. Drug-loaded PEGylated phospholipid micelles cannot cross normal vasculature but can easily extravasate out of circulation at the tumor site. This increases the therapeutic efficacy of the drug while decreasing the dose required for an effect and minimizing toxicity observed with the traditional mode of treatment. Reduced lymphatic drainage from the tumor decreases clearance of micelles from the tumor leading to gradual increase in accumulation of the drug at tumor tissues. Drug delivery through these nanocarriers therefore decreases dose as well as volume of distribution and clearance of the drug.

Active Targeting

Tumor targeting can also be achieved by modification of the surfaces of the drug carriers using ligands that are specific to receptors or antigens expressed on tumor cells. This enables efficient cellular drug uptake and prevents return of drug into circulation due to high intratumoral pressure. Through internalization of the drug-loaded carrier, delivery of high payloads of drug into cells can be achieved, which is greatly advantageous for killing both drug-sensitive and drug-resistant cancer cells. For targeting moieties that have low binding affinities with their epitopes, numerous ligands can be attached that will increase probability of binding to cells that have the target receptors.

Vasoactive intestinal peptide (VIP) is an example of targeting ligand whose receptors are overexpressed in many cancers. VIP can be chemically conjugated to DSPE-PEG and incorporated in the phospholipid micelle for intracellular delivery of anticancer drugs. To demonstrate efficiency of active targeting using internalizing receptors of VIP, we encapsulated hydrophobic CdSe/ZnS quantum dots (QD) in two micellar constructs: SSMM-VIP and SSMM and observed its internalization in MCF-7 breast cancer cells that overexpress VIP receptors (Rubinstein et al. 2008). To demonstrate specificity of interaction between micellar constructs and VIP receptor, a separate set of MCF-7 cells were pretreated with culture media containing human VIP, pituitary adenylate cyclase activating peptide$_{6-38}$ (PACAP$_{6-38}$) (a VIP receptor-1 antagonist) or galanin (unrelated peptide with similar molecular weight as VIP) prior to being replaced with culture media containing SSMM-QD or VIP-SSMM-QD and incubated as above. After 2 h incubation, cells treated with VIP-SSMM-QD exhibited higher normalized fluorescence signal than cells treated with SSMM-QD (Fig. 4). Over the period of 16 h, it was observed that VIP-SSMM-QD accumulated more rapidly than SSMM-QD alone in the cells. In cells treated with excess VIP, accumulation was significantly decreased. However, such observation was not made for cells treated with PACAP$_{6-38}$ or galanin, indicating specificity of interaction of VIP-SSMM-QD. After 16 h incubation, approximately 77% of MCF-7 cells that were incubated with VIP-SSMM-QD displayed QD in their cytoplasm, however only 17% of the cells that were incubated with SSMM-QD had QD (Fig. 5). Taken together, this study suggested that VIP-SSMM-QD was internalized into the cells using an active receptor-mediated endocytosis process. The same principle should also apply for *in vivo* targeted imaging of cancers with overexpressed VIP receptors using phospholipid micelles containing FDA-approved imaging agent such as gadolinium (Gd).

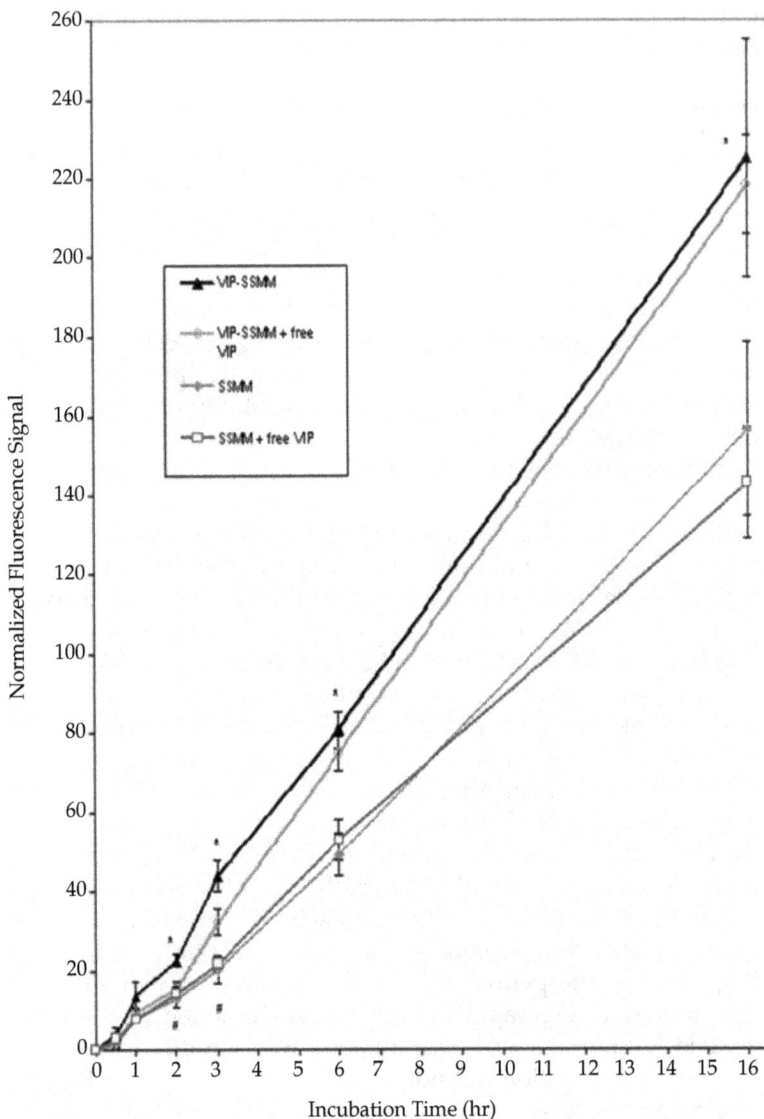

Fig. 4. Accumulation and specificity of uptake of quantum dot (QD) loaded sterically stabilized mixed micelles (SSMM) in human breast cancer cells. Normalized fluorescence signal of vasoactive intestinal peptide (VIP)-SSMM-QD and SSMM-QD in human MCF-7 cells with and without pre-treatment with free VIP (30 µM). Data represented as mean ± standard deviation (n = 5); $^*p < 0.05$ for VIP-SSMM-QD in comparison to SSMM-QD, and SSMM-QD with excess free VIP; $^\#p < 0.05$ for VIP-SSMM-QD in comparison to VIP-SSMM with excess free VIP. (Statistical analysis conducted using ANOVA followed by Tukey's post hoc test.)

Color image of this figure appears in the color plate section at the end of the book.

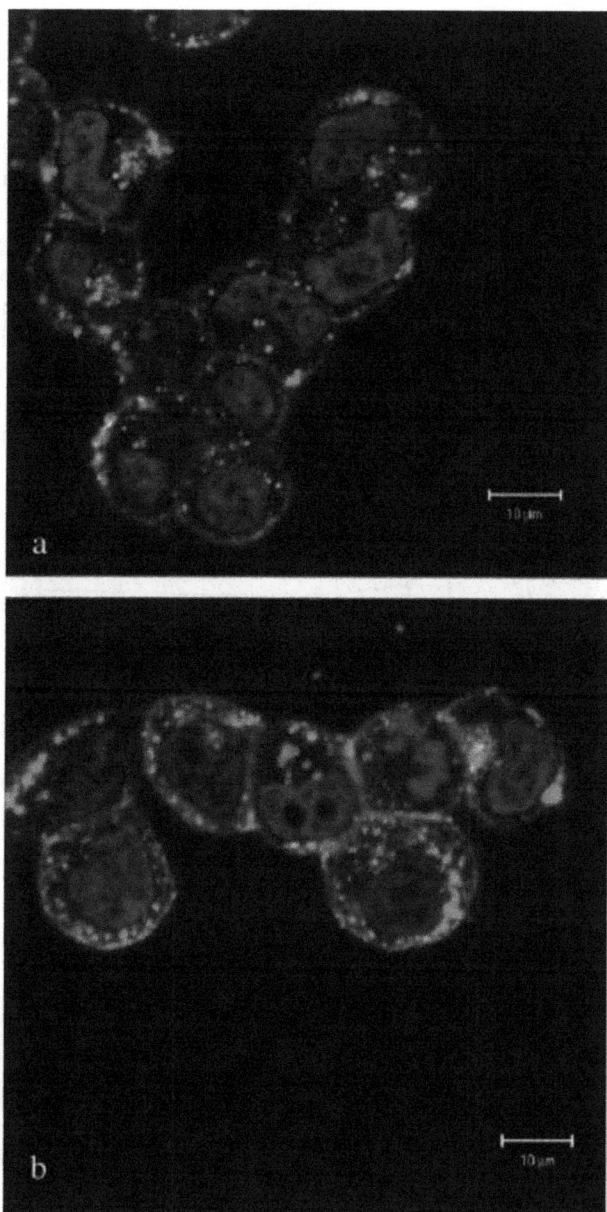

Fig. 5. Cellular uptake of sterically stabilized mixed micelles (SSMM) loaded with quantum dots (QD) in breast cancer cells. Representative confocal microscopy image of human breast cancer MCF-7 cells upon incubation with (a) SSMM-QD and (b) VIP-SSMM-QD for 16 h. (Figs. 4 and 5 reprinted from Rubinstein et al. 2008 with permission from Elsevier.)

Color image of this figure appears in the color plate section at the end of the book.

An advantage of using VIP as a targeting moiety over other targeting ligands is that VIP is not expressed in the vascular endothelium (Reubi 2003). Therefore, VIP-labeled micelles can only interact with its receptor once they extravasate out from circulation, thereby significantly reducing adverse effects due to interaction with normal tissues expressing VIP receptors. This has been shown experimentally when intravenous injection of VIP (a potent vasodilator) in healthy mice did not show any drop in systemic blood pressure when VIP was associated with SSM, while free VIP showed a significant drop in blood pressure upon injection (Fig. 6). Another benefit of using VIP for targeting is that VIP resides at the palisade region and adopts an alpha-helical conformation when associated with SSM (Lim et al. 2008). This conformation not only protects the peptide from enzymatic degradation but is the preferred conformation for receptor interaction that further enhances cellular uptake of the micellar construct.

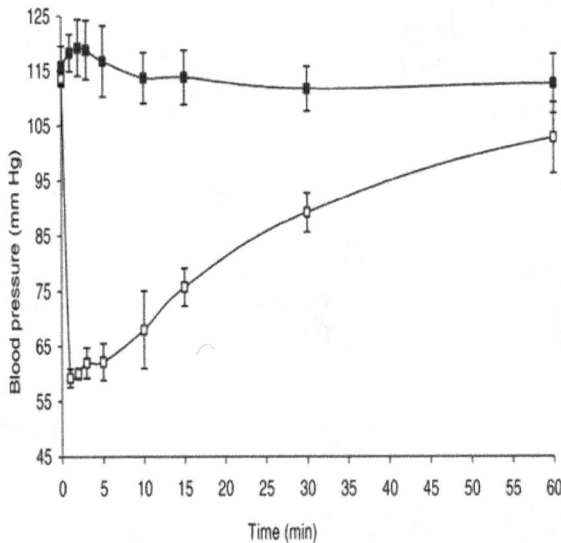

Fig. 6. Abrogation of vasoactive intestinal peptide (VIP) mediated decrease in mean systemic arterial blood pressure using VIP associated micelles. Effect on the mean arterial blood pressure in mice treated with VIP in micelles (■) and VIP alone (□). Data represented as mean ± standard deviation (n=4). (Reprinted from Koo et al. 2005 with permission from Elsevier.)

Examples of other active targeting agents used with phospholipid micelles for cancer therapy are transactivator of transcription (TAT) peptide, cyclic arginine-glycine-aspartic acid (cRGD) peptide, transferrin and monoclonal antibodies. TAT-peptide attached to PEG-PE micelles encapsulating paclitaxel showed significantly greater cytotoxicity in various cancer cells and significant apoptotic cancer cell death in tumor-bearing animals than non-targeted micelles (Sawant and Torchilin 2009). cRGD peptide functionalized PEGylated phospholipids have been used to solubilize quantum rods for imaging in mice bearing pancreatic cancer (Yong et al. 2009). Monoclonal antibody functionalized PEG-PE micelles, also known as immunomicelles, have also been developed and tested. Radiolabeled anticancer monoclonal antibody mAB2C5 immunomicelles encapsulating paclitaxel showed increased drug accumulation and antitumor efficacy in mice bearing lewis lung carcinoma as compared to formulation controls (Torchilin 2007).

APPLICATIONS TO AREAS OF HEALTH AND DISEASE

Phospholipid micelles find application in many areas of biomedical sciences, especially in the treatment of cancer. They are effective in the treatment of both drug-sensitive and drug-resistant cancers and have been used for imaging of tumor tissues. Apart from cancer, phospholipid micelles have been used for antifungal therapy (Vakil et al. 2008). They find application in numerous other diseases such as rheumatoid arthritis (Koo et al. 2005, 2010) and lung hyperinflammation in sepsis-induced acute lung injury (Lim et al. 2010). The carrier itself has potential to be used as a therapeutic moiety for treatment of Alzheimer's disease (Pai et al. 2006). Soybean phosphatidylcholine-mixed micelles have been used to deliver corticosteroids such as dexamethasone and hydrocortisone epicutaneously for suppression of arachidonic acid–induced murine ear edema (Cevc and Blume 2004). Phospholipid micelles have also been used to determine the absorption pattern of carotenoids by intestinal cells (Sugawara et al. 2001). Micelles made of polyvinyl alcohol substituted with oleic acid have been used for transdermal delivery of retinyl palmitate (Torchilin 2007).

However, since the focus of this chapter is the use of phospholipid micelles for the treatment of cancer, they are discussed more elaborately in the following sections.

CANCER THERAPY AND IMAGING

Drug-sensitive Cancer

Human mortalities due to cancer are on a constant rise partly due to lack of effective treatment. Of the various clinical approaches taken to treat cancer, chemotherapy is widely used. However, as discussed in preceding sections, chemotherapy has its own inadequacies.

Phospholipid micelles are very attractive to use in cancer therapy and may overcome many of the challenges faced during chemotherapy. Hydrophobic drugs can be easily solubilized in the core of a micelle. For example, the solubility of paclitaxel, which is an effective anticancer drug, can be greatly enhanced when encapsulated in a micellar carrier. Aqueous solubility of paclitaxel is 0.5 µg/ml (Kim and Bae 2009), while solubility of paclitaxel in 15 mM sterically stabilized mixed micelles (SSMM) was more than 1 mg/ml (Krishnadas et al. 2003). In terms of treatment efficacy, paclitaxel solubilized in SSMM showed significantly greater reduction in tumor size in rats with carcinogen-induced breast cancer compared to taxol (Krishnadas et al. 2004). Promising results were also obtained for tamoxifen, another poorly water-soluble anticancer drug, when loaded in PEG_{5000}-PE micelles (Torchilin 2007).

Apart from solubility enhancement, micelles also improve the stability of anticancer drugs such as camptothecin. Encapsulation of camptothecin in the hydrophobic environment of micelles prevented conversion of the active lactone form to the inactive carboxylate form by reducing the hydrolytic interaction of the molecule. This led to approximately three-fold increase in the stability of camptothecin as compared to camptothecin alone in buffer (Koo et al. 2005, 2010).

Photodynamic therapy for cancer has been investigated using PEG-lipid micelles encapsulating hydrophobic benzoporphyrin derivatives (Zhang et al. 2003) and chlorine e6 trimethyl ester (Torchilin 2007). Apart from this, carotenoids such as lycopene have been solubilized in phospholipid micelles and tested in human prostate cancer LNCaP cells and human lung cancer Hs888Lu cells. The micelles were readily taken up by the cells, reaching a plateau at approximately 12 h, and were shown to be stable for 96 h in standard cell culture media (Xu et al. 1999). Gao et al. tested a mixed pluronic-DSPE-PEG_{2000} micellar formulation encapsulating doxorubicin in mice for antitumor efficacy and biodistribution with and without local ultrasonic irradiation of the tumor (Gao et al. 2005). Ultrasound had no effect on micelle extravasation but it enhanced drug internalization. Ultrasound treatment also helped delay tumor growth in groups treated with micellar doxorubicin as compared to groups without ultrasound exposure.

Multidrug-resistant Cancer

One major reason for failure in cancer treatment is development of MDR. Exposure of cancer cells to low amounts of chemotherapeutic agents frequently results in the re-emergence of malignant cells that are now resistant to anticancer drugs. Occasionally the cancer cells become resistant to chemically or structurally related drugs and to unrelated drugs because of cross-resistance. Studies have indicated three major mechanisms of MDR in cells (Szakacs et al. 2006): (1) decreased permeability of water-soluble drugs that need specific transporters for cellular uptake; (2) intrinsic changes in the cells such as increased repair of DNA damage, changes in cell cycle, reduced apoptosis or alteration of binding sites and metabolic pathways; and (3) increased efflux of hydrophobic drugs that cross cell membranes through simple diffusion. Owing to these myriad mechanisms by which cells can develop MDR, the treatment of drug-resistant cancer becomes complicated.

Among all the above-mentioned mechanisms, the most extensively studied is the resistance imparted by active efflux of hydrophobic drug molecules mediated by membrane transporters known as P-glycoprotein (Pgp). Many studies have been conducted to overcome Pgp-mediated drug efflux and one promising approach is the use of PEGylated phospholipid micelles. We have recently shown that paclitaxel-loaded VIP-conjugated SSMM (P-SSMM-VIP) can overcome MDR by sensitizing drug-resistant breast cancer cells to anticancer drug (Onyuksel et al. 2009b). The construct was tested in drug-sensitive breast cancer MCF-7 cells as well as in MDR breast cancer cells BC19/3. In BC19/3 cells, the half-maximal inhibitory concentration (IC_{50}) of P-SSMM-VIP (59.76 ± 11.27 ng/ml) was found to be significantly lower than P-SSMM (136.7 ± 23 ng/ml) and paclitaxel solubilized in dimethyl sulfoxide (DMSO) (282.3 ± 83 ng/ml) (Fig. 7). The rationale behind this observation is that SSMM-VIP delivers high drug-load of anticancer drugs into cancer cells using internalizing VIP receptors; that prevents interaction of the drug molecules with cell surface efflux pumps and therefore sensitizes the cells to the anticancer drug.

In another study, the cytotoxic potential of bile salt-egg PC mixed micelle formulation of paclitaxel was reported to be greater than Taxol in MDR cells such as KB-V1 (a human cervical carcinoma cell line resistant to vinblastine). Median effective dose (ED_{50}) of drug-loaded mixed micelles in KB-V1 cells was 2.6 µg/ml compared to 12.3 µg/ml for Taxol (Onyuksel et al. 1994).

Cancer Imaging

In addition to therapy, phospholipid micelles have also been used for imaging purposes through different image enhancement techniques.

Silicon QD encapsulated in phospholipid micelles have been tested for imaging cancer cells (Erogbogbo et al. 2008). It was found that transferrin-conjugated PEGylated phospholipids were internalized by Panc-1 pancreatic cancer cells much better than non-functionalized QD-loaded micelles.

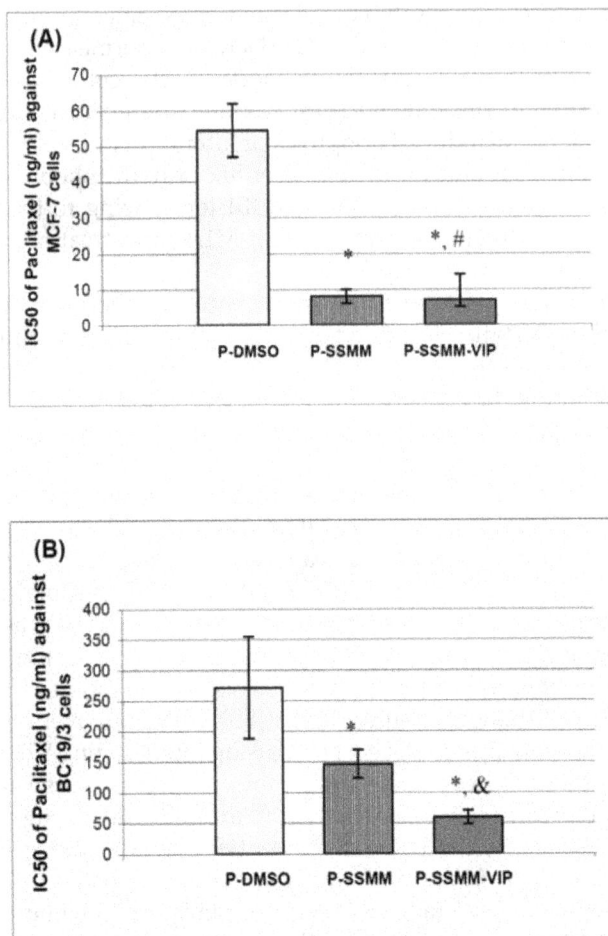

Fig. 7. Evaluation of cytotoxicity of paclitaxel formulations in drug-sensitive and drug-resistant cancers. Half-maximal inhibitory concentration (IC_{50}) of paclitaxel in dimethyl sulfoxide (DMSO), sterically stabilized mixed micelles (SSMM) and actively targeted SSMM (SSMM-vasoactive intestinal peptide). The formulations were tested in (A) drug-sensitive breast cancer MCF-7 cells and (B) multidrug-resistant BC19/3 cells (n=3). * and & $p < 0.05$ in comparison to paclitaxel in DMSO and paclitaxel in SSMM, respectively; #p > 0.5 in comparison to paclitaxel in SSMM. Data represented as mean ± standard deviation and statistical analysis conducted using ANOVA followed by Tukey's post hoc test. (Reprinted from Onyuksel et al. 2009b with permission from Elsevier.)

PEGylated phospholipids encapsulating quantum rods functionalized with cRGD have been used for targeting and imaging of pancreatic cancer in mice. Cytotoxicity studies demonstrated absence of toxicity in both cells and tissues in mice treated with cRGD functionalized micelles containing quantum rods (Yong et al. 2009).

Other probes such as Gd-diethylene triamine pentaacetic acid (DTPA) conjugated to PE (Gd-DTPA-PE) and [111]Indium (In)-DTPA conjugated to stearylamine (SA) [^{111}In-DTPA-SA] have also been encapsulated in PEG-PE micelles for magnetic resonance and gamma scintigraphy based lymphography and suggested for future application as lymphangiography agents (Torchilin 2007). Lymphangiography is often used to detect lymphoma or tumor metastasis. Phospholipid micelles have also been used for near infrared imaging (NIR) of tumor using double-labeled (Alexa and rhodamine) PEG-PE micelles (Papagiannaros et al. 2009). NIR suffers from the limitation of intense scattering of fluorescence signal by tissues that results in poor image quality. To visualize tumors with greater precision, Alexa750-PE and rhodamine-lysamine PE were encapsulated in PEG-PE micelles and tested in mice with breast cancer. Alexa-labeled micelles showed higher fluorescence and signal to noise ratio at the tumor site as compared to areas around the tumor, which is useful in pinpointing the location of tumor and its approximate size.

All the above examples clearly demonstrate the feasibility of using phospholipid micelles for imaging in cancer.

CHALLENGES AND FUTURE DIRECTION

Phospholipid micelles have useful applications in the treatment and imaging of cancer. However, to obtain effective targeted cancer nanomedicines certain challenges need to be considered.

One challenge is the release of the drug or imaging agent from micelles during circulation in the body. For targeted delivery, it is desirable that drug molecules remain in micelles during circulation until the target site is reached. However, not all poorly water-soluble drugs are stable in the micelles during circulation and a certain proportion of the loaded drug is released in blood during circulation. Stability of a drug in micelles depends on the drug's intrinsic hydrophobicity, amphiphilicity and location in the micelles. Drugs that reside in the hydrophobic core need higher energy to be released from the micelle than drugs that reside at the palisade region. Currently, we are using molecular dynamic simulation to predict the location of drug molecules in SSM (Vukovic et al. 2010). Another challenge is the issue of elevated interstitial pressure in tumor tissues. There is a possibility that carriers entering tumor tissues via leaky vasculature are thrown back to circulation because of high tumoral interstitial pressure. Actively targeted micelles will overcome this problem. It is also unclear whether pore sizes in tumor vasculature vary when tumor size decreases and whether EPR is observed in all the stages of tumor (Kim and Bae 2009). Smaller carriers will circumvent this issue. Another concern is the distance that a carrier traverses after it enters a tumor site and that may depend on the size and surface charge of the carrier. Tumor-penetrating peptides are being tested to overcome this problem (Feron 2010).

CONCLUSION

Phospholipid micelles are versatile drug delivery systems for treatment of cancer and enjoy distinct advantages over other drug delivery systems. Since phospholipids are natural components of the body, they are safe and biocompatible and therefore do not pose serious toxicity concerns. The micelles can be used to solubilize hydrophobic drugs that suffer delivery issues. Moreover, because they have lower CMC than other commonly used detergents, they are relatively stable to dilution. The micelles can be targeted both passively and actively to tumor tissues, which helps in improving the therapeutic efficacy while reducing systemic toxicity of the drug. They are effective in the treatment of both drug-sensitive and multidrug-resistant cancers and can also deliver imaging agents for diagnostic purpose. Furthermore, easy preparation and the ability to freeze-dry phospholipid micelles without the use of any lyo/cryoprotectant allow for easy transitioning of these micelles to the clinics. From these facts it can be summarized that phospholipid-based micelles have promising applications in the diagnosis and treatment of solid tumors.

Key Facts

- According to the American Cancer Society, about 1,529,560 new cancer cases are estimated to be diagnosed in 2011 and about 569,490 Americans are expected to die of cancer this year, which is more than 1,500 people a day.
- Lack of early diagnosis and chemotherapy-related treatment failure are the primary reasons for so many mortalities. Lipid nanocarriers such as phospholipid nanomicelles can be used to address these concerns. The patented technology of sterically stabilized micelles for delivery of poorly water-soluble drugs is one of the pioneering works on phospholipid micelles–based cancer therapy (Onyuksel and Rubinstein 2001).
- Phospholipids are essential components of cell membranes and therefore considered safe when used as delivery systems. PEGylated phospholipids are also part of FDA-approved products. PEGylation of the lipids reduces immunogenicity and imparts steric stability to micelles, making them long-circulating in the blood.
- The nanosized micelles alter the biodistribution and pharmacokinetics of chemotherapeutic drugs or imaging agents. Drug or imaging agent

loaded micelles extravasate out of circulation at tumor tissues through leaky vasculature in solid tumors, thus enhancing the efficacy of treatment or imaging.

- These micelles can be functionalized with targeting ligands specific to biomarkers overexpressed in the tumor. After reaching the tumor site, the micelles can be internalized through receptor-mediated endocytosis that further enhances treatment or imaging efficacy.
- Functionalized lipid micelles can also be used to overcome MDR in tumors. Receptor-mediated internalization of cytotoxic drugs prevents interaction of drugs with efflux pumps leading to sensitization of tumor cells to treatment drug.

Definitions

Active targeting: A process to enhance the recognition and interaction between a pathologic biomarker such as a receptor and its specific ligand. Attaching targeting ligands to nanocarriers also provides cellular uptake in cells expressing internalizing receptors for the ligand.

Enhanced permeation and retention (EPR): A phenomenon associated with tumor and inflammatory tissues. Because of rapid proliferation of cells at these sites, their vasculatures have large fenestrations or are essentially leaky. This enables the nanoparticles (usually less than 100 nm) in the blood to accumulate at the tumor/inflammation tissue.

Micelle: A self-aggregate of amphiphilic molecules arranged so that the hydrophilic parts of the molecule are directed towards water, while the hydrophobic parts are directed away from water. This reduces the free energy of the system.

Passive targeting: A process by which nanoparticles get accumulated at tissues that have leaky vasculature using the EPR effect.

Stealth: The ability of certain molecules to escape recognition by reticuloendothelial system of the body. Opsonins present in the bloodstream fail to identify such molecules as foreign components and the molecules are not immediately cleared from the circulation. PEG, being a hydrophilic polymer, forms a structured water layer around the micelle that prevents opsonization and makes them long-circulating.

Summary Points

- Major problems with cancer chemotherapy are the narrow therapeutic indices of anticancer agents, multidrug resistance and low aqueous solubility of many anticancer drugs.

- All these problems can be overcome by using phospholipid micelles as anticancer drug carriers.
- Poorly water-soluble drugs can be easily solubilized in the hydrophobic core of the micelles.
- Since phospholipids are natural components of the body, these micelles do not raise immunogenicity concerns as observed with other solvents and nanocarriers.
- Owing to their nanosize, the micelles can be passively targeted to disease site using enhanced permeation and retention effect.
- Attaching a targeting ligand such as vasoactive intestinal peptide to the micelles enables receptor-mediated internalization of the phospholipid micelles. This further enhances treatment efficacy in drug-sensitive cancers and helps overcome resistance in multidrug-resistant cancers.
- Phospholipid micelles generally have low critical micellar concentration compared to other detergents; therefore, they are more stable to dilution *in vivo*.
- These micelles are easy to prepare, are reproducible, can be lyophilized and also can be produced on a large scale.
- Polyethyleneglycolated phospholipid micelles are not recognized by reticuloendothelial system of the body and are therefore not cleared immediately, making them long-circulating.
- Phospholipid micelles have also been used to deliver imaging agents to tumor sites for use as diagnostic probes for cancer.
- In summary, even though there is no commercial product of phospholipid micelles yet, they have promising applications in treatment and diagnosis of cancer and have high potential to be developed in the near future as nanomedicines in clinics for cancer.

Abbreviations

17-AAG	:	17-allylamino-17-demethoxy-geldanamycin
ABC	:	ATP binding cassette
CMC	:	critical micellar concentration
cRGD	:	cyclic arginine-glycine-aspartic acid
DMSO	:	dimethyl sulfoxide
DSPE	:	distearoyl phosphatidyl ethanolamine
DTPA	:	diethylene triamine pentaacetic acid
ED_{50}	:	median effective dose
EPR	:	enhanced permeation and retention
FDA	:	Food and Drug Administration
Gd	:	gadolinium
IC_{50}	:	half-maximal inhibitory concentration

In   :   indium
LPC   :   lysophosphatidylcholine
LPE   :   lysophosphatidylethanolamine
MDR   :   multidrug resistance
NIR   :   near infrared
P   :   paclitaxel
$PACAP_{6-38}$:   pituitary adenylate cyclase activating peptide$_{(6-38)}$
PC   :   phosphatidyl choline
PE   :   phosphatidylethanolamine
PEG   :   polyethyleneglycol
Pgp   :   P-glycoprotein
QD   :   quantum dot
SA   :   stearylamine
SSM   :   sterically stabilized micelles
SSMM   :   sterically stabilized mixed micelles
TAT   :   transactivator of transcription
VIP   :   vasoactive intestinal peptide

References

Ashok, B., L. Arleth, R.P. Hjelm, I. Rubinstein, and H. Onyuksel. 2004. *In vitro* characterization of PEGylated phospholipid micelles for improved drug solubilization: Effects of PEG chain length and PC incorporation. J. Pharm. Sci. 93: 2476–2487.

Bummer, P.M. 2000. Interfacial phenomena. pp. 275–287. *In*: A.R. Gennaro, A.H.D. Marderosian, G.R. Hanson, T. Medwick, N.G. Popovich, R.L. Schnaare, J.B. Schwartz and H.S. White (eds.) 2000. Remington. The Science and Practice of Pharmacy. Lippincott Williams and Wilkins, Baltimore.

Cesur, H., I. Rubinstein, A. Pai, and H. Onyuksel. 2009. Self-associated indisulam in phospholipid-based nanomicelles: A potential nanomedicine for cancer. Nanomed.: Nanotechnol. Bio. Med. 5: 178–183.

Cevc, G., and G. Blume. 2004. Hydrocortisone and dexamethasone in very deformable drug carriers have increased biological potency, prolonged effect, and reduced therapeutic dosage. Biochim. Biophys. Acta. Biomemb. 1663: 61–73.

Erogbogbo, F., K.T. Yong, I. Roy, G. Xu, P.N. Prasad, and M.T. Swihart. 2008. Biocompatible luminescent silicon quantum dots for imaging of cancer cells. ACS Nano 2: 873–878.

Farmer, P., J. Frenk, F.M. Knaul, L.N. Shulman, G. Alleyne, L. Armstrong, R. Atun, D. Blayney, L. Chen, R. Feachem, M. Gospodarowicz, J. Gralow, S. Gupta, A. Langer, J. Lob-Levyt, C. Neal, A. Mbewu, D. Mired, P. Piot, K.S. Reddy, J.D. Sachs, M. Sarhan, and J.R. Seffrin. 2010. Expansion of cancer care and control in countries of low and middle income: A call to action. Lancet 376: 1186–1193.

Feron, O. 2010. Tumor-penetrating peptides: A shift from magic bullet to magic guns. Sci. Transl. Med. 2: 1–5.

Gao, Z., H.D. Fain, and N. Rapoport. 2005. Controlled and targeted tumor chemotherapy by micellar-encapsulated drug and ultrasound. J. Control. Release 102: 203–222.

Kim, D., and Y.H. Bae. 2009. Polymeric carriers for anticancer drugs. pp. 207–243. *In*: Y. Lu and R.I. Mahato (eds.). Pharmaceutical Perspectives of Cancer Therapeutics. Springer, New York.

Koo, O.M., I. Rubinstein, and H. Onyuksel. 2005. Role of nanotechnology in targeted drug delivery and imaging: A concise review. Nanomed.: Nanotechnol. Biol. Med. 1: 193–212.

Koo, O.M.Y., I. Rubinstein, and H. Onyuksel. 2010. Actively targeted low dose camptothecin as a safe, long-acting disease-modifying nanomedicine for rheumatoid arthritis. Pharm. Res. (in press).

Krishnadas, A., I. Rubinstein, and H. Onyuksel. 2003. Sterically stabilized phospholipid mixed micelles: *In vitro* evaluation as a novel carrier for water-insoluble drugs. Pharm. Res. 20: 297–302.

Krishnadas A., I. Rubinstein, M. Sekosan, and H. Onyuksel. 2004. Targeted delivery of paclitaxel to breast cancer by vasoactive intestinal peptide conjugated sterically stabilized phospholipid mixed micelles. AAPS J. 6: M1191.

Lim, S.B., I. Rubinstein, and H. Onyuksel. 2008. Freeze-drying of peptide drugs self-associated with long-circulating, biocompatible and biodegradable sterically stabilized phospholipid nanomicelles. Int. J. Pharm. 356: 345–350.

Lim, S.B., I. Rubinstein, R.T. Sadikot, J.E. Artwohl, and H. Onyuksel. 2010. A novel peptide nanomedicine against acute lung injury: GLP-1 in phospholipid micelles. Pharm. Res. (in press).

Onyuksel, H.A., S. Ramakrishnan, H. Chai, and J.M. Pezzuto. 1994. A mixed micellar formulation suitable for the parenteral administration of taxol. Pharm. Res. 11: 206–212.

Onyuksel, H., and I. Rubinstein. 2001. Materials and methods for making improved micelle compositions. U.S. Patent # 6,217,886.

Onyuksel, H., P.S. Mohanty, and I. Rubinstein. 2009a. VIP-grafted sterically stabilized phospholipid nanomicellar 17-allylamino-17-demethoxy-geldanamycin: A novel targeted nanomedicine for breast cancer. Int. J. Pharm. 365: 157–161.

Onyuksel, H., E. Jeon, and I. Rubinstein. 2009b. Nanomicellar paclitaxel increases cytotoxicity of multidrug resistant breast cancer cells. Cancer Lett. 274: 327–330.

Pai, A.S., I. Rubinstein, and H. Onyuksel. 2006. PEGylated phospholipid nanomicelles interact with beta-amyloid(1–42) and mitigate its beta-sheet formation, aggregation and neurotoxicity *in vitro*. Peptides 27: 2858–2866.

Papagiannaros, A., A. Kale, T.S. Levchenko, D. Mongayt, W.C. Hartner, and V.P. Torchilin. 2009. Near infrared planar tumor imaging and quantification using nanosized Alexa750-labeled phospholipid micelles. Int. J. Nanomed. 4: 123–131.

Peer, D., J.M. Karp, S. Hong, O.C. Farokhzad, R. Margalit, and R. Langer. 2007. Nanocarriers as an emerging platform for cancer therapy. Nature Nanotech. 2: 751–760.

Reubi, J.C. 2003. Peptide receptors as molecular targets for cancer diagnosis and therapy. Endocrine Rev. 24: 389–427.

Rubinstein, I., I. Soos, and H. Onyuksel. 2008. Intracellular delivery of VIP-grafted sterically stabilized phospholipid mixed nanomicelles in human breast cancer cells. Chem. Biol. Interactions 171: 190–194.

Sawant, R.R., and V. P. Torchilin. 2009. Enhanced cytotoxicity of TATp-bearing paclitaxel-loaded micelles *in vitro* and *in vivo*. Int. J. Pharm. 374: 114–118.

Sezgin, Z., N. Yuksel and T. Baykara. 2006. Preparation and characterization of polymeric micelles for solubilization of poorly-soluble anticancer drugs. Eur. J. Pharm. Biopharm. 64: 261–268.

Sugawara, T., M. Kushiro, H. Zhang, E. Nara, H. Ono and A. Nagao. 2001. Lysophosphatidylcholine enhances carotenoid uptake from mixed micelles by Caco-2 human intestinal cells. J. Nutr. 131: 2921–2927.

Szakacs, G., J.K. Paterson, J.A. Ludwig, C. Booth-Genthe, and M.M. Gottesman. 2006. Targeting multidrug resistance in cancer. Nat. Rev. Drug Discovery 5: 219–234.

Torchilin, V.P. 2007. Micellar nanocarriers: Pharmaceutical perspectives. Pharm. Res. 24: 1–16.

Vakil, R., K. Knilans, D. Andes, and G.S. Kwon. 2008. Combination antifungal therapy involving amphotericin B, rapamycin and 5-fluorocytosine using PEG-phospholipid micelles. Pharm. Res. 25: 2056–2064.

Vukovic, L., N. Shah, A. Madriage, P. Kral, and H. Onyuksel. 2010. Molecular dynamics of lipid-based nanomedicines: Drug solubilization in PEGylated phospholipid nanocarriers. 239th National ACS Meeting, San Francisco.

Xu, X., Y. Wang, A.I. Constantinou, M. Stacewicz-Sapuntzakis, P.E. Bowen, and R.B. Van Breemen. 1999. Solubilization and stabilization of carotenoids using micelles: Delivery of lycopene to cells in culture. Lipids 34: 1031–1036.

Yamanaka, T., N. Ogihara, T. Ohhori, H. Hayashi, and T. Muramatsu. 1997. Surface chemical properties of homologs and analogs of lysophosphatidylcholine and lysophosphatidylethanolamine in water. Chem. Phys. Lipids 90: 97–107.

Yong, K.T., R. Hu, I. Roy, H. Ding, L.A. Vathy, E.J. Bergey, M. Mizuma, A. Maitra, and P.N. Prasad. 2009. Tumor targeting and imaging in live animals with functionalized semiconductor quantum rods. ACS App. Mat. Interface 1: 710–719.

Zhang, J.X., C.B. Hansen, T.M. Allen, A. Boey, and R. Boch. 2003. Lipid-derivatized poly(ethylene glycol) micellar formulations of benzoporphyrin derivatives. J. Control. Release 86: 323–338.

17

Tumor-specific Liposomal Nanomedicines: Antitumor Antibody-modified Doxorubicin-loaded Liposomes

Tamer Elbayoumi[1], and *Vladimir Torchilin[2]*

ABSTRACT

The severe toxic side effects associated with the administration of the anticancer drug doxorubicin makes this drug an ideal candidate for tumor-targeted delivery. To achieve tumor targeting, doxorubicin is often encapsulated into pharmaceutical nanocarriers, such as liposomes. The use of liposomes as anticancer drug delivery systems was originally hampered by their fast removal from the blood; however, liposome longevity in the body was achieved by coating of its surface with polyethylene glycol. Doxorubicin-loaded long-circulating liposomes underwent a successful transfer from the bench to the clinic and demonstrated good properties based on their ability to passively accumulate in the tumor via the enhanced permeability and retention effect. Still, it is believed that the active targeting of drug-loaded liposomes to tumor with the

[1]Department of Pharmaceutical Sciences, Midwestern University, Glendale, AZ 85308, USA.
Email: telbayoumi@midwestern.edu
[2]Center for Pharmaceutical Biotechnology and Nanomedicine, Northeastern University, Boston, MA 02115, USA.
*Corresponding author

List of abbreviations after the text.

use of liposome-attached tumor-specific ligand can improve their therapeutic potential furthermore. Among a variety of methods employed to target liposomal anticancer drugs to tumors, the most successful results were achieved using antibody-mediated (immuno) targeting. Various schemes for attaching antibodies and their fragments on the surface of liposomes, either directly or via distal coupling or functionalized polymeric liposome coatings are well elaborated and characterized.

Currently, multiple tumor-homing strategies exist to guide doxorubicin liposomes to a broad variety of tumors via different tumor-associated targets and liposome-attached antibodies. Doxorubicin-loaded tumor-targeted immuno-liposomes, which are now under preclinical and early clinical development, clearly demonstrate improved tumor accumulation and enhanced anti-tumor activity.

INTRODUCTION

The anthracyclin antibiotic doxorubicin (adriamycin) is one of the most potent and widely used anticancer drugs. It can kill tumor cells via disseminated protein and lipid peroxidation as well as DNA and topoisomerase inactivation (Chapner et al. 2005). Because of the unrestricted oxidative damage of doxorubicin on benign tissues, it results in a number of undesirable and even-life threatening adverse effects. Severe myelosuppression, nausea and vomiting, alopecia, mucositis, and severe tissue necrosis at the injection site or at any site where the skin is exposed to considerable pressure may limit its use (Lorusso et al. 2007). The cardiotoxicity associated with doxorubicin (cardiomyopathy and congestive heart failure) is a major concern during doxorubicin therapy (Chapner et al. 2005). Although researchers have studied numerous strategies to target doxorubicin to cancer tissues or at least to diminish its side effects, until now, liposomal nanocarriers represent the most successful platform for delivery of doxorubicin. Liposomes, artificial phospholipid vesicles, can be obtained by various methods from lipid dispersions in water, and their preparation, physico-chemical properties, and biomedical application have already been extensively discussed in many papers and monographs (Lasic and Papahadjopoulos 1998; Gregoriadis 2007). To give them various required properties, liposomes can be surface modified with different functionalities, the most important of which are as follows:

(A) Certain water-soluble polymers (first of all, polyethylene glycol, PEG) for prolonged circulation in the blood allowing PEGylated liposomes to accumulate in various pathological areas (such as solid tumors) via the enhanced permeability and retention (EPR) effect.

(B) Targeting ligands to specifically recognize and bind required tissues or cells (monoclonal antibodies as well as their Fab' fragments and some other moieties, such as folate or transferrin).

(C) Stimuli-sensitive modifiers with the ability to respond to local stimuli characteristic of the pathological site, for example, by releasing an entrapped drug or specifically acting on cellular membranes under the abnormal pH or temperature in disease sites.

(D) Cell-penetrating moieties to facilitate intracellular delivery of liposomes and bypass the lysosomal degradation.

TUMOR ACCUMULATION OF LIPOSOMES

At present, the concept of "magic bullet" suggested by Paul Erlich more than 100 years ago includes a coordinated behavior of three components: (1) drug; (2) targeting moiety; and (3) pharmaceutical carrier to multiply the number of drug molecules per single targeting moiety. The recognition of the target can occur on the level of the organ, on the level of certain cells/tissues specific for a given organ or even the level of individual components specific for certain target cells, such as cell surface antigens.

Targeting ligands like antibody moieties, and their fragments, polysaccharides, peptides and small molecular entities could be attached to nanocarriers, i.e., liposomes, without affecting their integrity and specific properties by covalent binding to the liposome surface or by hydrophobic insertion into the liposomal membrane after modification with hydrophobic residues. So far, the antibodies are the most diverse and broadly used specific ligands for experimental targeted chemotherapy of various tumors with drug-loaded liposomes. For reviews see (Torchilin 2008).

A combination of liposome targetability with prolonged circulation seems to provide the optimal properties for the liposomal drugs. In general, long-circulating pharmaceuticals and pharmaceutical nanocarriers represent currently an important and still growing area of biomedical research. There are several important reasons for that. One is to maintain a required level of a pharmaceutical agent in the blood for extended time periods. Then, long-circulating drug-containing nanoparticulates or large macromolecular aggregates can slowly accumulate via the EPR effect in pathological sites with affected and leaky vasculature (such as tumors, inflammations, and infarcted areas), and facilitate drug delivery in those areas (Maeda et al. 2000). In addition, the prolonged circulation can help to achieve a better targeting effect for targeted preparations allowing for more time for their interaction with the target due to higher number of passages of targeted pharmaceuticals through the target.

Clinical applications of long-circulating (Stealth™) liposomes are multiple and well known; see for example our review (Elbayoumi and Torchilin 2010). Doxorubicin in PEG-coated liposomes (Doxil®/Caelyx®) alone or in combination with other chemotherapeutic agents are successfully used for the treatment of tumors in patients with different cancers (Soloman and Gabizon 2008).

In this context, it was feasible to additionally use certain target-(tumor)-specific "vector" molecules to further enhance tumor targeting of chemotherapeutic drug nanocarriers by combining both "active" and "passive" delivery strategies into one liposomal formulation. The use of specific vector molecules is especially important for tumors with immature vasculature, such as tumors in the early stages of their development, and for delocalized tumors.

In general, coupling the antibodies to the liposomes surface must meet the following important criteria: (1) Antibody specificity and affinity should not change upon coupling to liposomes. (2) A sufficient quantity of transport molecules should firmly bind to the liposome surface. (3) Liposomal integrity during the antibody coupling procedure must be preserved. (4) The antibody coupling should be simple and with high yield. Other conditions to be considered to achieve successful binding of immuno-liposomes with their target include the accessibility of a specific binding site in the target for immuno-liposomes and high specific and low non-specific immuno-liposome binding with the target.

BASIC CONSIDERATIONS FOR ENGINEERING TUMOR-TARGETED DOXORUBICIN-LOADED IMMUNO-LIPOSOMES

Clearly, tumor-specific liposomes have to meet certain requirements: (1) Such liposomes should accumulate in target tumors fast and effectively. (2) The quantity of the drug delivered into the tumor via such liposomes should be higher than in the case of other delivery systems. (3) Ideally, liposomal drugs should not only accumulate in the interstitial space inside tumors but also be internalized by the target cells, thus creating high intracellular drug concentration and allowing for bypassing multidrug resistance. To meet these requirements, a target should be identified that is present (over-expressed) on the surface of tumor cells to be targeted in sufficient quantity; the specific ligand (antibody or its fragment) should be attached to the surface of the drug-loaded liposome in a way that does not affect its specific binding properties (the method suitable for one antibody will not necessarily be suitable for another), and in sufficient quantity to provide the multipoint binding with the target; and in the case of PEGylated long-circulating liposomes the quantity of the attached antibodies should not strongly compromise the liposome longevity. It is also highly desirable that the targeting antibody be internalizable and facilitate the internalization of the liposome and liposome-incorporated drug. In addition, doxorubicin release from the liposome inside the tumor or inside the tumor cell should provide the therapeutic concentration of doxorubicin in the target and maintain it within a reasonable period of time (a few hours).

With all the promising data on antibody-targeted drug-loaded liposomes for cancer therapy, we must mention several biological and technological problems associated with these systems. Biologically, one can expect certain changes in normal pharmacokinetics and biodistribution of plain and long-circulating liposomes after their modification with antibodies. These changes could result in an increased uptake of antibody-bearing liposomes by the reticuloendothelial system especially upon repeated administration (Bendas et al. 2003). Evidently, the use of smaller antibody fragments (such as Fab) instead of whole antibodies can minimize protein-mediated liposome clearance (Fig. 1).

From the technological point of view, the addition of the surface-attached antibody to the liposomal preparation will certainly result in the additional production step and cost increase of the final product. At present, it is rather difficult to say how serious this problem could become; however, the quantity of the attached antibody can be minimized, and technologies can be used that allow for minimal loss of the antibody during the attachment procedure.

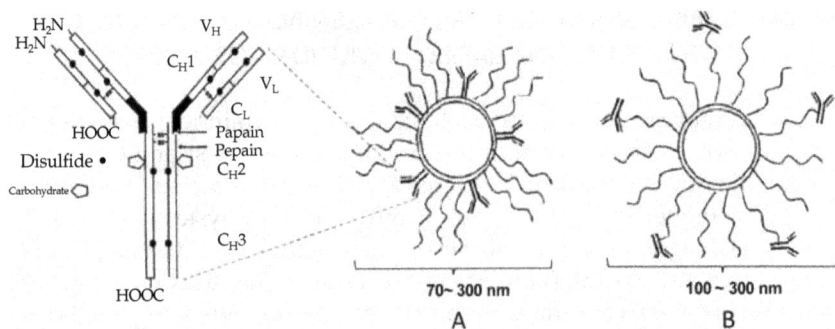

Fig. 1. Schematics of antibody attachment to liposomes. Left: Structure of IgG antibody, with possible sites of modifications. Right: Different ways to attach specific targeting ligands to PEGylated liposomes. (A) Co-immobilization of a ligand with PEG on the liposome surface. (B) Attachment of a ligand to an activated distal tip of a liposome-grafted PEG chain (unpublished).

POSSIBLE SCHEMES FOR ANTIBODY COUPLING TO LIPOSOMES

Direct Surface Modification of Liposomes

Early attempts to attach antibodies to liposomes for their targeting to certain cells and tissues in the body go back to the late 1970s and early 1980s, when antibody molecules and some model proteins were coupled to the surface of plain, non-PEGylated liposomes. Numerous procedures for the conjugation of antibodies to liposomes have been developed. These fall into four general categories defined by the particular functionality of the antibody being modified, namely amine modification, carbohydrate modification, disulfide modification, and non-covalent conjugation (Klibanov et al. 2003). One primary binding chemistry involves the use of bifunctional reagents (crosslinkers) to couple liposome-incorporated reactive groups with a protein, which is most widely employed in the reaction of sulflhydryl groups with maleimide groups, as detailed in figure 2. This reaction has the advantage of being relatively clean, fast, and efficient, and has been adapted to the modification of all of the antibody functional groups in the preparation of liposome conjugates. A protein (antibody) can also be modified with hydrophobic residues providing its efficient incorporation into the liposomal membrane (Torchilin et al. 1980). Protein attachment to liposome can also be performed via activated liposome-incorporated sugar moieties or via activated sugar moieties in the antibody. Some other approaches have also been developed.

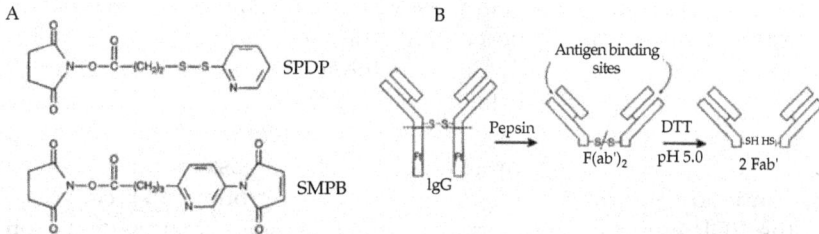

Fig. 2. Possible thiol group modifications for antibody conjugation reactions. (A) Structures of the two compounds most widely used for introducing thiol groups into moieties containing amino groups. SPDP, N-succinimidyl pyridyl dithiopropionate, and SMPB, N-succinimidyl-(-4-[p-maleimldophenyl]) butyrate. (B) Conversion of IgG to Fab' fragments containing free endogenous thiol groups. The Fab' fragments formed in this way contain free sulfhydryl groups that can react directly with liposomes containing PDP-PE, without the need for introduction of PDP residues into the protein (unpublished).

Distal Attachment on Long-Circulating Liposomes

Further development of liposomal carriers resulted in combination of properties of long-circulating liposomes and targeted liposomes in one preparation. To achieve better selectivity of PEG-coated liposomes, it is advantageous to attach the targeting ligand via a PEG spacer arm, so that the ligand is extended outside of the dense PEG brush, excluding steric hindrances for the ligand binding to the target. Various advanced technologies are used for this purpose, and the targeting moiety is usually attached above the protecting polymer layer by coupling it with the distal water-exposed terminus of activated liposome-grafted polymer molecule (Fig. 1).

For this purpose, several types of end-group functionalized lipo-polymers of general formula X-PEG-PE were introduced, where X represents a reactive functional group-containing moiety, while PEG-PE represents the conjugate of PE and PEG. Most of the end-group functionalized PEG-lipids were synthesized from heterobifunctional PEG derivatives containing hydroxyl and carboxyl or amino groups. Typically, the hydroxyl end-group of PEG was derivatized to form a urethane attachment with the hydrophobic lipid anchor, PE, while the amino or carboxyl groups were utilized for the conjugation reaction or further functionalization. To further simplify the coupling procedure and to make it applicable for single-step binding of a large variety of amino group-containing ligands (including antibodies, proteins and small molecules) to the distal end of nanocarrier-attached polymeric chains, amphiphilic PEG derivative p-nitrophenylcarbonyl-PEG-PE (pNP-PEG-PE) was introduced (Fig. 3) (Torchilin et al. 2001). pNP-PEG-PE readily adsorbs on hydrophobic nanoparticles or incorporates into liposomes and micelles via its phospholipid residue, and easily binds any amino group-containing compound via its water-exposed pNP group forming stable and non-toxic urethane (carbamate) bond. Other methods that could be used for the coupling of ligands to the distal tips of PEG chains include PEG activation with hydrazine group (in case of antibody attachment, hydrazine reacts with the oxidized carbohydrate groups in the oligosaccharide moiety of the antibody); pyridyldithiopropionate (PDP) group (after conversion of the PDP into the thiol, it reacts with maleimide groups of the pre-modified ligand); or maleimide group (reacts with thiol groups in pre-thiolated ligand) (Klibanov et al. 2003), (see Table 1 for various coupling techniques).

An attractive approach to couple various ligands, such as antibodies, to liposomes including PEGylated liposomes involves a "post-insertion" technique (Ishida et al. 1999). This technique is based on the preliminary activation of ligands with any reactive PEG-PE derivative and subsequent co-incubation of unstable micelles formed by the modified ligand-PEG-PE conjugates with pre-formed drug-loaded plain or PEGylated liposomes.

$$m \left(\text{Amine-reactive cross-linker (pNP-PEG}_{3400}\text{-PE)} \right)$$

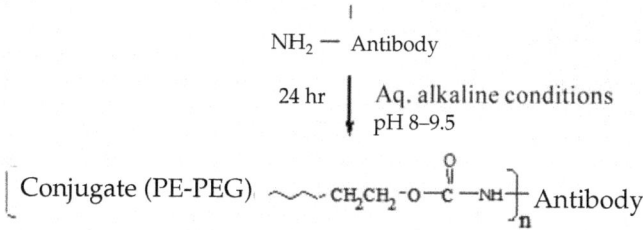

$$NH_2 - \text{Antibody}$$

$$24 \text{ hr} \quad \Bigg| \quad \text{Aq. alkaline conditions} \atop \text{pH 8–9.5}$$

$$\left[\text{Conjugate (PE-PEG)} \right. \sim\!\!\sim\!\! \text{-CH}_2\text{CH}_2\text{-O} \overset{\overset{\text{O}}{\|}}{\text{-C}} \left. -\text{NH} \right]_n \text{Antibody}$$

$$m = 10\text{–}40$$
$$n = 10\text{–}32$$

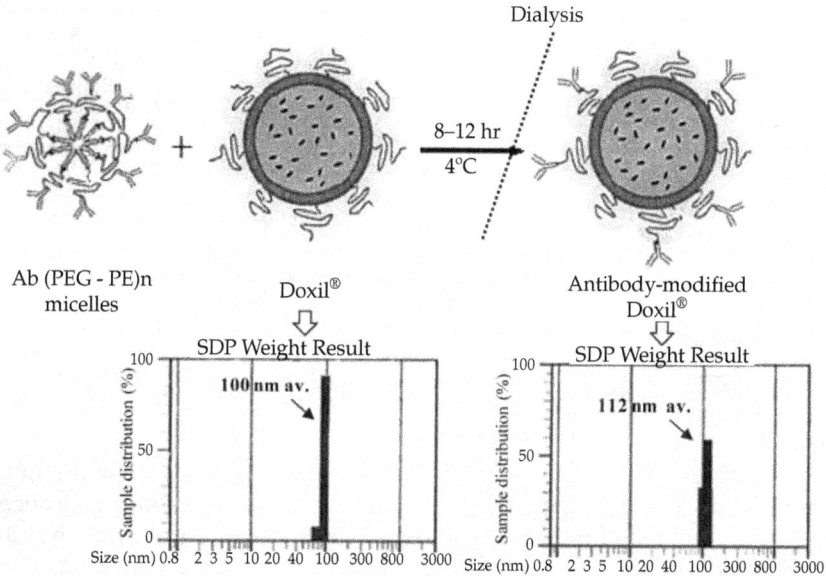

Fig. 3. Amine-based antibody conjugation, and coupling to PEGylated doxorubicin liposomes. Schematics of antibody anime conjugation reaction with pNP-PEG-PE (top), and the immuno-modification of Doxil® using post-insertion method (bottom), showing liposomal particle size measurement before (100 nm) and after the immumo-modification (112 nm) (unpublished).

Eventually, modified ligands spontaneously incorporate from their micelles into more thermodynamically favorable surrounding of the liposome membrane. This method was used, in particular, to prepare immuno-Doxil by modifying it with pNP-PEG-PE-modified anticancer 2C5 monoclonal antibodies (Lukyanov et al. 2004; Elbayoumi and Torchilin 2007).

Table 1. Summary of end group functionalized polymer lipids of general structure X-PEG-DSPE used for protein coupling to liposome surface.

X=	Functionalized polymer lipid	Application notes
–NH$_2$	**Amino-PEG-PE**	Amino group modification; forms long circulating liposomes with cationic charge characteristics.
–COOH	**Carboxy-PEG-PE**	Used for additional modification and conjugation via carbodiimide-mediated coupling, used for preparation of immunoliposomes.
	Hydrazide-PEG-PE	Used for conjugation of conjugation of antibodies modified with azide moieties on their carbohydrate residues, to PEGylated liposomes; used for N-terminal Ser/Thr peptide conjugation too.
	PDP-PEG-PE	Forms bi-sulfide linkage for binding thiol-containing ligands; precursor for HS-PEG-PE, for coupling maleimide and bromoacetyl-containing ligands to liposomes.
	Maleimide-PEG-PE	Used frequently for attachment of thiol-containing ligands, like Cys-containing peptides and Fab' antibody fragments.
	p-(nitrophenyl) carbonate-PEG-PE	Forms stable urethane attachment via binding amino-containing ligands and residues.

X represents the functional reactive group, present in the polymer lipid conjugate PEG-PE (unpublished).

A report by Hosokawa et al. (2003) demonstrated that while non-targeted doxorubicin-containing liposomes were toxic to various cancer cells to the extent of reflecting cell sensitivity to the drug, the cytotoxicity of antibody-targeted liposomes was proportional to the surface density of the surface antigen against which liposomes were targeted. The critical antigen surface concentration was about 4×10^4 sites per single cell, and after this value, the further increase in the antigen density was no longer important. Since cancer cells are often rather heterogeneous in respect to antigens they express, it was suggested (Sapra and Allen 2003) that a combination of antibodies against different antigens on a single liposome be used to provide better and more uniform targeting of all cells within the tumor.

POTENTIAL OF ANTIBODY-TARGETED DOXORUBICIN-LOADED LIPOSOMES IN CHEMOTHERAPY

Several antibodies have been used to target doxorubicin-loaded liposomes to tumors. For example, the monoclonal antibody against HER2, the antigen frequently over-expressed on various cancer cells, has been used to render drug-loaded liposomes specific for HER2-positive cancer cells (Park et al. 1995; Kirpotin et al. 2006), in particular, to deliver doxorubicin, both in plain and long-circulating liposomes, to breast tumor xenografts in mice, which resulted in significantly enhanced therapeutic activity of the drug. PEGylated liposomes decorated with anti-HER2 antibody were shown to undergo effective endocytosis by HER2-positive cancer cells allowing for better doxorubicin accumulation inside tumor cells (because of antibody-mediated internalization) with better therapeutic outcome. Another promising antibody to target tumors with drug-loaded liposomes is the monoclonal antibody against CD19 antigen, which is also frequently over-expressed on various cancer cells. Anti-CD19 antibody-modified liposomes loaded with doxorubicin demonstrated clearly enhanced targeting and therapeutic efficacy both *in vitro* and *in vivo* in mice with human CD19+ B lymphoma cells (Lopes de Menezes et al. 1998). Similar results have also been obtained with doxorubicin-loaded liposomes modified with antibodies against internalizable CD19 antigen and against non-internalizable CD20 antigen (Sapra and Allen 2004). Anti-CD19 antibodies have also been used to target doxorubicin-loaded liposomes with variable drug release rates to experimental tumors (Allen et al. 2005). Recently, a successful attempt was made to target doxorubicin-loaded long-circulating liposomes to CD19-expressing cancer cells with single chain Fv fragments of CD19 antibodies (Cheng and Allen 2008). Another viable target or therapeutic intervention is the CD22 antigen on B-cell lymphomas, which triggers antibody-mediated cell depletion therapy for treatment of lymphomas and leukemias. PEGylated liposomal doxorubicin was conjugated with HB22.7, an anti-CD22 monoclonal antibody, which was shown to improve cytotoxicity over non-targeted liposomes using non-Hodgkins lymphoma (NHL) xenograft-bearing mice, only in CD22+ cells (Tuscano et al. 2010). While myelotoxicity and pharmacokinetic profiles were comparable between anti-CD22 and plain PEGylated doxorubicin-loaded liposomes, in NHL xenograft-bearing mice, significant reduction in tumor volumes along with enhanced survival were demonstrated in immunoliposomal doxorubicin (DXR) group, versus all control treatments, especially at three-dose regimen (8 mg DXR/kg). ImmunoPEG-liposomes conjugated with S5A8 monoclonal antibody, an anti-idiotype antibody to 38C13 murine B-cell lymphoma, loaded with doxorubicin showed enhanced cell binding and internalization leading to greater *in vitro* cytotoxicity of drug encapsulated in immunoliposomes against idiotype-positive 38C13 lymphoma cells than for the idiotype-negative variant of this cell line. This was translated into prolonged survival of mice bearing 38C13 tumor treated with the specific immunoliposomes over non-targeted liposomal doxorubicin or free drug (Tseng et al. 1999). Since neuroblastoma cells usually over-express disialoganglioside GD2, it has been suggested that antibodies against GD2 and their Fab' fragments target drug-loaded liposomes to corresponding tumors (Pastorino et al. 2003, 2006). Fab' fragments of anti-GD2 antibodies covalently coupled to long-circulating liposomes loaded with doxorubicin allowed for increased binding and higher cytotoxicity against target cells both *in vitro* and *in vivo*, including in models of human tumors in nude mice and in metastatic models.

An interesting target for anti-tumor drug delivery by means of targeted liposomes is the membrane type-1 matrix metalloproteinase (MT1-MMP), playing an important role in tumor neoangiogenesis and over-expressed both on tumor cells and on neoangiogenic endothelium. The modification of doxorubicin-loaded long-circulating liposomes with anti-MT1-MMP antibody resulted in an increased uptake of the targeted liposomes by MT1-MMP-over-expressing HT1080 fibrosarcoma cells *in vitro* and in more effective inhibition of tumor growth *in vivo* compared to antibody-free doxorubicin-loaded PEGylated liposomes (Hatakeyama et al. 2007). Epidermal growth factor receptor (EGFR) and its variant EGFRvIII can serve as valuable targets for intracellular drug delivery into tumor cells over-expressing these receptors. Fab' fragments of the monoclonal antibody C225, which binds both EGFR and EGFRvIII, and scFv fragment of the monoclonal antibody, which binds only to EGFR, were coupled to drug-loaded liposomes and allowed for substantially enhanced binding of such targeted liposomes with cancer cell over-expressing corresponding receptors, such as glioma cells U87 and carcinoma cells A0431 and MDA-MB-468. The better binding resulted in enhanced internalization and increased cytotoxicity (Mamot et al. 2003). *In vivo* therapy with such targeted drug-loaded liposomes (doxorubicin, epirubicin and vinorelbine were used as drugs) always resulted in better tumor growth inhibition than therapy with non-targeted liposomal drugs (Mamot et al. 2005).

Various proteins of the extracellular matrix expressed on the surface of cancer cells have also been used as targets for the antibody-mediated delivery of the liposomal drugs. A possible target for antibody-mediated cancer therapy with drug-loaded liposomes is the epithelial cell adhesion molecule (EpCAM), which is expressed in many tumors but not in normal cells (Hussain et al. 2007). EpCAM-targeted immunoliposomes were generated by covalent attachment of the humanized scFv fragment of the 4D5MOCB monoclonal antibody to the surface of PEGylated doxorubicin-loaded liposomes and demonstrated significantly improved binding, internalization and cytotoxicity with EpCAM-positive cancer cells. Proliferating endothelial cells have been targeted with doxorubicin-loaded liposomes modified with scFv fragments of the antibody against endoglin over-expressed on such cells (Volkel et al. 2004).

Lipid-based drug nanocarriers have also been conjugated with antibodies (or their fragments) against transferrin receptor (TfR) frequently over-expressed on the surface of various cancer cells. This was recognized early to modulate doxorubicin resistance in a doxorubicin-resistant sub-line (K562/ADM) of human leukaemia K562 cells, by an anti-TFR monoclonal antibody, OKT9 coupled to liposomal surface. *In vitro* data indicated faster and increased intracellular temporal accumulation of doxorubicin in K562/ADM cells, compared to control formulations, with evident internalization into juxtanuclear vesicles in both adriamycin-resistant and sensitive cell lines (Suzuki et al. 1997).

Non-pathogenic antinuclear autoantibodies (ANAs), frequently detected in cancer patients and in healthy elderly individuals, represent a subclass of natural anti-cancer antibodies. Earlier, we have shown that certain monoclonal ANAs (such as mAbs 2C5 and 1G3) recognize the surface of numerous tumors, but not normal cells. Nucleosome-restricted specificity was shown for some of these monoclonal ANAs, and tumor cell surface-bound nucleosomes (NSs) have been shown to be their universal molecular target on the surface of the variety of tumor cells (Iakoubov and Torchilin 1998). Because these antibodies can effectively recognize a broad variety of tumors, they may serve as specific ligands to deliver other drugs and drug nanocarriers into tumors. These antibodies were used to prepare drug-loaded tumor-targeted long-circulating immunoliposomes (with doxorubicin), which demonstrated highly specific binding with various cancer cells (murine Lewis lung carcinoma, 4T1, C26, and human BT-20, MCF-7, PC3 cells) *in vitro* (Lukyanov et al. 2004; Elbayoumi and Torchilin 2007), significantly increased tumor accumulation in model tumors in mice including intracranial human brain U-87 MG tumor xenografts in nude mice, decreased side effects, and superior antitumor activity *in vivo* (Elbayoumi and Torchilin 2009a, b) (Fig. 4).

Doxorubicin-loaded PEGylated liposomes were also modified with Fab' fragments of an anti-CD74 antibody via a PEG-based heterobifunctional coupling reagent and demonstrated a significantly accelerated and enhanced accumulation in Raji human B-lymphoma cells *in vitro* (Lundberg et al. 2007). Anti-CD166 scFv attached to drug-loaded liposomes facilitated doxorubicin internalization by several prostate cancer cell lines (Du-145, PC3, LNCaP) (Roth et al. 2007). The immunoliposomal doxorubicin was targeted to human CD34(+) and MDR(+) human myelogenous leukemia KG-1a cell line showed a higher cytotoxicity against KG-1a cells than non-

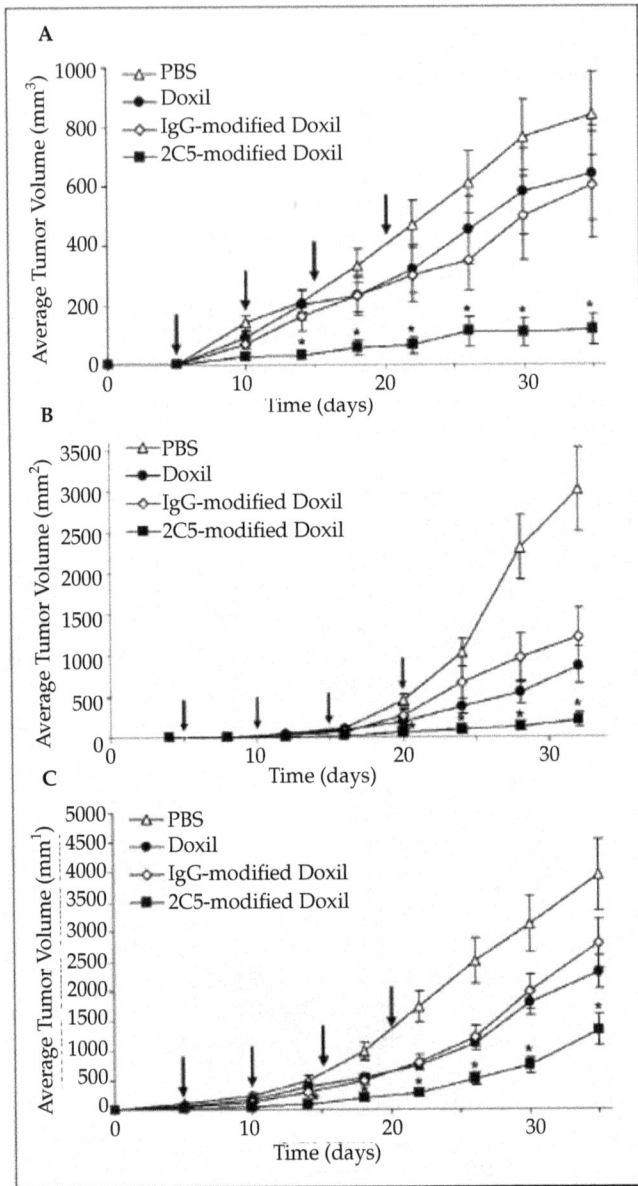

Fig. 4. Pre-clinical *in vivo* antitumor efficacy of mAb 2C5-modified Doxil®. Therapeutic activity, expressed as tumor volumes, of 2C5-modified Doxil against control preparations in mice implanted with 4T1 (A), C26 (B), and PC3 (C). Arrows, treatment schedule, 2 mg/kg every 5 d (n = 8–10). *$P \leq 0.05$ versus corresponding control, IgG-Doxil; non-parametric Kruskal-Wallis with Tukey's post hoc test (mean ± SD). (Reprinted from ElBayoumi and Torchilin 2009b, with permission from American Association for Cancer Research.)

targeted liposomal doxorubicin, but without improvement over that of free drug. The minor improvement in efficacy was attributed to stable surface binding of the immune-doxorubicin liposomes to the CD34 antigen, without internalization, and subsequently, no reversal of doxorubicin resistance was found. Thus, immunotargeting of liposomal doxorubicin to CD34(+) leukemic cells may only provide an *extra-cellular* strategy for site-selective CD34(+) leukemia cell killing (Carrion et al. 2004).

Doxorubicin-loaded liposomes were also successfully targeted to the kidney by using Fab' fragments of the monoclonal OX7 antibody directed against Thy1.1 antigen in rats (Tuffin et al. 2005). Tumor necrotic zones were effectively targeted by doxorubicin-loaded liposomes modified with chimeric TNT-3 monoclonal antibody specific towards degenerating cells located in necrotic regions of tumors and demonstrated enhanced therapeutic efficacy in nude mice bearing H460 tumors (Pan et al. 2008).

Early clinical trials of antibody-targeted drug-loaded liposomes have already demonstrated some promising results.Doxorubicin-loaded PEGylated liposomes (approx. 140 nm) modified with F(ab')$_2$ fragments of the GAH monoclonal antibody specific for stomach cancer were tested in Phase I clinical studies and demonstrated pharmacokinetics similar to that of Doxil® (Matsumura et al. 2004).

Thus, there exists a whole set of antibodies or their fragments used for targeting nanoscale liposomal anticancer drugs, starting with doxorubicin, to various tumors (Table 2).

APPLICATIONS TO AREAS OF HEALTH AND DISEASE

Doxorubicin remains one of the most commonly used anticancer drugs in chemotherapy treatment protocols, owing to its potency against a broad spectrum of solid and hematological neoplasms. With the inception of Doxil^{TM}, nanopreparation of doxorubicin loaded into long-circulating PEGylated liposomes, some important problems associated with doxorubicin side effects, cardiotoxicity and insufficient tumor accumulation have been resolved. Tumor-targeted nanomedicines, especially antibody-targeted liposomal doxorubicin, are the promising drug delivery systems being developed now in oncology to improve drug performance by overcoming most limitations of non-specific doxorubicin therapy. The most striking feature of antibody-targeted doxorubicin-loaded liposomes is their ability to bring the drug to the tumor site, thereby enhancing tumoral drug levels (site-specific delivery; aiming for enhanced anti-tumor activity), and/or to direct a drug away from those body sites that are particularly sensitive to the toxic effects of the drug (site-avoidance delivery; aiming for reduced damage to normal tissues). The variety of immuno-liposomal doxorubicin nanoformulations is believed to open the door of true "tailored" therapy, based on customization of both drug dose regimen and targeting antibody to the specific tumor marker identified at the single patient level.

Table 2. Examples of antibodies (or their fragments) used to target liposomal doxorubicin to tumors.

Targeting agent	Cell surface antigen	Model
AntiCD19	CD19	Namala hu-B-cell lymphoma human multiple myeloma, ARH and cell line
Anti-CD19, scFv	CD19	Raji human B-lymphoma
HB22.7 and anti-CD22-ScFv	CD22	non-Hodgkins lymphoma, BJAB and Raji lymphoma
Recombinant human anti-HER2-Fab' or scFv C6.5	HER2	HER2-overexpressing human breast cancer
Anti-GD$_2$ and anti-GD$_2$-Fab' GD$_2$		human neuroblastoma
Anti-idiotype mAb, S5A8 38C13		murine D-cell lymphoma
Anti-ganglioside G$_{M3}$(DH2) or anti-L^{Ex}(SH1)	Carbohydrate, ganglioside (G$_{M3}$); Lewis X(Le^x)	B16BL6 mouse melanoma and HRT-18 human colorectal adenocarcinoma
chTNT	TNT	human non-small lung carcinoma H460
Anti-β$_1$-integrin Fab'	human β$_1$ integrins	human non-small cell lung carcinoma
Anti CD74 LL1	CD74	Raji human B-lymphoma
Anti-nucleosome 2C5 mAb	nucleosome	human BT-20, MCF-7, PC3 murine LLC, 4T1, C26;
C225 mAb or Fab'	EGF receptor	human MDA-MB-468 adenocarcinoma, U87 glioblastoma
Anti-MT1-MMP-Fab'	metalloproteinase	human HT1080 fibrosarcoma MT1-MMP
anti-Thy-1.1 OX7 mAb	Thy-1.1	rat mesamgial cells
scFv A5	endothelin	endothelial cells HUVEC, HDMEC
OKT9 mAb	Transferrin receptor	human leukemia

Each immuno-targeting ligand in the first column was shown to specifically recognize and bind its target cell antigen, which is over-expressed on the test model tumor or cell line (unpublished).

From our growing knowledge on the liposome technology so far and the considerable simplicity and attainable efficiency in industrial manufacturing of liposomes (plain or PEGylated), plus the adequate stability of liposomal preparations of doxorubicin in storage and within the body, it is clear that antibody-targeted doxorubicin-loaded liposomes represent promising candidates for cancer chemotherapy, and we can expect their extensive clinical evaluation in the near future. Breast, colon and lung cancer, which have been used as targets in a vast majority of experiments, can be named as leading candidates for therapy with antibody-targeted liposomes.

Overall, the pros seem to overweigh the cons, and practical application of antibody-targeted liposomal doxorubicin for tumor therapy should become a reality within the next few years.

Definitions

Anti-idiotype (Id) antibodies: An antibody that is the mirror image of the original antibody, formed against a specific surface antigen.

Doxorubicin/Adriamycin (hydroxydaunorubicin): a classical drug used in cancer chemotherapy. It is an anthracycline antibiotic (derived from *Streptomyces* bacteria). It works by intercalating DNA and oxidative destruction of intracellular proteins and lipid membranes.

Endocytosis: The process in which a substance gains entry into a cell without passing through the cell membrane.

Endothelial cells: Thin layer of cells that lines the interior surface of blood vessels.

Epithelial cell adhesion molecule (EPCAM): A pan-epithelial differentiation antigen that is expressed on almost all carcinomas and is involved in central cellular pathways for intracellular signaling and polarity.

HER2/neu (ErbB-2): Human Epidermal growth factor Receptor 2, a cell membrane surface-bound receptor tyrosine kinase normally involved in the signal transduction pathways leading to cell growth and differentiation. It is a prognostic indicator of higher aggressiveness in breast cancers.

Matrix metalloproteinases (MMPs): Zinc-dependent endopeptidases with the ability to degrade all kinds of extracellular matrix proteins and involved in processing a number of bioactive molecules. MMPs are known to be involved in the cleavage of cell surface receptors, release of apoptotic ligands, and cell proliferation, migration, differentiation, angiogenesis, apoptosis and host defense.

Monoclonal antibodies (mAbs): Any of the highly specific antibodies produced in large quantity by the clones of a single hybrid cell formed in the laboratory by the fusion of a B cell with an immortal cell.

Vascular endothelial growth factor (VEGF): An important signaling protein involved in both vasculogenesis (the formation of the circulatory system) and angiogenesis (the growth of blood vessels from pre-existing vasculature).

Summary Points

- A significant number of monoclonal antibodies are now identified and engineered as chimeric or humanized antibodies, or full/partial antigen-binding fragments of the antibodies that can target various tumor-specific ligands.
- Liposomes, as lipidic nanocarriers, are reasonably easy to make and liposomal nanopreparations demonstrate a sufficient stability at storage and in the body, with a remarkable biocompatibility record.
- Several reasonably simple, highly effective and reproducible methods to couple antibodies or their fragments to the surface of plain or long-circulating drug-loaded liposomes are developed yielding antibody-modified liposomes with preservation of specific antibody affinity and capable of effective recognition of the target cells both *in vitro* and *in vivo*.
- Antibody attachment can decrease the plasma circulation time of liposomes because of whole antibody-mediated (primarily involving the antibody constant region) clearance of the modified liposomes and can also trigger immunological reactions. These effects could be minimized by using antibody fragments instead of whole antibodies.
- In many cases, antibody-targeted liposomes demonstrate better internalization by cancer cells and more effective intracellular drug delivery than other preparations, which could allow overcoming of multidrug resistance.
- Extensive data clearly demonstrate significant benefits of antibody-targeted liposomal doxorubicin in numerous animal models, such as accelerated target accumulation, increased quantity of the drug delivered to the target, decreased side effects associated with the administration of non-targeted liposomal drugs, and significantly enhanced therapeutic outcomes.
- A number of early clinical trials with antibody-targeted liposomal drugs have already yielded promising results, particularly against breast, ovarian and lung cancers.

Key Facts

- The hydrochloride salt of doxorubicin is the most commonly used form of the anthracycline antibiotic doxorubicin, with antineoplastic activity. Doxorubicin was first isolated in the 1950s from the bacterium

Streptomyces peucetius var. *caesius*, as the hydroxylated congener of daunorubicin.

- Doxorubicin is commonly used in chemotherapy regimens for treatment of some leukemias and Hodgkin's lymphoma, as well as cancers of the bladder, breast, stomach, lung, ovaries, thyroid, soft tissue sarcoma, multiple myeloma, and others.
- Doxorubicin anticancer action is produced via different mechanisms: drug intercalation between base pairs in the DNA helix, thus preventing DNA replication and protein synthesis; inhibition of topoisomerase II, leading to double-DNA strand breakage; cytotoxic action involving the formation of highly reactive oxygen free radicals, causing DNA and lipid peroxidation of cell membrane phospholipids.
- The formation of oxygen free radicals also contributes to the adverse toxicity of the anthracycline antibiotics, primarily the cardiac and cutaneous vascular side effects.
- The conventional non-PEGylated liposomes-encapsulating doxorubicin, Myocet®, is approved in Europe and Canada for treatment of metastatic breast cancer in combination with cyclophosphamide.
- Doxil®, the long-circulating (PEG-coated) liposome-encapsulated form of doxorubicin, was developed and approved by the FDA in 1996 to treat Kaposi's sarcoma.

Abbreviations

EPR	:	enhanced permeability and retention
IgG	:	Immunoglobulin G
Fab	:	fragment antigen-binding
ScFv	:	single-chain variable fragment
MDR	:	multidrug resistance

References

Allen, T.M., D.R. Mumbengegwi et al. 2005. Anti-CD19-targeted liposomal doxorubicin improves the therapeutic efficacy in murine B-cell lymphoma and ameliorates the toxicity of liposomes with varying drug release rates. Clin. Cancer Res. 11: 3567–3573.

Bendas, G., U. Rothe et al. 2003. The influence of repeated injections on pharmacokinetics and biodistribution of different types of sterically stabilized immunoliposomes. Biochim. Biophys. Acta. 1609: 63–70.

Carrion, C., M.A. de Madariaga et al. 2004. *In vitro* cytotoxic study of immunoliposomal doxorubicin targeted to human CD34(+) leukemic cells. Life Sci. 75: 313–328.

Chapner, B.A., P.C. Amrein et al. 2005. Antineoplastic agents. *In:* L.L. Brunton, J.S. Lazo and K.L. Parker. Goodman & Gilman's the Pharmacological Basis of Therapeutic. McGraw-Hill Medical Pub. Division, New York, 1: 1315–1403.

Cheng, W.W., and T.M. Allen 2008. Targeted delivery of anti-CD19 liposomal doxorubicin in B-cell lymphoma: a comparison of whole monoclonal antibody, Fab' fragments and single chain Fv. J. Control. Release 126: 50–58.

Elbayoumi, T.A., and V.P. Torchilin 2007. Enhanced cytotoxicity of monoclonal anticancer antibody 2C5-modified doxorubicin-loaded PEGylated liposomes against various tumor cell lines. Eur. J. Pharm. Sci. 32: 159–168.

Elbayoumi, T.A., and V.P. Torchilin 2009a. Tumor-specific anti-nucleosome antibody improves therapeutic efficacy of doxorubicin-loaded long-circulating liposomes against primary and metastatic tumor in mice. Mol. Pharm. 6: 246–254.

ElBayoumi, T.A., and V.P. Torchilin 2009b. Tumor-targeted nanomedicines: enhanced antitumor efficacy in vivo of doxorubicin-loaded, long-circulating liposomes modified with cancer-specific monoclonal antibody. Clin. Cancer Res. 15: 1973–1980.

Elbayoumi, T.A., and V.P. Torchilin 2010. Current trends in liposome research. Methods Mol. Biol. 605: 1–27.

Gregoriadis, G., (ed.). 2007. Liposome Technology: Liposome Preparation and Related Techniques. Taylor & Francis, London.

Hatakeyama, H., H. Akita et al. 2007. Tumor targeting of doxorubicin by anti-MT1-MMP antibody-modified PEG liposomes. Int. J. Pharm. 342: 194–200.

Hosokawa, S., T. Tagawa et al. 2003. Efficacy of immunoliposomes on cancer models in a cell-surface-antigen-density-dependent manner. Br. J. Cancer 89: 1545–1551.

Hussain, S., A. Pluckthun et al. 2007. Antitumor activity of an epithelial cell adhesion molecule targeted nanovesicular drug delivery system. Mol. Cancer Ther. 6: 3019–3027.

Iakoubov, L.Z., and V.P. Torchilin 1998. Nucleosome-releasing treatment makes surviving tumor cells better targets for nucleosome-specific anticancer antibodies. Cancer Detect Prev 22: 470–475.

Ishida, T., D.L. Iden et al. 1999. A combinatorial approach to producing sterically stabilized (Stealth) immunoliposomal drugs. FEBS Lett. 460: 129–133.

Kirpotin, D.B., D.C. Drummond et al. 2006. Antibody targeting of long-circulating lipidic nanoparticles does not increase tumor localization but does increase internalization in animal models. Cancer Res. 66: 6732–6740.

Klibanov, A.L., V.P. Torchilin et al. 2003. Long-circulating sterically protected liposomes. In: V.P. Torchilin and V. Weissig. Liposomes: A Practical Approach. Oxford University Press, Oxford, New York, pp. 231–265.

Lasic, D.D., and D. Papahadjopoulos (eds.). 1998. Medical Applications of Liposomes. Elsevier, Amsterdam, New York.

Lopes de Menezes, D.E., L.M. Pilarski et al. 1998. In vitro and in vivo targeting of immunoliposomal doxorubicin to human B-cell lymphoma. Cancer Res. 58: 3320–3330.

Lorusso, D., A. Di Stefano et al. 2007. Pegylated liposomal doxorubicin-related palmar-plantar erythrodysesthesia ('hand-foot' syndrome). Ann. Oncol. 18: 1159–1164.

Lukyanov, A.N., T.A. Elbayoumi et al. 2004. Tumor-targeted liposomes: doxorubicin-loaded long-circulating liposomes modified with anti-cancer antibody. J. Control. Release 100: 135–144.

Lundberg, B.B., G. Griffiths et al. 2007. Cellular association and cytotoxicity of doxorubicin-loaded immunoliposomes targeted via Fab' fragments of an anti-CD74 antibody. Drug Deliv. 14: 171–175.

Maeda, H., J. Wu et al. 2000. Tumor vascular permeability and the EPR effect in macromolecular therapeutics: a review. J. Control. Release 65: 271–284.

Mamot, C., D.C. Drummond et al. 2003. Epidermal growth factor receptor (EGFR)-targeted immunoliposomes mediate specific and efficient drug delivery to EGFR- and EGFRvIII-overexpressing tumor cells. Cancer Res. 63: 3154–3161.

Mamot, C., D.C. Drummond et al. 2005. Epidermal growth factor receptor-targeted immunoliposomes significantly enhance the efficacy of multiple anticancer drugs in vivo. Cancer Res. 65: 11631–11638.

Matsumura, Y., T. Hamaguchi et al. 2004. Phase I clinical trial and pharmacokinetic evaluation of NK911, a micelle-encapsulated doxorubicin. Br. J. Cancer 91: 1775–1781.

Pan, H., L. Han et al. 2008. Targeting to tumor necrotic regions with biotinylated antibody and streptavidin modified liposomes. J. Control. Release 125: 228–235.

Park, J.W., K. Hong et al. 1995. Development of anti-p185HER2 immunoliposomes for cancer therapy. Proc. Natl. Acad. Sci. USA. 92: 1327–1331.

Pastorino, F., C. Brignole et al. 2006. Targeting liposomal chemotherapy via both tumor cell-specific and tumor vasculature-specific ligands potentiates therapeutic efficacy. Cancer Res. 66: 10073–10082.

Pastorino, F., C. Brignole et al. 2003. Doxorubicin-loaded Fab' fragments of anti-disialoganglioside immunoliposomes selectively inhibit the growth and dissemination of human neuroblastoma in nude mice. Cancer Res. 63: 86–92.

Roth, A., D.C. Drummond et al. 2007. Anti-CD166 single chain antibody-mediated intracellular delivery of liposomal drugs to prostate cancer cells. Mol. Cancer Ther. 6: 2737–2746.

Sapra, P., and T.M. Allen 2003. Ligand-targeted liposomal anticancer drugs. Prog. Lipid Res. 42: 439–462.

Sapra, P., and T.M. Allen 2004. Improved outcome when B-cell lymphoma is treated with combinations of immunoliposomal anticancer drugs targeted to both the CD19 and CD20 epitopes. Clin. Cancer Res. 10: 2530–2537.

Soloman, R., and A.A. Gabizon 2008. Clinical pharmacology of liposomal anthracyclines: focus on pegylated liposomal Doxorubicin. Clin. Lymphoma Myeloma 8: 21–32.

Suzuki, S., K. Inoue et al. 1997. Modulation of doxorubicin resistance in a doxorubicin-resistant human leukaemia cell by an immunoliposome targeting transferring receptor. Br. J. Cancer 76: 83–89.

Torchilin, V. 2008. Antibody-modified liposomes for cancer chemotherapy. Expert Opin. Drug Deliv. 5: 1003–1025.

Torchilin, V.P., T.S. Levchenko et al. 2001. p-Nitrophenylcarbonyl-PEG-PE-liposomes: fast and simple attachment of specific ligands, including monoclonal antibodies, to distal ends of PEG chains via p-nitrophenylcarbonyl groups. Biochim. Biophys. Acta. 1511: 397–411.

Torchilin, V.P., V.G. Omel'yanenko et al. 1980. Incorporation of hydrophilic protein modified with hydrophobic agent into liposome membrane. Biochim. Biophys. Acta. 602: 511–521.

Tseng, Y.L., R.L. Hong et al. 1999. Sterically stabilized anti-idiotype immunoliposomes improve the therapeutic efficacy of doxorubicin in a murine B-cell lymphoma model. Int. J. Cancer 80: 723–730.

Tuffin, G., E. Waelti et al. 2005. Immunoliposome targeting to mesangial cells: a promising strategy for specific drug delivery to the kidney. J. Am. Soc. Nephrol. 16: 3295–3305.

Tuscano, J.M., S.M. Martin et al. 2010. Efficacy, biodistribution, and pharmacokinetics of CD22-targeted pegylated liposomal doxorubicin in a B-cell non-Hodgkin's lymphoma xenograft mouse model. Clin. Cancer Res. 16: 2760–2768.

Volkel, T., P. Holig et al. 2004. Targeting of immunoliposomes to endothelial cells using a single-chain Fv fragment directed against human endoglin (CD105). Biochim. Biophys. Acta. 1663: 158–166.

18

Nanomedicine in Blood Diseases

Denis B. Buxton

ABSTRACT

Nanotechnology offers a wide range of opportunities to impact a spectrum of blood diseases. The balance between thrombosis and thrombolysis represents a pivotal equilibrium that impacts many disease processes. Nanotechnology tools to diagnose and treat thrombosis are expected to contribute significantly to minimizing the harmful sequelae of disease processes such as stroke and deep vein thrombosis (DVT). Nanoformulation of anticoagulant drugs can facilitate oral or pulmonary administration, while prevention of thrombosis on surfaces of devices such as stents, valves, and ventricular assist devices could reduce the need for anticoagulation in patients. Conversely, nanotechnology approaches such as synthetic platelets to promote thrombosis could have broad applicability in surgery and trauma. In the field of hematological cancers, nanoparticle-based delivery offers opportunities for optimizing therapeutic delivery to cancerous cells while minimizing off-target systemic damage. Nanotechnology can also help to overcome multi-drug resistance and provide better control of combination therapy to enhance synergism between drugs. Nanoscaffolds show promise for the expansion of hematopoietic cells such as umbilical cord blood cells while

National Heart, Lung, and Blood Institute, 6701 Rockledge Drive, Suite 8216, Bethesda MD 20892-7940, USA; Email: db225a@nih.gov

List of abbreviations after the text.

maintaining engraftment potential, which could improve success rates for bone marrow transplants. Sepsis remains an area with limited therapeutic options, and nanotechnology offers significant potential for reducing mortality. Mesoporous microparticles show promise for rapid removal of inflammatory cytokines to damp down the hyperinflammatory response. Nanostructured peptides can efficiently remove bacterial lipopolysaccharide, a key sepsis mediator, from the bloodstream. Loss of lymphocytes from the circulation due to programmed cell death also contributes to sepsis; the use of targeted nanoparticles as a delivery agent for small interfering RNA (siRNA) can efficiently down-regulate key pro-cell death proteins in lymphocytes, helping to maintain B and T cell viability. Finally, nanofiltration provides a key step in the removal of viruses from blood products, helping to ensure the safety of treatments for hemophilia and other clotting disorders.

INTRODUCTION

The application of nanotechnology to diagnosing and treating blood diseases represents an exciting and rapidly growing area. Some technologies are already quite mature; the use of nanofiltration to remove viruses from blood products is now well established, providing a critical technology to ensure the safety of patients who need to receive blood proteins for clotting disorders. A number of nanoparticle therapeutics have reached clinical application for hematological cancers such as leukemias and are being tested in clinical trials to determine efficacy. Other technologies, such as the use of synthetic platelets as a hemostatic agent for treating trauma patients, are earlier in their developmental cycle, but show great promise for eventual clinical application. The goal of this chapter is to give an overview of this spectrum of nanomedicine as it is being applied to blood diseases, from very early-stage technologies to those already being applied for human use.

THROMBOSIS AND THROMBOLYSIS

Thrombus Detection

Thrombus formation represents a key biological process with potentially devastating pathophysiological effects. It is the final step in the initiation of almost all myocardial infarctions and strokes. Venous thromboembolism, including DVT and pulmonary embolism, results in more than 60,000 deaths and 600,000 hospitalizations each year in the US alone.

Incidence rises rapidly with age, reaching 500–600 per 100,000 in older patients (>75 yr) (Douma et al. 2010). Mortality from untreated venous thromboembolisms can be high, e.g., 25% for pulmonary embolisms; diagnosis can be difficult (Douma et al. 2010), but accurate diagnosis is important to avoid unnecessary complications of giving anticoagulants to patients who do not need them.

Molecular imaging agents that recognize intravascular thrombi would facilitate the diagnosis, risk stratification and treatment of thrombotic syndromes. McCarthy et al. developed multimodal cross-linked iron oxide nanoparticles directed through peptide targeting sequences to two thrombus constituents, fibrin and activated factor XIII (McCarthy et al. 2009). The targeted nanoparticles demonstrate high affinity for thrombi and can be detected through fluorescence and magnetic resonance imaging (McCarthy et al. 2009).

Pan et al. developed a manganese-based magnetic resonance theranostic imaging agent in order to avoid nephrogenic systemic fibrosis, a serious complication of gadolinium contrast agents occurring in some patients with renal disease or following liver transplant (Pan et al. 2008). The "bialy"-shaped nanoparticles are formed by self-assembly of amphiphilic branched polyethylenimine and targeted to fibrin using an anti-fibrin antibody. *In vitro* experiments using fibrin-rich clots demonstrated good contrast with the targeted agent (Pan et al. 2008). The nanoparticles can also be loaded with hydrophilic or hydrophobic drugs for targeted delivery. The same group has also developed manganese nanocolloids for fibrin-targeted imaging of intravascular thrombus (Pan et al. 2009).

Thrombolysis

Thrombolytic therapy plays a key role in preventing or mitigating tissue damage following occlusion of blood vessels in acute myocardial infarction. However, most ischemic stroke patients do not receive thrombolytics because of the danger of cerebral hemorrhage, and thrombolytic therapy is not always effective. One approach to improving the safety and efficacy of thrombolytic therapy is to use nanoparticles to target the thrombolytic to the clot. Targeting streptokinase-conjugated nanoparticles to fibrin using anti-fibrin antibody increases the effective dose of the thrombolytic at the clot surface. Marsh et al. showed using human plasma clots *in vitro* that targeted nanoparticles were orders of magnitude more effective than free drug (Marsh et al. 2007).

Another approach to targeting nanoparticles to clots uses magnetic nanoparticles in conjunction with a local magnetic field. Bi et al. conjugated urokinase to dextran-coated iron oxide magnetic nanoparticles and tested their efficacy in a rat arteriovenous shunt thrombosis model. The conjugate had a longer circulating half-life than urokinase, and the magnetic field enhanced the thrombolytic efficacy five-fold relative to free urokinase, allowing local thrombolysis with no systemic fibrinogen activation and little prolongation of bleeding time (Bi et al. 2009). Magnetic nanoparticles have also been used to target recombinant tissue plasminogen activator (rTPA) using an external magnet (Ma et al. 2009b). They demonstrated in a rat hind limb embolic model that administration of magnetic nanoparticles carrying rTPA, in conjunction with cyclical movement of the external magnet along the iliac artery, resulted in rapid restoration of blood flow, while free rTPA or magnetic nanoparticles without bound rTPA were ineffective (Fig. 1). Use of a stationary magnet with the rTPA nanoparticles was ineffective in restoring blood flow, consistent with a role for mechanical force in enhancing rTPA nanoparticle penetration into the clot (Fig. 1, inset) (Ma et al. 2009b) .

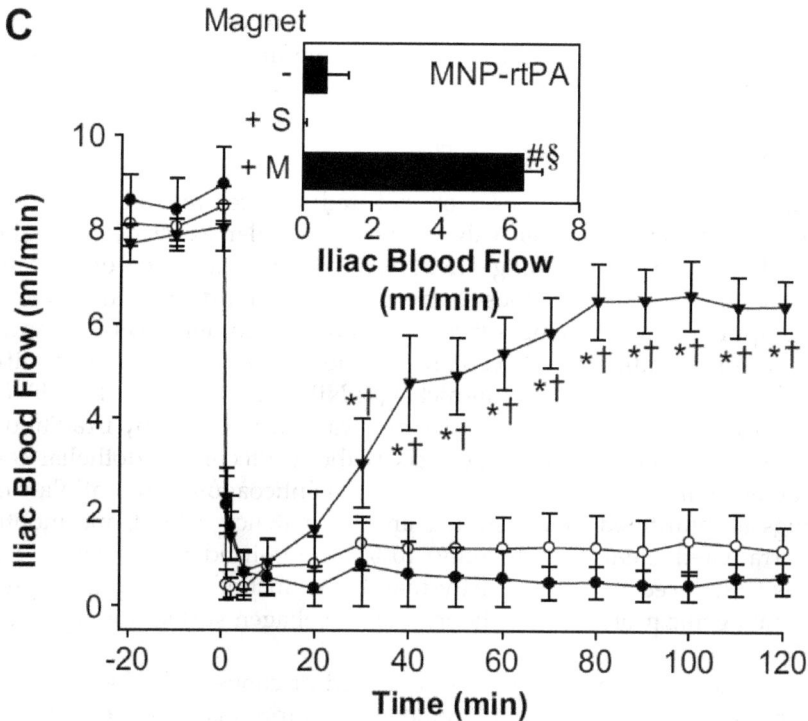

Fig. 1. Restoration of blood flow by magnetic nanoparticles in a rat embolic model. After introduction of a whole blood clot into the left iliac artery at time 0, animals were treated with: (o) rTPA; (●) magnetic nanoparticle; (▼) magnetic nanoparticle-rTPA. Results are means ± SEM for 8–10 animals. Inset: Comparison of animals treated with moving (M) versus stationary (S) magnet. Reproduced from Ma et al. (2009b) with permission of Elsevier.

Anticoagulation

Oral and Pulmonary Anticoagulant Administration

Low molecular weight heparin (LWMH) is the anticoagulant of choice for preventing DVT, but its poor oral bioavailability necessitates parenteral administration. Encapsulation of the LMWH tinzaparin into nanoparticles prepared from a blend of polyester and polycationic polymethacrylate resulted in good bioavailability following oral administration in rabbits, with an anticoagulant effect prolonged up to 8 h (Hoffart et al. 2006).

Nanoformulations also show promise for pulmonary administration of LMWH. Bai et al. used positively charged polyamidoamine dendrimers to deliver enoxaparin via intratracheal administration. Pulmonary administration was as effective as subcutaneous injection in a rat model of DVT (Bai et al. 2007). PEGylation of the dendrimers increased pulmonary absorption and plasma half-life; pulmonary administration of LMWH using the PEGylated dendrimer carrier at 48 h intervals resulted in similar efficacy to subcutaneous LMHW administered in saline at 24 h intervals (Bai and Ahsan 2009).

Prevention of Surface-Mediated Thrombosis

Acute and chronic thromboses on exposed surfaces remain a problem for devices such as stents, valves, and ventricular assist devices. The introduction of drug-eluting stents reduced the incidence of in-stent thrombosis, but late thrombosis remains an issue for drug-eluting stents. One approach to reducing surface thrombogenicity employs nanoscale surface modification. Sugita et al. used a He^+ ion beam to surface modify Type I collagen-coated titanium-nickel (Ti-Ni) stents (Sugita et al. 2009). *In vitro* studies showed platelet adherence was reduced 70% by irradiation of the collagen-coated Ti-Ni surface, without affecting endothelial cell adherence. *In vivo* studies in dogs with no anticoagulant or antiplatelet drugs administered demonstrated an 80% patency rate at one month in irradiated stents, while only 10% of non-irradiated stents were patent. Reduced thrombogenicity may be a function of binding of pro-thrombogenic proteins onto the irradiated collagen surface (Sugita et al. 2009).

The replacement of damaged, diseased or congenitally incompetent heart valves with artificial replacements is compromised by the tendency for thrombosis, and so patients receiving artificial valves must take anticoagulants for the rest of their lives. A number of nanotechnology-based approaches are under development to improve both the hemocompatibility and the durability of artificial replacement valves. These include the following (Ghanbari et al. 2009):

- Use of nanocomposites with improved biocompatibility and anti-thrombogenicity such as a composite between polyhedral oligomeric silsequiozane and polycarbonate urethane.
- Surface modification by biofunctionalization or plasma immersion ion implantation to improve surface endothelialization and render the surface antithrombogenic.
- Modification of surface topography at the nanoscale to control cell attachment and proliferation at the surface through contact "guidance", where cells sense and respond to nanoscale cues.

While these approaches show promise in the laboratory, they have yet to be translated into the clinic (Ghanbari et al. 2009).

COAGULATION

Hemostatic Agents

A number of topical hemostatic agents have been developed to control bleeding in surgery, trauma, and battlefield injury. However, the current generation suffer from a variety of shortcomings, leading to continued efforts to develop improved agents (Ellis-Behnke 2010). Synthetic platelets to stop bleeding in patients would be of enormous benefit in situations such as civilian trauma and battlefield injuries. Bertram et al. designed synthetic platelets based on poly(lactic-co-glycolic acid)-poly-L-lysine (PLGA-PLL) block copolymer cores with conjugated polyethylene glycol (PEG) arms terminated with arginine-glutamate-aspartate (RGD) functionalities (Fig. 2, top) (Bertram et al. 2009). The synthetic platelets are incorporated into clots (Fig. 2, bottom) and halved bleeding time in a rat model of major trauma. The nanoparticle platelets are stable at room temperature, cleared from the body within 24 h, and no significant complications were observed up to 7 d after administration. RGD, and the more active modified GRGDS moiety, are hypothesized to interact with activated platelet receptors glycoprotein IIb-IIIa and integrin $\alpha v \beta_3$, promoting adhesion between the polymers and platelets (Bertram et al. 2009).

Another approach to hemostasis employs self-assembling nanofiber peptides (Ellis-Behnke 2010). The peptides, which consist of repeating units of arginine-alanine-aspartate-alanine (RADA) peptides, self-assemble rapidly into well-ordered interwoven nanofibers with hydrophobic alanines inside and hydrophilic charged residues outside. In a variety of experimental wound settings, including brain, spinal cord, liver and femoral artery, regular and irregular wounds containing voids, and high pressure settings, hemostasis was obtained in 10–15 s. Additional advantages include optical transparency, ease of use, and safety. The material could be used in the surgical setting, facilitating the speed of surgery significantly, or in trauma settings; clinical testing in humans has yet to be performed, however (Ellis-Behnke 2010).

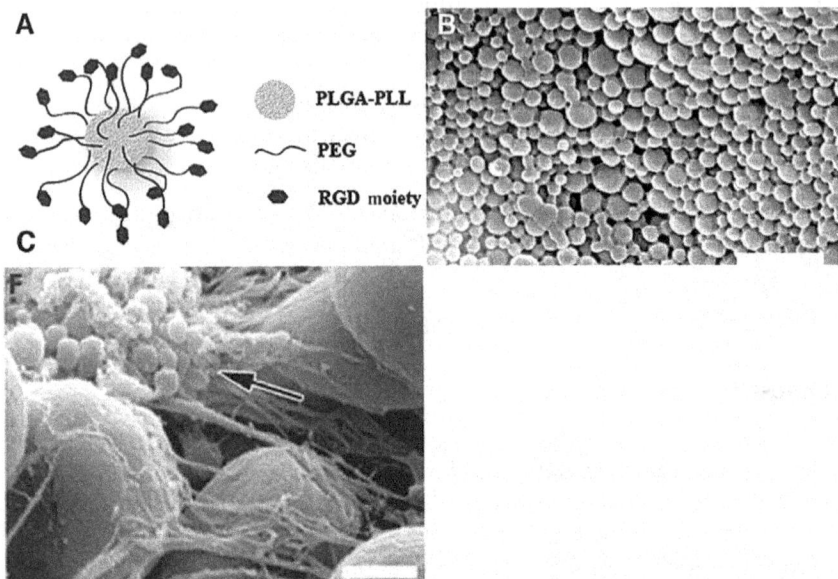

Fig. 2. Synthetic platelets. Top left: Schematic of synthetic platelet composed of PLGA-PLL core with PEG arms terminated with RGD peptide. Top right: SEM micrograph of synthetic platelets. Scale bar, 1 μm. Bottom: SEM micrograph of clot excised from injured artery after synthetic platelet administration. Arrow points to synthetic platelets intimately associated with clot and connecting fibrin mesh. The large (5 μm) spheres are blood cells. Scale bar, 1 μm. Reproduced from Bertram et al. (2010) with permission of AAAS.

Dai et al. have developed mesoporous silica xerogel beads coated with macroporous chitosan as a hemostatic agent that takes advantage of different modes of action (Dai et al. 2010). Chitosan provides hemostatic activity through electrostatic interactions between the positively charged material and negatively charged erythrocyte membranes; the polar environment may also provide a scaffold for autocatalytic activation of coagulation factors. In addition, the mesoporous silica xerogel, with high porosity and an average pore size of 3.6 nm, contributes high capacity for fluid absorption, leading to concentration of clotting factors at the surface. Calcium ions doped in the silica xerogel cores may also contribute to activation of intrinsic clotting cascades. The xerogel beads reduced coagulation time *in vitro*, and stopped bleeding in a rabbit femoral artery injury model. In addition, unlike some current hemostatic agents there is no exothermic reaction and hence no tissue thermal injury, and no obvious cytotoxicity was observed during a 7 d observation period (Dai et al. 2010).

Heparin Antagonism

Uncontrolled bleeding can also occur in the setting of an overdose of the anticoagulant heparin, leading to a need for compounds with anti-heparin activity. A novel approach to this problem takes advantage of engineered virus-like particles derived from bacteriophage Qβ (Udit et al. 2009). By generating point mutations to the coat protein and covalent attachment of poly-arginine peptides, the surface charge of the virus-like particles could be widely varied since the capsid consists of 180 coat proteins. Using an activated partial thrombin time clotting assay, they identified modified particles that inhibited heparin-mediated anticoagulation by > 95%. The inhibition is believed to be by presentation of regions with high positive charge density to bind heparin's extensively sulfated domains (Udit et al. 2009).

HEMATOLOGICAL CANCERS

The treatment of hematological cancers such as leukemia has improved dramatically in recent years, but problems such as development of multidrug resistance (MDR) and metastasis can decrease treatment efficacy. Nanotechnology offers a number of opportunities to improve drug efficacy for the treatment of hematological cancers. These include:

- Altering drug pharmacokinetics by providing sustained release and passive targeting.
- Inhibiting drug efflux pathways.
- Protecting drugs and nucleic acid-based therapeutics from metabolism.
- Targeting drugs to cancer cells to increase the effective concentration and decrease systemic effects.
- Delivering optimal concentrations of drug combinations for multi-drug therapy.

Sustained Release

For many antitumor drugs, the patient-tolerated dose is limited by significant systemic toxicities. Encapsulation of drugs into nanoformulations can help to reduce off-target toxicity by prolonging the circulation of active drug and targeting drug passively to sites with fenestrated vasculature, such as bone marrow, spleen, and solid tumors. Vincristine is an alkaloid drug with efficacy against acute lymphoblastic leukemia and non-Hodgkins lymphoma, causing cell cycle-specific apoptosis by binding to tubulin during active mitosis resulting in metaphase arrest (Thomas et al. 2009). Free vincristine has a very short plasma half-life due to rapid cellular uptake and tissue binding. Encapsulation of vincristine sulfate into the aqueous core of sphingomyelin and cholesterol liposomes, approximately 100 nm in size, extends drug deposition over several days, resulting in significantly greater antitumor activity in preclinical studies (Thomas et al. 2009). A Phase I multicenter study of vincristine sulfate liposomes given by injection and dexamethasone in adults with relapsed or refractory acute lymphoblastic leukemia showed potential as a salvage therapy (Thomas et al. 2009), and a Phase II trial is now underway (NCT00495079). The formulation has also shown promise in a Phase II study in refractory aggressive non-Hodgkins lymphoma (Rodriguez et al. 2009).

Inhibiting Drug Efflux Pathways

MDR occurs when a response to one drug results in the development of resistance to a variety of chemically unrelated drugs and represents a major reason for the failure of therapies for hematological and other cancers. Adenosine triphosphate-binding cassette (ABC) transporters are major contributors to MDR through ATP-dependent drug efflux. To address this problem, Batrakova et al. formulated doxorubicin into polymeric nanomicelles containing the block polymer Pluronic P85 (Batrakova et al. 2010). P85 inhibits the ABC transporter P-glycoprotein (Pgp) drug efflux system, in addition to other energy-related effects on cancer cells. In mice bearing T-lymphocyte leukemia-derived MDR solid tumors, the doxorubicin–P85 nanoformulation enhanced uptake of the Pgp-substrate ^{99m}Tc-sestamibi, induced ATP depletion in tumors, and increased tumor apoptosis (Batrakova et al. 2010).

Solid lipid nanoparticles containing doxorubicin have also shown increased efficacy relative to free doxorubicin in treating the Pgp-over-expression P388/ADR murine leukemia model (Ma et al. 2009a). Survival with doxorubicin nanoparticles was increased from a median of 11 d for untreated mice and 14.5 d for mice treated with free doxorubicin to 20 d. In this case the mechanism of action is less clear; increased net influx of doxorubicin into cancer cells mediated by the nanoparticle formulation is likely to contribute, while inhibition of Pgp by the surfactant Brij 78 and Vitamin E used in the nanoparticle formulation could also be factors.

Protection against Metabolism

Formulation of gemcitabine into nanoparticles by conjugation to squalene and self-assembly in water has shown promise in improving drug efficacy against murine leukemia (Reddy et al. 2007, 2008b). Protection of the drug against deamination and improved intracellular pharmacokinetics and retention are believed to be the mechanisms responsible (Reddy et al. 2008a). In leukemia-bearing rats, the squalene-gemcitabine formulation was much more potent than free gemcitabine at equitoxic doses, and was also more effective than cytarabine (Reddy et al. 2008a).

Targeted Delivery

GTI-2040, an antisense oligonucleotide (ODN) to the R2 subunit of ribonucleotide reductase, has been used in a Phase I trial in patients with acute myeloid leukemia (AML) to reduce the endogenous pool of deoxycytidine and improve the efficacy of high-dose cytarabine (Klisovic et al. 2008). While overall the ODN reduced bone marrow cell R2 protein expression by ~50%, there was marked heterogeneity in patients achieving complete remission and non-responders, with the latter seeing an increase rather than a decrease in R2 levels. Increased delivery of ODN to the cancer cells could potentially improve the efficiency of R2 down-regulation, and targeted delivery represents a strategy that is being pursued in many laboratories for enhancing delivery of therapeutics to cancer cells. Jin et al. have targeted the transferrin receptor, which is up-regulated in many cancer cells including AML cells, to deliver GTI-240 (Jin et al. 2009). They employed pH-sensitive lipopolyplex nanoparticles, which release ODNs at acidic endosomal pH facilitating cytoplasmic delivery after endocytosis, to deliver GTI-240 to Kasumi-1 cells and AML patient primary cells. They showed that transferrin-conjugated nanoparticles were more effective than non-conjugated at downregulating R2 mRNA and protein expression, and were effective at sensitizing AML cells to cytarabine. The combination of pH-sensitive lipopolyplex formulation and transferrin-mediated targeting may represent a promising strategy for antisense ODN delivery in leukemia therapy (Jin et al. 2009).

Delivering Optimal Concentrations of Drug Combinations

When cancers are targeted simultaneously with more than one drug, efficacy may be highly dependent on the ratio of effective drug concentrations at the cancer cell, and suboptimal concentrations can even be antagonistic (Mayer et al. 2006). Since different drugs may have very different pharmacokinetics, configuring multi-drug therapy to achieve a constant optimal ratio can be problematic. One approach to this problem is to encapsulate the drugs in liposomes. CPX-351, a liposomal formulation of cytarabine/daunorubicin at a ratio of 5:1, maintained circulating cytarabine/daunorubicin molar ratios near 5:1 for 24 h after i.v. administration and proved to be > 2000 times as effective as expected if the encapsulated drugs contributed only additively (Mayer et al. 2006). In mice with human T-lymphoblastic leukemia xenografts, CPX-351 administration resulted in dramatically higher peak bone marrow drug levels despite a 30-fold lower cytarabine dose than for the free drug cocktail (Fig. 3a,b); the liposomal preparation administration also resulted in prolonged maintenance of synergistic cytarabine:daunorubicin ratios in bone marrow (Fig. 3C) (Lim et al. 2010a). CPX-351 ablated bone marrow leukemic cells below detectable levels for multiple weeks, whereas leukemia growth was only transiently suppressed by free-drug cocktail. The improved *in vivo* efficacy of CPX-351 may arise from direct liposome-leukemia cell interactions in addition to the improved availability in bone marrow (Lim et al. 2010a). Survival of leukemia xenograft tumor-bearing mice was significantly improved by treatment with CPX-351 versus free drugs, and improved further by addition of consolidation therapy (Lim et al. 2010b). CPX-351 is currently being tested versus standard intensive salvage treatment in a Phase II clinical trial for adult patients with first relapse AML (NCT00822094).

Expansion of Hematopoietic Cells

The expansion of specific populations of hematopoietic stem or progenitor cells while maintaining the desired cell characteristics would be of significant benefit. An example of particular clinical relevance is umbilical cord blood transplantation for high-risk or relapsed hematological malignancies. Cord blood units contain 1–2 log fewer cells than bone marrow units, so transplantation with a single unit of cord blood may not provide enough cells capable of engraftment, leading to delayed or failed engraftment. This delay or failure to reconstitute the immune system results in significant mortality and morbidity from infections, particularly in adult patients. A number of culture methods are now being investigated for *ex vivo* expansion of cord blood and other hematopoietic stem cells. Taking a nanotechnology-based approach, Chua et al. demonstrated that electrospun polymer nanofiber scaffolds functionalized with amino groups improve the expansion of human umbilical cord hematopoietic stem/progenitor cells, increasing the number of $CD34^+$ $CD45^+$ cells from 50-fold expansion on tissue culture plates to ~200-fold (Chua et al. 2006). Scanning electron microscopy (SEM) showed that the aminated mesh encouraged extension of uropodia from cells to interact with the fibers (Fig. 4) (Chua et al. 2006, 2007). Spacer length was important, with 1,2-ethanediamine and 1,4-butanediamine providing optimal results (200-fold and 235-fold expansion of $CD34^+$ $CD45^+$ cells respectively), while 1,6-hexanediamine grafted meshes were less effective at $CD34^+$ $CD4^+$ expansion (86-fold) but maintained the highest fraction of $CD34^+$ $CD45^+$ to total cells (> 40%) (Chua et al. 2007). All three scaffolds gave similar results for colony-forming unit granulocyte-erythrocyte-monocyte-megakaryocyte and long-term culture-initiating cell maintenance, but engraftment efficiency in bone marrow of non-obese diabetic/severe combined immunodeficient mice was higher for the 1,2-ethanediamine and 1,4-butanediamine scaffolds (Chua et al. 2007).

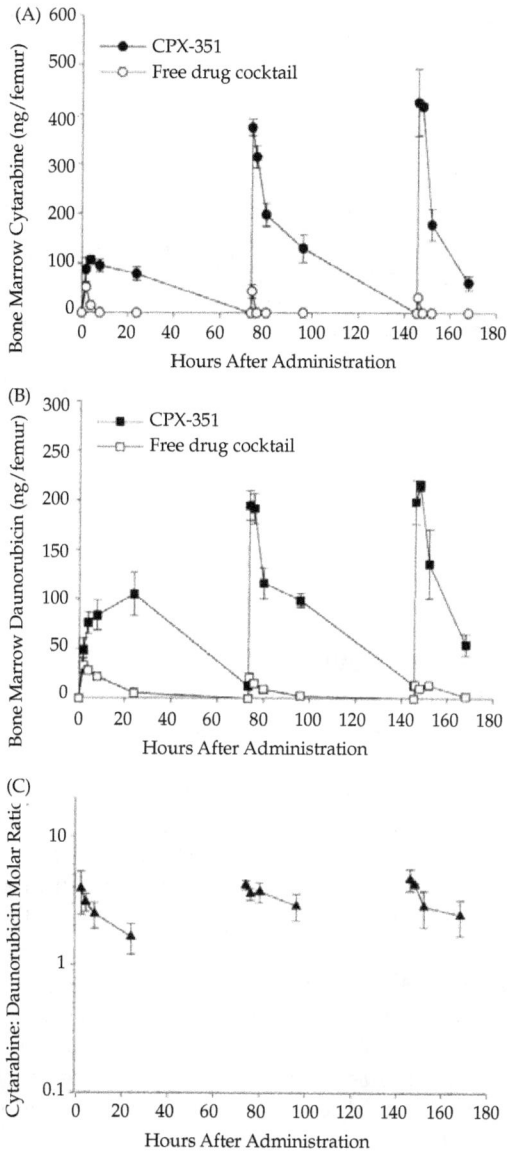

Fig. 3. Bone marrow pharmacokinetics of CPX-351 and free-drug cocktail. At 21 d after implantation, T-cell leukemia-engrafted Rag2-M mice were treated with drug cocktail (300:4.5 mg/kg) (open symbols) or CPX-351 (10:4.4 mg/kg) (closed symbols); the dosage was the maximum tolerated dose every 3 d for three doses. At 2, 4, 8, and 24 h after each treatment, bone marrow levels of (A) cytarabine (circles) and (B) daunorubicin (squares) were determined by HPLC analysis. Three femurs were analyzed for each time point. (C) Cytarabine:daunorubicin molar ratio over time as calculated from drug levels per femur. Reproduced from Lim et al. (2010a) with permission of Elsevier.

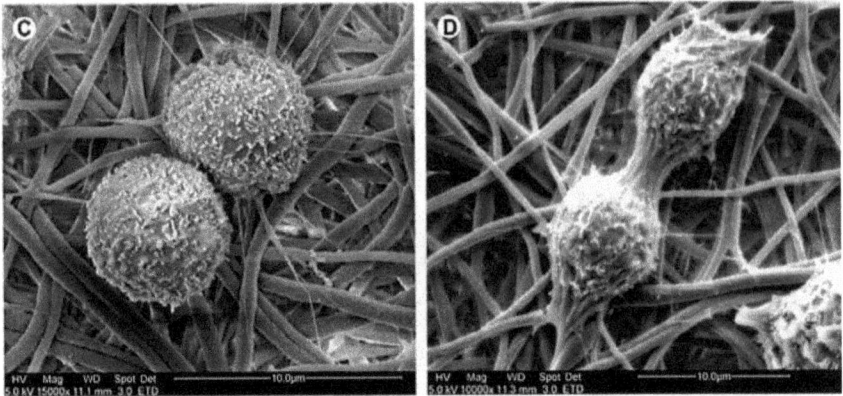

Fig. 4. SEM images of hematopoietic stem/progenitor cells after 3 d cultures on polyethersulfone 1,4-butanediamine nanofiber mesh. (Left) Cells exhibited numerous filopodia interacting with the aminated nanofibers. (Right) Cell division was observed occurring on the nanofiber surface. Reproduced from Chua et al. (2007) with permission of Elsevier.

SEPSIS

Sepsis involves a systemic inflammatory response to local infection. Severe sepsis, defined as acute organ dysfunction secondary to infection, remains a major health care issue, affecting millions of individuals around the world each year. Together with septic shock, where sepsis is compounded by hypotension that cannot be reversed with fluid resuscitation, these conditions kill at least one in four affected patients (Dellinger et al. 2008). Continuous exposure to bacterial lipopolysaccharide (LPS), a major cell wall component in Gram negative bacteria, stimulates the immune system causing a hyperinflammatory response to infection, characterized by production of excessive amounts of inflammatory cytokines such as tumor necrosis factor alpha (TNFα), interleukins, and high-mobility group box 1. Cytokine removal from blood represents an important strategy for halting the progression of sepsis, but requires rapid removal of a range of inflammatory cytokines in order to be effective (Yachamaneni et al. 2010). Yachamaneni et al. developed mesoporous carbide-derived carbon microparticles that form "molecular sieves" with pore sizes in the 1–50 nm range (Fig. 5). By controlling particle size, surface chemistry, synthesis temperature, and post-synthesis annealing, they could adsorb ~100% of TNFα from plasma within 1 h, while the smaller interleukins 1β and 6 could be removed completely within 5 min (Yachamaneni et al. 2010).

Another approach to damping down the septic response is to neutralize circulating LPS (Mas-Moruno et al. 2008). Mas-Maruno et al. employed peptides derived from three LPS-binding proteins of distinct origins:

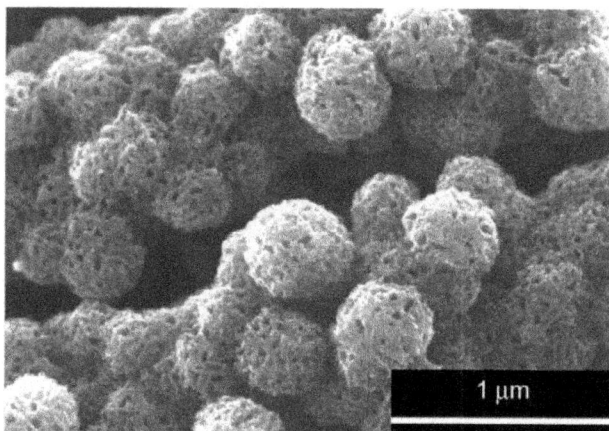

Fig. 5. SEM image of mesoporous carbide-derived carbon. Low magnification image (top) shows layered carbon structure. High magnification (bottom) shows edges of the lamella covered with mesoporous carbon spherules. Reproduced from Yachamaneni et al. (2010) with permission of Elsevier.

- *Limulus* anti-LPS factor (LALF), a small basic protein that inhibits the LPS-mediated coagulation cascade.
- Bactericidal permeability-increasing peptide (BPI), a 57kD protein from human neutrophils that binds to and neutralizes LPS.
- Serum amyloid protein (SAP), a serum glycoprotein that binds LPS and other molecules.

They demonstrated using a 14 amino acid cyclic peptide representing the minimal LPS binding domain that neutralization of LPS from *E. coli* 055:B5 was greatly enhanced by N-acetylation; efficacy increased with increasing acyl chain length from C_2 to C_{16}, and correlated with the ability to form micelles. A similar pattern of increasing efficacy with C_{16} acetylation was also observed for BPI and SAP; for these peptides, fibril-like nanostructures were formed. The results demonstrate that nanostructure formation can lead to anti-LPS peptides with higher LPS-neutralizing activities at cell-tolerable concentrations (Mas-Moruno et al. 2008).

Another factor contributing to the progression of sepsis is the loss of large numbers of lymphocytes and T cells through activation of apoptotic pathways, To target apoptotic cell death, Brahmamdam et al. employed small interfering RNA (siRNA) targeting two key cell death proteins, Bim and PUMA; to overcome the difficulty of transfecting lymphocytes, they delivered siRNA using Calando Pharmaceuticals RONDEL system (Brahmamdam et al. 2009). The delivery vehicle combines cyclodextrin polymer, stabilizing agent, and a targeting ligand, in this case targeted to the transferrin receptor. Administration of siRNA therapy to mice following cecal ligation and puncture (CLP), a model for sepsis, decreased lymphocyte apoptosis and prevented loss of splenic CD4 T and B cells (Fig. 6). Flow cytometry confirmed that Bim and PUMA protein levels were decreased in CD4 T and B cells in response to the siRNA administration (Brahmamdam et al. 2009).

Splenic B Cells

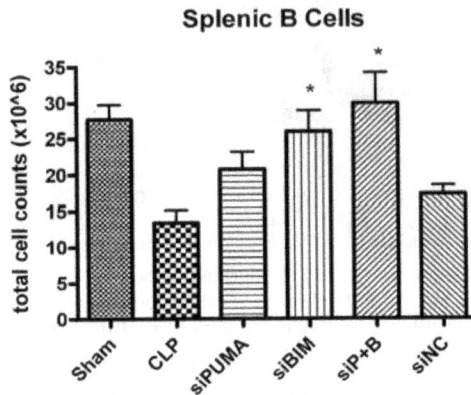

Fig. 6. Absolute splenic B cell counts in septic mice after siRNA treatment. Mice receiving siRNA to Bim and to PUMA and Bim in combination had a significantly higher number of surviving B cells after CLP-induced sepsis when compared with untreated septic controls (CLP) and CLP mice treated with non-specific siRNA (siNC) (*P > 0.01). Results are means ± SEM for 6–9 animals. Reproduced from Brahmamdam et al. (2009) with permission of Wolters Kluwer Health.

BLOOD PROTEIN PRODUCTION

Patients with inherited bleeding disorders such as hemophilia B depend on coagulation factor concentrates derived from pooled plasma of many donors, leading to a risk of transmission of viral diseases such as human immunodeficiency virus, hepatitis B virus, and hepatitis C virus. Solvent/detergent treatment has greatly reduced transmission of enveloped viruses such as human immunodeficiency virus, hepatitis B, and hepatitis C, but has no effect on non-enveloped viruses such as hepatitis A and human parvovirus B19, and so the Committee for Proprietary Medicinal Products mandated a second complementary inactivation/removal step to remove non-enveloped viruses (Chtourou et al. 2007; Kim et al. 2008). Pasteurization and dry heat treatment have the disadvantage that they can reduce biological activity of the product or form of neoantigens; any denaturation of the product that occurs can expose antigens that are normally buried, resulting in the generation of neutralizing antibodies in the patient and inhibiting blood product activity. Nanofiltration using pore sizes in the 15–20 nm range has proved effective for decreasing viral load by several logs while maintaining yield and biological activity. Examples include the Planova 15N filter, a hollow non-selective cuprammonium regenerated cellulose filter with nominal pore size of 15 nm used in preparing Factor VIII (Chtourou et al. 2007) and immunoglobulin; Planova 20N, manufactured from the same material but with a nominal 19 nm pore size, employed for purification of Rhesus (Rh) immunoglobulin, given intravenously to prevent Rh-negative mothers from developing antibodies to Rh-positive fetal red blood cells (Soluk et al. 2008); and the Viresolve NFP filter, a modified polyvinylidene fluoride filter with 20 nm pore size, used in preparing Factor IX (Kim et al. 2008). In each case the mechanism of virus removal is by size exclusion; nanofiltration has proved effective even for such large proteins as the 240kD Factor VIII, provided the production conditions are carefully controlled (Chtourou et al. 2007).

APPLICATIONS TO OTHER AREAS OF HEALTH AND DISEASE

The detection and treatment of blood clots is relevant to a wide range of diseases. Formation of blood clots blocking arteries is the final event that precipitates myocardial infarctions and ischemic strokes, while hypercoagulation is a common problem in many cancers. DVT in the lower limbs can cause sudden fatal pulmonary embolism or embolic stroke and can lead to chronic venous complications termed post-thrombotic syndrome. For oncology, many of the therapies being tested in hematological cancers also show promise for other types of tumor. Nanoscaffold-based methods being developed for expansion of hematopoietic stem cells are likely to be used or modified for the expansion and differentiation of other types of stem and progenitor cells, for example, for injection into patients with peripheral artery disease to promote angiogenesis and treat ischemia.

Key Facts

- When atherosclerotic plaques in coronary or carotid arteries rupture, this exposes blood to pro-thrombotic materials causing blood clot formation.
- When clots obstruct coronary arteries, this leads to myocardial infarction, while occlusion of carotid arteries leads to ischemic stroke.
- If clots are not lysed rapidly, either spontaneously by the body's defense mechanisms or by treatment with thrombolytic drugs and/ or mechanical disruption, permanent damage to the heart or brain occurs.
- Devices such as mechanical heart valves, ventricular assist devices and stents tend to form blood clots on their surfaces, so patients receiving

such devices must take anticoagulant therapy, in many cases for the rest of their lives.

- Blood clots that form in deep veins in the legs (DVT) or on devices such as ventricular assist devices can break free and travel to the lungs, heart or brain, causing arterial blockages that can be fatal.

Definitions

Multi-drug resistance: Resistance to a broad spectrum of drugs with different structures due to the induction of expression of energy-dependent transport channels that actively pump drugs out of the cancer cells, making it difficult to achieve an effective intracellular drug concentration to kill the cancer cells.

Sepsis: A hyperinflammatory response to bacterial infection, precipitated by extended exposure to a bacterial cell wall component, lipopolysaccharide. It is characterized by the overproduction of inflammatory cytokines such as TNFα, interleukins, and high-mobility group box 1.

Severe sepsis: A condition involving organ failure, low blood pressure (hypotension) and insufficient blood flow (hypoperfusion). Severe sepsis has a high rate of mortality, typically around 25%. The prognosis is worse for patients in whom treatment with fluid resuscitation is ineffective at reversing hypotension or hypoperfusion (termed septic shock); these patients have mortality rates of 50% or greater.

Thrombosis: Formation of blood clots (or thrombi) through an aggregation of platelets and fibrin to form a plug. Platelets are activated by exposure to subendothelial proteins such as von Willebrand factor that are normally kept from contact with platelets by healthy endothelium. Fibrin is formed in response to activation of the coagulation cascade, two pathways (extrinsic and intrinsic) that converge on the protease thrombin, which cleaves the inactive precursor fibrinogen to form fibrin.

Summary Points

- Targeting of nanoparticles to blood clots appears promising both for thrombus detection and for clot lysis.
- For anticoagulation, nanoformulations can facilitate effective oral and pulmonary delivery of anticoagulant, while nanostructured surfaces may reduce the need for anticoagulation in patients receiving devices.
- Nanotechnology-based hemostatic agents may help to curb bleeding in surgical patients and trauma victims.

- For treatment of hematological cancers, nanotechnology offers a number of opportunities for improving therapeutic efficacy, including targeted delivery, sustained release, and inhibition of drug efflux from cancer cells.
- Nanoscaffold-based expansion of hematopoietic stem cells may also help to improve engraftment in patients undergoing bone marrow transplants.
- In sepsis, nanostructured particles can remove inflammatory cytokines from blood quickly and efficiently, while nanostructured peptides can remove the initiating bacterial LPS.
- Nanoparticle-based targeted delivery of siRNA to lymphocytes can overcome barriers to transfection of these cells and protect them from targeted cell death.
- Nanofiltration of blood products helps to ensure the safety of patients with clotting disorders such as hemophilia by efficiently removing viruses from the preparation.

Abbreviations

ABC	:	adenosine triphosphate-binding cassette
AML	:	acute myeloid leukemia
BPI	:	bactericidal permeability-increasing peptide
CLP	:	cecal ligation and puncture
DVT	:	deep vein thrombosis
GRDGS	:	glutamine-arginine-glutamate-aspartate-serine
LALF	:	*Limulus* anti-LPS factor
LPS	:	lipopolysaccharide
LWMH	:	low molecular weight heparin
MDR	:	multi-drug resistance
ODN	:	oligodeoxynucleotide
PEG	:	polyethylene glycol
Pgp	:	P-glycoprotein
PLGA-PLL	:	poly(lactic-co-glycolic acid)-poly-L-lysine
RADA	:	arginine-alanine-aspartate-alanine
RGD	:	arginine-glutamate-aspartate
Rh	:	Rhesus
rTPA	:	recombinant tissue plasminogen activator
SAP	:	serum amyloid peptide
SEM	:	scanning electron microscope
siRNA	:	small interfering RNA
Ti-Ni	:	titanium-nickel

References

Bai, S., and F. Ahsan. 2009. Synthesis and evaluation of pegylated dendrimeric nanocarrier for pulmonary delivery of low molecular weight heparin. Pharm. Res. 26: 539–548.

Bai, S., C. Thomas, and F. Ahsan. 2007. Dendrimers as a carrier for pulmonary delivery of enoxaparin, a low-molecular weight heparin. J. Pharm. Sci. 96: 2090–2106.

Batrakova, E.V., S. Li, A.M. Brynskikh, A.K. Sharma, Y. Li, M. Boska, N. Gong, R.L. Mosley, V.Y. Alakhov, H.E. Gendelman, and A.V. Kabanov. 2010. Effects of pluronic and doxorubicin on drug uptake, cellular metabolism, apoptosis and tumor inhibition in animal models of MDR cancers. J. Control. Release 143: 290–301.

Benesch, M., and C. Urban. 2008. Liposomal cytarabine for leukemic and lymphomatous meningitis: recent developments. Expert Opin. Pharm. 9: 301–309.

Bertram, J.P., C.A. Williams, R. Robinson, S.S. Segal, N.T. Flynn, and E.B. Lavik. 2009. Intravenous hemostat: nanotechnology to halt bleeding. Sci. Trans. Med. 1: 11ra22.

Bi, F., J. Zhang, Y. Su, Y.-C. Tang, and J.-N. Liu. 2009. Chemical conjugation of urokinase to magnetic nanoparticles for targeted thrombolysis. Biomaterials 30: 5125–5130.

Brahmamdam, P., E. Watanabe, J. Unsinger, K.C. Chang, W. Schierding, A.S. Hoekzema, T.T. Zhou, J.S. McDonough, H. Holemon, J.D. Heidel, C.M. Coopersmith, J.E. McDunn, and R.S. Hotchkiss. 2009. Targeted delivery of siRNA to cell death proteins in sepsis. Shock 32: 131–139.

Chtourou, S., P. Porte, M. Nogre, N. Bihoreau, E. Cheesman, B. Samor, A. Sauger, S. Raut, and C. Mazurier. 2007. A solvent/detergent-treated and 15-nm filtered factor VIII: a new safety standard for plasma-derived coagulation factor concentrates. Vox Sang 92: 327–337.

Chua, K.N., C. Chai, P.C. Lee, S. Ramakrishna, K.W. Leong, and H.Q. Mao. 2007. Functional nanofiber scaffolds with different spacers modulate adhesion and expansion of cryopreserved umbilical cord blood hematopoietic stem/progenitor cells. ExpHematol 35: 771–781.

Chua, K.N., C. Chai, P.C. Lee, Y.N. Tang, S. Ramakrishna, K.W. Leong, and H.Q. Mao. 2006. Surface-aminated electrospun nanofibers enhance adhesion and expansion of human umbilical cord blood hematopoietic stem/progenitor cells. Biomaterials 27: 6043–6051.

Dai, C., C. Liu, J. Wei, H. Hong and Q. Zhao. 2010. Molecular imprinted macroporous chitosan coated mesoporous silica xerogels for hemorrhage control. Biomaterials 31: 7620–7630.

Dellinger, R.P., M.M. Levy, J.M. Carlet, J. Bion, M.M. Parker, R. Jaeschke, K. Reinhart, D.C. Angus, C. Brun-Buisson, R. Beale, T. Calandra, J.F. Dhainaut, H. Gerlach, M. Harvey, J.J. Marini, J. Marshall, M. Ranieri, G. Ramsay, J. Sevransky, B.T. Thompson, S. Townsend, J.S. Vender, J.L. Zimmerman, and J.L. Vincent. 2008. Surviving Sepsis Campaign: international guidelines for management of severe sepsis and septic shock: 2008. Crit. Care Med. 36: 296–327.

Douma, R.A., P.W. Kamphuisen, and H.R. Buller. 2010. Acute pulmonary embolism. Part 1: epidemiology and diagnosis. Nat. Rev. Cardiol. 7: 585–596.

Ellis-Behnke, R. 2010. At the nanoscale: nanohemostat, a new class of hemostatic agent. Wiley Interdiscip. Rev. Nanomed. Nanobiotechnol. DOI: 10.1002/wnan.1110.

Ghanbari, H., H. Viatge, A.G. Kidane, G. Burriesci, M. Tavakoli, and A.M. Seifalian. 2009. Polymeric heart valves: new materials, emerging hopes. Trends Biotechnol. 27: 359–367.

Hoffart, V., A. Lamprecht, P. Maincent, T. Lecompte, C. Vigneron, and N. Ubrich. 2006. Oral bioavailability of a low molecular weight heparin using a polymeric delivery system. J. Control. Release 113: 38–42.

Jin, Y., S. Liu, B. Yu, S. Golan, C.-G. Koh, J. Yang, L. Huynh, X. Yang, J. Pang, N. Muthusamy, K.K. Chan, J.C. Byrd, Y. Talmon, L.J. Lee, R.J. Lee, and G. Marcucci. 2009. Targeted delivery of antisense oligodeoxynucleotide by transferrin conjugated pH-sensitive lipopolyplex nanoparticles: a novel oligonucleotide-based therapeutic strategy in acute myeloid leukemia. Mol. Pharmaceut. 7: 196–206.

Kim, I.S., Y.W. Choi, Y. Kang, H.M. Sung, K.W. Sohn, and Y.S. Kim. 2008. Improvement of virus safety of an antihemophilic factor IX by virus filtration process. J. Microbiol. Biotechnol. 18: 1317–1325.

Klisovic, R.B., W. Blum, X. Wei, S. Liu, Z. Liu, Z. Xie, T. Vukosavljevic, C. Kefauver, L. Huynh, J. Pang, J.A. Zwiebel, S. Devine, J.C. Byrd, M.R. Grever, K. Chan, and G. Marcucci. 2008. Phase I study of GTI-2040, an antisense to ribonucleotide reductase, in combination with high-dose cytarabine in patients with acute myeloid leukemia. Clin. Cancer Res. 14: 3889–3895.

Lim, W.S., P.G. Tardi, N. Dos Santos, X. Xie, M. Fan, B.D. Liboiron, X. Huang, T.O. Harasym, D. Bermudes, and L.D. Mayer. 2010a. Leukemia-selective uptake and cytotoxicity of CPX-351, a synergistic fixed-ratio cytarabine:daunorubicin formulation, in bone marrow xenografts. Leuk. Res. 34: 1214–1223.

Lim, W.S., P.G. Tardi, X. Xie, M. Fan, R. Huang, T. Ciofani, T.O. Harasym, and L.D. Mayer. 2010b. Schedule- and dose-dependency of CPX-351, a synergistic fixed ratio cytarabine:daunorubicin formulation, in consolidation treatment against human leukemia xenografts. Leuk. Lymphoma 51: 1536–1542.

Ma, P., X. Dong, C.L. Swadley, A. Gupte, M. Leggas, H.C. Ledebur, and R.J. Mumper. 2009a. Development of idarubicin and doxorubicin solid lipid nanoparticles to overcome Pgp-mediated multiple drug resistance in leukemia. J. Biomed. Nanotechnol. 5: 151–161.

Ma, Y.H., S.Y. Wu, T. Wu, Y.J. Chang, M.Y. Hua, and J.P. Chen. 2009b. Magnetically targeted thrombolysis with recombinant tissue plasminogen activator bound to polyacrylic acid-coated nanoparticles. Biomaterials 30: 3343–3351.

Marsh, J.N., A. Senpan, G. Hu, M.J. Scott, P.J. Gaffney, S.A. Wickline, and G.M. Lanza. 2007. Fibrin-targeted perfluorocarbon nanoparticles for targeted thrombolysis. Nanomedicine (Lond.) 2: 533–543.

Mas-Moruno, C., L. Cascales, L.J. Cruz, P. Mora, E. Pérez-Payá, and F. Albericio. 2008. Nanostructure formation enhances the activity of LPS-neutralizing peptides. Chem. Med. Chem. 3: 1748–1755.

Mayer, L.D., T.O. Harasym, P.G. Tardi, N.L. Harasym, C.R. Shew, S.A. Johnstone, E.C. Ramsay, M.B. Bally, and A.S. Janoff. 2006. Ratiometric dosing of anticancer drug combinations: controlling drug ratios after systemic administration regulates therapeutic activity in tumor-bearing mice. Mol. Cancer Ther. 5: 1854–1863.

McCarthy, J.R., P. Patel, I. Botnaru, P. Haghayeghi, R. Weissleder, and F.A. Jaffer. 2009. Multimodal Nanoagents for the Detection of Intravascular Thrombi. Bioconjugate Chem. 20: 1251–1255.

Pan, D., S.D. Caruthers, G. Hu, A. Senpan, M.J. Scott, P.J. Gaffney, S.A. Wickline, and G.M. Lanza. 2008. Ligand-directed nanobialys as theranostic agent for drug delivery and manganese-based magnetic resonance imaging of vascular targets. J. Am. Chem. Soc. 130: 9186–9187.

Pan, D., A. Senpan, S.D. Caruthers, T.A. Williams, M.J. Scott, P.J. Gaffney, S.A. Wickline, and G.M. Lanza. 2009. Sensitive and efficient detection of thrombus with fibrin-specific manganese nanocolloids. Chem. Commun. (Camb.) 3234–3236.

Reddy, L.H., C. Dubernet, S.L. Mouelhi, P.E. Marque, D. Desmaele, and P. Couvreur. 2007. A new nanomedicine of gemcitabine displays enhanced anticancer activity in sensitive and resistant leukemia types. J. Control. Release 124: 20–27.

Reddy, L.H., H. Ferreira, C. Dubernet, S.L. Mouelhi, D. Desmaele, B. Rousseau, and P. Couvreur. 2008a. Squalenoyl nanomedicine of gemcitabine is more potent after oral administration in leukemia-bearing rats: study of mechanisms. Anticancer Drugs 19: 999–1006.

Reddy, L.H., P.E. Marque, C. Dubernet, S.L. Mouelhi, D. Desmaele, and P. Couvreur. 2008b. Preclinical toxicology (subacute and acute) and efficacy of a new squalenoyl gemcitabine anticancer nanomedicine. J. Pharmacol. Exp. Ther. 325: 484–490.

Rodriguez, M.A., R. Pytlik, T. Kozak, M. Chhanabhai, R. Gascoyne, B. Lu, S.R. Deitcher, and J.N. Winter. 2009. Vincristine sulfate liposomes injection (Marqibo) in heavily pretreated patients with refractory aggressive non-Hodgkin lymphoma. Cancer 115: 3475–3482.

Soluk, L., H. Price, C. Sinclair, D. Atalla-Mikhail, and M. Genereux. 2008. Pathogen safety of intravenous Rh immunoglobulin liquid and other immune globulin products: enhanced nanofiltration and manufacturing process overview. Am. J. Ther. 15: 435–443.

Sugita, Y., Y. Suzuki, K. Someya, A. Ogawa, H. Furuhata, S. Miyoshi, T. Motomura, H. Miyamoto, S. Igo, and Y. Nosé. 2009. Experimental evaluation of a new antithrombogenic stent using ion beam surface modification. Artificial Organs 33: 456–463.

Thomas, D.A., H.M. Kantarjian, W. Stock, L.T. Heffner, S. Faderl, G. Garcia-Manero, A. Ferrajoli, W. Wierda, S. Pierce, B. Lu, S.R. Deitcher, and S. O'Brien. 2009. Phase 1 multicenter study of vincristine sulfate liposomes injection and dexamethasone in adults with relapsed or refractory acute lymphoblastic leukemia. Cancer 115: 5490–5498.

Udit, A.K., C. Everett, A.J. Gale, J. Reiber Kyle, M. Ozkan, and M.G. Finn. 2009. Heparin antagonism by polyvalent display of cationic motifs on virus-like particles. Chembiochem 10: 503–510.

Yachamaneni, S., G. Yushin, S.-H. Yeon, Y. Gogotsi, C. Howell, S. Sandeman, G. Phillips and S. Mikhalovsky. 2010. Mesoporous carbide-derived carbon for cytokine removal from blood plasma. Biomaterials 31: 4789-4794.

Nanomedicine and Neurodegenerative Disorders

Ari Nowacek,[1] Gang Zhang,[1] JoEllyn McMillan,[1] Tomomi Kiyota,[1] Elena Batrakova[2] and Howard E. Gendelman[1,*]

ABSTRACT

Included in the broad range of degenerative disorders are those that affect the nervous system. These are particularly significant as they cause deficits in day-to-day activities and quality of life. Such disorders include, but are not limited to, Alzheimer's and Parkinson's diseases, amyotrophic lateral sclerosis, and Huntington's disease. Importantly, each is associated, at varying degrees, with genetic, environmental factors, and advancing age. Thus, as the average age of the world population increases, it is predicted that the incidence of age-related neurodegenerative disorders will grow. Over the past several decades much has been learned about disease pathobiology and the molecular mechanisms that underlie neurodegeneration. Some of these recent advances include knowledge about misfolded proteins

[1]Department of Pharmacology and Experimental Neuroscience, University of Nebraska Medical Center, 985800 Nebraska Medical Center, Omaha, NE 68198-5800.

[2]Department of Pharmaceutical Sciences, 985830 Nebraska Medical Center, Omaha, NE 68198-5830.

[3]Pharmacology and Experimental Neuroscience.

*Corresponding author

List of abbreviations after the text.

and protein aggregates, mitochondrial dysfunction, neuronal dysfunction, disease epidemiology, novel therapeutics that target disease-causing events, and drug delivery schemes. These works have yielded possible targets for therapeutic interventions. Most recently, harnessing the principles of nanomedicine has served to exploit disease-inciting targets. While the structure of the brain is secure, in that it contains both physical and chemical barriers to foreign molecules, nanotechnology has offered unique approaches towards overcoming blood-brain barrier limitations. The role that nanomedicine has played in providing improvements in diagnosis and therapeutic interventions is discussed in this chapter.

INTRODUCTION

Neurodegenerative disorders such as Alzheimer's and Parkinson's diseases (AD and PD), amyotrophic lateral sclerosis, multiple sclerosis, and Huntington's disease commonly leave people mentally and physically incapacitated. Neurodegeneration results in the death or malfunction of neurons and their connections to supportive glial cells. However, not all cells are equally affected and in some conditions only a specific sub-type of neuron in a defined anatomical region is affected. Even though significant therapeutic advances have been made, neurodegenerative disorders continue to present a significant challenge for modern medicine. They afflict more than 6 million people in the United States alone at an annual health care related cost of approximately US$ 180 billion/annum. Statistics like these highlight the urgent need to find effective methods for detecting, halting the progression of, or altogether curing these diseases. Regrettably, there are presently few therapeutic options available that can affect clinical outcomes. Therapies are palliative and do no more than temporarily manage symptoms. Another factor that makes it challenging to treat neurodegenerative disorders is that the conditions are difficult to detect in the early stages of disease. Commonly, a disease will progress for years before obvious clinical symptoms appear and a diagnosis can be made, at which time significant and permanent damage has occurred and degeneration is fully underway. Certainly, the management of disease may be more effective if a diagnosis was made early. Efforts are underway through biomarker discovery and bioimaging, in particular, to target this unmet need.

In this chapter, we focus on the two most common neurodegenerative disorders, AD and PD. However, it is important to mention that, regardless of their individual etiology or the particular region of the central nervous system (CNS) affected, all of the neurodegenerative disorders share common underlying molecular mechanisms, the most important of which is the presence of misfolded and oxidized proteins that incite neuronal loss through a range of mechanisms. Thus, most of the therapeutic approaches being studied for these two diseases can be applied, in measure, to the others.

The blood-brain barrier (BBB) protects the CNS from physical and chemical injuries and also serves to prevent passage of therapeutic compounds. Therefore, one of the primary goals of nanomedicine research for neurodegenerative disorders has been to find ways to circumvent these barriers and deliver drugs specifically to diseased regions of the brain. To this end, we explore a number of nanoparticulate compounds and delivery techniques. First, we review the structure and function of the CNS and describe why the diseases that affect it present particularly challenging obstacles to effective treatment strategies.

LIMITATIONS FOR DRUG DELIVERY: THE BBB

The BBB, as the name suggests, serves as both physical and chemical barrier between the blood and the brain. It evolved to protect the brain from toxic substances and microorganisms, but it also impedes the delivery of both diagnostic and therapeutic agents. In order to find ways to deliver medications to diseased regions of the brain, it is necessary to understand the structural and physiological features of the BBB.

On the most basic level, the BBB consists of small blood vessels known as capillaries, which separate the brain from circulating blood. The BBB is a massive structure that contains more than 100 billion capillaries and has a surface area of roughly 20 sq m. The smallest element of the BBB is called the neurovascular unit. This unit describes the general structure of the BBB and includes all of its cells and cellular components. It begins with the endothelial cells that line the blood vessels. Endothelial cells of the BBB, which are referred to as brain microvascular endothelial cells (BMVEC), have a number of important features that help them to act as effective barriers. First, BMVEC are stitched tightly together by tight intracellular junctions and contain fewer and smaller pores, called fenestrations, than peripheral endothelial cells. This helps prevent the free flow of material either through or around BMVEC. Second, BMVEC undergo low rates of pinocytosis, a process where the cells continually and non-specifically take up small samples of the blood. This means that in general if a substance is to enter the brain it must be selectively removed from the bloodstream by BMVEC. Third, BMVEC sit on top of a layer of extracellular matrix called the basement membrane, which is about 40–80 nm thick. This layer is not effective at inhibiting the movement of solutes, but it can prevent the crossing of small nanosized particles. The next layer of the barrier is made up of a different cell type, the pericyte. Pericytes do not actively participate in barrier function but rather serve to support BMVEC and secrete a number of factors that control the state of both the BMVEC and the cells of the next and final layer of the barrier, astrocytes. Astrocytes surround BMVEC with large cellular projections called astrocytic feet. These mainly serve a supportive role for BMVEC but can also serve as a biochemical barrier. Thus, the CNS is anatomically sealed off from the rest of the body and the transport of drugs to the brain must occur via a transcellular route through BMVEC.

Despite the physical obstructions, some molecules are still able to enter BMVEC by passive diffusion; however, they can still be prevented from entering the brain. BMVEC possess a number of passive and active efflux systems for a wide variety of compounds. A few of these transporters include the multiple drug resistance protein, P-glycoprotein, breast cancer resistance protein, and the multiple organic anion transporter. Thus, of the few molecules that do enter BMVEC, many are pumped out of the cell and directly back into the bloodstream.

In addition to efflux systems, BMVEC also possess an enzymatic barrier. These cells express many different types of drug-metabolizing enzymes that are capable of breaking down a wide array of compounds. These enzymes include gamma-glutamyl transpeptidase, alkaline phosphatase and aromatic amino acid decarboxylase. Thus, even if a molecule manages to avoid the efflux system, drug-metabolizing enzymes could degrade it before entering the brain.

However, the BBB does not totally prevent the passage of molecules and cells into the brain. The BBB must allow nutrients and exogenous compounds to enter the CNS if homeostasis is to be maintained. Thus, by first restricting passage of unwanted material and then carefully selecting compounds for transport, the BBB precisely regulates the homeostatic environment of the CNS. This is important because neurons are particularly sensitive to their surroundings and even small fluctuations can result in cellular dysfunction or death.

AD AND PD

AD and PD are the two most common neurodegenerative disorders. While both cause the degeneration of neurons, the cells affected are of different types and are in different anatomical locations. As a result, the symptoms of disease and the pathological hallmarks differ greatly. AD patients typically present with memory problems that steadily progress toward dementia. AD is the result of degeneration of neurons in the cerebral cortex particularly in regions involved with learning and memory such as the hippocampus. The two pathological hallmarks of AD are neurofibrillary tangles and senile plaques that consist of amyloid beta (Aβ) protein. PD patients, on the other hand, typically present with resting tremors, slowness of movement, and rigidity. This is because PD is the result of degeneration of neurons in the substantia nigra (SN), an area of the brain that is vital to the neural circuitry that controls movement. Pathologically, PD is characterized by the formation of Lewy bodies within neurons and the loss of pigmentation within the SN.

While AD and PD may differ, these two diseases have in common a number of important characteristics that are also shared by other neurodegenerative disorders. For example, both AD and PD involve the accumulation of misfolded protein and reactive oxygen species (ROS) production. This can lead to an imbalance in the oxidation state of the cell and may be a significant contributor to cellular stress. In AD, Aβ aggregates to form senile plaques and neurofibrillary tangles; in PD, Lewy bodies are formed by the aggregation of nitrated alpha-synuclein. Furthermore, the cause for disease probably involves a combination of genetic-predisposition and exposure to exogenous compounds; the risk of all neurodegenerative disorders increases with age. Owing to these and other similarities, findings from nanomedicine research into one neurodegenerative disorder can usually be applied to many others.

In the rest of this chapter, we focus on specific work for AD and PD. It is important to note that each method being used to target a particular molecular mechanism could potentially be modified to treat a different neurodegenerative disorder.

NANOMEDICINE AND AD

Metal Chelation

Metal ions such as Cu^{2+}, Fe^{3+} and Zn^{2+} have been implicated as key participants in the development of Aβ pathology in AD. The ions directly bind to both soluble and insoluble complexes of Aβ peptides and promote Aβ aggregation by precipitating the peptides into metal-enriched plaques. Moreover, Aβ aggregates and peptides can, through interactions with redox-active metal ions, generate ROS, which are the primary mediators of oxidative stress, suggesting that metal dysregulation may be a potential therapeutic target for AD (Faller 2009). Although studies on chelation therapy for AD using compounds such as desferrioxamine (DFO), ethylenediaminetetraacetic acid, and iodochlorhydroxyquin have shown promising results in the past, there were concerns about potential side effects and safety. However, nanomedicine has helped make chelation therapy à possibility again. A new nanoparticle (NP) delivery system was developed using the copper (I) chelator, D-penicillamine, which is covalently conjugated to NP via a disulfide bond or a thioether bond and is capable of solubilizing copper-Aβ aggregates under reducing conditions (Cui et al. 2005). Two other chelators, 2-methyl-*N*-(2'-aminoethyl)-3-hydroxyl-4 pyridinone (MAEHP) and DFO, have also been incorporated into NP. DFO conjugated to NP coated with the surfactant polysorbate were able to significantly chelate metal ions from brain sections of AD patients (Liu et al. 2005). Furthermore, these particles were also able to bind to apolipoproteins, which are proteins that have a specific transport system for crossing the BBB, suggesting that NP could be delivered to the brain in the same fashion. MAEHP-NP conjugates are synthesized with carboxylic functionalized polystyrene NP and MAEHP in an acid buffer solution. These particles can completely prevent Aβ aggregation and inhibit Aβ neurotoxicity (Fig. 1) (Liu et al. 2010), suggesting their therapeutic potential for AD.

Inhibition of Aβ Fibrillization

Since Aβ fibrillization is a seminal event in the development of AD, inhibition of this reaction is potentially a beneficial therapeutic strategy. Fibrillization occurs when monomers and oligomers of Aβ polymerize to form long, tangled fibrils of protein. One study used copolymeric *N*-isopropylacrylamide: *N-tert*-butylacrylamide nanoparticles to inhibit Aβ fibrillization by disturbing the balance of Aβ units. It was suggested that the mechanism worked by trapping monomers and oligomers of Aβ onto the surface of the NP, thus reducing the concentration of Aβ units available to form fibrils (Cabaleiro-Lago et al. 2008). Another study revealed that *N*-acetyl-L-cysteine capped quantum dots have strong molecular interactions with Aβ units and can inhibit fibrillization in a dose-dependent manner (Xiao et al.). In the two previous examples, NP inhibited fibrillization by physically interacting with various Aβ units; however, another potential strategy is to use the antioxidant enzyme catalase, which has a high binding affinity to a specific recognition sequence on the Aβ peptide, preventing polymerization (Milton and Harris 2010). It has been shown that catalase alone co-incubated with Aβ units can inhibit Aβ fibrillization (Fig. 2). Thus, the NP-catalase nanozyme complex discussed in the PD section may also be beneficial for AD therapy. This suggests a dual mechanism for NP-catalase nanozyme complexes as both an inhibitor of Aβ fibrillization and an antioxidant crucial to balancing the redox state of cells.

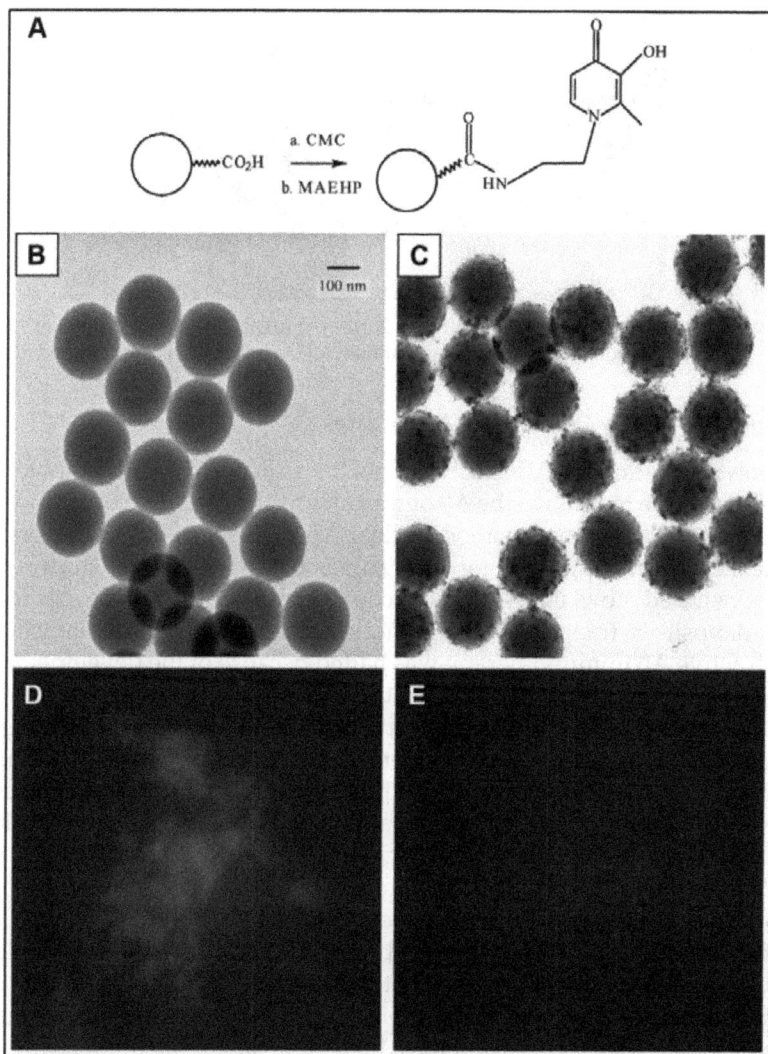

Fig. 1. Aβ NP–chelator conjugate. (A) Synthesis strategy for NP-chelator conjugates. (a) Reaction of carboxylic functionalized NP with N-cyclohexyl-N'-(2-morpholinoethyl) carbodiimide methyl-p-toluensulfonate in MES. (b) Conjugation of activated carboxylic nanoparticles with excessive MAEHP in MES. (B, C) Transmission electron microscope (TEM) images of nanoparticles without chelator conjugation and iron binding (B) or with both reactions (C). (D, E) Fluorescence microscopy images. Aβ aggregates (D) are inhibited in the presence of Nano-N2PY (E). Reproduced from Liu et al. (2009) with permission from Elsevier.

Color image of this figure appears in the color plate section at the end of the book.

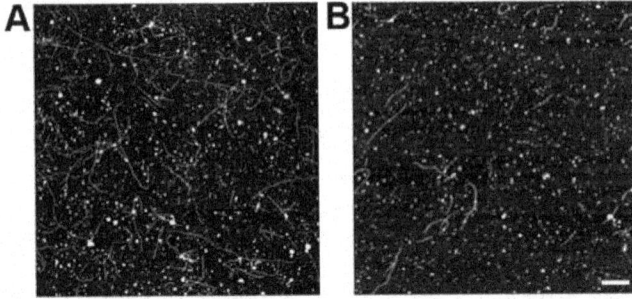

Fig. 2. Atomic force microscopy image of Aβ fibrils. (A) Aβ42 fibrils are generated in 10 mM HCl. (B) Co-incubation of Aβ with catalase inhibits Aβ42 fibrillization. Unpublished data.

Removal of Fully Formed Aβ Aggregates

If polymerization of Aβ fibrils cannot be prevented, another potential option may be to remove the Aβ aggregates as they form. Development of active or passive immunization to clear Aβ aggregates has been pursued as a therapeutic approach for AD. This is based on several studies that demonstrated how immunization could significantly reduce Aβ levels and deposits in transgenic-mouse models of AD. Although a clinical trial with active Aβ immunotherapy was halted because of the development of non-bacterial meningoencephalitis, trials using passive immunotherapies with antibodies are still ongoing. One type of NP that has been designed to remove Aβ aggregates is magnetic iron-oxide NP (Skaat and Margel 2009). These particles can bind to Aβ fibrils to form NP-Aβ complexes in the aqueous phase and then be separated out by using a magnetic field. Another type of particle is sialic acid-conjugated dendrimetric polymer NP. It has been shown that sialic acid residues have high affinity for Aβ and can attenuate Aβ-associated toxicity *in vitro* (Wang et al. 2001). A follow-up study using sialic acid residues conjugated to NP made of dendrimetric polymers showed that this complex can also attenuate Aβ-induced neurotoxicity *in vitro* by binding and removing Aβ aggregates (Patel et al. 2007). Unfortunately, there are no reports describing the effective use of these particles to remove Aβ aggregates *in vivo*; it is hoped, however, that the positive results from *in vitro* studies will encourage further investigation into the use of these particles in animal models of AD.

Improving Disease Diagnosis

Deposition of Aβ plaques and fibrillary tangles in neuronal cells are common pathological signs of AD. However, at present AD can only be officially confirmed via postmortem examination of the brain. Typically, diagnosis of AD is based on a comprehensive picture of a patient formed by including multiple data types such as clinical, behavioral, and biological analyses. It has been shown that certain AD-related biomarkers may exist at low concentrations in the early stage of AD. As with many diseases, early diagnosis means earlier treatment and better clinical outcomes. Thus, the development of a detection system with high sensitivity and specificity that could identify AD patients before obvious clinical symptoms set in could be very beneficial. A new diagnosis technique using a NP-based detection system with monoclonal antibodies specific for Aβ-derived diffusible ligands has been reported (Georganopoulou et al. 2005). The system can measure the concentration of pathogenic Aβ-derived diffusible ligands, which are soluble non-fibrillar Aβ protein assemblies that can induce neurotoxic effects, in the cerebrospinal fluid (CSF) using an ultrasensitive bio-barcode amplification assay at clinically relevant concentrations. Another recent technology to detect Aβ involves the use of gold-NP. The gold-NP are conjugated with Aβ antibodies and bind to Aβ in solution to form complexes, which can be successfully detected with scanning tunneling microscopy in concentrations as low as 1 fg/ml (Kang et al. 2009). In addition, another type of gold-NP coated with monoclonal tau antibodies was used to detect tau protein at the 1 pg/mL level in CSF using a two-photon scattering assay (Neely et al. 2009) (Fig. 3). It is hoped these new technologies will accurately diagnosis AD in patients early in the course of disease, thereby allowing for early intervention.

NANOMEDICINE AND PD

Anti-inflammatory Nanoparticles

Cerium, a rare earth element of the lanthanide series, and cerium oxide (CeO_2) are two inorganic compounds shown to be excellent oxygen buffers because of their redox capacities. The main oxidation and reduction processes occur through the active interface where the cerium oxide easily changes from a Ce^{4+} to a Ce^{3+} oxidation state. These compounds have been shown to be effective at reducing damage caused by ROS even at extremely low concentrations. For example, one study demonstrated that cerium oxide NP could extend cell and organism longevity through their actions as free radical scavengers (Rzigalinski et al. 2006); another found that the presence of just 5 nM cerium NP was sufficient to prevent the ROS-induced apoptosis in cultured retinal neurons exposed to exogenously added H_2O_2 (Chen et al. 2006). Also, in an animal model of disease, these particles have been shown to be effective at inhibiting the progression of ROS-induced cell death in a dose-dependent manner (Chen et al. 2006).

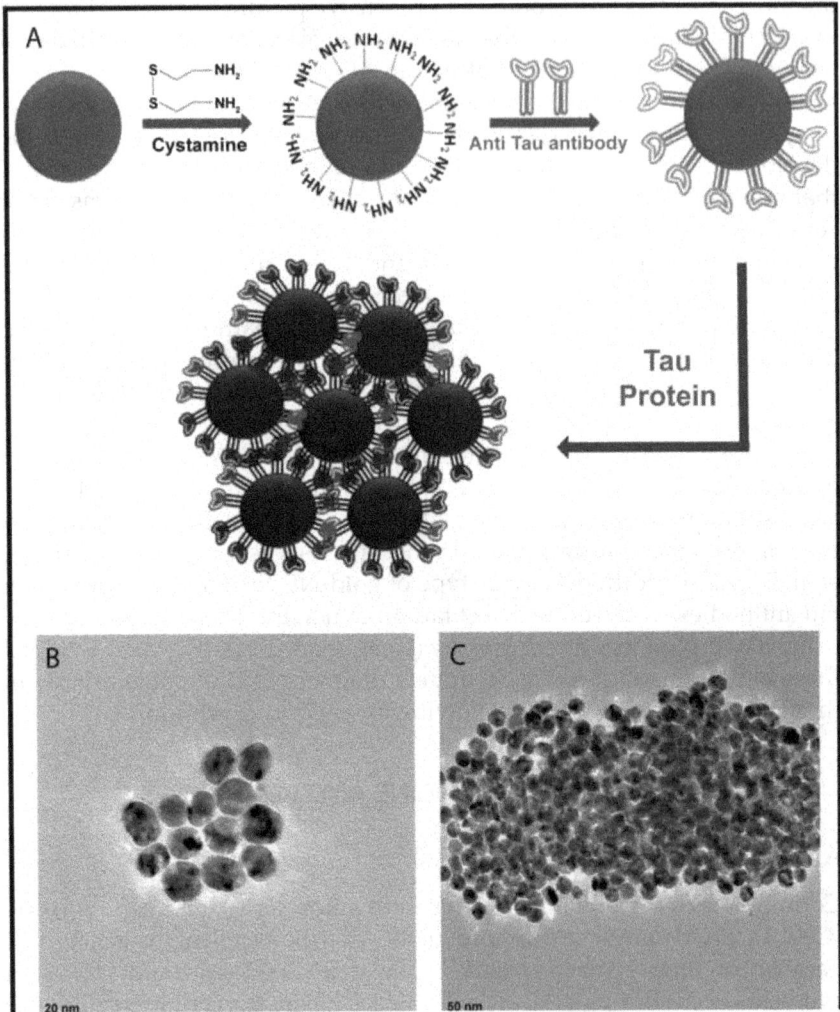

Fig. 3. Antibody-conjugated gold NP–antigen (tau protein) interaction. (A) Tau antibody–Au NP conjugation. First two steps show synthesis of monoclonal tau antibody-conjugated gold NP. Third step shows monoclonal tau antibody-conjugated gold NP-based sensing of tau protein. (B) TEM image of tau antibody-conjugated gold NP before addition of tau protein. (C) TEM image of tau antibody-conjugated gold NP after addition of tau protein. Reproduced from Neely et al. (2009) with permission from the American Chemical Society.

Since cerium NP have repeatedly been shown to be effective at minimizing the cellular damage caused by ROS, particularly in neurons, it has been suggested that they could be used to reduce the ROS-induced cell death seen in PD or in other neurodegenerative conditions. However, one major challenge that remains is a method by which to specifically deliver these particles to the SN. One potential method for accomplishing this, which is discussed in further detail later, is cell-mediated delivery.

Antioxidant C_{60} Fullerenes

Fullerenes, another type of potential ROS scavenger, are three-dimensional carbon networks that can be formed into a hollow sphere, ellipsoid, or tube ranging in size from 1 to 55 nm. Such structures have been shown to have high reactivity towards free radicals, which allows them to soak up ROS and act like free radical sponges. It has been suggested that direct reactions between free radical species and the highly conjugated double bond system of C_{60} structures are responsible for their antioxidant capabilities. Such compounds have been demonstrated to be effective at reducing the cellular damage caused by ROS. For example, a malonic acid C_{60} derivative (Fig. 4) showed dose-dependent levels of protection and at the highest concentration was almost fully protective in an *in vitro* model of PD (Ali et al. 2008).

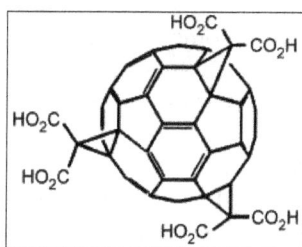

Fig. 4. Chemical structure of C3 malonic acid C_{60} fullerene. It has been suggested that direct reactions between ROS and the highly conjugated double bond system of C_{60} are responsible for their antioxidant properties, allowing them to act as a free radical scavenger. Reproduced from Ali et al. (2004) with permission from Elsevier.

Importantly, molecular modeling demonstrated that the reactivity of fullerenes toward ROS is sensitive to changes in dipole moment, which are dictated not only by the number of carboxyl groups but also by their distribution in the fullerene ball (Ali et al. 2008). However, the molecular mechanisms behind the antioxidant reactions of C_{60} compounds, especially for biologically important radicals such as superoxide, remain controversial. Owing to their success in reducing ROS-associated neuronal death both *in vitro* and *in vivo*, it has been suggested that fullerenes could be used as a treatment for PD. However, as with to the cerium NP discussed above, a major obstacle that still remains is the specific delivery of the particles to the SN. In addition, one point of strong contention with the use of fullerenes has been their potentially toxic side effects. Multiple studies have suggested that fullerenes may have a number of unanticipated toxic effects that need to be further explored. Thus, the future of fullerenes in the treatment of PD or other diseases is unsure.

Improved Pharmacokinetics of Traditional Drugs

A major problem with the treatment of PD and many other CNS diseases is that a majority of drugs used in these conditions have a low hydrolytic stability and are subject to degradation by blood proteins or by enzymes encountered in the BBB. Therefore, there is a need to develop a drug carrier that incorporates many drug molecules, increases their hydrolytic stability, defends against degradative enzymes or transporters and could considerably improve the efficacy of drug delivery to the brain. One way to accomplish this would be to transport the drugs within stable polymer coats. There are several fundamental properties of polymers, which are organic chains of a repeated molecular structure that contain both hydrophilic and hydrophobic regions, that make them useful for solving drug delivery problems. First, polymers can be designed to be intrinsically multifunctional and can, for example, be combined either covalently or non-covalently with drugs to overcome multiple problems such as solubility, stability, and permeability into tissues and cells. Second, polymers can be easily modified with various vectors to help direct drugs to specific sites in the body and thus allow for targeted delivery to a specific tissue or even cell type. Third, polymers can be designed to be environmentally responsive materials, allowing for the controlled and sustained release of a drug at its site of disease. Finally, polymers themselves can be biologically active, and this property can be exploited in order to modify the activity of various endogenous drug transport systems within the body to improve delivery and, therefore, drug performance.

One type of colloidal particle is the solid lipid NP (SLN). These solid particles consist of a nanosized droplet of polymer that contains within it suspended drug molecules in a very highly ordered fashion. Anti-parkinsonian activities of SLN loaded with one of the original dopamine agonists, bromocriptine, have been evaluated in an animal model of PD (Esposito et al. 2008). Bromocriptine crystals were suspended with tristearin/tricaprin lipid combination and coated with poloxamer-188. In animals, these bromocriptine SLN were able to improve performance in behavioral tests when compared to free drug. This suggested that simply repackaging a traditional drug into a nanocarrier might greatly improve its efficacy.

Another type of colloidal particle is the nanostructured lipid carrier (NLC). NLC are similar to SLN in that they both consist of a lipid matrix. However, while SLN are made up of only solid lipids that form a very organized nanostructure, NLC also consist of liquid lipids and thus have a disorganized nanostructure that allows for greater loading and slower release (Fig. 5).

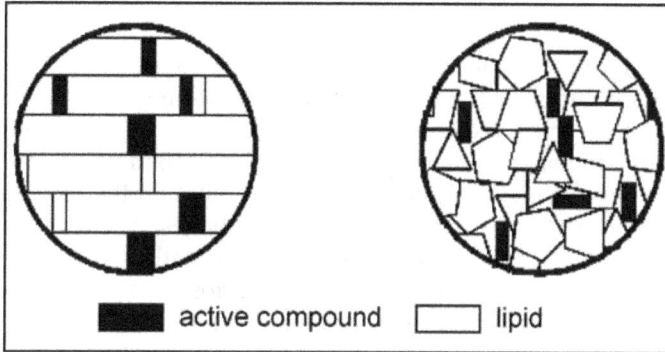

Fig. 5. Structural organization of SLN and NLC. The highly organized structure of SLN (left) are able to hold less active compound than the unorganized structure of NLC (right). Reproduced from Pardeike et al. (2009) with permission from Elsevier.

See Pardeike et al. (2009) for more detail on the differences between these two types of nanoparticles and their applications. This type of particle has also been explored for use in PD therapy. NLCs containing the anti-Parkinson's drug apomorphine have been tested for their abilities to improve drug delivery to the brain (Hsu et al. 2010). These particles were developed using cetyl palmitate as the lipid matrix, squalene as the cationic surfactant, and Pluronic-F68 combined with polysorbate-80 and polyethylene glycol as the interfacial additives. When used in an animal model, apomorphine-loaded NLC could indeed be targeted to selected brain regions (Hsu et al. 2010). This suggests that by repackaging drugs into nanocarriers specific regions of the brain, such as the SN can be targeted.

In recent years, multifunctional NP for treatment of neurodegenerative disorders have focused on vectorized carriers. Such NP are particles that have had their surfaces modified with particular molecules that bind to specific receptors on certain types of cells. Recently, a new gene delivery system using polyamidoamine NP vectorized with lactoferrin through bifunctional polyethyleneglycol (PEG) were used as a potential drug delivery system for human glial cell line–derived neurotrophic factor gene. In particular, the therapeutic efficiency of NP encapsulating this gene was evaluated in a rat model of PD (Huang et al. 2008; Xu et al. 2008). The results indicated that lactoferrin-modified NP could significantly improve locomotor activity, reduce dopaminergic neuronal loss, and enhance monoamine neurotransmitter levels. These results demonstrate the powerful neuroprotective effects of this type of particle and highlight their potential usefulness in PD.

Cell-mediated Drug Delivery

A promising approach for the targeted delivery of drugs to the CNS is the use of specific cells that can incorporate nano-containers loaded with drugs into their cytoplasm, acting as Trojan horses by migrating across the BBB and transporting the drugs to the brain (Batrakova et al. 2007; Brynskikh et al. 2010). The cell-based system relies on the biological and functional properties of blood-borne macrophages. These cells have a high rate of endocytosis that allows them to efficiently accumulate micro- and nanoparticles within intracytoplasmic endosomes and release them through processes that include exocytosis. In addition, they are also able to cross biological barriers in response to cytokine signaling from diseased or damaged tissue. CNS inflammation is characterized by chemokine- and cytokine-mediated leukocyte recruitment to the site of disease by processes involving macrophage diapedesis and chemotaxis. Once at the site of disease, the drug-loaded cells would be able, through processes including exocytosis, to maintain controlled drug release over a long period of time and affect ongoing disease (Jain et al. 2003). All these features make macrophages attractive candidates for cell-mediated delivery of drugs and therapeutic proteins. In particular, cell-based delivery of nanocarriers that contain redox enzymes, catalase, and super oxide dismutase to the brain would be of great therapeutic value in PD.

Using macrophages as carriers for therapeutic compounds offers several advantages to combat CNS disease, including (1) prolonged plasma drug levels, (2) time-controlled release of the loaded drug, (3) targeted drug transport to the site of disease, and (4) diminished drug immunogenicity. Recently, our laboratories developed novel CNS drug delivery systems using macrophages for delivery of the antioxidant enzyme catalase in a mouse model of PD (Batrakova et al. 2007; Brynskikh et al. 2010). For this system, nanoformulated catalase was obtained by coupling the enzyme to a cationic block copolymer, polyethyleneimine-poly(ethylene glycol), leading to a polyion complex micelle referred to as a nanozyme. It was demonstrated that macrophages loaded with nanoformulated catalase and injected into mice with 1-methyl-4-phenyl-1,2,3,6-tetrahydropyridine (MPTP)-induced PD reduced neuroinflammation and attenuated degeneration of the SN (Brynskikh et al. 2010) (Fig. 6). Subsequent studies examined relationships between the composition and structure of catalase, the physicochemical characteristics of the NP (including morphology, size, and zeta-potential), and their abilities to be loaded, retained, and released by macrophages (Zhao et al. 2010). Overall, these experiments indicated that cell-mediated delivery of catalase NPs could potentially be used as a therapy for PD.

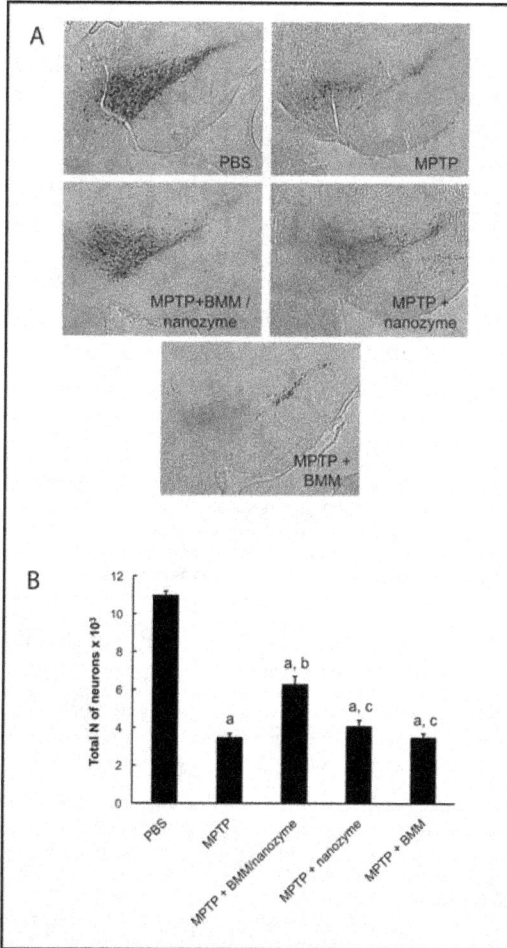

Fig. 6. Neuroprotective effect of nanozyme-loaded BMM in an MPTP mouse model of PD. MPTP-intoxicated C57Bl/6 mice (18 mg/kg) were *i.v.* injected with PBS (second bar), nanozyme alone (third bar), macrophages loaded with nanozyme (5×10^6 cells/mouse/100 μl) (fourth bar), or empty macrophages (fifth bar). Healthy non-intoxicated animals were used as control group (first bar). The animals were sacrificed 7 d after treatment; brain slices were stained for tyrosine hydroxylase-positive nigral dopaminergic neurons (A). Results from N = 5 animals per group demonstrating significant loss in MPTP-treated mice, which is prevented by adoptive transfer of macrophages loaded with nanozyme (B). No significant neuroprotective effect was detected after treatment with nanozyme alone, or empty macrophages. Values are means ± SEM, and $P < 0.05$ compared with [a]PBS; [b]MPTP; [c]MPTP+Macrophage/nanozyme. Reproduced from Brynskikh et al. (2010) with permission from Future Medicine Ltd.

DISCUSSION

The BBB has been studied in great detail for many years. Despite all the knowledge about its structure and function, it is still difficult to deliver drugs to a diseased brain. Nanotechnology and all the discoveries that have come along with it have provided many new tools for realizing this goal. Nanomedicine has helped to greatly advance drug delivery for neurodegenerative disorders. NP can be manufactured to have a wide range of unique physical and chemical properties that allow for loading, protection, and transport of drug molecules. This has helped to improve both the pharmacokinetics and targeting of old drugs and has shown promising results in disease models both *in vitro* and *in vivo*. NP are now being made that can target drugs to specific tissues, cells, and even subcellular compartments. This is helping to increase the efficacy of the drugs while simultaneously reducing untoward side effects. In addition, NP are being used to inhibit the mechanisms that lead to disease and neuronal death in a number of new ways. Gold-NP, with their ability to be conjugated to a wide range of different molecules, show promise in many diseases, particularly in AD. Finally, nanomedicine is also helping to increase the accuracy of diagnosis of neurodegenerative disorders, which is particularly important since many of these diseases are not diagnosed until the late stages.

Even though we only discussed AD and PD in detail, it is important to note that since there are common underlying mechanisms to many neurodegenerative disorders, such as ROS production and inflammation, the therapies designed for specific neurodegenerative disorders could potentially be applied to others. For example, methods that efficiently transport drugs across the BBB or decrease neuronal death could be applied to nearly every neurodegenerative disorder. Thus, nanomedical research into any neurodegenerative disorder will eventually benefit treatment strategies for them all.

The frequency of neurodegenerative disorders is predicted to continue to rise. The incidence of neurodegenerative disorders and the cost of care are already high and there is thus an increased urgency to find effective treatments. Nanomedicine has already produced a number of promising treatment methods and new ones are continually being introduced. Thus, it is not unreasonable to hope that in the next 10 years we will begin to see positive changes in clinical outcomes for those who suffer from neurodegenerative disorders.

APPLICATIONS TO OTHER AREAS OF HEALTH AND DISEASE

Many of the same nanomedical methods being applied to the treatment of neurodegenerative disorders are also being used to improve therapy in a number of other diseases. For example, cell-mediated drug delivery is being explored for its potential use in the treatment of HIV infection and for other diseases that are characterized by inflammation and macrophage infiltration such as rheumatoid arthritis and stroke. This method in particular holds a great deal of promise for direct targeting to areas of disease. Drug-containing NP are another tool being used to protect drugs from degradation and increase the circulation time in treatment of microbial infections and cancer. Gold-NP are versatile particles that can be conjugated to a variety of molecular compounds and as a result they have been used in treatment of a wide range of diseases. The antioxidant properties of fullerenes have been researched for their use in treating respiratory diseases and also cancer. Many of these approaches are relatively new; and as the use of these techniques improves, they can be effectively applied to more and more diseases.

Summary Points

- Neurodegenerative disorders are a set of diseases that affect the functioning and survival of neurons within the central nervous system.
- Despite efforts of modern medicine, there are still no cures for neurodegenerative disorders and therapy is usually only palliative.
- The blood-brain barrier separates the brain from the circulatory system and is extremely efficient at keeping drugs and other therapeutic compounds out of the brain.
- Nanomedicine is finding methods of packaging drug molecules into containers that can simultaneously protect them from degradation and deliver them to specific regions of the brain.
- One potential cause of aggregation of $A\beta$ protein in AD is an overabundance of metal ions; therefore, metal chelating nanoparticles may be a possible treatment option.
- Another potential treatment method for AD would be to eliminate $A\beta$ fibrils and aggregates by using NP that can bind and remove these structures.
- Early and accurate diagnosis has been a challenge for all neurodegenerative disorders. In AD, diagnostic methods using gold-NP have shown promising results.

- One potential cause of neuronal damage in PD is the presence of excess amounts of reactive oxygen species; therefore, antioxidant NP may provide a method for inhibiting this disease pathway.
- Cell-mediated drug delivery is a unique method for protecting and delivering drugs to the brain. In this method NP are loaded into phagocytic cells, which are then able to act as Trojan horses and transport drug into the brain.
- Many challenges to both diagnosing and treating neurodegenerative disorders remain. However, research using nanomedical approaches offers a number of promising methods to overcome these difficulties.

Definitions

Blood-brain barrier (BBB): A special type of small blood vessel in the CNS that acts as both physical and chemical barrier to molecules trying to enter the brain.

Brain microvascular endothelial cell (BMVEC): The major cell type of the BBB, which has both structural and physiological characteristics that help it prevent the transport of many types of molecules into the brain.

Efflux systems: Protein structures within BMVEC that pump drug molecules out of the cells before the compound can reach the brain.

Fullerenes: Molecular-sized cage-like structures made of carbon molecules (usually around 60) that have antioxidant properties and have the ability to transport drug molecules.

Gold nanoparticles: Nanosized particles of gold used in a variety of ways and currently a major focus for nanomedicine and neurodegenerative disorders therapy.

Nanozyme: NP of the antioxidant enzyme catalase coupled to polymeric compounds.

Quantum dots: Nanosized semiconductors that have unique electronic properties that allow them to have power fluorescence and are commonly being using in nanomedical research.

Solid lipid nanoparticles (SLN) and nanostructured lipid carriers (NLC): Nanoparticles that consist of liquid or solid lipids alone or in combination that are able to transport drug molecules.

Key Facts

- AD, first described in 1901 by the German psychiatrist Alois Alzheimer, is the most common neurodegenerative disorder, affecting over 26 million people worldwide.
- The pathological hallmarks of AD are neurofibrillary tangles and amyloid plaques, which are aggregations of insoluble hyperphosphorylated microtubule-associated tau protein and amyloid-beta respectively.
- AD causes neurons in the cerebral cortex to malfunction or die, resulting in impaired memory and dementia.
- PD, first described in 1817 by the French neurologist James Parkinson, is the second most common neurodegenerative disorder, affecting over 6 million people worldwide.
- The pathological hallmarks of PD are abnormal accumulations of insoluble alpha-synuclein protein called Lewy bodies and depigmentation of the substantia nigra.
- PD causes the death of dopaminergic neurons in the substantia nigra, which results in motor impairments.
- There are no cures for either AD or PD and the therapies that are currently available are only palliative.

Abbreviations

AD	:	Alzheimer's disease
Aβ	:	amyloid beta
BBB	:	blood-brain barrier
BMVEC	:	brain microvascular endothelial cell
CNS	:	central nervous system
CSF	:	cerebral spinal fluid
DFO	:	desferrioxamine
MAEHP	:	2-methyl-N-(2'-aminoethyl)-3-hydroxyl-4 pyridinone
MPTP	:	1-methyl-4-phenyl-1,2,3,6-tetrahydropyridine
NLC	:	nanostructured lipid carriers
NP	:	nanoparticles
PD	:	Parkinson's disease
PEG	:	polyethylene glycol
ROS	:	reactive oxygen species
SLN	:	solid lipid nanoparticles
SN	:	substantia nigra

References

Ali, S.S., J.L. Hardt, and L.L. Dugan. 2008. SOD activity of carboxyfullerenes predicts their neuroprotective efficacy: a structure-activity study. Nanomedicine 4: 283–294.

Ali, S.S., J.I. Hardt, K.L. Quick, J.S. Kim-Han, B.F. Erlanger, T.T. Huang, C.J. Epstein, and L.L. Dugan. 2004. A biologically effective fullerene (C60) derivative with superoxide dismutase mimetic properties. Free Radic. Biol. Med. 37: 1191–1202.

Batrakova, E.V., S. Li, A.D. Reynolds, R.L. Mosley, T.K. Bronich, A.V. Kabanov, and H.E. Gendelman. 2007. A macrophage-nanozyme delivery system for Parkinson's disease. Bioconjug. Chem. 18: 1498–1506.

Brynskikh, A.M., Y. Zhao, R.L. Mosley, S. Li, M.D. Boska, N.L. Klyachko, A.V. Kabanov, H.E. Gendelman, and E.V. Batrakova. 2010. Macrophage delivery of therapeutic nanozymes in a murine model of Parkinson's disease. Nanomedicine 5: 379–396.

Cabaleiro-Lago, C., F. Quinlan-Pluck, I. Lynch, S. Lindman, A.M. Minogue, E. Thulin, D.M. Walsh, K.A. Dawson, and S. Linse. 2008. Inhibition of amyloid beta protein fibrillation by polymeric nanoparticles. J. Am. Chem. Soc. 130: 15437–15443.

Chen, J., S. Patil, S. Seal, and J.F. McGinnis. 2006. Rare earth nanoparticles prevent retinal degeneration induced by intracellular peroxides. Nat. Nanotechnol. 1: 142–150.

Cui, Z., P.R. Lockman, C.S. Atwood, C.H. Hsu, A. Gupte, D.D. Allen, and R.J. Mumper. 2005. Novel D-penicillamine carrying nanoparticles for metal chelation therapy in Alzheimer's and other CNS diseases. Eur. J. Pharm. Biopharm. 59: 263–272.

Esposito, E., M. Fantin, M. Marti, M. Drechsler, L. Paccamiccio, P. Mariani, E. Sivieri, F. Lain, E. Menegatti, M. Morari, and R. Cortesi. 2008. Solid lipid nanoparticles as delivery systems for bromocriptine. Pharm. Res. 25: 1521–1530.

Faller, P. 2009. Copper and zinc binding to amyloid-beta: coordination, dynamics, aggregation, reactivity and metal-ion transfer. Chembiochem. 10: 2837–2845.

Georganopoulou, D.G., L. Chang, J.M. Nam, C.S. Thaxton, E.J. Mufson, W.L. Klein, and C.A. Mirkin. 2005. Nanoparticle-based detection in cerebral spinal fluid of a soluble pathogenic biomarker for Alzheimer's disease. Proc. Natl. Acad. Sci. 102: 2273–2276.

Hsu, S.H., C.J. Wen, S.A. Al-Suwayeh, H.W. Chang, T.C. Yen, and J.Y. Fang. 2010. Physicochemical characterization and in vivo bioluminescence imaging of nanostructured lipid carriers for targeting the brain: apomorphine as a model drug. Nanotechnology 21: 405101.

Huang, R., W. Ke, Y. Liu, C. Jiang, and Y. Pei. 2008. The use of lactoferrin as a ligand for targeting the polyamidoamine-based gene delivery system to the brain. Biomaterials 29: 238–246.

Jain, S., V. Mishra, P. Singh, P.K. Dubey, D.K. Saraf, and S.P. Vyas. 2003. RGD-anchored magnetic liposomes for monocytes/neutrophils-mediated brain targeting. Int. J. Pharm. 261: 43–55.

Kang, D.Y., J.H. Lee, B.K. Oh, and J.W. Choi. 2009. Ultra-sensitive immunosensor for beta-amyloid (1–42) using scanning tunneling microscopy-based electrical detection. Biosens. Bioelectron. 24: 1431–1436.

Liu, G., M.R. Garrett, P. Men, X. Zhu, G. Perry, and M.A. Smith. 2005. Nanoparticle and other metal chelation therapeutics in Alzheimer disease. Biochim. Biophys. Acta. 1741: 246–252.

Liu, G., P. Men, W. Kudo, G. Perry, and M.A. Smith. 2009. Nanoparticle-chelator conjugates as inhibitors of amyloid-beta aggregation and neurotoxicity: a novel therapeutic approach for Alzheimer disease. Neurosci. Lett. 455: 187–190.

Liu, G., P. Men, G. Perry, and M.A. Smith. 2010. Nanoparticle and iron chelators as a potential novel Alzheimer therapy. Methods Mol. Biol. 610: 123–144.

Milton, N.G., and J.R. Harris. 2010. Human islet amyloid polypeptide fibril binding to catalase: a transmission electron microscopy and microplate study. ScientificWorldJournal 10: 879–893.

Neely, A., C. Perry, B. Varisli, A.K. Singh, T. Arbneshi, D. Senapati, J.R. Kalluri, and P.C. Ray. 2009. Ultrasensitive and highly selective detection of Alzheimer's disease biomarker using two-photon Rayleigh scattering properties of gold nanoparticle. ACS Nano 3: 2834–2840.

Pardeike, J., A. Hommoss, and R.H. Muller. 2009. Lipid nanoparticles (SLN, NLC) in cosmetic and pharmaceutical dermal products. Int. J. Pharm. 366: 170–184.

Patel, D.A., J.E. Henry, and T.A. Good. 2007. Attenuation of beta-amyloid-induced toxicity by sialic-acid-conjugated dendrimers: role of sialic acid attachment. Brain. Res. 1161: 95–105.

Rzigalinski, B.A., K. Meehan, R.M. Davis, Y. Xu, W.C. Miles, and C.A. Cohen. 2006. Radical nanomedicine. Nanomedicine 1: 399–412.

Skaat, H., and S. Margel. 2009. Synthesis of fluorescent-maghemite nanoparticles as multimodal imaging agents for amyloid-beta fibrils detection and removal by a magnetic field. Biochem. Biophys. Res. Commun. 386: 645–649.

Wang, S.S., D.L. Rymer, and T.A. Good. 2001. Reduction in cholesterol and sialic acid content protects cells from the toxic effects of beta-amyloid peptides. J. Biol. Chem. 276: 42027–42034.

Xiao, L., D. Zhao, W.H. Chan, M.M. Choi, and H.W. Li. 2009. Inhibition of beta 1–40 amyloid fibrillation with N-acetyl-L-cysteine capped quantum dots. Biomaterials 31: 91–98.

Xu, S., M. Chen, Y. Yao, Z. Zhang, T. Jin, Y. Huang, and H. Zhu. 2008. Novel poly(ethylene imine) biscarbamate conjugate as an efficient and nontoxic gene delivery system. J. Control. Release 130: 64–68.

Zhao, Y., M. Haney, N. Klyachko, S. Li, S. Booth, S. Higginbotham, J. Jones, M. Zimmerman, R. Mosley, A. Kabanov, H. Gendelman, and E. Batrakova. 2011. Polyelectrolyte complex optimization for macrophage delivery of redox enzyme nanoparticles. Nanomedicine 6: 25–42.

Nanotechnological Applications in Tissue and Implant Engineering

Viswanathan S. Saji[1] and Han-Cheol Choe[2],*

ABSTRACT

Nanobiotechnology, often described as the hybrid science resulting from the merging of two powerful scientific realms, biotechnology and nanotechnology, reflects science's growing ability to conduct investigations beyond molecular level. The applications of nanotechnology in medicine and biomedical engineering are extensive, spanning areas such as implant and tissue engineering, diagnosis and therapy. Research and development in different areas of nanobiotechnology aspires to develop highly functional biosensors, nanosized microchips, molecular switches and tissue analogs for skin, bones, muscles and other organs. This chapter discusses recent developments in nanobiotechnology applicable in the areas of load-bearing orthopedic and dental implants as well as novel tissue engineered scaffolds. Major research outputs in this area are discussed with reference to metallic and composite biomaterials. Many studies have demonstrated that implants

[1]School of Energy Engineering, Ulsan National Institute of Science and Technology, Ulsan 689-805, Republic of Korea.

[2]Department of Dental Materials, School of Dentistry, Chosun University, Gwangju 501-825, Republic of Korea; Email: hcchoe@chosun.ac.kr

*Corresponding author

List of abbreviations after the text.

having nanotopography stimulate more positive cellular responses than conventional materials. Nanomaterialistically manipulating the surface chemistry of implants is an advantageous approach to enhance implant performance. Nanostructured tissue engineered scaffolds are attractive as they mimic natural bone materials and can enhance the mechanical properties without diminishing bioactivity. Despite these advantages, there exist several unresolved questions concerning the safety of engineered nanomaterials, raising challenges to their practical applications.

INTRODUCTION

The major goal in tissue and implant engineering is the development of biomaterials that accelerate adaptation of artificial implants to living tissues, thereby extending implant life. A biomaterial can be defined as a substance or a combination of substances, other than drugs, derived either naturally or synthetically and used to repair or replace the function of living human tissues (Williams 1987). In accordance with application in body systems, examples of biomaterials include bone plates, total joint replacements (skeletal system), sutures (muscular), artificial heart valves (circulatory), and artificial skin (integumentary). An ideal biomaterial should be highly biocompatible, bioactive, easy to fabricate, sterilizable, non-toxic, non-corrosive and non-immunogenic.

Load-bearing biomaterials such as orthopedic and dental implants face the challenge of a coupled effect between structural requirements and the aggressive environment of the body. Metallic, ceramic, polymeric and composite biomaterials have been used extensively as orthopedic and dental implants (Park and Bronzino 2003). Biodegradable scaffolds are attractive for less load-bearing applications and it helps to eliminate additional surgery to remove the implant after it has served its function. The series of complex post-implantation interactions at the tissue/implant interface depend on many parameters including surface chemistry, elasticity and topography of the implant. In the past, bioactive surface coatings based on hydroxyapatite (HA) were mainly investigated to selectively improve the desirable tissue formation on implants.

In recent times, biomaterials engineered at the nanometer scale have gained recognition in various biomedical applications because of their potential advantages over conventional biomaterials. A nanostructured biomaterial is expected to interact more effectively with biological molecules normally present in the natural extracellular matrix, as most of these molecules are also in the nanometer scale dimensions. Nanomaterials and structures such as nanoparticles, nanosurfaces, nanofibres, nanocoatings, and nanocomposites are being considered for various applications in orthopedics and dentistry. Reports show that nanometer scale surface features of implants influence cellular attachment, differentiation, and alignment significantly (Karlsson et al. 2003; Richert et al. 2008; Saji et al. 2010). The enhanced material properties at the nanometer scale dimension such as increased surface area and finer surface roughness are expected to yield better biological responses of osteogenic cells and effective tissue-implant mechanical interlocking. Also, the enhanced hardness and strength of nanomaterials in comparison with their coarser counterparts are attractive in making highly wear-resistant implants. This chapter discusses recently projected applications of nanotechnology in the area of load-bearing orthopedic and dental implants.

APPLICATIONS TO AREAS OF HEALTH AND DISEASE

Orthopedic implants are mainly used for bone repairs (e.g., bone fracture) and joint replacements (e.g., hip joint, spine). The treatment of a bone fracture can be non-surgical (e.g., immobilization with plaster) or surgical (external or internal fracture fixation). All internal fracture fixation devices should meet the requirement of biomaterials. Stainless steel (SS), Co-Cr alloys, Ti and Ti alloys are the most suitable metallic biomaterials used for the purpose. Among these, Ti-based materials are the best choice and have excellent biocompatibility and osseointegration properties. Biodegradable polymeric scaffolds have been used to treat minimally loaded fractures, thereby eliminating the need for a second surgery to remove the implant

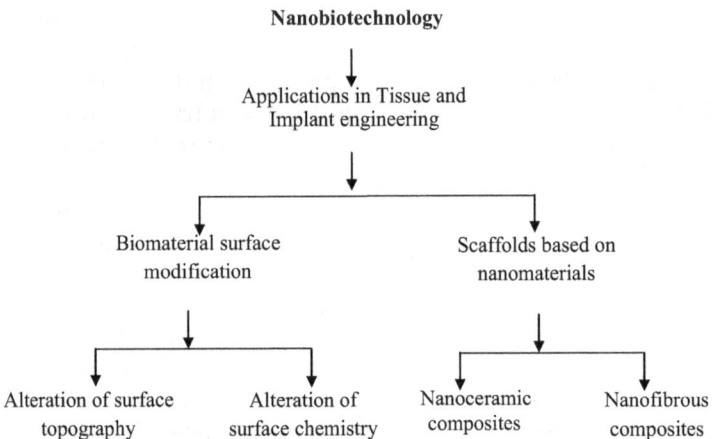

Fig. 1. Diagram of the application areas of nanobiotechnology in tissue and implant engineering.

after the fracture heals. The major issues to be considered in a tissue scaffold fabrication are the selection of appropriate matrix materials, control of porosity, proper mechanical strength, scaffold degradation properties and bioactivity. Widespread investigations are going on for the effective use of nanomaterials in fabrication of tissue scaffolds. Total joint replacements (e.g., hip, knee, shoulder, finger) are permanent implants, unlike those used to treat fractures. Metals (316 L SS, CoCrMo cast alloy, Ti6Al4V alloy), ceramics (alumina, bioglass, HA) and polymeric biomaterials (UHMWPE, PMMA bone cement) are employed in joint implants. Commercially pure Ti and Ti alloys, Co-Cr cast alloys, HA and bioglass are the major dental implant materials. The initial biological response of host tissue to the implant (healing phase) is critical for implant success. During the healing phase the initial biological processes of protein and molecular deposition on the implant surface are followed by cellular attachment, migration and differentiation. These tissue responses primarily lead to the cellular expression and maturation of extracelllar matrix and ultimately to the development of bony interfaces with the implant material. The surface topography (micro or nano) of the implant can have significant impact on these cellular interactions.

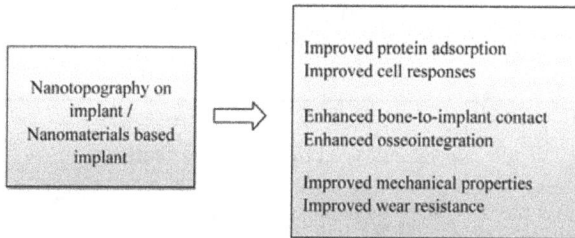

Fig. 2. Expected advantages (Saji et al. 2010).

APPLICATIONS IN TISSUE AND IMPLANT ENGINEERING

Metallic Implants

The current research in this area is striving to uncover novel and economical surface modification strategies for metallic implants to achieve nanometer scale surface roughness. Bone cells are naturally accustomed to interact with surfaces having a large degree of nanometer roughness. Studies show that nanotopography alters cellular response through controlled cell growth, protein deposition, increased osteoblast adhesion and proliferation. This enhanced protein and cell binding capacity is attributed to the larger surface areas/roughness degrees and altered surface energies (Jager et al. 2007). Different nanoscale surface modification strategies investigated for metallic implants can be classified into two groups: implants that alter a surface topographically and implants that introduce nanoscale chemical molecules on a surface.

Implants that Alter a Surface Topographically

Researchers have attempted various techniques on different biomaterials (especially titanium and their alloys) to explore the hypothesis that nanoscale surface modification could improve the tissue response. Although problems do exist (e.g., in solvent casting and embossing, distortion during transfer), lithographic technologies continue to be the most widely used technique for producing nanotopography for biological applications. Colloidal lithography is especially promising. By functionalizing the surface with poly-L-lysine and dip coating it in a colloidal sol, one can impart nanotopography to even a three-dimensional structure such as dental screws (Wood 2007). New lithography techniques such as "schematic overview of template-directed assembly on an ordered microsphere array" may limit the possible irregularities occurring in in-plane topographies (Grego et al. 2005). Imprint lithography is emerging as an alternative nanopatterning technology to traditional photolithography that permits the fabrication of two-dimensional and three-dimensional structures with high resolution and with operational ease (Van and Watts 2006). Other than the lithographic techniques, there are diverse

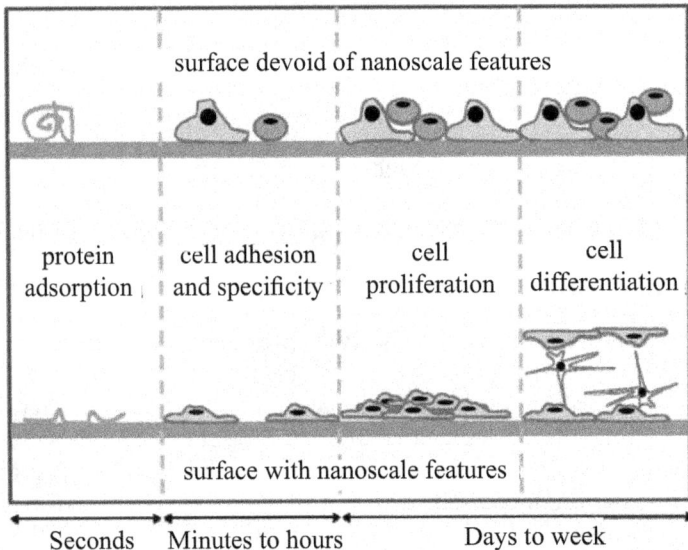

Fig. 3. A schematic representation, showing the expected positive effect of surface nanotopography on cell interactions at surfaces when compared to conventional-sized topography (Mendonca et al. 2008).

physical and chemical approaches investigated to create nanopatterning on implant surfaces (Mendonca et al. 2008). Self-assembled monolayers can change the topography as well as chemistry of a surface to impart novel physical and/or biochemical properties. Other physical methods comprise ion beam deposition and nanoparticles deposition. By suitably controlling the experimental parameters, chemical methods such as acid etching and anodization can produce nanopatterned surfaces. Methods such as peroxidation and alkali treatment alter both the surface chemistry and the topography.

In this direction, a few novel approaches have been reported. Among them, self-organized nanotubular oxide layer formation through controlled anodization is particularly attractive (Macak et al. 2007; Saji et al. 2009). It was shown that the cells cultured on nanotubular surfaces showed higher adhesion, proliferation, alkaline phosphate activity and bone matrix deposition than those grown on flat titanium surfaces (Popat et al. 2007). The bioactivity of such nanotubular surfaces can be enhanced through growth of nanoscale HA (Kar et al. 2006). Bond strength was further improved by annealing the HA-coated nanotubular titanium alloy in argon atmosphere. Bauer et al. showed that self-assembled monolayer (octadecylphosphonic acid) formation on TiO_2 nanotubular layers creates diameter-dependent wetting behavior ranging from hydrophobic ($108 \pm 2°C$) up to super-hydrophobic ($167 \pm 2°C$). Mesenchymal stem cells adhesion and proliferation are strongly affected in the super-hydrophobic range (Bauer et al. 2008). Inherent bioactive nanotubular titanium alloys can be fabricated through incorporation of bioactive elements within the nanotubes. Such a surface will be attractive in terms of both topography and bioactivity. It has been shown that a low temperature heat treatment is desirable in terms of biocompatibility of nanotubular titanium alloys (Saji and Choe 2009). Analogous to the nanotube formation, a few alternative novel approaches were reported (Liu et al. 2008; Ogawa et al. 2008). Nanoscale rod arrays were fabricated on Ti surface by glass phase topotaxy growth method by an interfacial reaction between sodium borate glass coating and the pre-heated titanium substrates at elevated temperature (Liu et al. 2007). Ogawa et al. reported formation of titanium nanonodular self-assembly through electron beam PVD of Ti onto specifically conditioned microtextured titanium surfaces (Ogawa et al. 2008). Uniform nanonodular structures were developed only on the pre-textured surfaces by either sand-blasting or acid-etching with HCl and/or H_2SO_4, but not on relatively smooth surfaces, including machined and HF-treated surfaces. It was suggested that helical rosette nanotubes (HRN)-coated Ti surface may simulate an environment that bone cells are accustomed to interact with. HRN are a new class of self-assembled organic nanotubes possessing biologically inspired nanoscale dimensions (Chun et al. 2004).

Implants that Introduce Nanoscale Chemical Molecules on a Surface

Nanoparticles can be deposited onto implant surfaces to alter their surface chemistry. For example, Ag nanoparticles can be used in coating orthopedic pins to prevent bacterial colonization and dispersed silver nanoparticles can be used in PMMA bone cement. Nanoparticles promoting osseointegration can be deposited on pre-roughened surface though exposure to nanoparticles solution (Berckmans et al. 2007). Heidenau et al. used nanoscale sol-gel titania layers with embedded metal salts of Ag, Zn, Hg, Cu, Co and Al deposited on Ti surfaces and the highest bacterial-reduction rate was achieved with Cu (Heidenau et al. 2005). Bone implants with modified titania and zirconia nanocrystal coatings were investigated for osseointegration and antibacterial effects (Bignozzi et al. 2008). The coating made of nanocrystalline material comprising nanoparticles of formula $AO_{x-}(L-M^{n+})_i$, where AO_x represents TiO_2 or ZrO_2; M^{n+} is a metallic ion having antibacterial activity (Ag, Cu etc); $n = 1$ or 2; L is a bifunctional organic molecule that can simultaneously bind to the metal oxide and to the metallic ion; and i is the number of $L-M^{n+}$ groups bound to one nanoparticle of AO_x. Introduction of biomolecules on the surface is another strategy. Biocompatibility and performance of an implant can be significantly improved by localization of such biomolecules (peptides, proteins and genes) on the surface through DNA hybridization, ligand/receptor, or antigen/antibody interaction.

Ceramic and Polymeric Implants: Nanocomposite Scaffolds

The majority of fabrication methods used for nanotopographies on planar metallic substrates is unlikely to be capable of providing controlled topographical cues to cells seeded within three-dimensional scaffolds. An alternative approach is the use of nanometric building blocks to fabricate scaffolds. *Nanoscale bioceramics* play a pivotal role in this direction because of their large surface area and ultrafine structure, which resembles that of natural components of the bone. Examples of this category include nanoalumina, nanozirconia and nano calcium phosphates (HA, tricalcium phosphate, biphasic calcium phosphate). Bioinert nanoscale alumina and zirconia are employed in coroneal replacements, couplings of knee prosthesis and other implants. Studies showed that there exists an increasing trend in osteoblast adhesion on the nanostructure of alumina phase (Webster et al. 1999). The poor mechanical properties of HA, which restrict their uses in load-bearing applications, can be minimized by nanotechnology. It was found that nanocrystalline HA powders improve sinterability and densification, which could improve the fracture toughness and other mechanical properties. Nano HA is also expected to have better bioactivity than coarser crystals. Novel methods for production of high quality nanoparticles with extremely high purity and crystallinity improved the quality of the already existing HA-based implants. Ergun et al. investigated osteoblast adhesion on nanoparticulate calcium phosphates of various Ca/P ratios. Their study showed that the average nanometer grain size, porosity and average pore size decreased with increasing Ca/P ratios; the osteoblast adhesion increased on calcium phosphates with higher Ca/P ratios (Ergun et al. 2008).

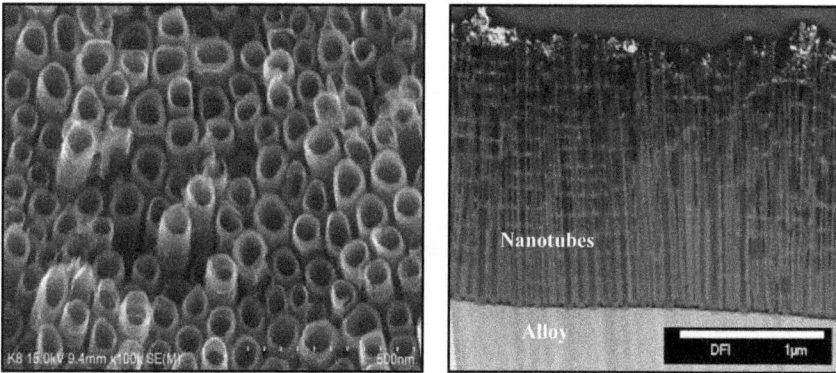

Fig. 4. Electron microscopy images of nanotubes formed on a β titanium alloy (Saji et al. 2009).

The interest in *polymers* for various biomedical applications is mainly due to their design flexibility and biodegradability. Polymers offer the benefit of being intrinsically resistant to environmental attack; however, load-bearing applications often put them at direct risk of fatigue, wear and fracture. Polymeric systems that have been investigated for hard tissue generation include PEG, PLA, and polyurethanes. UHMWPE is widely used in total hip replacements. Recent advances in the fabrication of nanofibers, which are capable of mimicking the size and scale of natural collagen fibers, offer an advanced form of building blocks for scaffolds. For the past few years, there has been growing interest in novel fiber production technology especially by electrospinning (Balasundaram and Webster 2007). Nanofibrous scaffolds designed to elicit specific cellular responses through the incorporation of signaling ligands (e.g., growth factors, adhesion peptides) or DNA fragments are viewed as particularly promising. Properly applying right polymer with the considerations of porous size, degradation rate and surface morphology as a parent material by mixing with high-strength fibers can indeed provide a good alternative for existing materials for scaffold applications (Cheung et al. 2007).

The main drawback of nano HA for use in load-bearing implants is their poor mechanical properties. Compared to the strength of metals and ceramics, the strength of biodegradable polymers is also very low. If the mechanical property of an implant does not match that of natural bone, it can cause stress shielding effect and can result in implant failure (Sumner and Galante 1992). In order to seek better scaffolds for bone tissue engineering, a *composite* strategy is attracting great interest. The advantage of composite systems is that they combine the desired mechanical properties of each of constituent phases into one material system, without diminishing its bioproperties. Many investigations on polymer/ceramic composite systems for bone replacement have been conducted during the last decade. Biocompatible macromolecules, like biopolymers, proteins and polysaccharides, have been widely used and incorporated into nanoceramics to form novel bone-like composites (Xu et al. 2007). The ideal reinforcement material would impart mechanical integrity to the composite at high loading without diminishing its bioactivity. Among the nanoceramic protein composites, nano HA/collagen composites are widely investigated. Studies showed that incorporation of higher elastic modulus biopolymers such as Ca-cross-linked alginate into nano HA/collagen composite systems improved the mechanical properties considerably (Zhang et al. 2003). It was reported that after mixing biodegradable poly(lactic-co-glycolic acid) with nanophase alumina, the bending properties of the composites match more closely with the value of human femur bone (Kay et al. 2002). In the presence of chitosan, some of the limitations of nano HA, such as bioresorption and particle migration, can be alleviated. When chitosan was added into HA systems, the fracture strength and microhardness of the resulting nanocomposites were found enhanced (Rhee and Tanaka 2001). A novel bone-tissue engineering scaffold design should mimic the macro- and micro-structure of natural bone. A nano HA-biodegradable polymer scaffold of a multi-channel configuration can produce an ideal structure that can replace the natural extracellular matrix until host cells can repopulate and resynthesize a new natural matrix (Fig. 5). Polymer nanofiber-based composites are attracting considerable attention and can be used in either hard or soft tissue applications depending on their mechanical and structural properties.

Carbon nanotubes and nanofibers have been investigated as reinforcement material. The versatility of these fibers suggests that there are a large number of possibilities for future designs that could enhance the efficiency of medical implants (Webster et al. 2004). For example, carbon nanofibers-polyurethane composites can be employed in non-loading orthopedic applications. Carbon nanotubes with their high aspect ratio and excellent mechanical properties have the potential to strengthen and toughen HA without offsetting its bioactivity. PMMA-modified nano

HA with multi-walled carbon nanotubes as reinforcement was projected as new generation biomedical bone cement and implant coatings (Singh et al. 2008). Employing freeze-granulation technique, it was possible to increase material homogeneity and also enhance the dispersion of the nanotubes in the composite matrix.

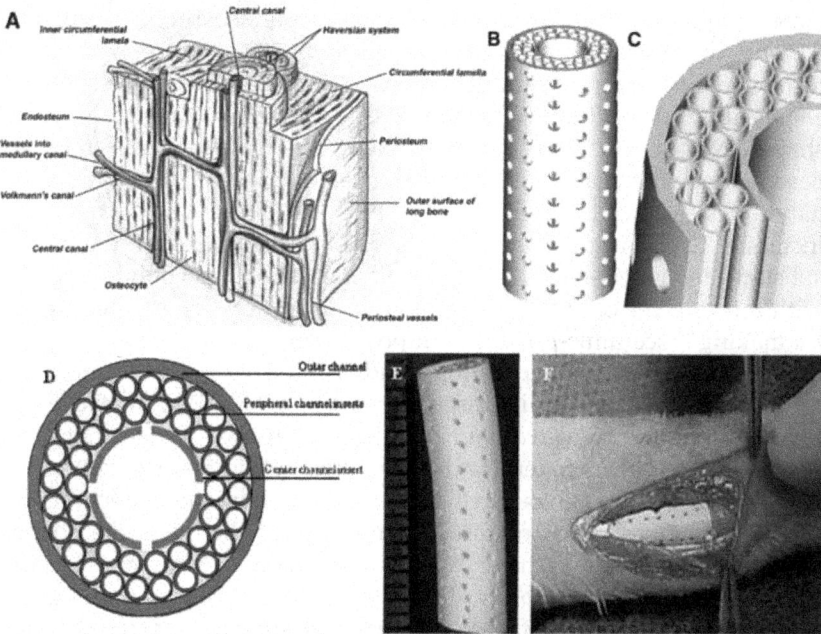

Fig. 5. (A) Schematic drawing of long bone microstructure. (B–D) Schematic drawing of honeycomb scaffold for guided and biomimetic long bone tissue regeneration. (E) Nano HA and degradable polymer-based honeycomb scaffold. (F) A honeycomb scaffold was implanted into radius defect rabbits to regenerate long bone (Xu et al. 2007).

Color image of this figure appears in the color plate section at the end of the book.

Numerous proteins and peptides have been emerging as novel biomimetic nanomaterials because of their ability to self-assemble into nanoscale structures like nanotubes, nanovesicles, helical ribbons and three-dimensional fibrous scaffolds (Chun et al. 2004). There is a growing interest in the design of nanotube-nanoparticle hybrid materials. Biomineralization represents a particularly promising approach in this direction.

Nanoceramic Coatings

Nanoceramic coatings are attractive in terms of enhancement of mechanical properties (hardness, toughness, friction coefficient) and/

or bioactivity. Nano HA-based bioactive coatings are expected to have improved mechanical and osteoconductive properties for dental and orthopedic implant applications. Different methods such as sol-gel, electrophoretic and electrolytic deposition, high velocity oxy-fuel process, electrohydrodynamic spray deposition, ion implantation, cathodic arc plasma deposition, RF magnetron sputtering and pulse laser deposition were investigated for fabricating HA-based nanocoatings on implants. Among these, sol-gel derived coatings demonstrated promise owing to their relative ease of production and ability to form a physically and chemically pure and uniform coating especially over complex geometric shapes. Many efforts are currently going on in making different HA-based composite coatings. Kaya et al. coated Ti6Al4V alloy with carbon nanotube-reinforced-HA employing EPD. Addition of carbon nanotubes increased the bonding strength of the EPD-formed layers to the metallic substrate and is cost-effective (Kaya et al. 2008). Coating a polymer scaffold with a thin layer of HA through biomimetic process is another strategy for making bioceramic-polymer composite scaffolds. It will help to impart bioactivity on the polymer scaffolds and also strengthen the scaffold.

Nanostructured metalloceramic coatings (Cr-Ti-N) were extensively investigated as wear-resistant coatings. Various surface-hardening processes such as nitrogen ion implantation, plasma ion nitriding, and PVD of various coatings have been evaluated in this direction. It is well known that nanostructured diamond and diamond-like coatings are biocompatible and have extreme hardness, wear resistance, low friction and biocompatibility. Nanostructured diamond coatings and metalloceramic coating increase the corrosion resistance of stainless steels and cobalt-based alloys (Saji and Thomas 2007).

MAJOR CHALLENGES

Despite the excitement associated with the tremendous progress of nanotechnology applications, evaluation of potential hazards related to nanoscalar-material exposure and its subsequent outcomes have become an important area of study in toxicology and health risk assessment. Those special properties that make nanoscale materials useful are the same properties that make them potentially hazardous in certain respects. The high specific surface area materials have high interfacial chemical and physical reactivity that translates to biological reactivity. The term "nanotoxicity" refers to potential adverse human health effects resulting from exposure to novel nanomaterials.

In orthopedics, particle is often associated with artificial joint wear, a major factor in aseptic loosening of orthopedic implants. Nanosized abrasion particles especially are known to be highly inflammatory, leading to osteolysis, and are also able to migrate everywhere in the body with potentially adverse effects. Nanoparticles can escape from the normal phagocytic defenses and can modify the structure of proteins. The size, shape, surface chemistry, and degree of aggregation of nanomaterials influence the production of free radicals and subsequent oxidative stress. Oxidative stress causes inflammation, which can lead to genotoxicity. There are reports that nanoparticles cause toxicities in liver, spleen, kidneys, lymph nodes, heart, lungs, and bone marrow (Chen et al. 2007; Sadik et al. 2009). Most of the reported studies on nanotoxicity are short-term studies and need to be prolonged to predict potential long-term effects. New standards and expectations for analyzing the interfacial properties of nanoparticles and nanofabricated technologies are required.

Summary Points

- Current research trends in nanotechnological applications in tissue and implant engineering are discussed.
- Advanced investigations are currently in progress to explore novel nanotechnological surface modifications on implants as well as to design nanomaterials-based composite scaffolds.
- Although not yet truly realized, surface nanopatterning of functional medical devices offers great potential in implant engineering.
- Novel nanocomposites-based scaffold can constitute a major breakthrough in this field.
- These nanotechnology-based approaches can significantly enhance tissue/implant interactions.
- More authentic studies are needed to understand the toxicological effects of nanoparticles for their widespread clinical use.

Key Facts

- Private and public research efforts are increasing worldwide in developing nanoproducts for improving health care. Three applications of nanotechnology are particularly suited to biomedicine: diagnostic techniques, drugs, and prostheses and implants.
- The ability to replace damaged joints and repair bone fractures and dental disorders with implants brings relief to millions of patients every year. According to Straumann implant company estimates, in the global medical device market, 17% goes to orthopedics and 7% goes to dental implants. The global dental device market in 2004 was estimated to be US$ 5.5–6.6 billion.
- With the advancement of science and technology, the field of implants underwent revolutionary changes. The trend is shifting towards

nanotechnology to improve the biological responses. For example, nano HA is a constituent of bone and teeth (bone is a natural composite of nano HA with collagen fibers). The use of nano HA in orthopedics is therefore considered to be very promising, owing to its dimensional similarity with bone crystals.

- It is generally accepted that the three-dimensional surface topography (size, shape, surface texture) of implants is one of the most important parameters that influence cellular reactions. Typically applied techniques to enhance the degree of roughness and promote the osteointegrative properties of metallic biomaterials are chemical etching, anodization, sand-blasting, and sputter coating. Significant efforts are being taken to produce implants with suitable surface nanotopography (created by nano-grains, pores, particles, ridges or valleys). However, most implants currently in clinical use are micrometer-scale surface roughened.

Definitions

Bioactivity: Favoring a chemical bonding between bone and the material after implantation. Examples of bioactive materials include HA and glass-ceramics.

Biocompatibilty: A measure of the acceptance of an artificial implant by the surrounding tissues and by the body as a whole. Biocompatible materials do not irritate the surrounding structures, do not provoke an abnormal inflammatory response, and do not incite allergic or immunologic reactions.

Biodegradable implant: An implant that can be broken down by nature either through hydrolytic mechanisms without the help of enzymes and/or through enzymatic mechanism. Other terms such as adsorbable, erodible, resorbable have also been used to indicate biodegradation. A biodegradable polymeric-based implant is gradually adsorbed by the body and does not permanently leave traces of residual in the implantation sites.

Biomaterial: A synthetic material used to replace part of a living system or to function in intimate contact with living tissue. Depending on the problem area, biomaterials have different uses: for example, they replace diseased or damaged part, assist in healing, improve function, correct functional abnormality, correct cosmetic problem, or help in diagnosis and treatment.

Nanocomposite: A multiphase solid material in which one of the phases has dimensions of less than 100 nm. Nanocomposites differ from conventional composite materials in the exceptionally high surface to volume ratio of the reinforcing phase and/or its exceptionally high aspect ratio. Nanocomposites may be biological, synthetic or hybrid.

Nanomaterial: Materials with nanostructures (up to 100 nm in size, atleast in one dimension) embedded in them. Examples include quantum dots, nanoparticles, nanorods, nanowires, nanotubes thin films, self-assemblies and composites. Owing to the small size of the building blocks and the high surface to volume ratio, these materials are expected to demonstrate unique mechanical, optical, electronic and magnetic properties.

Nanotopography: Surface topography of an implant in nano scale.

Osseointegration: Ability of host tissues to form a functional interface with implant surface without an intervening layer of fibrous connective tissue. The concept is originally described for titanium alloys by P.I. Branemark.

Tissue engineering: An interdisciplinary field that applies the principles of engineering and life sciences toward the development of biological substitutes that restore, maintain or improve tissue function. As a potential medical treatment, tissue engineering holds promises of eliminating re-operations by using biological substitutes, using biological substitutes to solve problems of implant rejection, and providing long-term solutions in tissue repair or treatment of diseases.

Abbreviations

DNA	:	deoxyribonucleic acid
EPD	:	electrophoretic deposition
HA	:	hydroxyapatite
HRN	:	helical rosette nanotube
PEG	:	poly(ethylene glycol)
PLA	:	poly(L-lactic acid)
PMMA	:	polymethylmethacrylate
PVD	:	physical vapor deposition
RF	:	radio frequency
SS	:	stainless steel
UHMWPE	:	ultra high molecular weight polyethylene

References

Balasundaram, G., and T.J. Webster. 2007. An overview of nano-polymers for orthopedic applications. Macromol. Biosci. 7: 635–642.

Bauer, S., J. Park, K. von der Mark, and P. Schmuki. 2008. Improved attachement of mesenchymal stem cells on super hydrophobic TiO$_2$ nanotubes. Acta Biomater. 4: 1576–1582.

Berckmans, B., R.W. Towse, and R.L. Mayfield. 2007. Deposition of discrete nanoparticles on an implant surface. U.S. Pat. Appl. US 2007110890.

Bignozzi, C.A., F. Carinci, S. Caramori, and V. Dissette. 2008. Nanomaterial coatings for osteointegrated biomedical prosthesis. PCT Int. Appl. WO 2008020460.

Chen, Z., H. Meng, G. Xing, C. Chen and Y. Zhao. 2007. Toxicological and biological effects of nanomaterials. Int. J. Nanotechnol. 4: 179–196.

Cheung, H.Y., K.T. Lau, T.P. Lu, and D. Hui. 2007. A critical review on polymer-based bioengineered materials for scaffold development. Composites 38B: 291–300.

Chun, A.L., J. Moralez, H. Fenniri, and T.J. Webster. 2004. Helical rosette nanotubes: a more effective orthopedic implant material. Nanotechnology 15: S234–S239.

Ergun, C., H. Liu, T.J. Webster, E. Olcay, S. Yilmaz, and F.C. Sahin. 2008. Increased osteoblast adhesion on nanoparticulate calcium phosphates with higher Ca/P ratios. J. Biomed. Mater. Res. 85A: 236–241.

Grego, S., T.W. Jarvis, B.R. Stoner, and J.S. Lewis. 2005. Template directed assembly on an ordered microsphere array. Langmuir 21: 4971–4975.

Heidenau, F., W. Mittelmeier, R. Detsch, M. Haenle, F. Stenzel, G. Ziegler and H. Gollwitzer. 2005. A novel antibacterial titania coating: metal ion toxicity and in vitro surface colonization. J. Mater. Sci-Mater. M. 16: 883–888.

Jager, M., C. Zilkens, K. Zanger, and R. Krauspe. 2007. Significance of nano- and microtopography for cell-surface interactions in orthopedic implants. J. Biomed. Biotech. 69036 (1–19).

Kar, A., K.S. Raja, and M. Misra. 2006. Electrodeposition of hydroxyapatite onto nanotubular TiO$_2$ for implant applications. Surf. Coat. Technol. 201: 3723–3731.

Karlsson, M., E.P. Pålsgård, P.R. Wilshawb, and L. Di Silvio. 2003. Initial in vitro interaction of osteoblasts with nano-porous alumina. Biomaterials 24: 3039–3046.

Kay, S., A. Thapa, K.M. Haberstroh, and T.J. Webster. 2002. Nanostructured polymer/ nanophase ceramic composites enhance osteoblast and chondrocyte adhesion. Tissue Eng. 8: 753–761.

Kaya, C., I. Singh, and A.R. Boccaccini. 2008. Multi-walled carbon nanotube-reinforced hydroxyapatite layers on Ti6Al4V medical implants by electrophoretic deposition. Adv. Eng. Mater. 10: 131–138.

Liu, Y., W. Chen, Y. Yang, J.L. Ong, K. Tsuru, S. Hayakawa, and A. Osaka. 2008. Novel fabrication of nano-rod array structures on titanium and *in vitro* cell responses. J. Mater. Sci-Mater. M., 19: 2735–2741.

Macak, J.M., H. Tsuchiya, A. Ghicov, K. Yasuda, R. Hahn, S. Bauer, and P. Schmuki. 2007. TiO$_2$ nanotubes: self organized electrochemical formation, properties and applications. Curr. Opin. Solid St. Mater. Sci. 11: 3–18.

Mendonca, G., D.B.S. Mendonca, F.J.L. Aragao, and F.L. Cooper. 2008. Advancing dental implant surface technology—From micron to nanotopography. Biomaterials 29: 3822–3835.

Ogawa, T., L. Saruwatari, K. Takeuchi, H. Aita, and N. Ohno. 2008. Ti nano-nodular structuring for bone integration and regeneration. J. Dent. Res. 87: 751–756.

Park, J.B., and J.D. Bronzino. 2003. Biomaterials—Principles and Applications. CRC Press, Boca Raton.

Popat, K.C., L. Leoni, C.A. Grimes, and T.A. Desai. 2007. Influence of engineered titania nanotubular surfaces on bone cells. Biomaterials 28: 3188–3197.

Rhee, S.H., and J. Tanaka. 2001. Synthesis of a hydroxyapatite/collagen/chondroitin sulfate nanocomposite by a novel precipitation method. J. Am. Ceram. Soc. 84: 459–461.

Richert, L., F. Vetrone, J.H. Yi, S.F. Zalzal, J.D. Wuest, F. Rosei, and A. Nanci. 2008. Surface nanopatterning to control cell growth. Adv. Mater. 20: 1488–1492.

Sadik, O.A., A.L. Zhou, S. Kikandi, N. Du, Q. Wanga, and K. Varner. 2009. Sensors as tools for quantitation, nanotoxicity and nanomonitoring assessment of engineered nanomaterials. J. Environ. Monit. 11: 1782–1800.

Saji, V.S., and J. Thomas. 2007. Nanomaterials for corrosion control. Curr. Sci. 92: 51–55.

Saji, V.S., H.C. Choe, and W.A. Brantley. 2009. An electrochemical study on nanoporous and nanotubular Ti-35Nb-5Ta-7Zr alloy for biomedical applications. Acta. Biomater. 5: 2303–2310.

Saji, V.S., and H.C. Choe. 2009. Electrochemical corrosion behavior of nanotubulat Ti-13Nb-13Zr alloy in ringer's solution. Corros. Sci. 51: 1658–1663.

Saji, V.S., H.C. Choe, and K.W.K. Yeung. 2010. Nanotechnology in biomedical applications—a review. Int. J. Nano and Biomaterials 3: 119–139.

Singh, M.K., T. Shokuhfar, J.J. de Almeida Gracio, A.C.M. de Sousa, J.M.D.F. Fereira, H. Garmestani, and S. Ahzi. 2008. Hydroxyapatite modified with carbon nanotube reinforced poly(methyl methacrylate): a nanocomposite material for biomedical applications. Adv. Funct. Mater. 18: 694–700.

Sumner, D.R., and J.O. Galante. 1992. Determinants of stress shielding: design versus materials versus interface. Clin. Orthop. Relat. R. 274: 202–212.

Van, N.T., and M.P.C. Watts. 2006. Trends in imprint lithography for biological applications. Trends Biotechnol. 24: 312–317.

Webster, T.J., R.W. Siegel, and R. Bizios. 1999. Osteoblast adhesion on nanophase ceramics. Biomaterials 20: 1221–1227.

Webster, T.J., M.C. Waid, J.L. McKenzie, R.L. Price, and J.U. Ejiofor. 2004. Nano-biotechnology: carbon nanofibers as improved neural and orthopedic implants. Nanotechnology 15: 48–54.

Williams, D.F. 1987. Definition of Biomaterials. Elsevier, Amsterdam.

Wood, M.A. 2007. Colloidal lithography and current fabrication techniques producing in-plane nanotopography for biological applications. J.R. Soc. Interface 4: 1–17.

Xu, T., N. Zhang, H.L. Nichols, D. Shi, and X. Wen. 2007. Modification of nanostructured materials for biomedical applications. Mater. Sci. Eng. C. 27: 579–594.

Zhang, S.M., F.Z. Cui, S.S. Liao, Y. Zhu, and L. Han. 2003. Synthesis and biocompatibility of porous nano-hydroxyapatite/collagen/alginate composite. J. Mater. Sci-Mater. Med. 14: 641–645.

Index

About the Editors

Ross J. Hunter MD BSc MRCP trained in medical sciences at King's College London (Times University ranking 11th in UK). He spent a further year at Imperial College London (Times University ranking 3rd in UK) and was awarded his BSc in Cardiovascular medicine in 1998. Since returning to his medical training at King's College School of Medicine, he has remained an honorary research fellow at The Department of Nutritional Sciences, researching the effect of different nutritional states and alcoholism on the cardiovascular system. He was awarded his Bachelor of Medicine and Surgery (MBBS) with distinction in 2001. He trained in general medicine in London and Brighton and was made a member of the Royal College of Physicians (UK) in 2005. He trained as a Registrar in the London Deanery from 2005-2008. Since 2008 he has been a research fellow at the Department of Cardiology & Electrophysiology at St Bartholomew's Hospital London, conducting clinical research and clinical trials in cardiology and electrophysiology. He has published over 60 scientific articles of various kinds.

Victor R. Preedy BSc, PhD, DSc, FIBiol, FRCPath, FRSPH is Professor of Nutritional Biochemistry, King's College London, Professor of Clinical Biochemistry, Kings College Hospital (Hon) and Director of the Genomics Centre, King's College London. Presently he is a member of the King's College London School of Medicine. Professor Preedy graduated in 1974 with an Honours Degree in Biology and Physiology with Pharmacology. He gained his University of London PhD in 1981 when he was based at the Hospital for Tropical Disease and The London School of Hygiene and Tropical Medicine. In 1992, he received his Membership of the Royal College of Pathologists and in 1993 he gained his second doctoral degree, i.e. DSc, for his outstanding contribution to protein metabolism in health and disease. Professor Preedy was elected as a Fellow to the Institute of Biology in 1995 and to the Royal College of Pathologists in 2000. Since then he has been elected as a Fellow to the Royal Society for the Promotion of Health (2004) and The Royal Institute of Public Health (2004). In 2009, Professor Preedy became a Fellow of the Royal Society for Public Health. In his career Professor Preedy has carried out research at the National

Heart Hospital (part of Imperial College London) and the MRC Centre at Northwick Park Hospital. He is visiting lecturer at University College London (UCL) and has collaborated with research groups in Finland, Japan, Australia, USA and Germany. He is a leading expert on the pathology of disease and has lectured nationally and internationally. He has published over 570 articles, which include over 165 peer-reviewed manuscripts based on original research, 90 reviews and numerous books.

Color Plate Section

Chapter 1

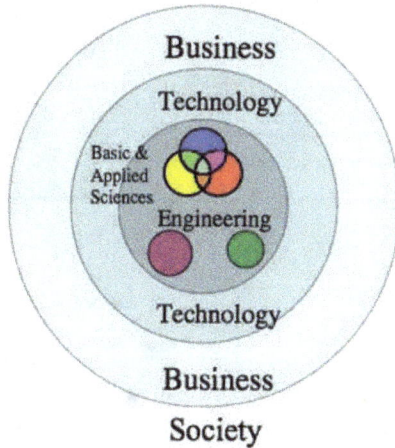

Fig. 1. The {{{{Science}Engineering}Technology}Business} hierarchy. Progress on these fields could be driven by hypothesis, need, and/or opportunity. Efficient communication between these layers is of utmost importance for successful commercialization.

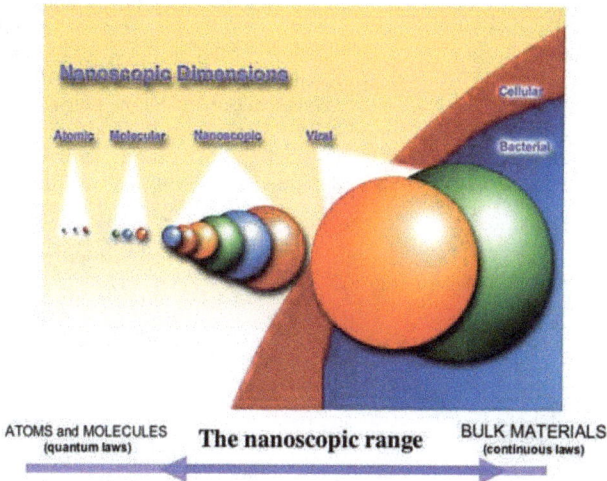

Fig. 2. The nanoscopic range is between "molecular" and "bulk" regimes, where molecular, "nano", and materials characteristics are present simultaneously.

Chapter 3

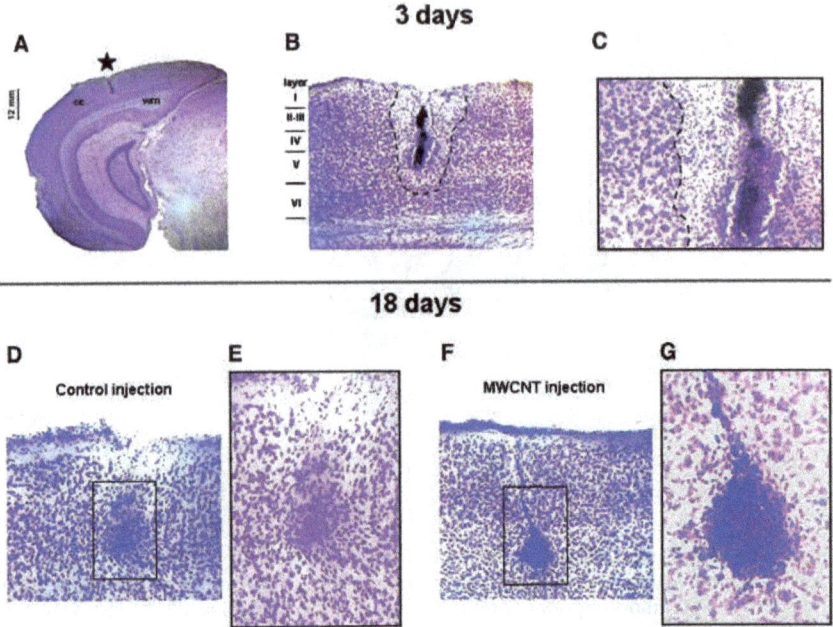

Fig. 1. *In vivo* testing of carbon nanotubes: injection in mouse brain. Sections of mouse brain at the site of CNT injection. (a) The injection site is labeled with a star. *cc*: cerebral cortex; *wm*: white matter. (b) The injection site in mice treated for 3 d at higher magnification. Normal neuronal density and tissue layering is present outside the lesioned site (broken line). (c) Higher magnification of the transition between the lesioned and the unlesioned area (on the right, broken line). Control mice (d, e) and MWCNT-injected mice (f, g) brain cortices present a scar of glial cells surrounding the injection site. Higher magnifications of (e) the control and (g) the MWCNT-injected mice are shown. Reprinted from Bardi et al. (2009) with permission from Elsevier.

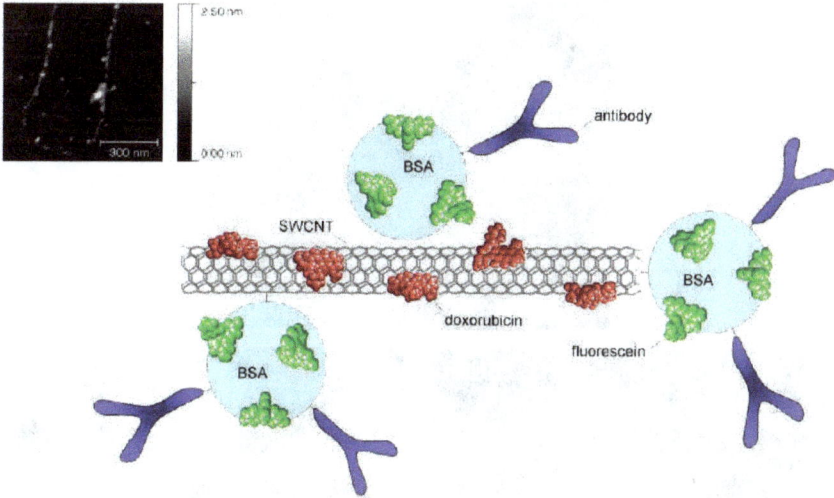

Fig. 2. Example of carbon nanotubes multiple functionalization. Schematic illustration of the doxorubicin–fluorescein–BSA–antibody-SWCNT complexes (red = doxorubicin, green = fluorescein, light blue = BSA, dark blue = antibodies). Inset: AFM image of doxorubicin–fluorescein–BSA–SWCNT complexes (without antibodies). Reprinted from Heister et al. (2009) with permission from Elsevier.

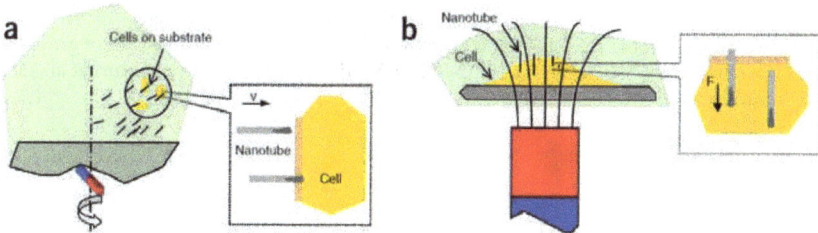

Fig. 3. DNA transfection via carbon nanotubes spearing. A two-step procedure of CNT spearing. (a) In the first step, a rotating magnetic field drives nanotubes (short black lines) to spear the cells (yellow) on a substrate. Inset: Close-up of one cell with the CNT penetrating the membrane. The plasma membrane is illustrated as an assembly of red circles. (b) In the second step, a static field persistently pulls CNT into the cells. *F*: magnetic force; *v*: velocity of CNTs. Reprinted from Cai et al. (2005) with permission from Macmillan Publishers Ltd.

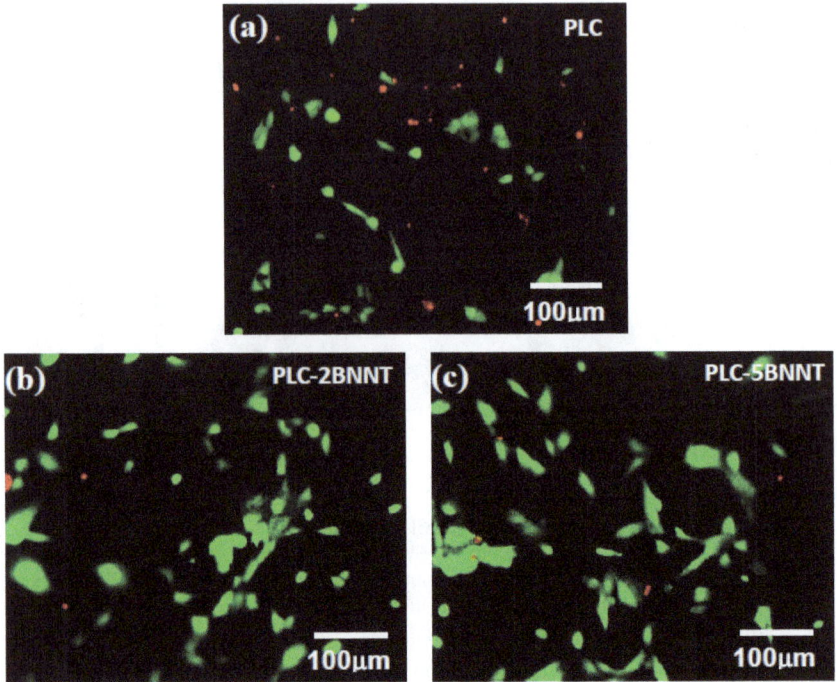

Fig. 5. Testing of boron nitride nanotube doped scaffolds. Fluorescent images of live (green) and dead (red) osteoblast cells obtained through Live/Dead staining after 2.5 d of growth on (a) PLC, (b) PLC-BNNT 2% and (c) PLC-BNNT 5% films. Reprinted from Lahiri et al. (2010) with permission from Elsevier.

Chapter 6

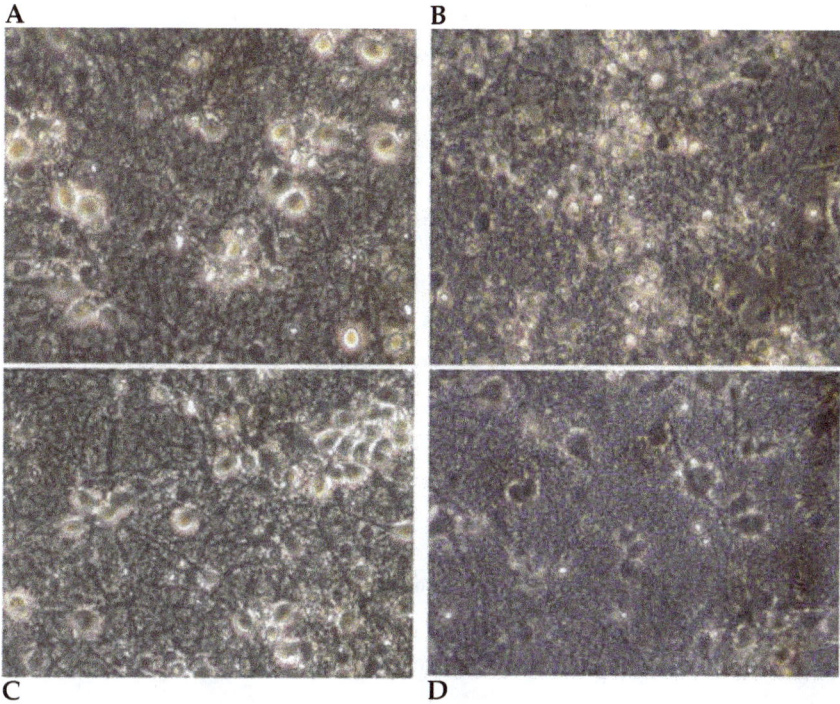

Fig. 3. Cerium oxide nanoparticles prevent neuronal damage associated with exposure to Aβ(1–42). Panel A shows a healthy culture of rat cortical neurons. Panel B shows a similar culture, treated with Aβ(1–42) for 24 h. Note fragmentation of neuronal processes and overall degradation of culture heath. In panels C and D, cultures were pretreated with 10 and 100 nM CeONP respectively, followed by Aβ(1–42) exposure. Note protective effects of cerium oxide nanoparticles (previously unpublished data).

Chapter 7

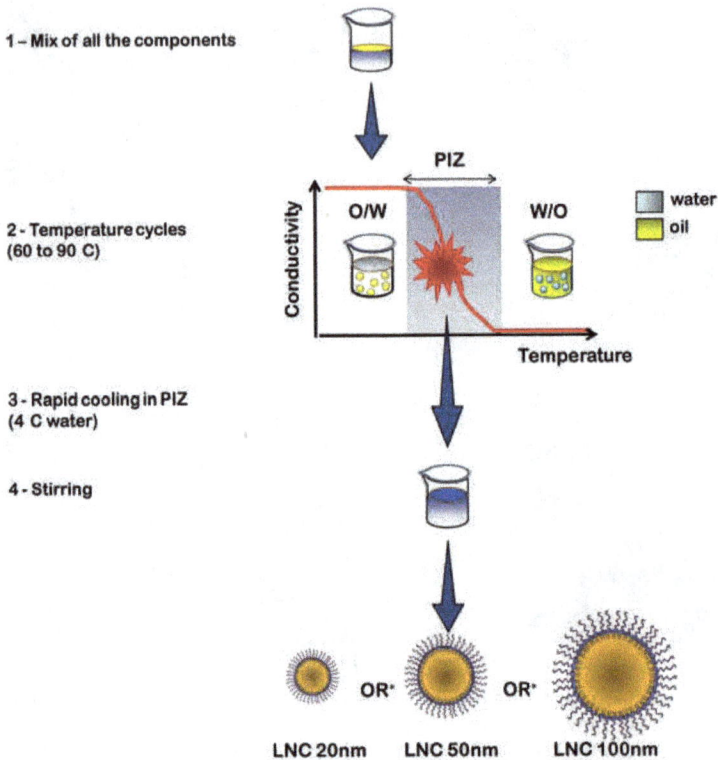

1 – Mix of all the components

2 - Temperature cycles (60 to 90 C)

3 - Rapid cooling in PIZ (4 C water)

4 - Stirring

PIZ

O/W   W/O

Conductivity

Temperature

water
oil

OR*   OR*

LNC 20nm   LNC 50nm   LNC 100nm

Fig. 3. Formulation of LNCs by the phase-inversion temperature method. Firstly, all the components are mixed. The suspension is then heated and cooled between 60°C and 90°C to obtain conductivity variation, and therefore phase inversion between the oil and water phase. After several temperature cycles, and in the phase-inversion zone, where the conductivity is much lower, a rapid dilution with cold water is performed to fix the suspension. The LNCs obtained are then stirred for 5 min. *The size of the obtained particle depends on the composition of non-ionic surfactants and oil.

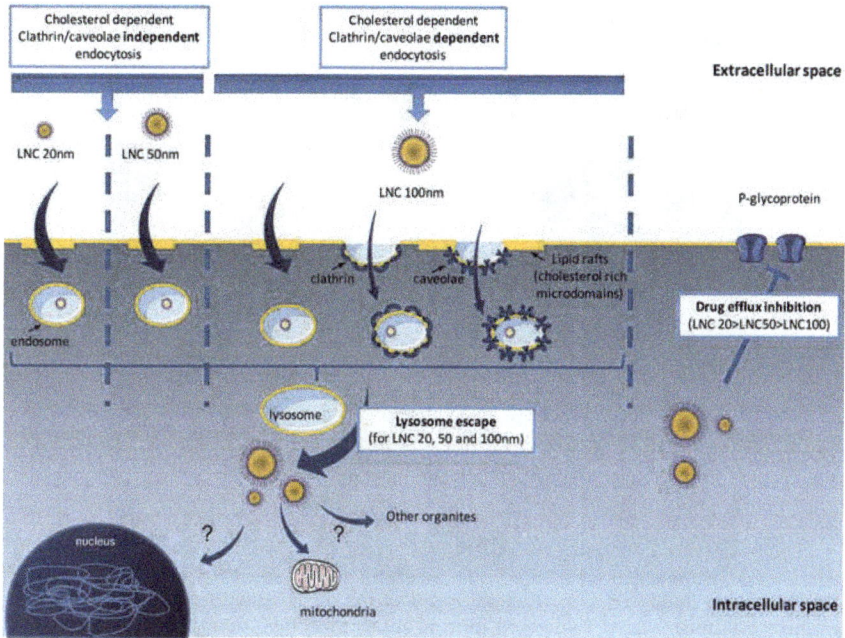

Fig. 8. Schematic representation of known LNC interactions with cells. 20 nm and 50 nm LNCs seem to be exclusively endocytosed by the cholesterol-dependent pathway. 100nm LNCs are also internalized by this pathway, as well as by clathrin-dependent mechanisms. LNCs (20, 50, and 100 nm) are then supposed to escape from lysosomal degradation. LNCs are also known to inhibit drug efflux pumps, such as P-glycoprotein.

Chapter 10

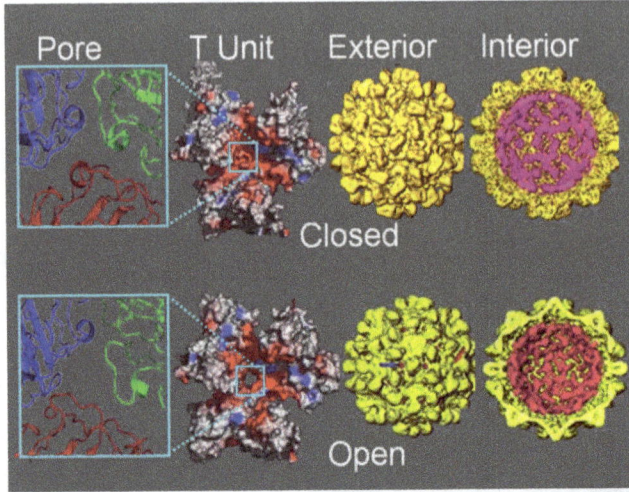

Fig. 1. Representation of two structural forms of the plant virus RCNMV determined by cryo-electron microscopy. The closed form is observed at high divalent ion concentration (Ca^{2+} and Mg^{2+}) and is likely the form of the virus as it moves through the soil from one host to another. The open form of the virus consists of 60 holes that lead from the exterior of the capsid to the hollow interior of the "cargo chamber". The open form is observed when the divalent ion concentration is low. This form is present in the cytosol of cells.

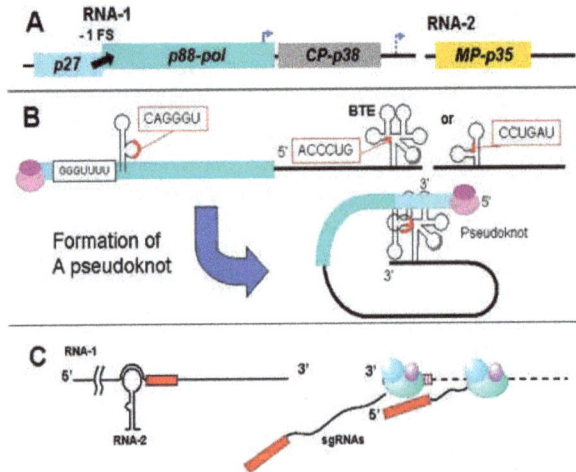

Fig. 2. RNA genome of RCNMV and long-range interactions. A. The four genes are shown on RNA-1 and RNA-2. B. The ribosomal frame shift element interacts with a distal structure to form a long-distance pseudoknot that stimulates ribosomal frameshifting. C. RNA-2 and RNA-1 interact to create the transactivator (TA) region.

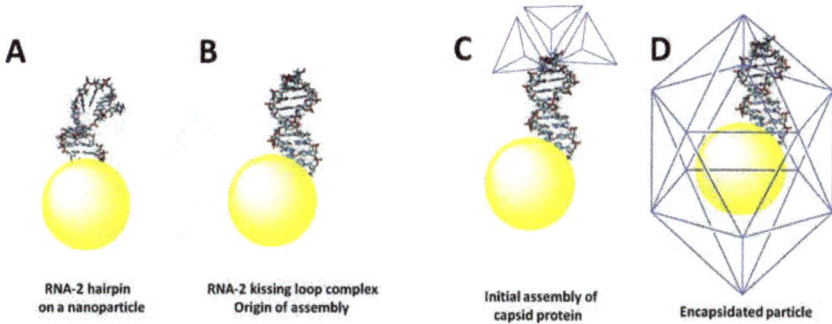

Fig. 3. Schematic illustration of a method for encapsidation of a solid (e.g., gold) nanoparticle inside a plant virus nanoparticle. A. The target nanoparticle is conjugated to RNA-2. B. The RNA-2 is permitted to react with genomic RNA-1 to form the origin of assembly. C. The origin of assembly is exposed to capsid proteins. D. The capsid proteins assemble around the particle.

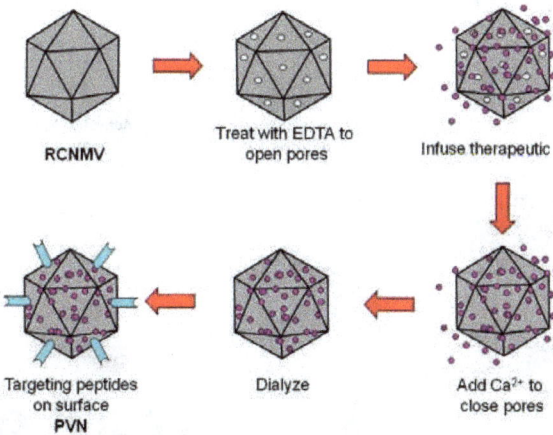

Fig. 4. Schematic representation of the approach for loading a cargo inside the plant virus RCNMV. The process is initiated by addition of excess EDTA to sequester the divalent ions, Ca^{2+} and Mg^{2+}. The cargo is then loaded under these conditions. Subsequently, Ca^{2+} is added back and the excess drug is dialyzed away. Then targeting peptides are added to the surface of the virus to complete the PVN formulation.

A

Lys 219

Lys 331

Lys 71

Lys 208

Lys 102/103

Lys 153

B

Cys 267

Cys 328/330

Cys 154

Cys 203

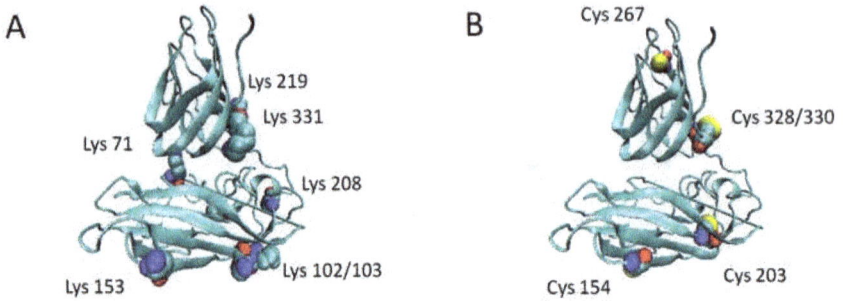

Fig. 5. Capsid structure with locations of lysines and cysteines specified by number. The coordinates were determined using a combination of electron diffraction and homology modeling as described in the text. A. The lysines in the structure are shown. B. The cysteines in the structure are shown.

Chapter 11

NP released transition metals

NP uptake

Endocytosis

Activation of receptor

NP surface caused oxidative stress

Increased oxidative stress

Diffusion

Signaling pathways

Oxidised lipids, thiols, ROS

Mitochondria

Nucleus

Lysosome

Respiratory chain

Cathepsins

Surface markers

Proteins RNA

Cytokines

Genotoxicity

Pyrimidine dimers

O_6-methyl guanine

Bulky adducts

Inflammatory mediators

Mismatches

Strand breaks

Inflammation

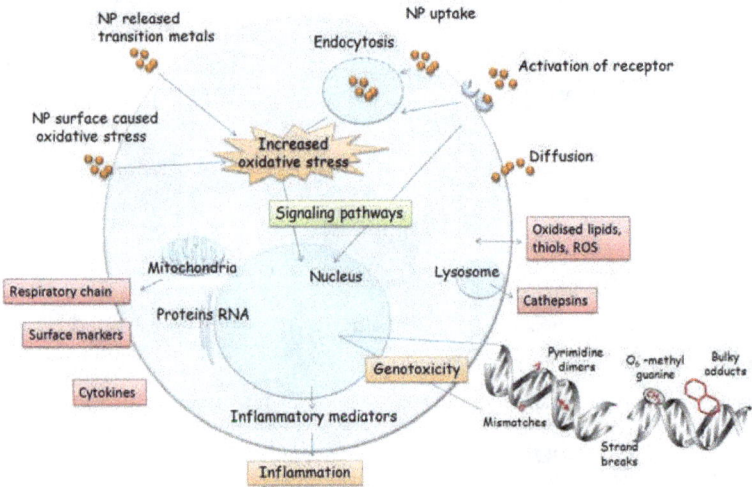

Fig. 2. Schematic illustration of cellular uptake and possible mechanisms of action of NMs. Unpublished.

Chapter 13

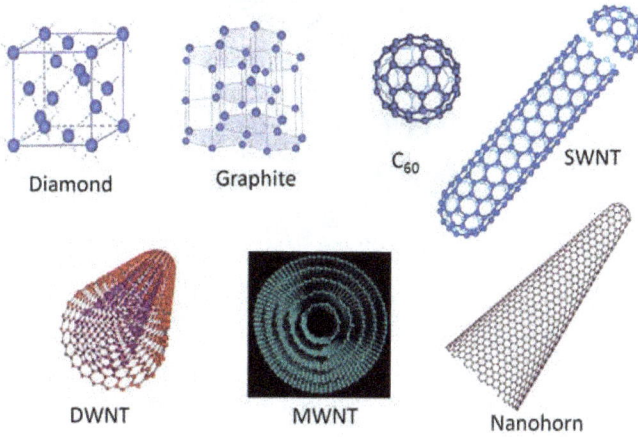

Fig. 1. Schematic representation of allotropes of carbon. These include a variety of entities that manifest as different physical structures as indicated in this figure.

Chapter 14

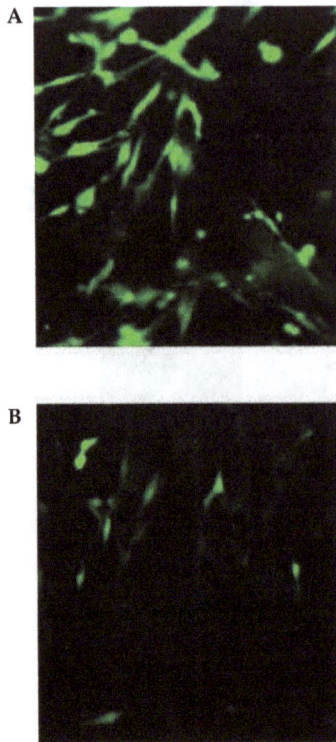

Fig. 5. ROS expression in irradiated normal lung fibroblasts. 4 hours post radiation, the levels of ROS were detected in (A) irradiated normal lung fibroblasts and (B) irradiated normal lung fibroblasts pretreated with CeO_2. Unpublished data from C.H. Baker.

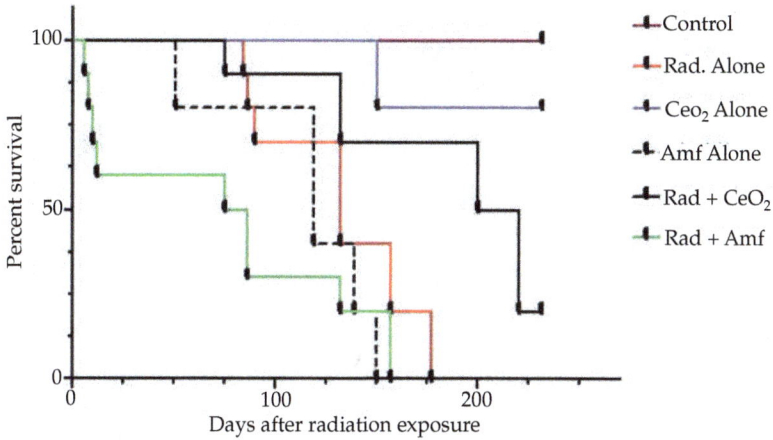

Fig. 6. Tolerability of CeO_2 nanoparticles in mice. CeO_2 were well tolerated by mice and the median survival of radiated mice was significantly increased in mice pretreated with 15 nM (0.00001 mg/kg) CeO_2 (50% alive on day 225) as compared to mice treated with radiation alone (50% alive on day 132) or pretreated with 150 mg/kg Amifostine before radiation (50% alive on day 81). Note that 20% of mice treated with CeO_2 alone were terminated on day 150 for histology analysis.. With permission from C.H. Baker 2009. Protection from radiation-induced pneumonitis using cerium oxide nanoparticles. Nanomedicine 5: 225–231.

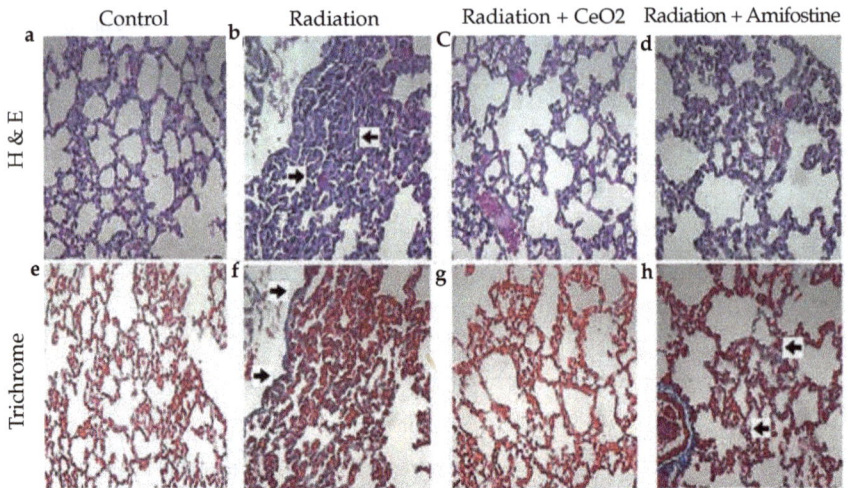

Fig. 7. CeO$_2$ nanoparticles protect lungs from radiation-induced pneumonitis. Hematoxylin and Eosin (H&E) stains to assess lung damage in normal lungs (a), lungs from mice treated with radiation alone (b), lungs from mice treated with radiation plus CeO$_2$ (c) and lungs from mice treated with radiation plus Amifostine (d). The H&E stains show significant lung damage in mice treated with radiation (b). Radiation-induced cell damage is protected in lungs of mice treated with radiation in combination with CeO$_2$ (c) and these lungs appear normal in control (a). The amount of fibrosis and collagen deposition (indicative of chronic lung conditions) was measured by using Masson's Trichrome stain. Results show that fibrosis and collagen deposition (indicated by arrows) were common in the lungs of those mice given radiation alone (f) and in lungs of those mice given a pretreatment of Amifostine (h). The amount of fibrosis and collagen deposition in lungs of mice treated with radiation in combination with CeO$_2$ (g) was minimal and these lungs appeared normal (e). With permission from C.H. Baker 2009. Protection from radiation-induced pneumonitis using cerium oxide nanoparticles. Nanomedicine 5: 225–231.

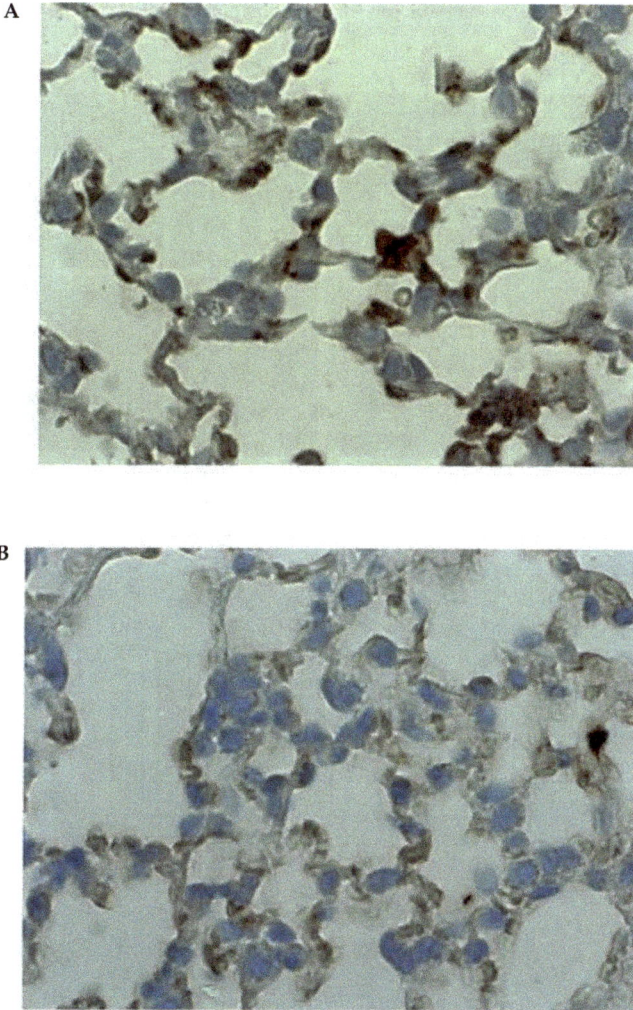

Fig. 8. CeO$_2$ nanoparticles reduce TGF-β expression post radiation. 120 d after XRT(30 Gy) fractionated over 5 doses and 2 wk, mice that received nanoceria treatment had significantly less TGF-β deposition. A. Lung tissue from untreated animal. B. Lung tissue from treated animal (0.005 mg/kg). Unpublished data from C.H. Baker.

A. Production of ROS post radiation treatment on normal colon CRL 1541 cells

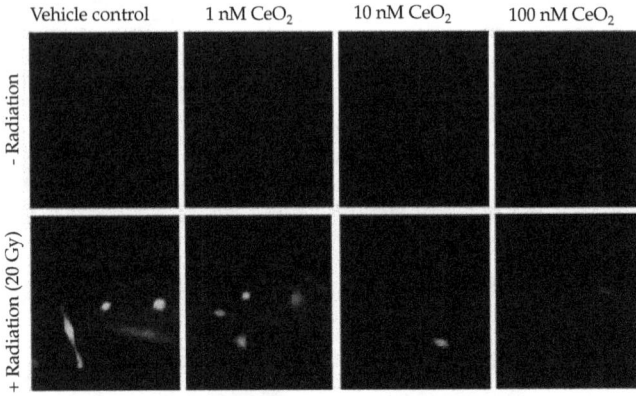

B. ATP assay 96 hrs post radiation treatment on normal colon CRL 1541 cells

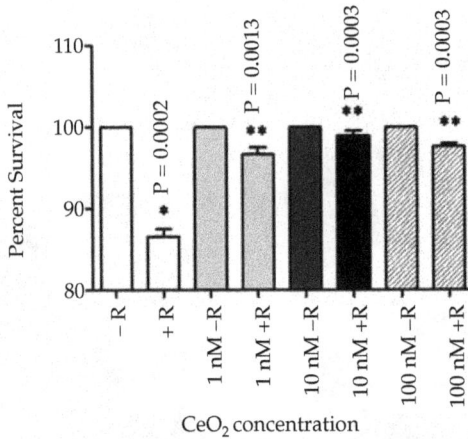

*Control (0 Gy) vs. Radiation (20 Gy)

** Radiation (20 Gy) vs. Radiation + CeO$_2$

Fig. 9. CeO$_2$ nanoparticles protect normal colon cells against radiation-induced cell damage. A. ROS production of normal human colon cells (CRL 1541) immediately following 20 Gy radiation exposure with pretreatment of 1, 10, or 100 nM CeO$_2$ nanoparticles was significantly reduced as compared to cells exposed to radiation alone. B. CRL 1541 cells were exposed to 20 Gy radiation in the absence or presence of 1, 10, or 100 nM CeO$_2$ and 96 h after exposure cell viability was measured by Cell Titer-Glo Luminescent Cell Viability Assay (cell number correlates with luminescent output (RLU). With permission from C.H. Baker 2010. Cerium oxide nanoparticles protect gastrointestinal epithelium from radiation-induced damage by reduction of reactive oxygen species and upregulation of superoxide dismutase 2. Nanomedicine 5: 698–705.

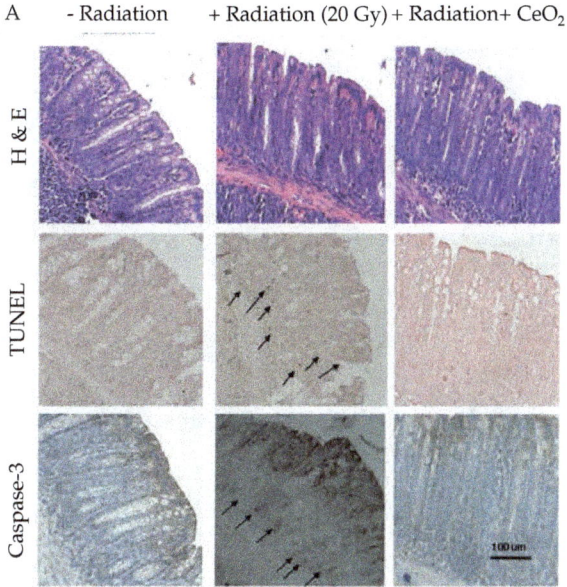

Fig. 11. CeO$_2$ nanoparticles protect normal human colon tissue from radiation-induced cell death. Hematoxlin and eosin (H&E) stains of murine colons 4 h post a single dose of 20 Gy radiation. Radiation was administered to the bowel of non–tumor-bearing athymic nude mice pretreated with four i.p. treatments of CeO$_2$ nanoparticles. Results show a significant decrease in apoptotic colon cryptic cells (as measured by TUNEL) and Caspase-3 expression as compared to the colonic crypt cells from mice treated with radiation alone. With permission from C.H. Baker 2010. Cerium oxide nanoparticles protect gastrointestinal epithelium from radiation-induced damage by reduction of reactive oxygen species and upregulation of superoxide dismutase 2. Nanomedicine 5: 698–705.

A.

B.

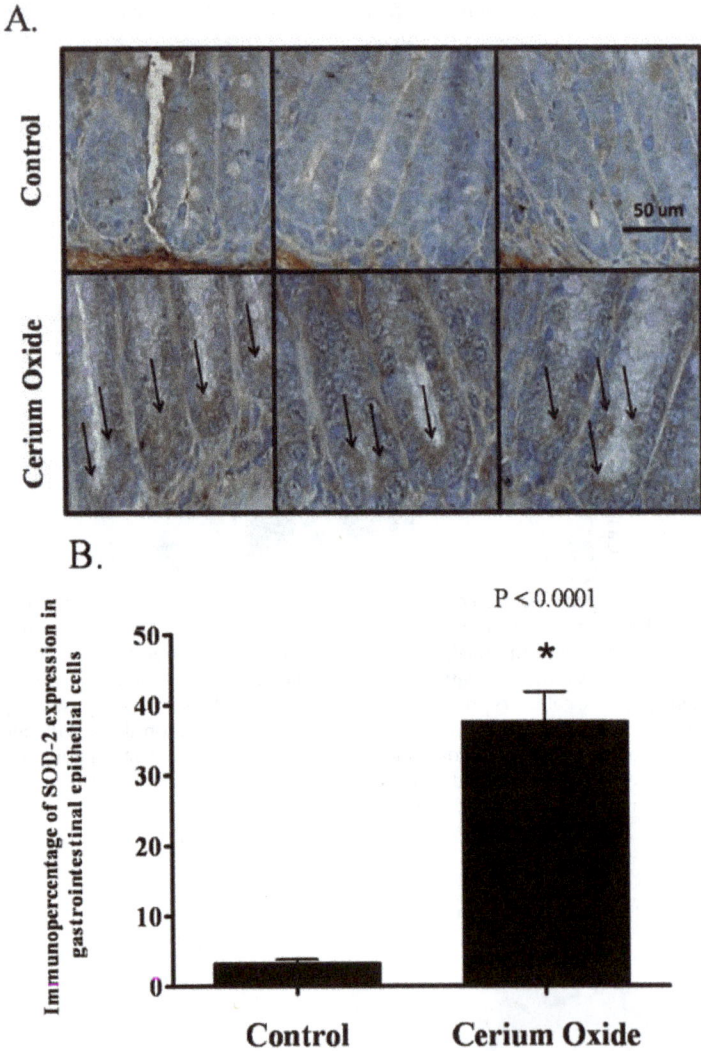

Fig. 12. CeO$_2$ nanoparticles induce SOD-2 expression in normal colon. A. Representative sections of SOD-2 expression (brown staining) in colonic crypts in mice treated with CeO$_2$ nanoparticles or in normal (control) mice. Colons were collected 24 h post a single injection of CeO$_2$ nanoparticles. B. The immunopercentage of SOD-2 expression increased by 40% in mice treated with CeO$_2$ nanoparticles as compared to control mice. Each data point represents the mean +/- SEM from analyzing 10 random crypts per mouse from five different mice, which has been expressed as percentage of crypt cells staining positive for SOD-2. With permission from C.H. Baker 2010. Cerium oxide nanoparticles protect gastrointestinal epithelium from radiation-induced damage by reduction of reactive oxygen species and upregulation of superoxide dismutase 2. Nanomedicine 5: 698–705.

Chapter 16

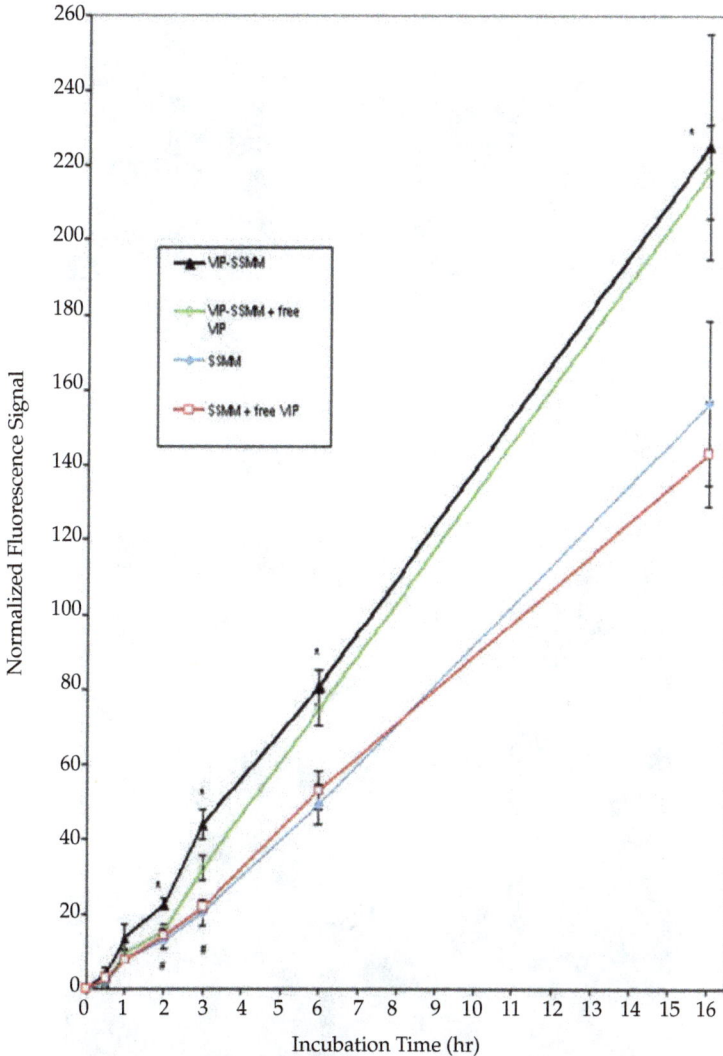

Fig. 4. Accumulation and specificity of uptake of quantum dot (QD) loaded sterically stabilized mixed micelles (SSMM) in human breast cancer cells. Normalized fluorescence signal of vasoactive intestinal peptide (VIP)-SSMM-QD and SSMM-QD in human MCF-7 cells with and without pre-treatment with free VIP (30 μM). Data represented as mean ± standard deviation (n = 5); $^*p < 0.05$ for VIP-SSMM-QD in comparison to SSMM-QD, and SSMM-QD with excess free VIP; $^\#p < 0.05$ for VIP-SSMM-QD in comparison to VIP-SSMM with excess free VIP. (Statistical analysis conducted using ANOVA followed by Tukey's post hoc test.)

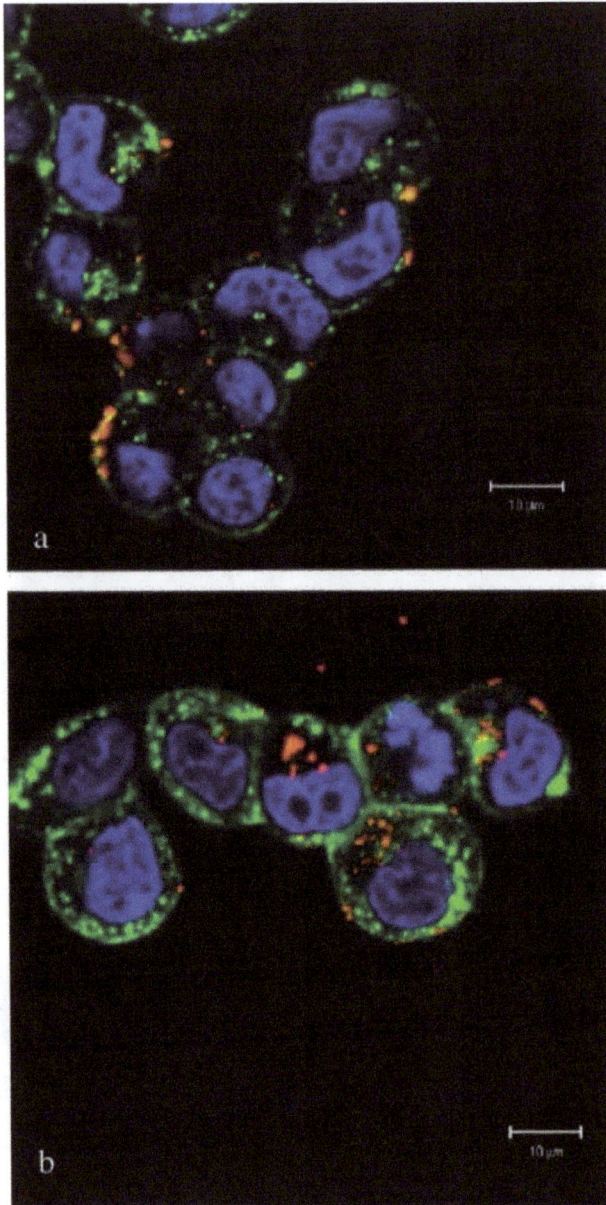

Fig. 5. Cellular uptake of sterically stabilized mixed micelles (SSMM) loaded with quantum dots (QD) in breast cancer cells. Representative confocal microscopy image of human breast cancer MCF-7 cells upon incubation with (a) SSMM-QD and (b) VIP-SSMM-QD for 16 h. (Figs. 4 and 5 reprinted from Rubinstein et al. 2008 with permission from Elsevier.)

Chapter 19

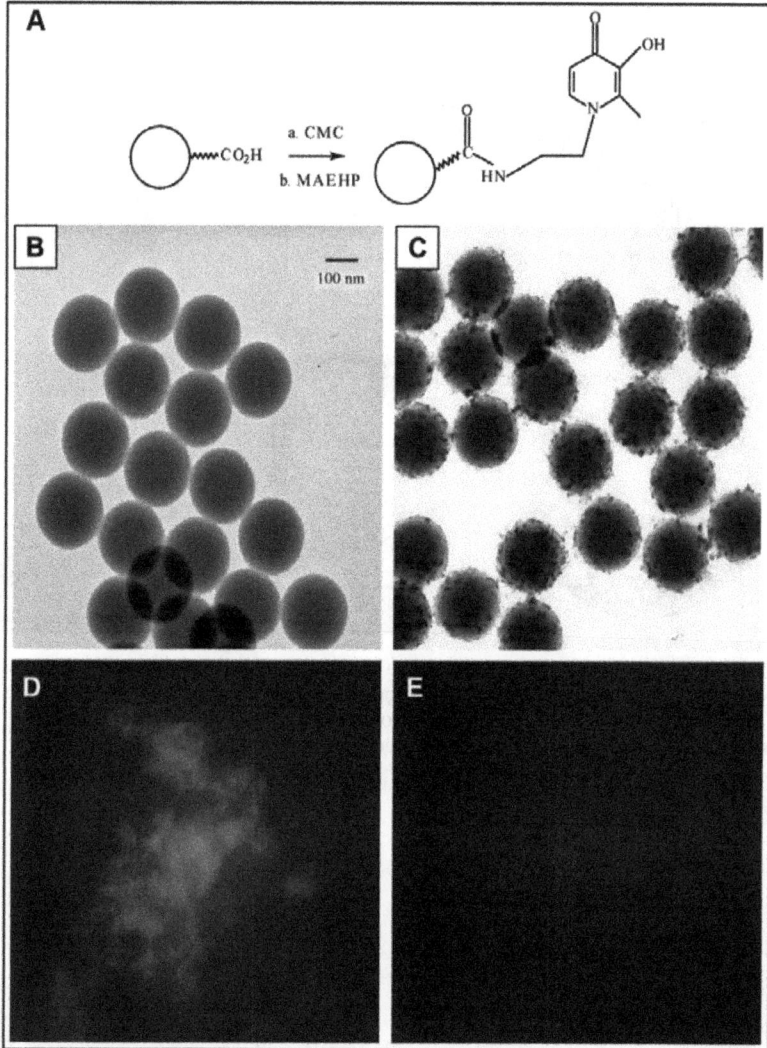

Fig. 1. Ab NP–chelator conjugate. (A) Synthesis strategy for NP-chelator conjugates. (a) Reaction of carboxylic functionalized nanoparticles with N-cyclohexyl-N′-(2-morpholinoethyl) carbodiimide methyl-p-toluensulfonate in MES. (b) Conjugation of activated carboxylic nanoparticles with excessive MAEHP in MES. (B, C) Transmission electron microscope (TEM) images of nanoparticles without chelator conjugation and iron binding (B) or with both reactions (C). (D, E) Fluorescence microscopy images. Ab aggregates (D) are inhibited in presence of Nano-N2PY (E). Reproduced from Liu et al. (2009) with permission from Elsevier.

Chapter 20

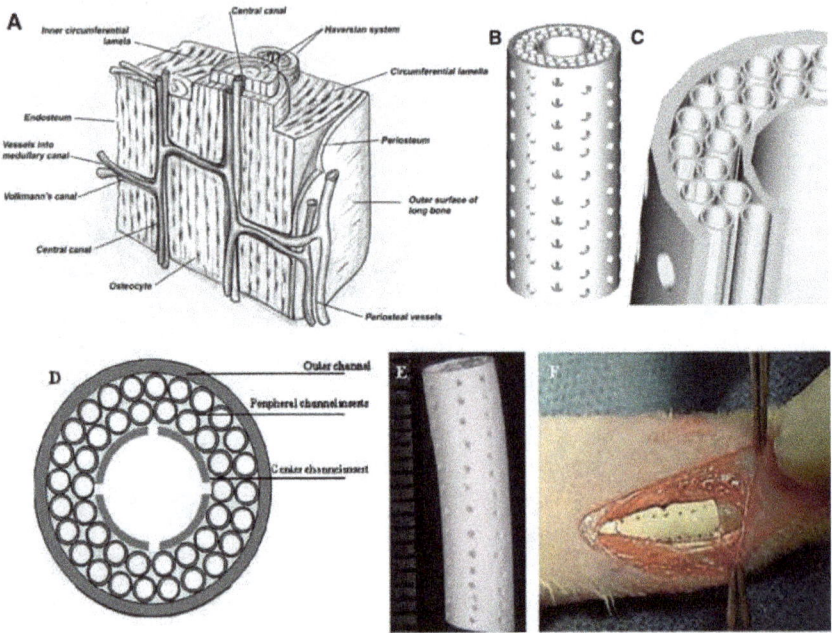

Fig. 5. (A) Schematic drawing of long bone microstructure. (B-D) Schematic drawing of honeycomb scaffold for guided and biomimetic long bone tissue regeneration. (E) Nano HA and degradable polymer-based honeycomb scaffold. (F) A honeycomb scaffold was implanted into radius defect rabbits to regenerate long bone (Xu et al. 2007).

For Product Safety Concerns and Information please contact our EU
representative GPSR@taylorandfrancis.com
Taylor & Francis Verlag GmbH, Kaufingerstraße 24, 80331 München, Germany

www.ingramcontent.com/pod-product-compliance
Lightning Source LLC
Chambersburg PA
CBHW060744220326
41598CB00022B/2323